ENVIRONMENTAL STUDIES

STUDIES

The Earth as a Living Planet

DANIEL B. BOTKIN
and
EDWARD A. KELLER

University of California, Santa Barbara

Charles E. Merrill Publishing Company
A Bell & Howell Company
Columbus • Toronto • London • Sydney

Cover photo:
Chimney Peak and Courthouse Mountain.
Uncomphahgre Wilderness Area near Ridgway, Colorado.
Courtesy of David Sumner Photography.

Published by
Charles E. Merrill Publishing Company
A Bell & Howell Company
Columbus, Ohio 43216

This book was set in Serifa.
Cover coordination and text design: Ann Mirels
Production coordination: JoEllen Gohr

Library of Congress Catalog Card Number: 82–80262
International Standard Book Number: 0–675–09813–0
Printed in the United States of America

2   3   4   5   6   7   8   9   10—87 86 85 84 83

for Erene and Jackie

Preface

The field of environmental studies integrates many disciplines in the study of our environment. It includes some of the most important applied topics of modern civilization as well as some of the oldest philosophical concerns of human beings—that is, the nature of our relationship to our surroundings. Both applied and basic aspects require a solid foundation in the natural sciences (including biology, geology, geography, oceanography, soil science, hydrology and climatology). We must also be aware of the cultural and historical context in which we make decisions about our environment and understand the ways in which choices are made and implemented. Thus, the field of environmental studies integrates the natural sciences with environmental ethics, environmental economics, environmental law, and planning. *Environmental Studies: The Earth as a Living Planet* attempts to examine the entire spectrum of relationships between people and environment. Because this text is an introduction to environmental studies, you do not need any prerequisites in the sciences or humanities.

The text is divided into three parts. Part One introduces the fundamental principles of environmental studies and examines the physical and biological processes so important to global, regional, and local ecology and geology. In Part Two we discuss the Earth's major resources, including the atmosphere, water, minerals, energy, and biological materials. Part Three examines aspects related to people in the environment—natural hazards, environmental ethics, environmental planning, the urban environment, and environmental law.

Successful completion of this text would not have been possible without the cooperation of many individuals. To all those who so freely offered their help in this endeavor, we offer our sincere appreciation. We wish to thank the following people for their constructive comments in reviewing the manuscript: Stanley Awramik, University of California, Santa Barbara; Charles Beveridge, Frederick Law Olmstead Papers, Department of History, American University; John B. Conway, Washington State University; Ira C. Darling, University of Maine; Margaret B. Davis, University of Minnesota; Anthony Dominski, University of California, Santa Barbara; Lon D. Drake, University of Iowa; Robert N. Ford, Millersville State College; Charles A. S. Hall, Cornell University; Robert E. Hennigan, State University of New York, Syracuse; Jean Hitzeman, State University of New York, Brockport; Sally J. Holbrook, University of California, Santa Barbara; Katherine Keating, Rutgers University; Lynn Margulis, Boston University; Eugene W. McArdle, Northeastern Illinois University; Mark McGinnes, Environmental Defense Center, Santa Barbara; Harold J. Morowitz, Yale University; Roderick Nash, University of California, Santa Barbara; Erene V. Pecan, Santa Barbara Institute for Environmental Studies; Richard C. Pleus, University of Minnesota; Tad E. Reynales, University of California, Santa Barbara; Donald L. Rice, State University of New York, Binghampton; Robert J. Robel, Kansas State University; Dorothy Rosenthal, University of Rochester; Alfred Runte, University of Washington; Jon Sonstelie, University of California, Santa Barbara; Oscar Soulé, Evergreen State College, Olympia, Washington; Alfred Suskie, Mohawk Valley Community College, Utica, New York; Frederick R. Swan, West Liberty State College; Laurence C. Walkter, Stephen F. Austin State University.

We are indebted to JoEllen Gohr for editing the text as well as making suggestions for improvement. Special thanks are extended to Marsha Sato for her work in bringing the manuscript to final completion. Without her able assistance, the project would have taken much longer. Penn Chu and Robin Panza provided invaluable assistance in literature reviews. We want to thank Mrs. Dorothy Fujii and Roberta M. Fujii for assistance in obtaining permissions for the manuscript.

The Environmental Studies Program and the departments of geological sciences and biological sciences at the University of California, Santa Barbara, provided the stimulating atmosphere necessary for writing. A fellowship to D. B. Botkin from the Woodrow Wilson International Center for Scholars provided time and the environment to integrate science and humanities.

v

Contents

6

WHO HAS BEEN WHERE, WHEN? THE GEOGRAPHY OF LIFE 123

7

AN ECOLOGICAL VIEW OF HUMAN BEINGS: HUMAN POPULATION DYNAMICS 149

PART 2: RESOURCES FOR HUMAN SOCIETY

8

THE ATMOSPHERE 169

9

THE WATERS 193

10

MINERAL RESOURCES: USING, CONSERVING, AND RECYCLING WHAT IS NOT RENEWABLE 227

11

ENERGY RESOURCES: FINITE, MEASURABLE, AND EXHAUSTIBLE 249

12
BIOLOGICAL RESOURCES: WHAT IS RENEWABLE AND HOW TO KEEP IT THAT WAY 289

PART 3: PEOPLE AND THE ENVIRONMENT

13
FIRES, STORMS, AND FLOODS: NATURAL HAZARDS AS PART OF THE ENVIRONMENT 325

14
THE HAZARDS WE PRODUCE: POLLUTANTS AND THEIR EFFECTS 357

15
PUTTING A VALUE ON THE ENVIRONMENT: ENVIRONMENTAL ETHICS AND ENVIRONMENTAL ECONOMICS 389

16
THINKING AHEAD WITH LESSONS FROM THE PAST: ENVIRONMENTAL PLANNING 411

17
ARTIFICE AS ENVIRONMENT: AN ENVIRONMENTAL PERSPECTIVE ON CITIES 439

18
WHOSE RIGHT TO DO WHAT? AN INTRODUCTION TO ENVIRONMENTAL LAW 459

APPENDICES 479

The study of our environment has expanded rapidly in recent years. Although this study is one of the oldest interests of human beings, it can also be considered one of the newest.

Like other living creatures, human beings live in, depend on, and influence the environment. Thus every human society has beliefs, myths, and attitudes about the environment. Our rapidly expanding technological civilization creates new demands on all aspects of the environment. Because more people are alive today than have ever lived on the Earth at one time, we are using more resources more rapidly than any civilization has used before. We also produce more wastes more rapidly than any previous civilization.

Human beings affect the environment in many ways. Chemicals produced in the United States have

PART 1

Environment and Life: An Introduction to Natural Processes in the Biosphere

been found throughout the oceans and in Antarctic creatures; burning of fossil fuels has increased the carbon dioxide concentration of the atmosphere; industries spew smoke that makes lakes acid, kills fish, and leaves toxic metals on the soil, killing trees. As we change our use of the land, we alter the nature of soils, sediments, and waters, and we increase the rate of extinction of species.

Along with these negative effects, modern civilization has made the environment more liveable in many ways. Since the invention of soap and the first understanding of modern medicine, we have developed better health care, and consequently people are healthier. We have learned to feed more people better than ever before, and more of us can travel further to see national parks and enjoy outdoor recreation than was possible in the past. In recent years, we have learned to live in closer harmony with our environment. For example, we are learning to control pests in a more benign manner than before.

Whether the positive benefits of technology outweigh the negative ones in the long run is an open question. We have many choices, but these choices can lead us in one of two directions. We can move forward to a future in which we live in harmony with our environment, maintaining our renewable resources and conserving and re-using our nonrenewable ones. Or, we can act in ways that will lead to an impoverished landscape, with its problems of pollution, the loss of resources, the exhaustion of soils, forests, and fisheries, and the extinction of many important species. Our choice of direction depends in part on our knowledge and understanding of the environment and in part on our *values*. This is why environmental studies must be seen as a broad and interdisciplinary field encompassing a range of activities—from biological and geological research to environmental ethics, planning, and environmental law. All of our social actions and decisions in regard to the environment require an understanding of basic principles. In Part One we will introduce these basic principles.

Chapter 1 sets down 13 fundamental principles of environmental studies which provide a set of themes for the entire book. You will find it useful to refer to these as you read later chapters.

Chapters 2 through 7 introduce fundamental scientific concepts necessary for the study of the environment. Chapters 2 and 3 consider the biosphere—the place where life exists on our planet. In Chapter 2 we examine climate, weather, geology, and soils, and in Chapter 3 we consider the living part of the biosphere.

In Chapters 4, 5, and 6 we present the basic principles of ecology, the study of the relation between living organisms and their environment. Chapter 4 examines populations, and Chapter 5 covers ecological communities and ecosystems—collections of populations functioning together in an environment. In Chapter 6 we consider the geography of life and the factors that influence the distribution of living creatures on the Earth's surface. There are somewhere between 3 and 10 million species on our planet. Our inability to estimate this number with any greater accuracy shows the complexity of the biosphere and our current state of knowledge.

Chapter 7 introduces the study of human populations. We examine the history of human beings on Earth and the species *Homo sapiens* as an ecological factor. The principles of ecology discussed in Chapters 4, 5, and 6 help us to understand the issues that confront our species within an environment.

Part One provides a foundation on which you can build your own ability to decide what is important in the environment and what we should do to best live within our environment, both local and global.

1

Fundamental Principles

In this chapter we introduce concepts basic to the study of the environment. Although these concepts do not constitute a complete list, they do provide the philosophical framework of this book. The concepts are not to be memorized. Rather, you should understand the general thesis of each concept to help you comprehend the material throughout the remainder of the text.

CONCEPT 1

The Earth is the only suitable habitat we have, and its resources are limited.

The Earth began approximately 4.6 billion years ago when a cloud of interstellar gas known as a solar nebula collapsed, forming protostars and planetary systems. Life on Earth began approximately 2 billion years later and since that time has profoundly affected our planet. Since the evolution of first life, many kinds of organisms have evolved, prospered, and died out, leaving only their fossils to record their place in the Earth's history. Several million years ago, our ancestors set the stage for the eventual dominance of the human race on Earth. It is certain, however, that our sun will eventually die. Then we too will disappear, and the impact of humanity on Earth history may not be particularly significant. However, to us living now and to our children and theirs, our environment is very important.

The Earth is a dynamic, evolving system in which material and energy are constantly being transferred and changing in form. The Earth, with respect to energy, is nearly in a steady state, receiving energy from the sun and releasing it to space. With respect to matter, however, the Earth is almost a closed system. There are a number of natural Earth cycles such as water and rock cycles in which earth materials are continually recycled. For example, the rain that falls today and erodes sediment to be washed to the sea will eventually return to the atmosphere, while the sediment that is deposited will eventually be transformed into solid rock.

We must understand the magnitude and frequency of processes that maintain dynamic Earth cycles. For example, if we wish to manage a region's water resources, we must be able to evaluate the nature and extent to which natural processes supply groundwater and surface water. Or, if we are concerned with the disposal of dangerous chemical or radioactive materials in the geologic environment, we must know how the disposal procedure will interact with natural cycles to insure that we or future generations will not be exposed to hazardous materials. Therefore, it is imperative that we recognize Earth cycles and determine the length of time involved in various parts

TABLE 1.1
Residence times of some natural cycles.

Earth Materials	Some Typical Residence Times
Atmosphere circulation	
Water vapor	10 days (lower atmosphere)
Carbon dioxide	5 to 10 days (with sea)
Aerosol particles	
Stratosphere (upper atmosphere)	Several months to several years
Troposphere (lower atmosphere)	One week to several weeks
Hydrosphere circulation	
Atlantic surface water	10 years
Atlantic deep water	600 years
Pacific surface water	25 years
Pacific deep water	1300 years
Terrestrial groundwater	150 years [above 760 m depth]
Biosphere circulation[a]	
Water	2,000,000 years
Oxygen	2000 years
Carbon dioxide	300 years
Seawater constituents[a]	
Water	44,000 years
All salts	22,000,000 years
Calcium ion	1,200,000 years
Sulphate ion	11,000,000 years
Sodium ion	260,000,000 years
Chloride ion	Infinite

SOURCE: *Earth and Human Affairs* by the National Academy of Sciences. Copyright ©1972 by the National Academy of Sciences. Reprinted by permission of Harper & Row, Publishers, Inc.
[a]Average time it takes these materials to recycle with the atmosphere and hydrosphere.

of these cycles. Tables 1.1 and 1.2 list characteristic residence times for some natural cycles and rates of several natural processes.

Two rather fundamental truths face us: First, the Earth is the only habitable place we have that is now accessible; and second, our resources, even renewable ones, are limited. If we consume the last of our grain and foul our water and air, then these so-called renewable resources may not be as renewable as we would wish. On the other hand, nonrenewable resources such as metals will eventually require large-scale recycling and/or extensive conservation to insure an adequate supply. Furthermore, we must recognize that many materials that are considered pollutants or waste are really resources that are out of place.

TABLE 1.2
Rates of some natural cycles.

Earth Processes	Some Typical Rates
Erosion	
Average U.S. erosion rate[a]	6.1 cm per 1000 years
Colorado River drainage area	16.5 cm per 1000 years
Mississippi River drainage area	5.1 cm per 1000 years
N. Atlantic drainage area	4.8 cm per 1000 years
Pacific slope (Calif.)	9.1 cm per 1000 years
Sedimentation[b]	
Colorado River	281 million metric tons per year
Mississippi River	431 million metric tons per year
N. Atlantic coast of U.S.	48 million metric tons per year
Pacific slope (Calif.)	76 million metric tons per year
Tectonism	
Sea-floor spreading	
N. Atlantic	2.5 cm per year
E. Pacific	7 to 10 cm per year
Faulting	
San Andreas (Calif.)	1.3 cm per year
Mountain uplift	
Cajon Pass, San Bernardino Mts. (Calif.)	1 cm per year

SOURCE: *Earth and Human Affairs* by the National Academy of Sciences. Copyright ©1972 by the National Academy of Sciences. Reprinted by permission of Harper & Row, Publishers, Inc.
[a]Thickness of the layer of surface of the continental United State per 1000 years.
[b]Includes solid particles and dissolved salts.

CONCEPT 2

Solutions to environmental problems often involve an understanding of systems and rates of change.

Analysis of many environmental problems and possible solutions often involves an understanding of systems, feedback cycles, and rates of change. For example, to properly manage renewable resources, whether it be timber or whales, one must be familiar with the system and feedback responsible for maintaining that resource as well as the current and past rates of exploitation and projected future harvest. The ocean, for example, is an open system in which there are few boundaries for energy and material. While we may consider many Earth systems such as the oceans or a volcanic system as open, the Earth is probably best regarded as a closed system in terms of resources, particularly renewable resources that are recycled on a rather frequent basis.

There are two types of feedback cycles in systems: positive feedback and negative feedback. Positive feedback is often known as the vicious cycle, whereas negative feedback is self-regulating, inducing the system to approach an equilibrium or steady state. Figure 1.1 illustrates the concepts of positive and negative feedback for the urban environment. Many processes in nature also exhibit feedback cycles. For example, off-road vehicle use may be a positive cycle because as vehicle use increases the number of plants that are uprooted increases, which

increases erosion. As this occurs, still more plants are damaged, which further increases erosion until eventually an area intensively used by off-road vehicles may be completely denuded of all vegetation and have a very high

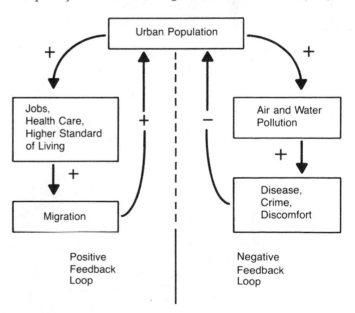

FIGURE 1.1
Idealized diagram of positive and negative feedback loops for an urban population. (Modified after Maruyama, 1963.)

erosion rate. On the other hand, systems such as rivers often display a negative feedback such that a rough steady state is formed. That is, as rivers change in response to an increase in regional rainfall or urbanization, the channel and floodplain system will make changes to accommodate the new, increased amount of water or sediment and within a relatively short time, a new steady state may be established.

Growth rates are important in changes that take place in systems. Exponential growth is particularly significant. Figure 1.2 shows an idealized diagram of an exponential growth curve. Notice that it is shaped like "J"; in the early stages the growth may be quite slow, but then it increases rapidly and then very rapidly. Many systems, both human-induced and natural, may approach the "J" curve for some lengths of time. Involved in exponential growth are two important factors: the rate of growth measured as a percentage, and the doubling time in years, which is the time necessary for the quantity of whatever is being measured to double. A general rule of thumb is that the doubling time is approximately equal to 70 divided by the growth rate. This rule applies to growth rates up to approximately 10 percent. Beyond that, the errors may become quite large.

There are some interesting consequences of exponential growth. For example, if we consider the growth rate of energy consumption, which in the United States has approached 7 percent, then the doubling time will be 10 years. This means that in one decade we will consume as much energy as we have consumed up until that time. A hypothetical story by Albert Bartlett will illustrate this [1]. Consider a hypothetical strain of bacteria that grows with a division time of 1 minute. The growth rate, then, is 1 divided by 60 or 1.66 percent per second, and the doubling time is 60 seconds or 1 minute. Assume that our hypothetical bacterium is put in a bottle at 11:00 A.M., and it is observed that the bottle is full at 12:00 noon. An important question is, When was the bottle half-full? The answer is 11:59 A.M. If you were an average bacterium in the bottle, at what time would you realize that you were running out of space? There is certainly no unique answer to this question, but at 11:58 A.M. the bottle was 75 percent empty, and at 11:57 A.M. it was 88 percent empty. Now assume that at 11:58 A.M. some farsighted bacteria realized that they were running out of space and started looking around for new bottles. Let's suppose that they were able to find three more bottles. How much time did they buy? The answer is 2 additional minutes. They will run out of space at 12:02 P.M.

The preceding example, while hypothetical, illustrates the power of exponential growth. Many systems in nature display exponential growth for some periods of time, so it is important that we be able to recognize it. In particular, it is important to recognize exponential growth with positive feedback as it may be very difficult to stop positive feedback cycles. Negative feedback, on the other hand, tends toward a steady state and thus is easier to control.

Changes in natural systems may be predictable and should be recognized by anyone looking for solutions to environmental problems (Fig. 1.3). Where the input into the system is equal to the output (Fig. 1.3a), a rough steady state is established and no change occurs. Examples of an approximate steady state may be on a global scale—the balance between incoming solar radiation and outgoing radiation from the Earth, or the system of plate tectonics in which new lithosphere is being created and destroyed at about the same rate—or on the smaller scale of a duck farm in which ducks are brought in and harvested at a constant rate. Another example of change is where the input into the system is less than the output (Fig. 1.3b). Examples of this would be the use of resources such as groundwater or the harvest of certain plants or animals. If the input is much less than the output, then the groundwater may be completely used or the plants and animals may die out. In a system where input exceeds output (Fig. 1.3c), positive feedback may occur, and the stock of whatever is being measured will increase. Examples are the buildup of heavy metals in lakes or the pollution of soil water. By using rates of change or input/output analysis of systems, we can derive an average *residence time* for such factors. The average residence time is a measure of the time it takes for the total stock or supply of a particular material, such as a resource, to be cycled through the pool. To compute the av-

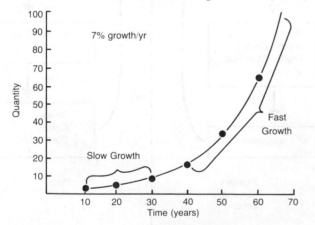

FIGURE 1.2
Idealized diagram of the "J" curve for exponential growth. Notice that in the lower part of the "J" the growth is slow, but past the bend, growth becomes extremely rapid.

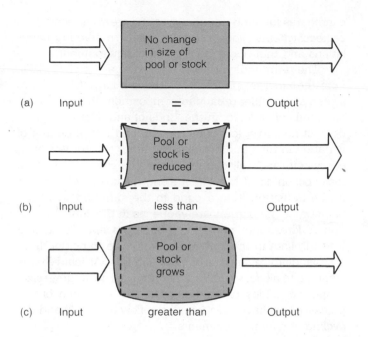

FIGURE 1.3
Idealized diagram of the major ways that a pool or stock of some material may change. (Modified after Ehrlich, Ehrlich, and Holdren, 1977.)

erage residence time, we simply take the total size of the stock or pool and divide it by the average rate of transfer through that pool or stock.

An understanding of changes in systems is primary in many problems in environmental studies. We conclude from our discussion of growth rates and exponential growth that very small growth rates may yield incredibly large numbers in modest periods of time. On the other hand, with other systems it may be possible to compute a residence time for a particular resource and, knowing this, apply the information to develop sound management principles. Recognizing positive and negative feedback systems as well as calculating growth rates and residence times, then, enable us to make predictions concerning resource management.

CONCEPT 3

The Earth, as a planet, has been profoundly altered by life; its atmosphere, oceans, and sediments are strongly modulated by life and are very different from what they would be on a lifeless planet.

The modern atmosphere of the Earth has some peculiar attributes. For example, in it oxygen and methane occur together in concentrations that are close to an explosive combination. That is, a little more methane or a little more oxygen and any spark could produce a large ex-

plosion. This means that methane and oxygen occur in concentrations that are far from a chemical equilibrium, an equilibrium that would be reached if the chemical constituents of the oceans, lands, and atmosphere were left together in a closed transparent container and placed out in the sun and under the stars, subject to ordinary Earth days and nights for a very long time. In such a closed container, the methane would inevitably combine with the oxygen to form carbon dioxide and water [2]. This does occur in our present Earth's atmosphere, so there must be a source producing methane as well as oxygen, replenishing what is converted in the air to carbon dioxide and water.

This curious chemical disequilibrium does not occur on Mars and Venus, the two planets in our solar system most like the Earth in size and distance from the sun. Space probes to Mars and Venus, as well as Earth-based observations of these planets with telescopic devices, reveal that the atmospheres of these two planets are similar in composition but are very different from the Earth's. The atmospheres of Mars and Venus are predominantly carbon dioxide, which exists only as a trace constituent (0.03 percent) of the Earth's atmosphere.

The peculiar chemical disequilibrium of the Earth's atmosphere is a result of life acting over long periods of geological time. All of the gases in the Earth's atmosphere are profoundly altered by life, and have been so for several billion years. Moreover, the atmospheric constituency of the Earth appears to have been relatively stable for millions of years, suggesting that life regulates the conditions of its existence by stabilizing the atmosphere as well as by profoundly altering its constitution. This is why we say *the Earth is a living planet.*

Life is thus intimately tied to the cycling of elements through the atmosphere, not only using the gases in the air but controlling the air's constitution. Life is responsible for the kinds of compounds and their concentration, for the amounts that are there at any one time, and the rates at which the compounds enter and leave the atmosphere.

The impact of life on the atmosphere extends beyond the atmosphere's chemistry. Life affects the physical characteristics of the atmosphere and the Earth's surface as well as the rate of heating and cooling—and therefore the temperature—of the Earth's surface and its weather and climate.

The effects of life on the Earth's chemistry are apparent in the soils and rocks and the fresh waters and oceans as well as in the air. As the famous ecologist G. Evelyn Hutchinson observed in 1954, the effects of life on geological processes is indicated in ''the presence of great thicknesses of limestone in the sedimentary column'' which indicate that "an enormous quantity of carbon dioxide had left the atmosphere" through the process of photosynthesis [3].

Locally, life has caused water to be much more varied than it would have been on a lifeless planet: basic-alkaline in some locales, highly acidic in others. Globally, life adds organic acids such as humic acids from decaying tree leaves; the biota also change the proportion of dissolved carbon dioxide, sulphur dioxides, and nitrogen oxides.

Other evidence suggests that many of the important mineral deposits are the result, directly or indirectly, of biological processes. For example, the deposition of copper and other metals can take place in shallow waters lacking oxygen and within a narrow range of acidity. It is life that removes the oxygen from these waters and creates the necessary conditions of acidity.

The effect of life on the Earth is all the more remarkable because living organisms make up only a tiny fraction of the Earth's mass. If the mass of all living things were evenly mixed with the rest of the material of the Earth, the concentration of living material would be two tenths of one part in a trillion, which is a concentration at the border of detectability of our most sophisticated chemical instruments. (The mass of living things is 1.2×10^{12} kilograms; the mass of the Earth is 6×10^{24} kilograms.) Even in relation to the atmosphere, the biota make up a small fraction of the total mass. The mass of the biota equals 0.02 percent of the mass of the atmosphere.

CONCEPT 4

Sustained life on the Earth is a characteristic of ecosystems, not of individual organisms or populations.

We know of no single organism, population, or species that both produces all of its own food and completely recycles all of its own metabolic products. Green plants in light produce sugar from carbon dioxide and water, and from sugar and inorganic compounds make many organic compounds, including protein, and woody tissue. But no green plant alone can degrade woody tissue back to its original inorganic compounds. Those living things that degrade woody tissue, such as bacteria and fungi, do not produce their own food, but instead obtain their energy and chemical nutrition from the dead tissues they feed on. From these observations, we know that for complete recycling of chemical elements to take place there must be several species.

Minimal systems that could have the property for a flow of energy and a complete chemical cycling are apparently composed of at least several interacting populations and their nonbiological environment. The smallest candidates for such minimal systems are what ecologists call *ecosystems*—local communities of interacting populations and their local, nonbiological environment.

The term **ecosystem** is applied to areas of all sizes, from the smallest puddle of water to a large forest. Ecosystems also differ greatly in composition: in the number and kinds of species, in the kinds of and relative proportions of nonbiological constituents, and in the degree of variation in time and space. Sometimes the borders of an ecosystem are well defined, as in the transition from a rocky ocean coast to a forest along the coast of Maine, or in the transition from a pond to the surrounding forest. Sometimes the borders are vague, as in the subtle gradation of forest into prairie in Minnesota and South and North Dakota in the United States, or the transition from grasslands to savannahs or forests in East Africa. What is common to all ecosystems is not physical structure—size, shape, variations of borders—but the existence of the processes we have mentioned, the flow of energy and the cycling of chemical elements.

Ecosystems can be natural or artificial. An artificial pond that is part of a waste treatment plant is an example of an artificial ecosystem. Ecosystems can be natural or managed, and the management can vary over a large range of actions. Agriculture can be thought of as partial management of certain kinds of ecosystems.

Natural ecosystems carry out many "public service" functions for us. Waste water from houses and industries is often converted to drinkable water by passage through natural ecosystems, and pollutants, like those in the smoke from industrial plants or the exhaust from automobiles, are often converted to harmless compounds by forests.

CONCEPT 5

Individual populations are capable of rapid exponential growth, but this is rarely achieved in nature; control of populations is the norm.

Although maximum population size is limited by resources, biological populations are capable of extremely rapid growth. When the increase in the size of a population is a constant percentage of the current population size, the growth is called **exponential** or **geometric**. Laboratory cultures, like yeast in a sugar solution, can increase exponentially for brief periods. Occasionally, natural populations of more complex multicellular organisms are also observed to increase exponentially for brief periods. A pheasant population introduced onto a small island in Puget Sound, Washington, grew exponentially for more than 2 decades (Fig. 1.4); a population of 8 introduced in 1937 grew to 1898 in less than 6 years [4].

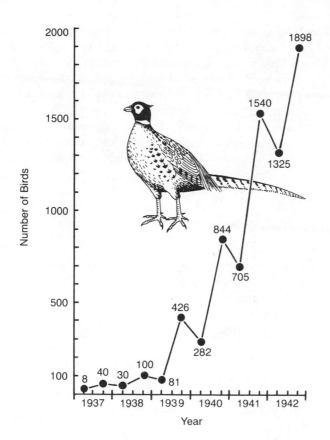

FIGURE 1.4

Growth of a pheasant population on a small island in Puget Sound, Washington, over a 6-year period.

Since the Industrial Revolution, the world's human population has increased at a rate more rapid than an exponential increase; that is, the annual increase in the total number of humans has been an increasingly larger fraction of the current population.

A biological population is characterized by three rates: a rate of birth (reproduction), a rate of growth, and a rate of death. When rapid population growth occurs, eventually a point is reached when the resources available per individual become limiting, and a change occurs in the population rates: births may decrease and/or deaths increase, and/or individual growth decreases.

When a population is self-regulated, it is said to exhibit **density-dependent population control.** Density-dependent population control implies that one or more of the three population rates changes with population size. That is, birth, survival, and growth rates may decrease as the population size increases.

One population may be regulated by another. Trees of different species compete with each other for light, and one species limits the abundance of another. An increase in the size of one population may cause an increase in the

incidence of parasitism, such as disease, or an increase in the incidence of predation.

The sizes of populations are also changed by factors that are unrelated to the population's density. For example, a tornado can sweep through a forest and knock down all the trees in its path. The incidence of the tornado is completely independent of the number of trees in the forest. If all trees in the path of the tornado are killed, the death rate is also independent of the size of the population. In this case, the size of a population may be restricted to a relatively narrow range by factors completely unrelated to the population's own size or rates of birth, growth, and death. This is called **density-independent population regulation.**

To summarize, nothing can increase forever. The Earth and the known universe are finite in space, matter, and energy. In a finite universe, there is an upper bound to the size of everything. So, too, are populations limited to a finite range.

CONCEPT 6

Today's physical and biological processes are maintaining and modifying our Earth and have operated throughout much of geologic time. However, the magnitude and frequency of these processes are subject to natural and artificially induced changes.

The concept that present physical and biological processes which are forming and modifying our Earth will help to explain the geologic and evolutionary history of the Earth is known as the doctrine of **uniformitarianism.** Simply stated as "the present is the key to the past," uniformitarianism was first suggested by James Hutton in 1785. Because Charles Darwin was impressed by the concept of uniformitarianism, it pervades his ideas on biological evolution. Today the doctrine is heralded as one of the fundamental concepts of the Earth and biological sciences.

Uniformitarianism does not demand or even suggest that the magnitude and frequency of natural processes remain constant with time. Obviously some processes do not extend back through all of geological time. For example, the processes operating in the oxygen-free environment during the first billion years or so of Earth history must certainly have been quite different from processes we observe today. However, as long as the early continents, oceans, and atmosphere were like modern ones, and so long as the basic factors that rule biological evolution have not changed, then we can infer that present processes also operated in the past. For example, if we study present-day stream channels and learn something about the types of deposits associated with streams, then we can infer that similar-looking deposits in the geologic record are most likely stream deposits. Similarly, we can study modern organisms and relationships between their morphology and biologic functions and interpret the fossil record. For example, the study of predator-to-prey ratios along with the bone structure of fossils and other information are being used to argue that dinosaurs, instead of being cold-blooded and close relatives of reptiles, may have been warm-blooded animals and more closely related to birds.

Human activities increase and decrease the magnitude and frequency of natural Earth processes. For example, rivers will emerge periodically from their banks and flood the surrounding countryside regardless of human activities, but the magnitude and frequency of flooding may be greatly increased or decreased by human activity. Therefore, in order to predict the long-range effects of certain processes such as flooding, we must be able to determine how our future activities will change the rate of a geological process.

From a biological point of view, we also know that the ultimate fate of every species is extinction. However, the effects of human use and interest in the land may greatly change the rates of extinction. Figure 1.5 shows that as the human population has increased there has been a parallel increase in the extinction of species. These extinctions are closely related to land-use change—to agricultural and urban uses that influence the ecological conditions of an area. Some species are domesticated or cultivated while others are removed as pests. The parallel rise of human population and extinction of nonhuman species should be of little surprise.

To be useful from an environmental standpoint, the doctrine of uniformitarianism will have to be more than a

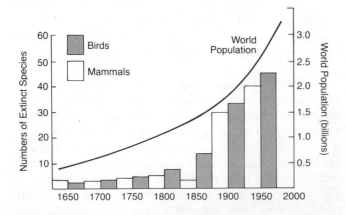

FIGURE 1.5
Increase in human population (a) paralleled by (b) an increase in the extinction of birds and animals. (From Ziswiler, 1967.)

key to the past. We must turn it around and say that a study of past and present processes may be the key to the future. That is, we can assume that in the future the same physical and biological processes will operate, but at rates that will vary as the environment is influenced by human activity. Geologically ephemeral landforms such as beaches and lakes will continue to appear and disappear in response to moderate and catastrophic natural processes; extinctions of animals and birds will also continue to some extent in spite of people's activities.

CONCEPT 7

There have always been natural Earth processes that are hazardous to people. These natural hazards must be recognized, and avoided where possible, and their threat to human life and property minimized.

We have established in our discussion of uniformitarianism that present physical and biological processes have been operating a good deal longer than people have been on Earth. Therefore, people have always had to contend with processes that make their lives difficult. Surprisingly, however, *Homo sapiens* appears to have evolved during recent Ice Ages, a period of harsh climates and environments.

Many physical processes continue to cause loss of life and property damage, including flooding of coastal or flood plain areas, earthquakes, and landslides. In addition, biological processes often mix with physical events to produce hazards. For example, after earthquakes and floods, water may be contaminated by bacteria and the rate of the spread of diseases increased. The magnitude (intensity of energy released) and frequency (recurrence interval) of natural, hazardous processes depend on such factors as the region's climate, geology, and vegetation. In general, there is an inverse relationship between the magnitude of an event and its frequency. That is, the larger the flood, the less frequently it occurs. Studies have demonstrated generally that most of the work in forming the Earth's surface is done by processes of moderate magnitude and frequency rather than by processes with low magnitude and high frequency or by extreme events of high magnitude and low frequency. As an analogy to the magnitude-frequency concept, consider the work in logging a forest done by termites, people, and elephants. The termites are small but work quite steadily; the people work less often than termites but are stronger. Given enough time, they are able to fell most of the trees in the forest and therefore do a great deal of work. Imagine several elephants that rarely visit the forests, but when they do are capable of knocking down many trees. We can see from this analogy that most of the work is done by people, who work at a rather moderate expenditure of energy and time, rather than by the termites' frequent but low expenditure of energy or the elephants' infrequent, high expenditure of energy.

Natural hazards are nothing more than natural processes. They only become hazards when people live or work in areas where these processes occur naturally. The naturalness of these hazards is a philosophical barrier that we encounter when we try to minimize their adverse effects. It is the environmental scientist's job, therefore, to identify potentially hazardous processes and make this information available to planners and decision makers so that they can formulate various alternatives to avoid or minimize the threat to human life or property.

CONCEPT 8

All human activities, from the most primitive farming to the most recent technological innovations, cause some changes in our environment.

Since the Industrial and Scientific revolutions, there have been more human-induced changes in the environment, but all human cultures have altered the environment in some way. Rarely, if ever, do undisturbed wildernesses provide an abundance of easily obtained food for people. Occasionally, shellfish and fish along ocean shores have provided primitive people with food that could be taken with little effect on the environment, as long as the human population was small. The American Indians who lived on Long Island in what is now New York State were able, at least during parts of the year, to obtain a plentiful amount of oysters. Peter Kalm, a Swedish naturalist who came to North America in 1749 to collect new species of shrubs and trees for European gardens, wrote in his journal about this abundance of food. He wrote that one could easily fill a cart full of oysters from what washed onshore between the tides, and that in New York City, then a small town, the poor people "lived on oysters and a little bread" [5]. There is also evidence that some of the American Indians of coastal California obtained much of their food from the ocean coast with little effect on the coastal environment, but these examples are exceptions to the rule. In general, wilderness areas, like the vast forests that once covered the temperate zones, provide little food for people.

To sustain a high human population, large areas of the landscape must be altered. We will see, in our discussion of Concept 9, that the effects of land use are cumulative. Since our goals cannot include an elimination of the effects of human beings on the Earth, we must attempt to seek a wise use of the environment.

CONCEPT 9

The effects of land use tend to be cumulative, and, therefore, we have an obligation to future generations to minimize its negative effects.

When early hominids roamed the broad grasslands, marshy deltas, and adjacent forests of ancient Lake Rudolph along the Great Rift Valley system of East Africa several million years ago, these prehistoric people were completely dependent upon their immediate environment. Their numbers were small and their effect on the environment was small as they hunted game and in turn were hunted by predators. This relationship between people and the environment probably existed until about 800,000 years ago, when people developed the skill of making fire. The use of fire affected the environment in new ways. First, fire can affect large areas of forest or grasslands. Second, fire is a repetitive process, occurring in the same area at rather frequent intervals. Third, fire tends to be selective; that is, certain species may be locally exterminated, while others that exhibit a resistance to or rapid recovery from fire are favored. Fire was the first instance of artificial land use capable of significantly modifying the natural environment.

The next major change occurred approximately 10,000 years ago when agriculture first began. The emergence of agriculture was the second instance of artificial land use which significantly modified the natural environment. Agriculture also fostered development of more or less continously occupied village sites or clusters of sites, which introduced further modifications of the environment such as shelters, primitive toilets, and protective barriers against predators and other people. These early villages probably became the first areas to experience pollution from disposal of waste and soil erosion from removal of indigenous vegetation. As agriculture supplied the necessary food for an increasing population, additional land had to be cleared. This activity certainly influenced the area's ecological conditions as some species were domesticated or cultivated and others were removed as pests. Thus, again the increase in human population is paralleled by an increase in the number of extinctions of other species.

By tracing the technological development of *Homo sapiens* through time, we see that as cities and farms increase, there is an increased demand for diversification of land use, the effects of which also tend to be cumulative with time. Since the beginning of civilization, between 6000 years ago and perhaps as far back as 15,000 years ago, virtually the entire surface of the Earth has been altered by human activity. Little, if any, land can be considered original or untouched. Furthermore, our ability to manipulate the environment and thus cause further changes is increasing rapidly. This being the case, we need to examine the effects of land use in a historical framework, if only to insure that our children and their children can survive in the environment they inherit. In defense of human activity, we should note that in comparison to the energy involved in the natural processes of mountain building, volcanic activity, and even the erosional power of streams, human impact is rather small. However, in large metropolitan areas the impact of people and urbanization is quite pronounced and is frequently detrimental to the health and well-being of people.

The lessons learned from past experiences with land use and people are quite explicit. Where sound conservation practices have been used, there have been successful adjustments of population. However, the negative results of wasteful exploitation of resources include gullied fields on alluvial plains, silted-up irrigation reservoirs and canals, and ruins of once prosperous cities. For example, about 5300 years ago the Phoenicians migrated from the desert to settle along the eastern coast of the Mediterranean Sea and established the towns of Tyre and Sidon, Beyrouth and Byblos. The area is mountainous with a relief of about 3000 meters and, at that time, was heavily forested with the famous cedars of Lebanon. These trees became the timber supply for shipbuilding and construction in the alluvial plains of the Nile and Mesopotamia. In many areas the amount of flat land was limited, so as population increased the people moved to the slopes. As the slopes were cleared and cultivated, they were subject to soil erosion. Today there are a large number of terrace walls in various states of repair, indicating that the ancient Phoenician farmers attempted to control erosion with rock walls across the slope 40 to 50 centuries ago [6].

The cedars of Lebanon retreated under the ax, until today very little of the original forest of approximately 2600 square kilometers is left. Evidence suggests that given present climatic conditions the cedars would still grow where soil has escaped the process of erosion. Unfortunately, today the bare limestone slopes strewn with remnants of former terrace walls are testimony to the results of erosion and the decline and loss of a country's resources [6].

Other examples could be cited, but we can conclude that in areas long populated by people soil erosion is often a serious problem. Furthermore, in several instances erosion has destroyed the land and retarded the progress of civilization; therefore, conservation of soils must remain a national interest for, as stated by W. C. Lowdermilk, "One generation of people replaces another, but productive soils destroyed by erosion are unrestorable and never replaceable" [6].

CONCEPT 10

In our environment, every action has more than one effect.

This concept is sometimes loosely referred to as the principle of environmental unity; that is, everything affects everything else. Of course, this cannot be absolutely true. For example, the extinction of a species of snails in North America is hardly likely to change the flow characteristics of the Amazon River. On the other hand, the principle does emphasize that many aspects of the natural environment are closely related and that perturbations or changes in one part of the system will often have secondary and tertiary effects within the system or even affect adjacent systems.

For example, consider a major change in land use from forests and grassland to a large urban complex. The progressive clearing of the land for urban uses will change the amount of sediment eroded from the land, which will significantly affect the form and shape of the stream channel; the sediment will also affect the stream's biology, changing the composition and structure of the aquatic biological system. Eventually, as more land becomes paved or otherwise made impervious, the amount of sediment eroded from the land will decrease and the streams will further adjust to this lesser sediment load. The urbanization process is also likely to pollute the streams or otherwise change water quality and consequently affect the biological systems in the stream and adjacent banks. Finally, as the city continues to develop and skyscrapers are constructed, these may form local rain shadows that modify the climate of the area, affecting both the amount and pattern of precipitation. Furthermore, as power plants for the cities burn fossil fuels, contaminants enter the atmosphere along with those produced by automobiles and other industries to cause air pollution which may change the air quality for the entire region around the urban center. Problems such as acid rain might also result in adjacent areas. Thus, the land-use conversion sets off a whole series of changes in the local and regional environment, and each change is likely to precipitate still others.

CONCEPT 11

An understanding of complex environmental problems requires a team approach involving several disciplines.

The serious student of the environment must be aware of the contributions to environmental research from biology, conservation, atmospheric science, chemistry, environmental law, architecture, engineering, and geology as well as from physical, cultural, economic, and urban geography and philosophy and history. Environmental studies, by its very nature, is the domain of the generalist with strong interdisciplinary interests. This description in no way refutes the specialist's significant contribution in various aspects of environmental work or the importance of the generalist's consideration of specific problems or specialty areas for research. It merely suggests that although our research interests may be specialized, we should be generalists by being aware of other disciplines and their contributions to environmental problems. Many projects may be studied best by an interdisciplinary team, provided that team is well coordinated.

CONCEPT 12

Every human society has a set of beliefs about nature, the effects of nature on human beings, and the effects of human beings on their natural surroundings.

An unknown and disordered world is discomforting. Every human society has its own beliefs about the history, structure, and purpose of the Earth and the creatures who live on it. Throughout the history of Western civilization, philosophers have sought to explain why the world existed. There was an appearance of order. What was the purpose for this order, and who was the object of the design of the world and the universe? The ancient philosophers believed on the whole that the universe was one of order, harmony, and balance, that there was a purpose behind the existence of everything, that there was a design and a designer, and that human beings were the object of this design [7]. "Who cannot wonder at this harmony of things, at this symphony of nature which seems to will the well-being of the world?" asked Cicero in *The Nature of the Gods,* written in the first century B.C. [8].

Concern with our "environmental crisis" is the most recent manifestation of ancient issues. Again, we are trying to discover the proper role of human beings in nature and the nature of our natural world. Again we are asking, Have we upset the crucial balance of nature in some way that may lead to our own destruction? Today we view these questions against a background of modern technology's effects on the entire Earth and against an immense accumulation of scientific information, but the questions themselves are ancient. The answers that we find may be based on scientific information, but they are set within a context of our history and culture and our view of aesthet-

ics and ethics. **Environmental ethics,** an active field in philosophy, attempts to bring our modern concerns and modern scientific knowledge into a philosophical framework so that we may ask these ancient questions again.

CONCEPT 13

What is necessary to sustain life is a scientific question; what effects specific actions have on the life around us and our environment are applied scientific questions; but what we choose is ultimately an ethical question beyond the reach of science.

Through physics, chemistry, geology, biology, meteorology, ecology, and the other Earth-related sciences, we can understand how life is sustained on the Earth. We can discover what is required for an individual organism, a population, or a single species. We can learn how ecological communities—groups of interacting populations—are sustained over long periods of time and the minimum characteristics for a functioning ecosystem. These are scientific questions requiring fundamental research. Although an immense amount of information has been gathered since the beginning of modern science about these questions, much yet remains to be learned.

Scientific knowledge enables us to understand the effects of specific actions on life around us: on individuals, on populations, on communities, on ecosystems, and on the Earth's entire **biosphere**—the region of the Earth where all life exists. The study of how our activities and how changes in our activities may change our environment is a rapidly growing area of applied science.

What we choose to do about our actions is an ethical decision that reaches beyond the realm of the sciences. Concept 12 showed that the concern with the role of human beings in nature is an ancient one and is part of all human culture. Science provides us with new knowledge and new tools that may lead to new principles of conservation and environmental management, but only when they are embedded within a set of ethical precepts. Our use of the environment never has been and never will be free from our cultural and ethical views. The field of environmental studies sharpens our understanding of our physical environment and the effects of our actions, and at the same time maintains an awareness of the historical, cultural, and ethical overtones of the environmental issues.

REFERENCES

1 BARTLETT, A. A. 1980. Forgotten fundamentals of the energy crisis. *Journal of Geological Education* 28: 4–35.

2 LOVELOCK, J. E., and MARGULIS, L. 1974. Homeostatic tendencies of the Earth's atmosphere. *Origins of Life* 5: 93–103.

3 HUTCHINSON, G. E. 1954. The biochemistry of the terrestrial atmosphere. In *The Earth as a planet*; ed. G. P. Kuiper, pp. 371–433. Chicago: The University of Chicago Press.

4 EINARSEN, A. S. 1942. Specific results from ring-necked pheasant studies in the Pacific Northwest. *Transactions of the North American Wildlife Conference* 7: 130–45.

5 KALM, P. 1963. *Travels in America*. New York: Dover.

6 LOWDERMILK, W. C. 1943. Lessons from the Old World to the Americans in land use. In *Smithsonian report for 1943*, pp. 413–28.

7 GLACKEN, C. J. 1967. *Traces on the Rhodian Shore: Nature and culture in Western thought from ancient times to the end of the eighteenth century*. Berkeley: University of California Press.

8 CICERO, Marcus Julius. 1972. *The nature of the gods*. (Translated by H. C. P. McGregor.) Aylesbury, England: Penguin.

FURTHER READING

1 CLOUD, P. 1978. *Cosmos, Earth, and man*. New Haven, Conn.: Yale University Press.

2 DASMANN, R. F. 1976. *Environmental conservation*. 4th ed. New York: Wiley.

3 DAY, J. A.; FOST, F. F.; and ROSE, P., eds. 1971. *Dimensions of the environmental crisis*. New York: Wiley.

4 LEOPOLD, A. 1949. *A Sand County almanac*. New York: Oxford University Press.

5 NASH, R. 1967. *Wilderness and the American mind*. New Haven, Conn.: Yale University Press.

6 ———, ed. 1972. *Environments and Americas*. New York: Holt, Rinehart and Winston.

7 SKINNER, B. J., ed. 1981. *Use and misuse of the Earth's surface*. Los Altos, Calif.: William Kaufmann.

8 STRONG, D. H. 1971. *The conservationists*. Reading, Mass.: Addison-Wesley.

STUDY QUESTIONS

1 Discuss the difference between positive and negative feedback cycles. Provide one example of each that has environmental significance.

2 Why is the "J" curve so important in understanding environmental problems?

3 Input/output analysis of systems is important in evaluating many environmental problems. Discuss potential problems where input is less than output. Provide two examples, one each from biological and physical systems.

4 What physical evidence supports the notion that life has greatly affected global physical processes?

5 Discuss how natural ecosystems perform "public service" functions for us. Which functions are most important? Why?

6 What is the environmental significance of knowing if a population is density-dependent or density-independent?

7   Why is uniformitarianism an important principle in Earth and biological sciences? Has this principle held throughout the entire 4 to 6 billion years of Earth history? Explain.

8   What is the magnitude-frequency principle, and how can it be used in environmental management or planning?

9   Explain how the effects of land use are cumulative. Provide one example from the past and one possible prediction for the future to illustrate the concept of cumulative effects of land use.

10 Discuss the principle of environmental unity. Provide several examples *not* discussed in the text.

11 What is the difference between a scientific question and an ethical question? Why is the difference important to society?

2 How Fit Is the Environment? The Biosphere's Physical Processes

FIGURE 2.1
St. Francis Dam prior to
failure (a), site geology (b),
and after failure (c). (Photos
courtesy of Los Angeles
Department of Water and
Power.)

(a)

ST. FRANCIS DAM

West   East

DAM

Sedimentary Rocks

CANYON WALLS

Metamorphic Rocks (Schist)

FOLIATION PLANES

Fault

(b)

(c)

18

The study of the environment involves the study of life and life's environment. At the largest scale—the scale of our entire planet—the study of the environment is the study of the **biosphere.** The biosphere is simply the place where life exists and where its influence extends. Living things are found from the depths of the oceans to the summits of the tallest mountains, but most life exists within a few meters of the Earth's surface.

The environment influences life, and life influences its environment. In this chapter we will examine the environment and its basic attributes important to environmental studies. In Chapter 3 we will consider the influence of life on the environment.

The Earth's surface is usually divided into three major physical systems: the solid portion—the rocks, soils, and other sediments—known as the **lithosphere;** the oceans and fresh waters, called the **hydrosphere;** and the gaseous portion, familiar to us as the **atmosphere.** Each is linked to the others and to life in many ways, and each undergoes change. In this chapter we will consider briefly the flow of energy and matter through these three systems. They are important not only at the large scale of the entire planet, but also in the environment immediately around us.

INTRODUCTION

The effects of the Earth's physical systems on structures built by people was illustrated by the failure of the St. Francis Dam near Saugus, California, on March 12, 1928. The resulting flood claimed over 500 lives and accounted for millions of dollars in property damage. The dam failed because of adverse geologic conditions at the site (Fig. 2.1). The dam was 63 meters high, with a main section 214 meters long, and held 47 million cubic meters of water. The east canyon wall was composed of a kind of rock that is made up of parallel layers of minerals (foliation). Because the foliation was parallel to the valley slope, it formed a zone of weakness likely to fail by landslide. Prior to the failure, both recent and ancient landslides had indicated the instability of this rock. The rocks of the west side of the dam formed prominent ridges, suggesting that they were very strong and resistant. Under dry conditions this was true; however, when the rocks became wet, they disintegrated. This was not discovered and tested until after the dam had failed. The two rock types meet along a fault or fracture zone with an approximately 1.5-meter thick zone of crushed and altered rock. Although it was shown on the 1922 fault map of California, the fault was either not recognized or was ignored.

The causes of the St. Francis Dam failure were a combination of slipping and disintegration of the rocks, and leakage of water along the fault zone, which washed out the crushed rock. From this tragedy, however, came some good; namely, we learned that it is necessary to investigate carefully the properties of earth materials prior to constructing large engineering structures. Such investigations are now standard procedure.

From the St. Francis Dam failure and other events we recognize that physical systems significantly affect the use of our environment. Thus, in assessing environmental impacts of dams, housing, roads, or energy delivery, we must consider how they will interact with a broad spectrum of physical processes in the atmosphere, hydrosphere, and lithosphere. We will examine these processes in the following sections on solar radiation, atmosphere and climate, and the geologic cycle.

SOLAR RADIATION, ATMOSPHERE, AND CLIMATE

SOLAR RADIATION

Although the Earth intercepts only a very tiny fraction of the total energy emitted by the sun, it is solar energy that sustains life on Earth and drives the surface portion of the geologic cycle. Energy emitted from the sun in the form of electromagnetic radiation travels to Earth at the speed of light through the vacuum of space. Different forms of electromagnetic energy may be distinguished by their wavelengths, and the collection of all the possible wavelengths may be grouped into what is known as the **electromagnetic spectrum** (Fig. 2.2). A modest appreciation of the spectrum is fundamental to understanding many environmental problems.

The energy that is radiated from a body such as the sun or the Earth varies with the fourth power of the temperature of the body. Thus, if a body's temperature doubles, the energy radiated increases by 16 times. This explains why the sun, with a temperature of 5800°C radiates a tremendously greater amount of energy compared with the Earth at 15°C (Fig. 2.3). Figure 2.3 illustrates another important point: The sun emits relatively short-wave radiation, whereas the Earth emits relatively long-wave radiation.

Essentially all the energy available at the Earth's surface comes from the sun; a small additional amount comes from the deep-seated geologic processes within the Earth (Fig. 2.4). Of the solar energy that reaches the

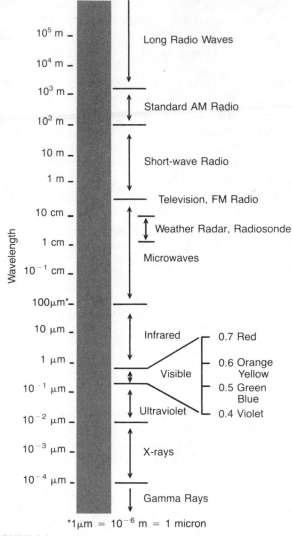

*1μm = 10^{-6} m = 1 micron

FIGURE 2.2
The electromagnetic spectrum.

THE ATMOSPHERE

The atmosphere is composed of gas molecules held close to the Earth's surface by gravitation. The major gases in the atmosphere are nitrogen (78%), oxygen (21%), argon (0.9%), and carbon dioxide (0.03%). The atmosphere also contains minor or trace amounts of numerous elements in compounds, including methane, ozone, hydrogen sulphide, carbon monoxide, oxides of nitrogen and sulphur, hydrocarbons, and various dust particles (called *particulates*). In addition, water vapor in the atmosphere varies from zero to approximately 4 percent by volume.

Two important properties of the atmosphere at any particular elevation above sea level are pressure and temperature. Pressure is a force per unit area, and atmospheric pressure increases as altitude decreases because there is a greater weight from the overlying air at lower altitudes. This relationship is shown in Figure 2.5. The average temperature through the first 50 kilometers of atmosphere is also shown.

Winds, the movement of clouds, and the passing of storms and clear days show us that the atmosphere changes rapidly and continually. This atmospheric circulation takes place at a variety of scales. The lower atmosphere is heated by the surface of the Earth. As warm air rises, it cools by expansion, and winds are produced as cooler surface air is drawn in to replace rising warm air. The lower atmosphere is therefore said to be "unstable" because it tends to circulate and mix, particularly in the lower 4 kilometers or so. On a global scale, circulation results primarily from differential heating of the atmosphere and the Earth's rotation (Fig. 2.6). On a regional scale, Figure 2.7 shows some of the air masses that cross North America on a yearly basis. Finally, on a local scale, differential heating of the air causes such phenomena as sea and land breezes.

CLIMATE AND CLIMATIC CHANGE

The term *climate* refers to the representative or characteristic conditions of the atmosphere at particular places on the Earth. The climate of a particular location may depend upon extreme and infrequent conditions and therefore is more than a simple average temperature and precipitation regime. Figure 2.8 shows some of the major climatic types for selected areas on Earth in terms of characteristic rainfall and temperature conditions throughout the year. Figure 2.9 is a world map showing the distribution of major climatic types. These figures illustrate the tremendous variability of climates on the Earth.

From the global or regional scale down to local conditions, the changes in climate can be quite remarkable. For example, Figure 2.10 shows an east-west trending

Earth's atmosphere, only about half ever arrives at the Earth's surface. The rest is either absorbed by the atmosphere or reflected from the atmosphere back into space. Much of the harmful radiation, such as X-rays and ultraviolet radiation, is filtered out by the upper atmosphere. An ozone layer from approximately 15 to 32 kilometers above the Earth's surface is particularly important in absorbing ultraviolet radiation, which is potentially hazardous to all forms of living things (Fig. 2.5). The small amount of ultraviolet radiation that does manage to get through the ozone layer is responsible for sunburning people as well as probably causing some skin cancers.

Examination of Figure 2.4 shows that energy from the sun each year is responsible for fueling the hydrologic cycle, generating winds which produce water waves, eroding and depositing sediment on land, and allowing plant growth through photosynthesis [1].

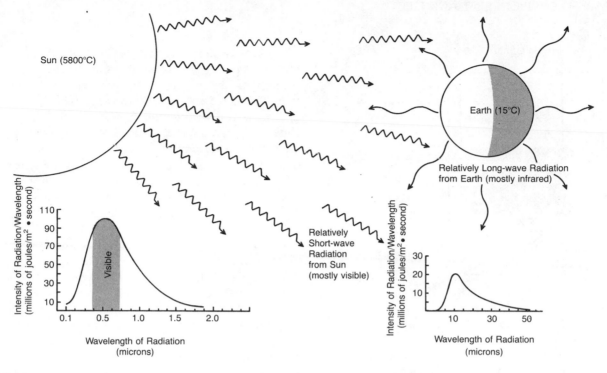

FIGURE 2.3

Idealized diagram comparing the emission of energy from the sun with that from the Earth. Notice that the solar emissions have a relatively short wavelength, whereas those from the Earth have a relatively long wavelength. (Modified after Marsh and Dozier, 1981.)

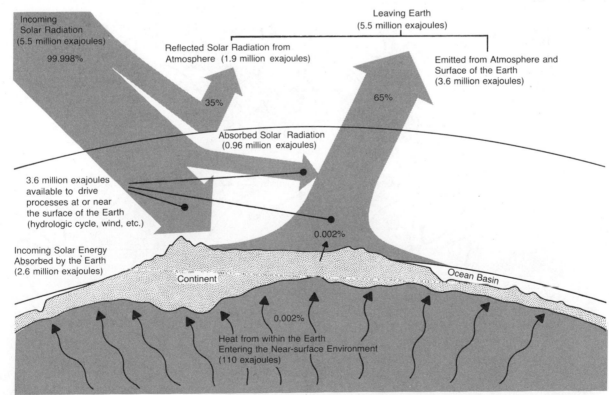

FIGURE 2.4

Annual energy flow to the Earth from the sun. Also shown is the relatively small component of heat from the Earth's interior to the near-surface environment. (Modified after Marsh and Dozier, 1981.)

21

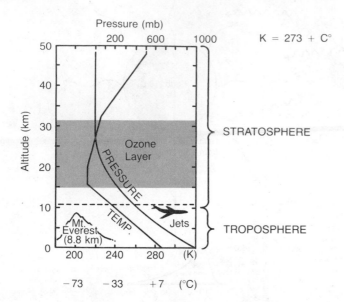

K = 273 + C°

FIGURE 2.5
Idealized diagram of the lower 50 kilometers of the Earth's atmosphere. Illustrated are the changes in temperature and pressure with elevation as well as the approximate location of the stratospheric ozone layer.

FIGURE 2.6
Generalized circulation of the atmosphere. (From Williamson, *Fundamentals of air pollution,* © 1973, Figure 5.5. Reprinted with permission of Addison-Wesley, Reading, MA.)

valley in New Mexico with a very steep south-facing slope and a relatively gentle north-facing slope. The south-facing slope is dry much of the year and has little vegetation. The northern slope, however, retains water longer and has a soil with abundant vegetation. Such local effects are called *microclimatic.* On an even smaller scale, microclimate may vary from one side of a small rock to another or from one side of a tree to another. Organisms often take advantage of these different conditions. Furthermore, as we shall discuss later, urban areas produce a characteristic microclimate with important environmental consequences.

Another important aspect of climate is climatic change. During the last million years there have been major climatic changes involving swings in the world's mean annual temperature of several degrees Centigrade. The present Ice Age began approximately 2 million years ago. Since then there have been numerous changes in the Earth's mean annual temperature (Fig. 2.11). On a scale of change over 1 million years, the times of high temperature reflect ice-free periods (called *interglacial periods*) over much of the Earth, whereas the low-temperature points reflect the glacial events. As one goes

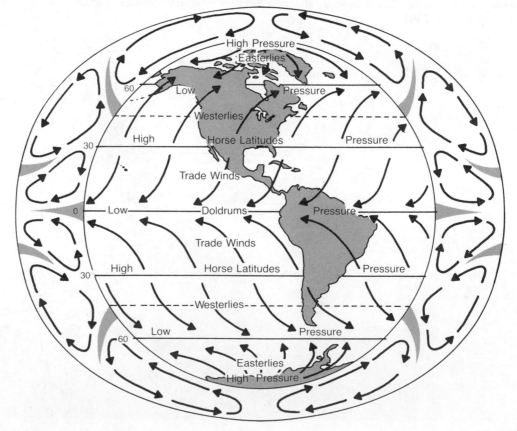

FIGURE 2.7
Warm and cold air masses that flow across North America. (From Strahler, 1973.)

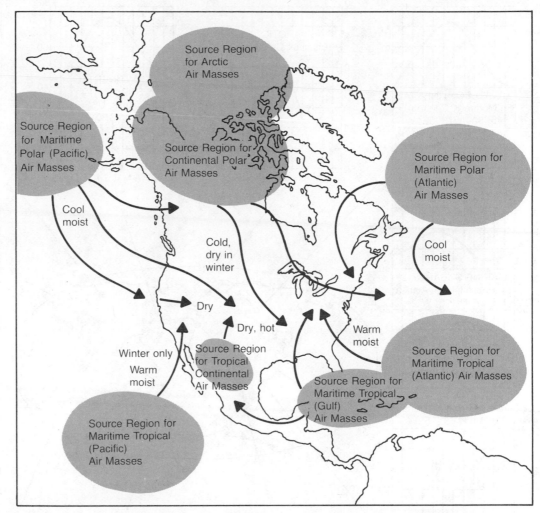

to a scale of 150,000 years, then additional minor glacial and interglacial events are more prominent. Finally, the change over the last 1500 years reflects several warming and cooling trends that have greatly affected people on Earth. For example, during the major warming trend which was at a maximum during the period from 800 to 1200 A.D. (Fig. 2.11d), the Vikings colonized Iceland, Greenland, and North America. When glaciers advanced during the cold period around 1400 A.D., the Viking settlements were abandoned in North America and parts of Greenland. Since approximately 1750 A.D. a warming trend has been apparent, lasting until the 1940s when temperatures again began to cool [1]. These major changes in climate are complex, and their causes are not well understood. Nevertheless, the consequences of climatic change are important from an environmental standpoint and will influence future conditions for people on Earth.

THE GEOLOGIC CYCLE

Throughout the nearly 5 billion years of Earth history, the materials at or near the Earth's surface have been more or less continuously created, maintained, and destroyed by numerous physical, chemical, and biological processes. Except during the very early history of our planet, the processes which produced many earth materials necessary for human survival have recurred periodically. Collectively, the processes responsible for the formation of new earth materials are referred to as the **geologic cycle** (Fig. 2.12), which is actually a group of subcycles: the hydrologic cycle, the tectonic cycle, the rock cycle, and the geochemical cycle.

The **hydrologic cycle** (Fig. 2.13) encompasses the movement of water from the ocean to the atmosphere and back to the ocean again by way of evaporation, runoff in streams and rivers, and groundwater flow. Only a very

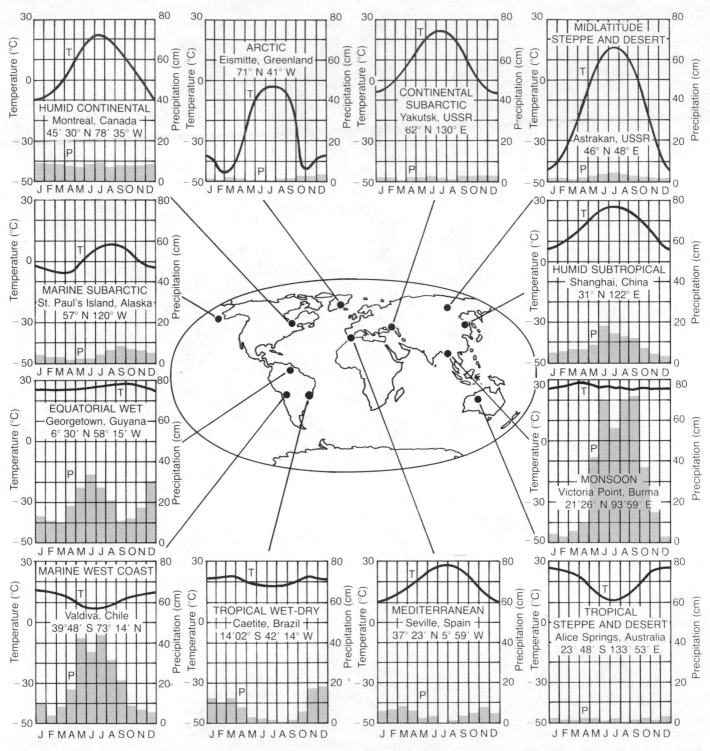

FIGURE 2.8

Some of the major climate types for selected areas on Earth in terms of characteristic precipitation and temperature conditions throughout the year. (From Marsh and Dozier, *Landscape,* © 1981, Figure 9.1. Reprinted with permission of Addison-Wesley, Reading, MA.)

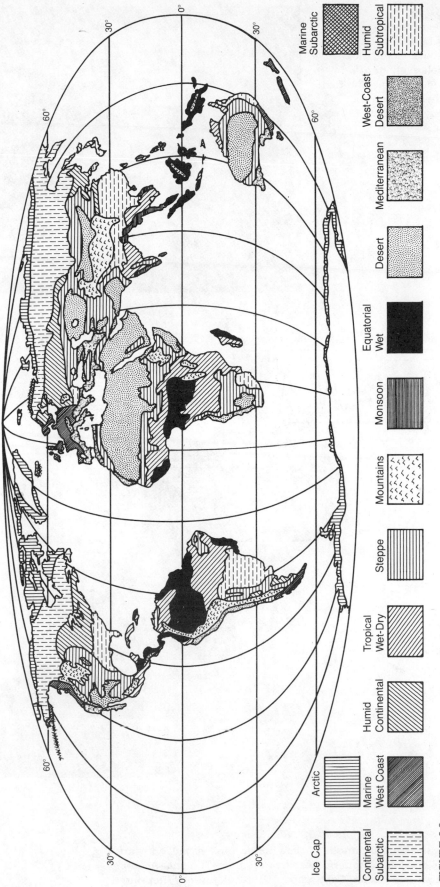

Marine Subarctic

Humid Subtropical

West-Coast Desert

Mediterranean

Desert

Equatorial Wet

Monsoon

Mountains

Steppe

Tropical Wet-Dry

Humid Continental

Arctic

Marine West Coast

Continental Subarctic

Ice Cap

FIGURE 2.9

The climates of the world. (From Marsh and Dozier, *Landscape,* © 1981, Figure 9.2. Reprinted with permission of Addison-Wesley, Reading, MA.)

FIGURE 2.10
East-west trending valley
in New Mexico with a very
steep south-facing slope
and a relatively gentle
north-facing slope. See text
for further explanation.
(Photo by E. A. Keller.)

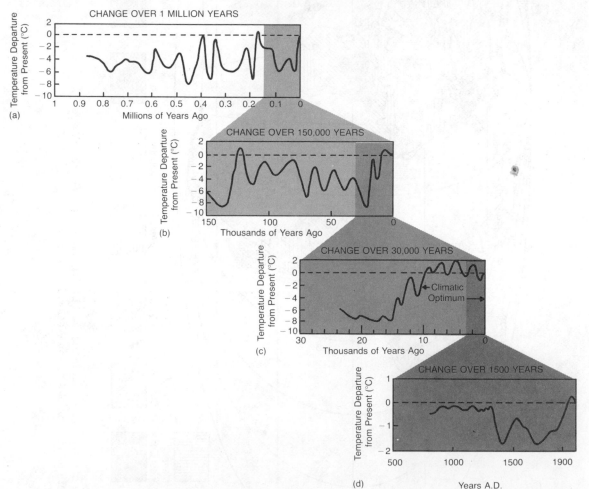

FIGURE 2.11
Change in temperature over different periods of time during the last million years.
(From Committee for the Global Atmospheric Research Program, National Academy
of Sciences, 1975, and Marsh and Dozier, *Landscape,* © 1981, Figure 9.3. Reprinted
with permission of the National Academy of Sciences and Addison-Wesley,
Reading, MA.)

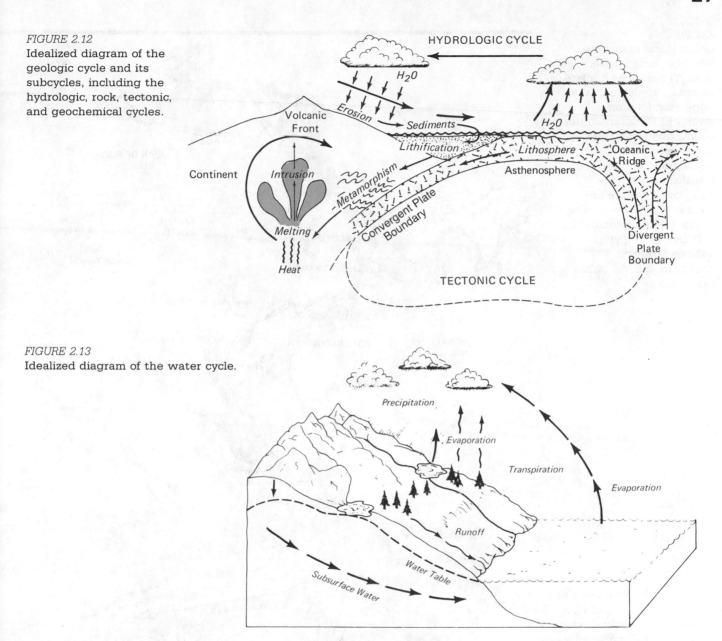

FIGURE 2.12
Idealized diagram of the geologic cycle and its subcycles, including the hydrologic, rock, tectonic, and geochemical cycles.

FIGURE 2.13
Idealized diagram of the water cycle.

small amount of water in the ocean is active in the hydrologic cycle at any one time, and yet this small amount of water is very important in the movement and sorting of chemical elements in solution (the geochemical, or biogeochemical, cycle), sculpturing the landscape, weathering rocks, transporting and depositing sediments, and providing our water resources. Therefore, we should learn more about how the hydrologic cycle operates to better understand and utilize our water resources.

Tectonic processes are driven by forces originating deep within the Earth. They deform the Earth's crust, producing external forms such as ocean basins, continents, and mountains. These processes are collectively known as the **tectonic cycle.** We now know that the Earth's outer layer, containing the continents and oceans,

is about 100 kilometers thick. This layer, known as the **lithosphere,** is not a continuous, uniform layer. Rather, the lithosphere is broken into several large segments known as *plates,* which are moving relative to one another (Fig. 2.14). As the lithospheric plates move over the **asthenosphere,** which is thought to be a more or less continuous layer of little strength below the lithosphere, the continents also move (Fig. 2.15). This movement of continents is known as *continental drift.* It is believed that the most recent episode of drift started approximately 200 million years ago when a supercontinent known as *Pangaea* broke up.

The boundaries between plates are geologically active areas, and most volcanic activity and earthquakes occur there. Three main types of boundaries are known:

FIGURE 2.14
Lithospheric plates that form the Earth's outer layer. Three types of plate junctions are shown: the ridge axis, forming divergent boundaries; the subduction zones, forming convergent plate boundaries; and the transform fault plate boundaries, where one plate is sliding by another. (From "Plate tectonics" by J. F. Dewey. Copyright© 1972 by Scientific American, Inc. All rights reserved.)

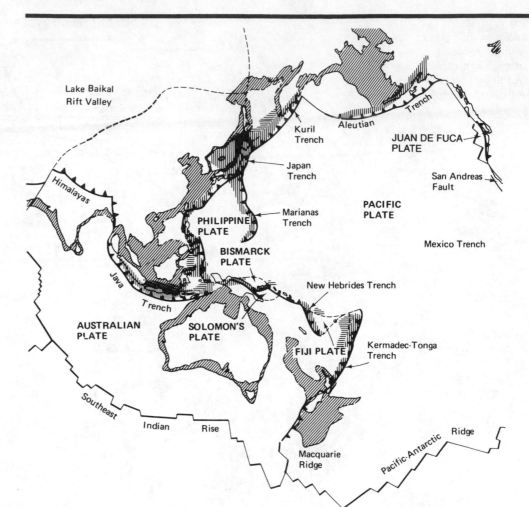

FIGURE 2.15
Idealized diagram showing the model of sea-floor spreading which is thought to drive the movement of the lithospheric plates. New lithosphere is being produced at the oceanic ridge (divergent plate boundary). The lithosphere then moves laterally and eventually returns down to the interior of the Earth at a convergent plate boundary (subduction zone). This process produces ocean basins and provides a mechanism that moves continents.

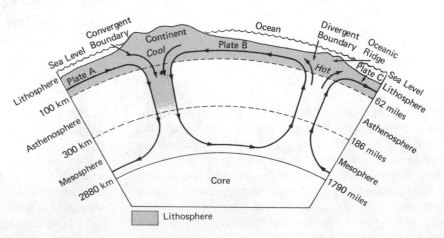

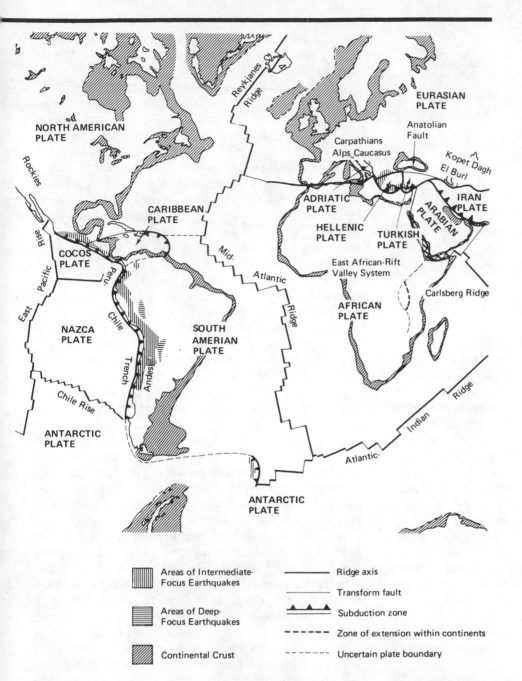

Areas of Intermediate-Focus Earthquakes

Areas of Deep-Focus Earthquakes

Continental Crust

Ridge axis

Transform fault

Subduction zone

Zone of extension within continents

Uncertain plate boundary

divergent, convergent, and transform fault. *Divergent boundaries* occur at oceanic ridges where plates are moving away from one another and new lithosphere is produced. *Convergent boundaries* (subduction zones) occur when one plate dives or moves beneath the leading edge of another plate. However, if both leading edges are composed of relatively light continental material rather than the heavier oceanic crust, it is more difficult for subduction to start, and a special type of convergent plate boundary known as a *collision boundary* may develop.

This may produce linear mountain systems such as the Alps and the Himalayas. *Transform fault boundaries* occur where one plate slides past another. A good example of this type of boundary occurs in California, where the San Andreas fault (a transform fault) is the boundary between the North American and Pacific plates for several hundred kilometers. There is a very close relationship between the tectonic cycle and the other cycles. The tectonic cycle provides water from volcanic processes as well as energy to form and change many earth materials.

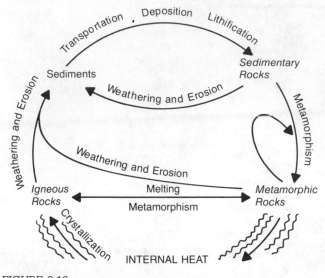

FIGURE 2.16
Idealized diagram of the rock cycle.

The other two subcycles of the geologic cycle are the geochemical and rock cycles. **Geochemistry** is the study of the distribution and migration of elements in Earth processes. The **geochemical cycle** is the path of elements during geologic processes and involves the chemistry of the lithosphere, atmosphere, hydrosphere, and biosphere. The **rock cycle** (Fig. 2.16) is a sequence of processes that produce sedimentary and metamorphic rocks. The geochemical and rock cycles are closely related to one another and intimately related to the hydrologic cycle, which provides the water necessary for the many chemical and physical processes.

In summary, earth materials such as minerals, rocks, soil, and water, as well as landforms such as ocean basins and mountain ranges, are constantly being created, maintained, and destroyed by internal and external (including biologic) Earth processes in numerous parts of the geologic cycle. Thus we see that the Earth is a dynamic, evolving system, changing in a consistent pattern according to the fundamental principle of uniformitarianism (Chapter 1).

MINERALS AND ROCKS

Minerals are naturally occurring inorganic crystalline substances with physical and chemical properties that vary within prescribed limits (see pp. 32–33). **Rocks** are aggregates of a mineral or minerals. Although there are over 2000 minerals, only a few are necessary to identify most rocks. Nearly 75 percent by weight of the Earth's crust is oxygen and silicon, and these two elements in combination with aluminum, iron, calcium, sodium, po-

tassium, and magnesium account for the minerals that make up about 95 percent of the Earth's crust. These minerals, called the **silicates,** are the most important rock-forming minerals.

Minerals and rocks are the foundation materials for the soils upon which we build most of our structures, grow all of our crops, and stand to use, enjoy, and appreciate our surroundings. The strength of rocks, and thus their ability to form mountains and their utility for building sites and other human uses, is affected by their composition, texture, and structure. **Composition** refers to the minerals in a rock; **texture** describes the size, shape, and arrangement of the mineral grains; and **structure** refers to such aspects as the nature of the forces holding the mineral grains together or the size and number of fractures.

All rocks are fractured, and the presence of fractures affects the strength of a rock. For example, water may concentrate and move along fractures, facilitating weathering and reducing rock strength. Furthermore, rocks tend to move along fractures and thereby create geological faults. The movement may cause earthquakes or may grind or pulverize the rocks, producing zones of weak rocks. The orientation of fractures is especially important, as fractures that dip down a slope may form potential *slip planes* along which landslides might occur (Fig. 2.17).

The *rock cycle* is truly the largest of the Earth cycles. The cycle depends on the tectonic cycle for energy, the geochemical cycle for materials, and the hydrologic cycle for water. There are three rock families (igneous, metamorphic, and sedimentary) involved in a worldwide recycling process, the elements of which are shown in Figure 2.16. Internal heat from the tectonic cycle drives the rock cycle and produces *igneous rocks* from molten material as it crystallizes beneath or on the Earth's surface. Rocks located at or near the surface break down chemically and physically by *weathering processes* to form the sediments that are transported by wind, water, and ice. The sediments accumulate in depositional basins such as the ocean, where they are eventually transformed into *sedimentary rocks*. After the sedimentary rocks are buried to sufficient depth, they may be altered by heat, pressure, or chemically active fluids to produce *metamorphic rocks* which may then melt to start the cycle again. Possible variations of this idealized cycle are indicated by the arrows in Figure 2.16.

Recycling of rock and mineral material is the most important aspect of the rock cycle. Our interest in the cycle is more than academic because it is upon these earth materials that we build our homes, industries, roads, and other structures, and it is from them that we obtain many mineral resources. Understanding the vari-

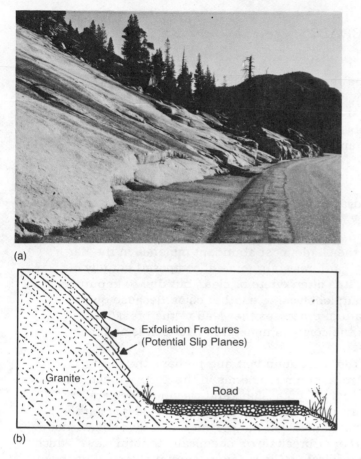

(a)

(b)

FIGURE 2.17
Fractures in granite rock dipping toward a highway in Yosemite National Park (a), and idealized diagram showing some of these fractures in cross section (b). (Photo by E. A. Keller.)

ous aspects of the rock cycle will facilitate the best use of the resources which are produced by the cycle. For example, the rock cycle concentrates as well as disperses materials—an extremely important aspect in the mining and use of minerals. We take resources that are concentrated by one aspect of the rock cycle, transform these resources through industrial activity, and then return them to the geologic cycle in a diluted form, where they are further dispersed by Earth processes. Once this dispersion process has taken place, the resource may not be concentrated again within a useful frame of time. For example, the lead in automobile fuel is mined in a concentrated form, transformed and diluted in fuel, and further dispersed by traffic patterns, air currents, and other processes. Eventually, it may become abundant enough to contaminate soil and water but never sufficiently to be efficiently recycled. Similar examples may be cited for other resources used in paints, solvents, and other industrial products.

Types of Rocks Our discussion of the generalized rock cycle has established that there are three rock families: igneous, sedimentary, and metamorphic. Table 2.1 lists the common rock types. Although rocks are primarily classified according to both mineral makeup and texture, it is *texture* (the size, shape, and arrangement of mineral grains) which is most significant in environmental studies. Rock texture along with discontinuities in rocks, such as fractures, determine the strength and utility of a particular rock.

Igneous rocks are rocks that have crystallized from a naturally occurring mobile mass of quasiliquid earth material known as *magma*. If magma crystallizes below the Earth's surface, the resulting igneous rock is called *intrusive*. One of the most common of the intrusive rocks is granite, which is generally a strong rock suitable for many engineering purposes.

Extrusive igneous rock forms when magma reaches the surface and is blown out of a volcano as pyroclastic debris or flows out as lava. The variety and composition of extrusive rocks are considerable, and it is difficult to generalize about their suitability for a specific purpose. One of the more common extrusive igneous rocks is basalt, which often forms extensive flows. Pyroclastic debris physically blown from a volcano produces a variety of extrusive rocks, some of which are listed in Table 2.1.

Because they are so variable in composition as well as in texture and structure, extrusive igneous rocks have to be carefully evaluated for any purpose. This was tragically emphasized on June 5, 1976, when the Teton Dam in Idaho failed, killing 14 and inflicting approximately $1 billion in property damage. The causes of the failure had strong geologic aspects, namely highly fractured volcanic rocks over which the dam was constructed and highly erodable wind-deposited clay-silts used in the construction of the dam interior or core. Open fractures in the volcanic rocks were probably not completely filled with a cement slurry (grout) in construction, and while the reservoir was filling, the water began moving under the foundation area of the dam. When the water came into contact with the highly erodable material of the core, it quickly eroded a tunnel through the base of the dam, explaining the observed whirlpool several meters across which formed in the reservoir just prior to failure. In other words, the development of the vortex of water draining out of the reservoir near the dam strongly suggested the presence of a subsurface tunnel of free-flowing water below the dam. The final failure of the dam came minutes later, sending a wall of water up to 20 meters high downstream, destroying homes, farms, equipment, animals, and crops along a 160-kilometer reach of the Teton and Snake rivers.

TYPES OF MINERALS

To understand the Earth's lithosphere as part of the environment, we should be familiar with some of the physical and chemical qualities of the major minerals as they relate to soil fertility, their ability to hold water, and their potential as a resource or a pollutant.

SILICATES

Silicates are the most important rock-forming minerals. The three most important rock-forming silicate minerals or mineral groups are *quartz, feldspar,* and *ferromagnesium.*

Quartz, one of the single most abundant minerals in the Earth's crust, generally is a hard, resistant mineral composed entirely of silicon and oxygen. It is often white or clear, but due to impurities it may also be rose, purple, black, or another color. Because quartz is very resistant to natural processes that lead to the breakdown of most minerals, it is the common mineral in river sands and most beach sands.

The *feldspars,* the most abundant and perhaps the most important group of rock-forming minerals in the Earth's crust, are aluminosilicates of sodium, potassium, and calcium. They are generally white, gray, or pink and are fairly hard. Feldspars are important commercial minerals in the ceramics and glass industries.

Feldspars *weather* or break down chemically to form clays, which are hydrated aluminosilicates. Clays cause many problems, but they are also very useful. For example, some clay minerals expand and contract greatly upon wetting and drying, and the resulting changes in volume may damage structures such as houses, streets, and sidewalks. On the other hand, clays are extremely important to life. Some scientists believe that clays played a crucial role in the origin of life. Their fine internal structure, where chemical elements collect and are held, may have provided sites where some large compounds that were precursors to life first formed. Today the ability of clays to hold onto chemical elements necessary for life is an important factor in soil fertility.

The *ferromagnesium minerals* are a group of silicates in which the silicon and oxygen combine with iron and magnesium. Generally dark minerals in most rocks, they are not particularly resistant to weathering and erosional processes and therefore tend to be altered or removed relatively quickly. Common weathering products may be oxides such as limonite (rust), clays, and soluble salts. Ferromagnesium minerals, when abundant, may produce weak rocks, so caution must be used in evaluating construction sites for highways, tunnels, and reservoirs when these rocks are encountered.

CARBONATES

From an environmental point of view, the most important carbonated mineral is *calcite,* which is calcium carbonate. Most calcium carbonate is formed by biological activity. Some organisms use calcium carbonate directly in bones and shells, which are then deposited. Other organisms, such as algae, induce changes in water chemistry that facilitate the precipitation of calcium carbonate. Calcite is the major constituent of limestone and marble, two very important rock types. Because water can weather calcite by dissolving it (putting it in *solution*), both limestone and marble may have solution pits or caverns associated with them. Cavern systems may carry groundwater, and water pollution problems in urban areas over limestone bedrock are well known. Furthermore, construction of highways, reservoirs, and other structures is a problem where caverns are likely to be encountered.

SULPHIDES

The sulphide minerals, such as pyrite or iron sulphide (fool's gold), are sometimes associated with serious environmental problems, particularly when roads, tunnels, or mines cut through rocks such as coal which contain sulphide minerals. When in contact with surface water or air, the minerals oxidize to form compounds such as ferric hydroxide and sulphuric acid. The acid water thus produced is a major problem in the coal regions of the Appalachian Mountains and other areas.

METAL OXIDES

Metallic elements react with free oxygen in the atmosphere to form metal oxides. Some of our most important mineral resources occur in this form. For example, iron and aluminum, the most important metals in our industrial society, are both mined from deposits which are oxides of these metals.

NATIVE ELEMENTS

Our last group of minerals include the uncommon *native elements,* such as gold, silver, copper, and diamonds, that have long been sought as valuable minerals. The native elements generally occur in rather small accumulations, but occasionally are found in sufficient quantities to justify mining. As we continue to mine these valuable minerals in lower-grade deposits, the environmental impact will continue to increase.

TABLE 2.1
Common rocks
(engineering geology
teminology).

Type	Texture	Materials
Igneous		
Intrusive		
Granitic[a]	Coarse[b]	Feldspar, quartz
Ultrabasic	Coarse	Ferromagnesians, ± quartz
Extrusive		
Basaltic[c]	Fine[d]	Feldspar, ± ferromagnesians, ± quartz
Volcanic breccia	Mixed—coarse and fine	Feldspar, ± ferromagnesians, ± quartz
Welded tuff	Fine volcanic ash	Glass, feldspar, ± quartz
Metamorphic		*Parent material*
Foliated		
Slate	Fine	Shale or basalt
Schist	Coarse	
Gneiss	Coarse	Shale, basalt, or granite
Nonfoliated		
Quartzite	Coarse	Sandstone
Marble	Coarse	Limestone
Sedimentary		*Materials*
Detrital		
Shale	Fine	Clay
Sandstone	Coarse	Quartz, feldspar, rock fragments
Conglomerate	Mixed—very coarse and fine	Quartz, feldspar, rock fragments
Chemical		
Limestone	Coarse to fine	Calcite, shells, calcareous algae
Rocksalt	Coarse to fine	Halite

[a]Textural name for a group of coarse-grained, intrusive igneous rocks, including granite, diorite, and gabbro.
[b]Individual mineral grains can be seen with naked eye.
[c]Textural name for a group of fine-grained, extrusive igneous rocks, including rhyolite, andesite, and basalt.
[d]Individual mineral grains cannot be seen with naked eye.

Sedimentary rocks form when sediments are weathered, transported, deposited, and then formed into rock by natural cement, compression, or other mechanism. Physical processes are obviously important in forming sedimentary rocks. Perhaps less well known is that the Earth's biota also play an important role in all stages of the sedimentary rock formation. Biological processes are especially significant in weathering, depositing, and cementing sediment, but also may be directly involved in the transport of sediment at a variety of scales. For example, many substances carried in chemical solution in river water are taken up by marine organisms and thus end up in sedimentary rocks when the organisms die and become part of the sedimentary record. As another example, kelp beds along the California coast are responsible for the transport of large (several centimeters in diameter) blocks of rock. Each main strand of kelp is attached to the bottom by a "hold-fast," which may be one or several rocks. In storms the kelp bed may be torn up, and as the kelp stems drift they move the hold-fast. In one instance near Santa Barbara, California, a substantial pad of gravel constructed to support an offshore pipe carrying sewage was eroded by this mechanism.

The two major types of sedimentary rocks are *detrital sedimentary rocks,* which form from broken parts of previously existing rocks, and *chemical sedimentary rocks,* which form from chemical or biochemical processes that remove material carried in chemical solution. The detrital sedimentary rocks include shale, sandstone, and conglomerate. Of the three, shale is by far the most abundant and causes most of the environmental problems associated with sedimentary rocks. Two types of shale are generally recognized: *compaction shale* and *cementation shale.* Compaction shale is an extremely weak rock that often is characterized by a propensity to expand and contract on wetting and drying and a high erosion potential. On the other hand, cementation shale may be a very hard rock, depending on the cementing agent. The presence of shale rock indicates the need for detailed evaluation before any land use is planned.

Sandstones and conglomerates are coarse-grained sedimentary rocks which make up about 25 percent of all sedimentary rocks. Depending upon the type of cementing material, these rocks may be very strong and suitable for many engineering purposes. However, because they may have an extremely variable chemistry, they vary considerably in their fertility as a base for soils.

Limestone makes up about 25 percent of all sedimentary rocks and is by far the most abundant of the chemical sedimentary rocks. Limestone is composed almost entirely of the mineral calcite, which weathers readily in the presence of water and forms open cavities or caverns within the limestone. In addition, large surface pits known as *sink holes* may form. Limestones tend to form highly fertile soils, as shown by adjacent valleys in Virginia near the North Carolina border. One, known as Rich Valley, has good soils and prosperous farms—the bedrock is limestone. The other, where the soils are less productive and the farms less prosperous, is called Poor Valley—the bedrock is shale.

Another important chemical sedimentary rock is salt, which is composed primarily of the mineral halite (sodium chloride) or common table salt. Rock salt forms when shallow seas or lakes dry up. As water evaporates, a series of salts, one of which is halite, is precipitated. The salts may later be covered up by other types of sedimentary rocks as the area again becomes a center of sediment deposition. It is from these sedimentary basins that salt is often mined. Rock salt is of particular environmental significance because it is one of the rock types that is being seriously considered as a storage medium for high-level radioactive waste.

Metamorphic rocks are changed rocks. Heat, pressure, and chemically active fluids produced in the tectonic cycle may change the mineralogy and texture of rocks. This, in effect, produces new rocks. Two types of metamorphic rocks are recognized: *foliated* (that is, occurring in layers like pages in a book), which are formed of elongated or flat mineral grains having a preferential alignment and forming parallel layers; and *nonfoliated,* those without preferential alignment or segregation of the mineral grains. Foliated metamorphic rocks such as salte, schist, and gneiss have a variety of physical and chemical properties; it is therefore difficult to generalize about their usefulness for engineering purposes or their qualities as a base for soils. Foliation planes in metamorphic rocks, however, are potential planes of weakness affecting the strength of the rock and in particular its potential to slide. This results because the foliation planes produce zones of weakness along which the strength of the rock is less than across the foliation planes (Fig. 2.18). Examples of nonfoliated metamorphic rocks

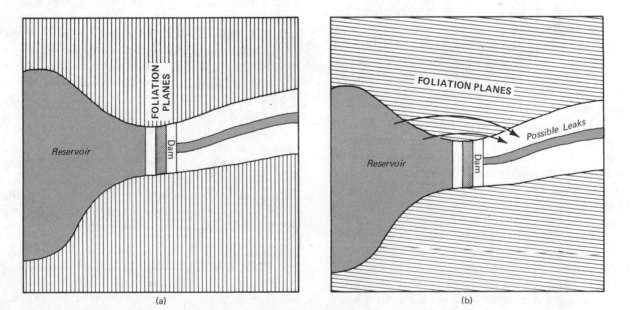

(a) (b)

FIGURE 2.18
Idealized diagram showing two possible orientations of foliation in metamorphic rocks at a dam and reservoir site. The most favorable orientation of the foliation is shown in part (a), where the foliation is parallel to the axis of the dam. The least favorable orientation is where the foliation planes are perpendicular to the axis of the dam, as shown in (b).

include quartzite and marble. Quartzite, a metamorphosed sandstone, is generally a hard, durable rock suitable for many engineering purposes. Marble, formed from the metamorphism of limestone, consists of recrystallized calcite. Marbles have many of the problems associated with limestones, namely the development of solution pits, caverns, or sink holes.

SOILS

Soils are made of weathered rock material and organic matter and thus are produced by interactions between geologic (rock and hydrologic cycles) and biologic processes. **Weathering,** the physical and chemical breakdown of rocks, is the first step in the soil-forming process. The more insoluble weathered material may remain essentially in place and be further modified by organic processes to form a residual soil as, for example, the red soils of the Piedmont in the southeastern United States. If the weathered material is transported by water, wind, or ice, and then deposited and further modified by organic processes, then a transported soil forms. The exceptionally fertile soils formed from the glacial deposits in the midwestern United States are an example.

Soils may be defined in two ways in environmental studies. A soil scientist defines soil as a solid earth material that has been altered by physical, chemical, and organic processes such that it can support rooted plant life. An engineer, on the other hand, defines soil as any solid earth material that can be removed without blasting.

The formation of soil is affected by climate, topography, parent material (material from which the soil is formed), maturity (time), and biological activity. Most of the differences we observe in soils are due to the effects of climate and topography. The type of parent rock, the organic process, and the length of time soil-forming processes have operated are of secondary importance. Given enough time, similar soils may develop on different parent materials if climate and topography are the same.

The vertical and horizontal movement of material in soils often produces a distinctive soil layering, or *soil profile,* divided into zones, or *horizons* (Fig. 2.19). An important process is *leaching* (downward movement of soluble material) from the upper zone, called the *A horizon,* to the intermediate zone, known as the *B horizon,* where the leached material is deposited. Depending upon climate and other variables, the B horizon may contain a "hardpan" layer of compacted clay, or calcium carbonate cemented materials known as *caliche.* In addition, materials may also move upward in the soil in response to a natural rise of the water table or in response to human use, such as extensive irrigation or removal of vegetation (timber

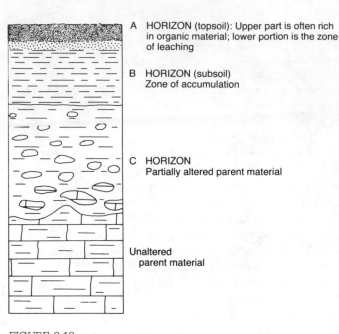

A HORIZON (topsoil): Upper part is often rich in organic material; lower portion is the zone of leaching

B HORIZON (subsoil) Zone of accumulation

C HORIZON Partially altered parent material

Unaltered parent material

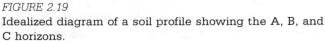

FIGURE 2.19
Idealized diagram of a soil profile showing the A, B, and C horizons.

harvesting) which induce a rise in the level of groundwater. When accompanied by extensive evaporation, this upward movement may cause salts to move up and be deposited in the upper soil layers or even on the ground surface. This happened in the Indus River valley in Pakistan, where approximately 93,000 square kilometers were irrigated. The water table rose and extensive evaporation deposited salts, making the soils unsuitable for crops. The problem can be, and has been in some places, corrected by installing drainage systems for the land and then providing an abundance of water to leach the salts out. A careful analysis and special system of wells to lower the water table were used with success in Pakistan to reclaim soils damaged by salts. In the Great Valley of California salts deposited on the surface by evaporation of irrigation water are regularly leached out of the soil by periodic applications of a large amount of water.

Soil Classification Soil scientists have developed a comprehensive and systematic classification of soils based on physical and chemical properties known as *Soil Taxonomy* (informally called *the Seventh Approximation*). The major kinds of soil in this classification are given in Table 2.2. This classification is especially useful for agriculture. However, it lacks sufficient information about soil texture to be of much use in a specific site evaluation for land-use planning.

TABLE 2.2
General properties of soil orders used with the Seventh Approximation.

Order	General Properties
Entisols	No horizon development (azonal); many are recent alluvium; synthetic soils are included; are often young soils.
Vertisols	Include swelling clays (greater than 35 percent) that expand and contract with changing moisture content. Generally form in regions with a pronounced wet and dry season.
Inceptisols	One or more of horizons have developed quickly; horizons are often difficult to differentiate; most often found in young but not recent land surfaces, have appreciable accumulation of organic material; most common in humid climates but range from the Arctic to Tropics; native vegetation is most often forest.
Aridisols	Desert soils; soils of dry places; low organic accumulation; have subsoil horizon where gypsum, caliche (calcium carbonate), salt, or other materials may accumulate.
Mollisols	Soils characterized by black, organic rich "A" horizon (prairie soils); surface horizons are also rich in bases. Commonly found in semiarid or subhumid regions.
Spodosols	Soils characterized by ash-colored sands over subsoil, accumulations of amorphous iron-aluminum sesquioxides and humus. They are acid soils that commonly form in sandy parent materials. Are found principally under forests in humid regions.
Alfisols	Soils characterized by: a brown or gray-brown surface horizon, an argillic (clay rich) subsoil accumulation with an intermediate to high base saturation (greater than 35 percent as measured by the sum of cations, such as calcium, sodium, magnesium, etc.). Commonly form under forests in humid regions of the mid-latitudes.
Ulfisols	Soils characterized by an argillic horizon with low base saturation (less than 35 percent as measured by the sum of cations); often have a red-yellow or reddish-brown color; restricted to humid climates and generally form on older landforms or younger, highly weathered parent materials.
Oxisols	Relatively featureless, often deep soils, leached of bases, hydrated, containing oxides of iron and aluminum (laterite) as well as kaolinite clay. Primarily restricted to tropical and subtropical regions.
Histosols	Organic soils (peat, muck, bog).

SOURCE: Soil Conservation Service, Soil Survey Staff, 1975.

Another soil classification known as the *unified soil classification system* (Table 2.3) has been devised for engineering purposes. In this system soils are divided according to the predominant particle size or the abundance of organic material. A useful way of estimating the size of soil particles in the field is as follows: It is *sand* or larger (coarse soil) if you can see the individual grains, *silt* (fine soil) if you can see the grains with a 10-power hand lens, and *clay* (fine soil) if you cannot see the grains with such a hand lens.

Because soils greatly affect the best use of the land, a soil survey to identify and map soils is an important part of planning. The important ecological soil properties for planning are physical structure, fertility, and organic content, all of which affect the movement of water and nutrients. The more important properties of soils for planning are strength, sensitivity, compressiblity, erodibility, permeability or drainage potential, corrosion potential, ease of excavation, and a shrink/swell potential.

It is difficult to generalize about the *strength* of a soil. The actual strength of a soil depends upon cohesive and frictional forces within the soil itself. In general, cohesion is most important in the fine-grained soils. Tree roots or moisture films between coarse grains may provide some cohesion, explaining the ability of wet sand (which is cohesionless when dry) to stand in vertical walls in children's sand castles on the beach [2]. For important projects the strength of a soil may be measured in the laboratory.

A soil's *sensitivity* reflects changes in its strength due to disturbances such as vibration or excavation. Fine-grained soils may lose up to 75 percent or more of their strength following disturbance and thus are considered quite sensitive. Coarser soils of sand and gravel, with little or no clay, are the least sensitive [3].

Compressibility of soils is a measure of the tendency to consolidate or decrease in volume. It is partly a function of the elastic nature of the soil particles. Excessive compression may crack foundations and walls. In general, it is the finer soils which have a high compressiblity.

Erodibility of soils is another important property. Generally, soils that are easily eroded include unprotected

TABLE 2.3
Unified soil classification system.

Major Divisions				Group Symbols	Soil Group Name
COARSE-GRAINED SOILS — Over half of material larger than 0.074 mm	GRAVELS	CLEAN GRAVELS	Less than 5% fines	GW	Well-graded gravel
				GP	Poorly graded gravel
		DIRTY GRAVELS	More than 12% fines	GM	Silty gravel
				GC	Clayey gravel
	SANDS	CLEAN SANDS	Less than 5% fines	SW	Well-graded sand
				SP	Poorly graded sand
		DIRTY SANDS	More than 12% fines	SM	Silty sand
				SC	Clayey sand
FINE-GRAINED SOILS — Over half of material smaller than 0.074	SILTS Non-Plastic			ML	Silt
				MH	Micaceous silt
				OL	Organic silt
	CLAYS Plastic			CL	Silty clay
				CH	High plastic clay
				OH	Organic clay
Predominantly Organics				PT	Peat and muck

silts, sands, and other loosely consolidated materials. Cohesive soils with a relatively high clay content or naturally cemented soils are not easily moved and therefore have a low erosion factor.

Permeability is the measure of the ease with which water can move through a material. The percentage of void or empty space in a soil or rock is known as the *porosity*. As the size of the material decreases, the permeability generally decreases. Clean gravels or sands have the highest permeabilities, transmitting several hundred cubic meters of water per day through a cross-sectional

TABLE 2.4
Porosity and permeability of selected earth materials.

Material	Porosity (%)	Permeability (m^3/day/m^2)
Unconsolidated		
Clay	45	0.041
Sand	35	32.8
Gravel	25	205.0
Gravel and sand	20	82.0
Rock		
Sandstone	15	28.7
Dense limestone or shale	5	0.041
Granite	1	0.0041

SOURCE: Modified after Linsley, Kohler, and Paulhus, 1975.

area of one square meter. Some of the most porous materials, such as clay, have a very low permeability (Table 2.4) because the small, flat clay particles can have a good deal of pore space, but the individual small openings tenaciously hold the water. A typical cross-sectional area of one square meter of clay may transmit only one tenth or less of a cubic meter of water per day (Table 2.4).

The *shrink/swell potential* reflects the tendency of a soil to gain or lose water. Soils that tend to increase or decrease in volume with water content are called *expansive*. The swelling is caused by the chemical attraction and addition of layers of water molecules between the flat submicroscopic clay plates of certain clay minerals. *Expansive soils cause significant environmental problems (Fig. 2.20) and in terms of total property damage to structures are our most costly natural hazards* [4].

This discussion of soil properties establishes that certain materials are more suitable for particular types of land uses. The information summarized in Table 2.5 should be considered carefully prior to any project that involves the use of soils.

WATER

The hydrologic or water cycle (Fig. 2.13) is driven by solar energy and supplies nearly all of our water resources. Of

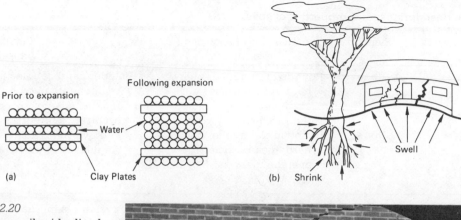

FIGURE 2.20

Expansive soils: idealized diagram showing an expansive clay as layers of water molecules are incorporated between clay plates (a); and effects of a soil's shrinking and swelling at a home site (b). (After Mathewson and Castleberry.) Cracked wall resulting from expansion of clay soil under the foundation (c). (Photo courtesy of U.S. Department of Agriculture.)

the water on Earth, approximately 97 percent is in the oceans, about 2 percent is locked up in glaciers and the icecaps, and only a small fraction of 1 percent (0.001) is in the entire atmosphere. Nevertheless, the fresh water phase of the hydrologic cycle is dependent upon this small portion of the Earth's total water. We shall now discuss in some detail some selected aspects of groundwater and the water in streams and rivers.

GROUNDWATER

Rain that falls on the land either evaporates, runs off along the surface, or moves below the surface and is transported underground. In the conterminous United States, approximately 30 percent of the rainfall enters the surface-subsurface flow system. Of this, approximately 1 percent reaches the ocean by way of groundwater flow. Water that moves into the ground first enters a belt of soil moisture known as the *zone of aeration* (Fig. 2.21). It may then move through the intermediate belt, which is sel-

dom saturated, and enter the groundwater system in the *zone of saturation*. The upper surface of this zone is called the *water table*. The area just above the water table is a narrow belt called the *capillary fringe*, where water is drawn up by surface tension. The rate of movement of groundwater depends upon the permeability of the soil or rock and upon the *hydraulic gradient*, which is related to the slope of the water table.

Groundwater is an important source of useful water for drinking, industry, and irrigation. A zone of earth material capable of producing groundwater at a useful rate from a well is called an *aquifer*. Unconsolidated gravel and sand as well as sandstone, granite, and metamorphic rocks with high porosity or open fractures generally make good aquifers.

Water under pressure may rise to the surface of a well without pumping. Water confined by rocks and soils under pressure is called an *artesian system*. In artesian systems water tends to rise to the height of the recharge

TABLE 2.5
Generalized sizes, descriptions, and properties of soils.

Soil	Soil Component	Symbol	Grain Size Range and Description	Significant Properties
Coarse-grained components	Boulder	None	Rounded to angular, bulky, hard, rock particle; average diameter more than 25.6 cm.	Boulders and cobbles are very stable components used for fills, ballast, and rip-rap. Because of size and weight, their occurrence in natural deposits tends to improve the stability of foundations. Angularity of particles increases stability.
	Cobble	None	Rounded to angular, bulky, hard, rock particle; average diameter 6.5–25.6 cm.	
	Gravel	G	Rounded to angular, bulky, hard, rock particles greater than 2 mm in diameter.	Gravel and sand have essentially the same engineering properties, differing mainly in degree. They are easy to compact, little affected by moisture, and not subject to frost action. Gravels are generally more pervious, stable, and resistant to erosion and piping than are sands. The well-graded sands and gravels are generally less pervious and more stable than those which are poorly graded and of uniform gradation. Irregularity of particles increases the stability slightly; finer, uniform sand approaches the characteristics of silt.
	Sand	S	Rounded to angular, bulky, hard, rock particles 0.074–2 mm in diameter.	
Fine-grained components	Silt	M	Particles 0.004–0.074 mm in diameter; slightly plastic or nonplastic regardless of moisture; exhibits little or no strength when air dried.	Silt is inherently unstable, particularly with increased moisture, and has a tendency to become quick when saturated. It is relatively impervious, difficult to compact, highly susceptible to frost heave, easily erodible, and subject to piping and boiling. Bulky grains reduce compressibility, whereas flaky grains (such as mica) increase compressibility, producing an elastic silt.
	Clay	C	Particles smaller than 0.004 mm in diameter; exhibits plastic properties within a certain range of moisture; exhibits considerable strength when air dried.	The distinguishing characteristic of clay is cohesion or cohesive strength, which increases with decreasing moisture. The permeability of clay is very low; it is difficult to compact when wet and impossible to drain by ordinary means; when compacted is resistant to erosion and piping; not susceptible to frost heave; and subject to expansion and shrinkage with changes in moisture. The properties are influenced not only by the size and shape (flat or platelike), but also by their mineral composition. In general, the montmorillonite clay mineral has the greatest and kaolinite the least adverse effect on the properties.

TABLE 2.5 (continued)

Soil	Soil Component	Symbol	Grain Size Range and Description	Significant Properties
	Organic matter	O	Organic matter in various sizes and stages of decomposition.	Organic matter present even in moderate amounts increases the compressibility and reduces the stability of the fine-grained components. It may decay, causing voids, or change the properties of a soil by chemical alteration; hence, organic soils are not desirable for engineering purposes.

SOURCE: After Wagner, 1957.
NOTE: The United Soil Classification does not recognize cobbles and boulders with symbols. The size range for these as well as the upper limit for clay are according to Wentworth (1922).

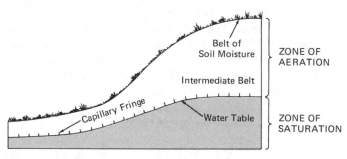

FIGURE 2.21
Idealized diagram showing the zones in which groundwater occurs.

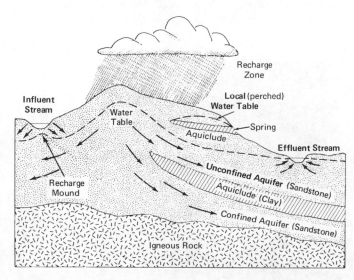

FIGURE 2.22
Idealized diagram showing an unconfined aquifer, a local (perched) water table, and influent and effluent streams. An aquiclude is a zone that holds water but does not release it at a useful rate.

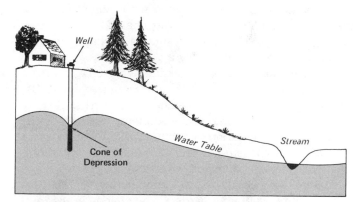

FIGURE 2.23
Idealized diagram of the cone of depression in a water table resulting from pumping water from a well.

zone, a condition analogous to the function of a water tower producing water pressure for houses. Both confined and unconfined aquifers may be found in the same area (Fig. 2.22).

Water pumped from a well may produce a *cone of depression* in the water table (Fig. 2.23). Overpumping of an aquifer may lower the water level continuously, necessitating lowering the pump setting or drilling deeper wells or possibly exhausting the available supply.

STREAMS AND RIVERS

Streams and rivers are the basic transportation systems in the part of the rock cycle involved with erosion and deposition of sediment. There are two types of streams: effluent and influent (Fig. 2.22). *Effluent streams* tend to be perennial, and stream flow is maintained during the dry season by groundwater seepage into the channel. On the other hand, *influent streams* are everywhere above

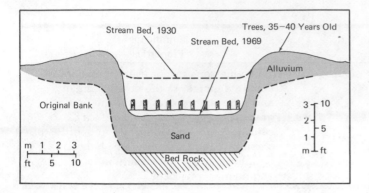

FIGURE 2.24
Accelerated sedimentation and subsequent erosion due to land-use changes at the Mauldin Millsite on the Piedmont of middle Georgia. (After Trimble, 1969.)

the groundwater table and flow only in direct response to precipitation. Water from the channel moves down through the water table, forming a recharge mound. These streams may flow only a part of the year.

The amount and kind of sediment moved by a stream or river depends on the velocity and amount of water. Because there is a delicate balance between the flow of water and the movement of sediment, changes in the sediment load entering a stream lead to changes in the shape of the stream channel—its slope and cross-sectional shape—which effectively change the velocity of the water. Changes in velocity may increase or decrease the amount of sediment carried. Any change in the sediment load or discharge will initiate slope changes to bring the system into balance again.

For example, in the southeastern United States farmland is being converted back to forest. When this occurs, the sediment load tends to decrease and the stream reacts by eroding the channel to lower the slope, which in turn lowers the velocity of the water. This process continues until an equilibrium between the load imposed and the work done is achieved again. In parts of the Piedmont of the southeastern United States, this change can actually be observed and recorded (Fig. 2.24). Land that has reverted to pine forest in conjunction with soil conservation measures has reduced the sediment load delivered to some of the Piedmont streams. Thus, some once muddy streams choked with sediment are now clearing slightly and eroding their channels. Whether this trend continues depends upon continuing land-use change and conservation measures. This example illustrates the important effect living things can have on geologic and hydrologic processes.

Local stream conditions may have environmental significance. For example, stream and river channels often contain a series of regularly spaced pools and riffles (Fig. 2.25). *Pools* at low flow are deep areas with slow-moving water and are produced by scour at high flow. *Riffles* at low flow are shallow areas with fast-moving water. Riffles are produced by depositional processes at high flow. Streams with well-developed pools and riffles provide a variety of flow conditions characterized by deep, slow-moving water alternating with shallow, fast-moving water, which fosters desirable biological activity. Such streams also produce landscape variations that are visually pleasing, and increase what has become known as the "aesthetic amenity" as well as the recreational poten-

FIGURE 2.25
Well-developed pool-riffle sequence in Sims Creek near Blowing Rock, North Carolina. (Photos by E. A. Keller.)

tial by providing better fishing and boating conditions. Unfortunately, land-use change that increases sediment input to streams, such as urbanization or timber harvesting, may degrade a stream's pool-riffle environment.

WIND AND ICE

Wind and ice-related processes are responsible for the erosion, transport, and deposition of vast quantities of materials on the Earth's surface. These processes create and modify a substantial number of landforms in environmentally sensitive areas such as coasts, deserts, and arctic and subarctic terrains.

WIND
Windblown deposits are generally divided into two major groups: *sand deposits,* mainly dunes; and *loess,* or windblown silt. Extensive deposits of windblown sand and silt cover thousands of square kilometers in the United States (Fig. 2.26). Sand dunes and related deposits are found along the coast of the Atlantic and Pacific oceans and the Great Lakes. Inland bodies of sand are found in areas of Nebraska, southern Oregon, southern California, Nevada, and northern Indiana and along large rivers flowing

through semiarid regions such as the Columbia and Snake rivers in Oregon and Washington. The majority of loess is located adjacent to the Mississippi Valley but is also found in the Pacific Northwest and Idaho.

Sand dunes are formed from sand moving close to the ground. They are of many sizes and shapes and develop under a variety of conditions. Regardless of where they are located or how they form, actively advancing sand dunes cause environmental problems. Stabilization of sand dunes is a major problem in the construction and maintenance of highways and railroads that cross sandy areas of deserts. The complex group of dunes shown in Figure 2.27 is encroaching on Highway 95 near Winnemucca, Nevada, at a rate of approximately 12 meters per year. The sand is removed about three times per year, and approximately 1500 to 4000 cubic meters are removed each time. Building and maintaining reservoirs in sand dune terrain are even more troublesome and extremely expensive. Such reservoirs can only be constructed if a very high water loss can be tolerated. Canals in sandy areas should be lined to hold water while controlling erosion.

In contrast to sand, which seldom moves more than a meter or so off the ground, windblown silt and dust can

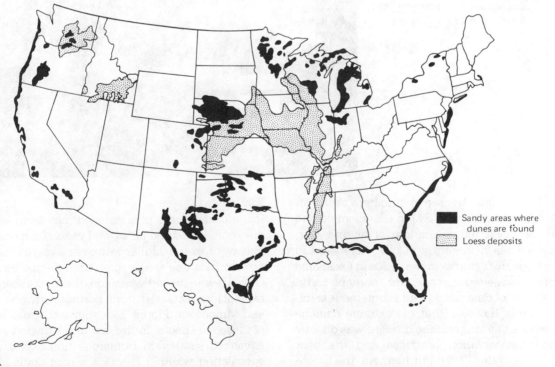

Sandy areas where dunes are found

Loess deposits

FIGURE 2.26
Distribution of windblown landforms within the United States. (From Way, Douglas, *Terrain analysis,* 2nd Edition, Fig. 9.3, p. 267. Copyright © 1973 and 1978 by Dowden, Hutchinson & Ross, Inc., Strondsburg, Pa. Reprinted by permission of the publisher.)

(a)

(b)

FIGURE 2.27
Complex group of sand dunes encroaching on Highway 95 near Winnemucca, Nevada (a). Removal of sand from the highway is a continuous problem (b). See text for further explanation. [Photos courtesy of J. O. Davis (a) and D. T. Trexler (b).]

FIGURE 2.28
Dust storm caused by cold front at Manter, Kansas, in 1935. (Photo courtesy of Environmental Science Services Administration.)

be carried in huge clouds thousands of meters in the air (Fig. 2.28). A dust storm may be 500 to 600 kilometers in diameter and carry over 100 million tons of silt and dust, sufficient to form a pile 30 meters high and 3 kilometers in diameter. The huge dust storms in the 1930s in Oklahoma and nearby areas exceeded these figures, perhaps carrying over 58,000 tons of dust per square kilometer. It is important to recognize, however, that the extreme damage caused by these dust storms and the drought was directly related to poor conservation practices and the poor weather conditions of the 1930s. Furthermore, the hatching of grasshoppers was accelerated by favorable conditions of heat and dryness. The insects destroyed crops worth several hundred million dollars from 1934 to 1938.

ICE

Ice, a cold-climate phenomenon, has been an important environmental topic for several years. As more people live and work in the higher latitudes, we will have to learn how to best use these sometimes fragile environments. Only a few thousand years ago, the most recent continental glaciers retreated from northern Indiana, Michigan, and Minnesota. Figure 2.29 shows the maximum extent of these ice sheets. In the last 2 to 3 million years the ice advanced southward numerous times, and we are still speculating whether it will advance again. Recent evidence indicates that we are still in the Ice Ages.

The effects of recent glacial events are easily seen in the landscape. The flat, nearly featureless ground moraine

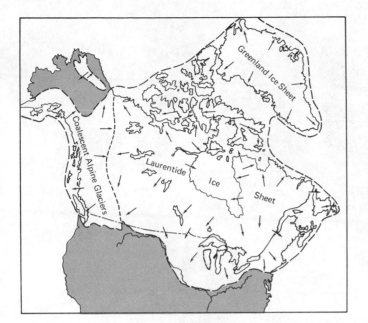

FIGURE 2.29
Maximum extent of ice sheets during the Pleistocene glaciation. (From Foster, 1979.)

or till plains of central Indiana are composed of material carried and deposited by continental glaciers, called *till*. The till buried preglacial river valleys, so beneath the glacial deposits is topography formed by running water, much like the hills and valleys of southern Indiana where the glaciers did not reach. The deposits of till and other material associated with glaciers create a varied landscape. Small lakes called *kettle lakes,* formed where ice blocks remained and melted, may fill up rather quickly with sediment and organic material, and thus are different from the surrounding glacial deposits. Furthermore, sands and gravels from streams in, on, or under and in front of the ice provide further diversity in materials. Because of the wide variety of deposits in glaciated areas, an area's physical properties must be evaluated in detail for planning, designing, and constructing large structures such as dams, highways, and buildings.

Glacial topography also affects natural ecosystems. For example, glaciated areas have many lakes and bogs, which provide habitat for fish and wildlife. Glacial till often supports rich soil that provides a good foundation for the development of forests.

In the higher latitudes permanently frozen ground, called **permafrost,** is a widespread natural phenomenon underlying about 20 percent of the world's area (Fig. 2.30). Special environmental problems are associated with the design, construction, and maintenance of struc-

tures such as roads, railroads, pipelines, airfields, and buildings in permafrost areas. Lack of knowledge about permafrost has led to very high maintenance costs and relocation or abandonment of highways, railroads, and other structures. For example, Figure 2.31 shows a gravel road that has sunk because of the thawing of permafrost. Another example is a tractor trail constructed by bulldozing off the vegetation cover over the permafrost on the North Slope of Alaska (Fig. 2.32). Small ponds found on the abandoned trail during the first summer following trail construction will continue to grow deeper and wider as the permafrost continues to thaw. In general, most problems are encountered where permafrost occurs in fine-grained, poorly drained, frost-susceptible materials such as silt. These generally contain a lot of ice, which melts if the temperature increases. Melting produces unstable materials, resulting in settling, subsidence, and lateral or downslope flow of saturated sediment. It is the thawing of permafrost and subsequent frost heaving and subsidence caused by freezing and thawing of the upper layer that is responsible for most of the environmental problems in the permafrost areas of the arctic and subarctic regions.

LANDFORMS

Landforms are very important to us. Every human activity on the surface of the Earth, whether it be agriculture or land development for urban or industrial uses, involves the use of landforms. Land uses on river floodplains should be different from those on adjacent valley slopes, and these in turn may be different from the uses in the higher areas of greater relief in adjacent uplands. Geological processes create and change the shape and makeup of the land surface. Biological processes, by changing and modifying erosion and depositional patterns, have influenced the development of these landforms. The kind of landform that is found in any particular area therefore depends on the underlying material and on geological and biological processes. For example, the particular set of landforms associated with river and stream processes are distinctly different from landforms produced by volcanic processes or coastal processes.

Because each combination of geological and biological processes produces rather distinctive landforms, it is possible to make a *genetic classification of landform assemblages* and, on a larger scale, to outline physiographic regions and provinces. For example, Figure 2.33 shows the physiographic regions for the conterminous United States. An understanding of these physiographic regions and the types of landforms found in each is essential in

FIGURE 2.30
Extent of permafrost zones
in the Northern
Hemisphere. (From
Ferrians, Kachadoorian, and
Greene, 1969.)

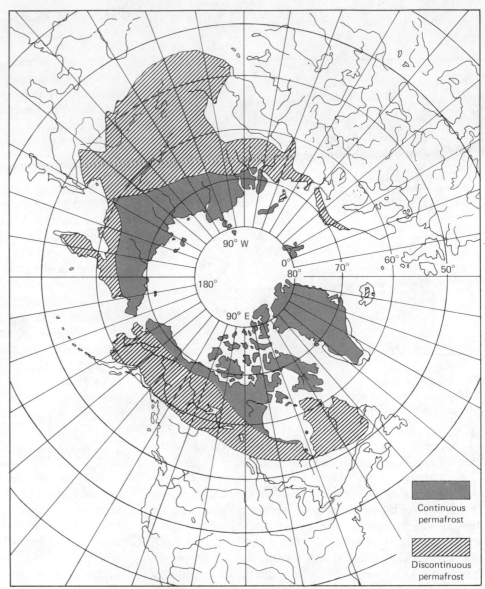

90° W

0°

80°

70°

60°

50°

180°

90° E

Continuous
permafrost

Discontinuous
permafrost

FIGURE 2.31
Gravel road in Alaska
showing severe differential
subsidence due to thawing
of permafrost. (Photo by O.
J. Ferrians, Jr., courtesy of
the U.S. Geological Survey.)

FIGURE 2.32
Tractor trail on the North Slope of Alaska. The small ponds are due to thawing of the permafrost in the roadway. (Photo by O. J. Ferrians, Jr., courtesy of the U.S. Geological Survey.)

environmental planning. The characteristic assemblages of plants and animals in these physiographic regions are also important in planning. The plants affect the soil, the erosion potential, and other factors which eventually feed back to change the landform assemblages themselves.

SUMMARY AND CONCLUSIONS

Although the Earth intercepts only a very tiny fraction of the energy emitted from the sun, it is solar energy that sustains life on Earth and drives two of the Earth's physical systems—the hydrologic cycle and atmospheric circulation—and ultimately drives the Earth's climate.

Climate and climatic change are important topics in environmental studies for two main reasons. First, many human endeavors are dependent on favorable climatic conditions. Second, human-induced processes are now capable of modifying local, if not global, climate.

Processes that create, maintain, change, or destroy earth materials such as minerals, rocks, soil, and water, as well as entire landforms, are collectively referred to as the geologic cycle. More correctly, the geologic cycle is a set of subcycles that include the tectonic, rock, geochemical, and water cycles. The geologic cycle performs many service functions for people, and people use earth materials found in different parts of the cycle. These materials, whether water or minerals, are initially uncontaminated or in a concentrated state. Once dispersed or used, these materials may not be so available for human use.

The physical properties of earth materials such as minerals, rocks, and soils are important to recognize in environmental studies. These materials behave predict-

ably but differently for various land uses. Thus, while expansive soils have undesirable properties for construction sites, well-drained soils over strong rock are more desirable and will cause fewer problems. Heavy clay soils may be poor for most agricultural purposes, and well-drained soils support different natural vegetation than poorly drained soils.

Continental landforms, even in arid and semiarid regions, are formed almost exclusively by processes associated with running water. Therefore, understanding both surface and subsurface hydrologic processes is important in environmental studies. Hydrologic systems tend to establish a rough steady state in which various parts of the systems adjust to each other. Changes, whether artificial or natural, cause readjustments and produce a new steady state. These readjustments may have adverse impacts on human systems.

Windblown sand, silt, and dust as well as glacial deposits from the recent Pleistocene glaciation cover many thousands of square kilometers of the Earth's surface. Understanding the physical properties of these materials and how they are deposited is necessary if environmental problems are to be avoided or minimized. In addition, the permafrost which underlies about 20 percent of the world's land area produces a fragile and sensitive environment. Special engineering procedures are necessary to minimize adverse effects from artificial thawing of frozen ground.

In dealing with the Earth's physical systems, we conclude that various physical processes produce characteristic materials and landforms, and that the specific properties associated with each must be considered in environmental problems.

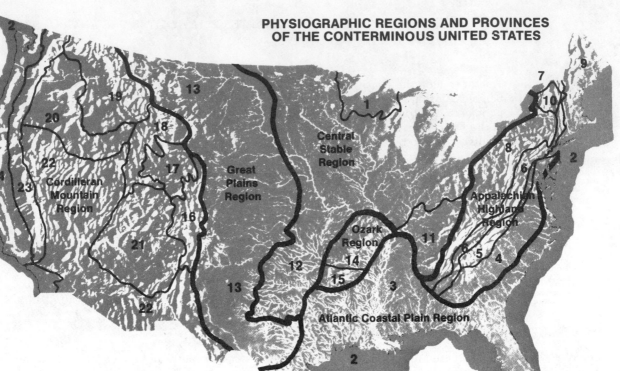

Topographical map courtesy of the United States
Department of the Interior, Geological Survey.

EXPLANATION

1. Superior Upland — Hilly area of erosional topography on ancient crystalline rocks.
2. Continental Shelf — Shallow sloping submarine plain of sedimentation.
3. Coastal Plain — Low, hilly to nearly flat terraced plains on soft sediments.
4. Piedmont Province — Gentle to rough, hilly terrain on belted crystalline rocks becoming more hilly toward mountains.
5. Blue Ridge Province — Mountains of crystalline rock 3,000 to 6,000 feet high, mostly rounded summits.
6. Valley and Ridge Province — Long mountain ridges and valleys eroded on strong and weak folded rock strata.
7. St. Lawrence Valley — Rolling lowland with local rock hills.
8. Appalachian Plateaus — Generally steep-sided plateaus on sandstone bedrock, 3,000 to 5,000 feet high on the east side, declining gradually to the west.
9. New England Province — Rolling hilly erosional topography on crystalline rocks in southeastern part to high mountainous country in central and northern parts.
10. Adirondack Province — Subdued mountains on ancient crystalline rocks rising to over 5,000 feet.
11. Interior low plateaus — Low plateaus on stratified rocks.
12. Central Lowland — Mostly low rolling landscape and nearly level plains. Most of area covered by a veneer of glacial deposits, including ancient lake beds and hilly lake-dotted moraines.

13. Great Plains — Broad river plains and low plateaus on weak stratified sedimentary rocks. Rises toward Rocky Mountains at some places to altitudes over 6,000 feet.
14. Ozark Plateaus — High, hilly landscape on stratified rocks.
15. Ouachita Province—Ridges and valleys eroded on up-turned folded strata.
16. Southern Rocky Mountains — Complex mountains rising to over 14,000 feet.
17. Wyoming Basin — Elevated plains and plateaus on sedimentary strata.
18. Middle Rocky Mountains — Complex mountains with many intermontane basins and plains.
19. Northern Rocky Mountains — Rugged mountains with narrow intermontane basins.
20. Columbia Plateau — High rolling plateaus underlain by extensive lava flows; trenched by canyons.
21. Colorado Plateau — High plateaus on stratified rocks cut by deep canyons.
22. Basin and Range Province — Mostly isolated ranges separated by wide desert plains. Many lakes, ancient lake beds, and alluvial fans.
23. Cascade-Sierra Nevada Mountains — Sierras in southern part are high mountains eroded from crystalline rocks. Cascades in northern part are high volcanic mountains.
24. Pacific Border Province — Mostly very young steep mountains; includes extensive river plains in California portion.

FIGURE 2.33

Physiographic regions and landforms of the conterminous United States. (Diagram and explanation courtesy of the U.S. Geological Survey.)

REFERENCES

1 MARSH, W. M., and DOZIER, J. 1981. *Landscape*. Reading, Mass.: Addison-Wesley.

2 PESTRONG, R. 1974. *Slope stability*. New York: McGraw-Hill.

3 KRYNINE, D. P., and JUDD, W. R. 1957. *Principles of engineering geology and geotechnics*. New York: McGraw-Hill.

4 MATHEWSON, C. C.; CASTLEBERRY, J. P., II; and LYTTON, R. L. 1975. Analysis and modeling of the performance of home foundations on expansive soils in central Texas. *Bulletin of the Association of Engineering Geologists* 17: 275–302.

FURTHER READING

1 GATES, D. M. 1972. *Man and his environment: Climate*. New York: Harper & Row.

2 GEIGER, R. 1965. *The climate near the ground*. Cambridge, Mass.: Harvard University Press. (English translation by Scripla Technica, Inc.)

3 GREGORY, K. J., and WALLING, D. E., eds. 1979. *Man and environmental processes*. London: William Dawson.

4 GRIBBEN, J., ed. 1978. *Climatic change*. Cambridge, Mass.: Cambridge University Press.

5 HENDERSON, L. J. 1913. *The fitness of the environment*. New York: Macmillan.

6 KELLER, E. A. 1982. *Environmental geology*. 3rd ed. Columbus, Ohio: Charles E. Merrill.

7 LE ROY LADURIE, E. 1971. *Times of feast, times of famine: A history of climate since the year 1000*. Garden City, N.Y.: Doubleday.

8 MARSH, W. M., and DOZIER, J. 1981. *Landscape*. Reading, Mass.: Addison-Wesley.

9 MATHEWSON,, C. C. 1981. *Engineering geology*. Columbus, Ohio: Charles E. Merrill.

10 SELLERS, W. D. 1965. *Physical climatology*. Chicago: The University of Chicago Press.

11 UTGARD, R.O.; MCKENZIE, G. D.; and FOLEY, D., eds. 1978. *Geology in the urban environment*. Minneapolis: Burgess.

STUDY QUESTIONS

1 Discuss the importance of the geological cycle to the maintenance of the biosphere.

2 What were the lessons learned from the St. Francis Dam failure? How could they be useful in planning the site for a nuclear power plant?

3 Distinguish among climate, microclimate, and weather. Which is most influenced by a major city?

4 What has been the relationship between human exploration of new regions and (a) climate; (b) weather; (c) microclimate?

5 How can the tectonic cycle influence the pattern of rainfall on the Earth's surface?

6 Why is a city sidewalk built over a clay soil more likely to crack and break than a sidewalk built over a sandy soil?

7 Why does groundwater move more slowly than surface water in streams? What is the implication of this for the management of water pollution?

8 What were the causes of the "dust bowl" in the United States in the 1930s? What is the likelihood of this phenomenon occurring again?

9 Why is it difficult to build a city on permafrost?

10 Your neighbor contends that glaciers could return quickly—in 10 years. Would you agree or disagree? Explain.

3

The Earth as an Unusual Planet: The Biosphere's Biological Processes

INTRODUCTION

The Earth as a planet has been profoundly altered by life. Its air, oceans, soils, and rocks are very different from what they would be on a lifeless planet. In some ways, life on the Earth controls the makeup of the air, oceans, and sediments. It has greatly changed the Earth's surface during the last 3 billion years and continues to control and to modify it.

The influence of life on global Earth processes is illustrated by the story of the source of phosphorus used as an agricultural fertilizer. Historically, one of the major sources of phosphorus fertilizer are deposits on certain peculiar islands found only in dry regions and in low latitudes, such as off the coast of Africa, in the Caribbean, in the mid- and western Pacific, and along the coast of Peru. The deposits are as much as 40 meters deep and lie on otherwise flat-topped islands (Fig. 3.1). Valuable and plentiful, the fertilizer is part of an international trade, which in Peru began in earnest about 1840 when ships carried the material to London—as much as 9 million tons per year—where it was sold for 28 British pounds per ton.

What is the source of the fertilizer, and why does it occur only in certain kinds of islands in certain climates and latitudes? The fertilizer is *guano*, the excrement of birds who nest on the islands in thousands and have done so for thousands of years, at least since the end of the last glacial age. The three principal species—the guanay (*Phalacrocorax bougainvillei*), the piquero (*Sula varie-*

gata), and the Chilean pelican (*Pelecanus occidentalis thagus*)—feed on anchovy, schooling fish that live in regions rich in tiny, floating ocean life called *plankton*. The plankton, in turn, thrive where nutritionally essential chemical elements like nitrogen and phosphorus occur. Nutrients are abundant in areas of oceanic *upwelling*, where ocean currents rise and carry elements to the surface.

The guano is deposited in great amounts only where there are large colonies of nesting, colonial birds, and these must be near a vast source of food. The birds nest on islands free from predators. The guano hardens into a rocklike mass only in a relatively dry climate [1].

Thus one of the world's major sources of phosphorus depends on a peculiar combination of biological and geological processes. There are few inorganic ways that phosphorus is brought to the surface of the land, as we will see later in this chapter. A change in the upwelling or the climate or a loss of species of plankton, fish, or birds could greatly alter the supply of phosphorus. In the past this has happened during periods called *El Niño* when climatic changes reduced the upwelling, reducing the growth of plankton and fish, and causing the birds to starve in huge numbers. In this way, the Earth's biota has had an important effect on a global cycle of an important element. Without the plankton, fish, and colonial birds, the phosphorus would have remained in the oceans, and the world's agricultural production would have been significantly reduced.

(a)

(b)

FIGURE 3.1
A guano island (a) and guano deposits (b). (Photos from Hutchinson, 1950.)

Human activities now also affect the Earth on a global scale, and these effects are increasing with technological advances. Our civilization has a great potential to alter our climate, the chemistry of our atmosphere, and even the chances that life will continue to persist on the Earth. Any study of human effects on the environment must examine them at the level of the whole Earth, or the *biosphere.*

The **biosphere** is that region of the Earth where life exists. It extends from the depths of the ocean to the summits of mountains, but most life is concentrated within a few meters of the Earth's surface. To understand human effects on the biosphere we must understand the Earth as a living planet, studying the global cycling of chemical elements necessary for life and the exchange of energy on the Earth's surface. The purpose of this chapter is to introduce this global perspective, to discuss how life has greatly altered the Earth in the past and how it continues to do so, and how the persistence of life depends on global characteristics of the Earth.

LIFE AFFECTS THE BIOSPHERE

Life has altered the Earth's atmosphere, oceans, and sediments for a very long time—more than 3 billion years. Ancient biologically induced changes greatly affect our modern world, the current makeup of the air, oceans, and rocks, the creatures who live on the Earth, and our lives.

All the gases in the atmosphere (except the noble ones—neon, argon, xenon, and their chemical relatives) are greatly affected by life. The circulation of chemicals in bodies of water, large and small, is altered in important ways by life. Many kinds of mineral deposits are of biological origin or greatly affected by life; not only oil and coal are organic products of organisms, but limestone rock and some metallic ones occur in abundance only because of life's impact on the Earth.

How do we know that life has greatly changed the Earth in the past? Some of the evidence can be found in ancient rocks. About 3.5 billion years ago dense mats of photosynthetic bacteria grew and died in shallow ocean bays. The bacteria trapped sediments which had washed down in ancient streams and rivers to the oceans. These sediments were fossilized into structures called *stromatolites,* which are rocks with layers like leaves in a wrinkled book (Fig. 3.2). By reading these leaves, we can learn the ancient history of the Earth [2].

These ancient stromatolites are similar to structures forming today in shallow ocean bays and estuaries from blue-green algae (*Cyanobacteria*)—with one important difference. The sediments found in modern environments are oxidized: that is, the chemical elements on the surface of these sediments have combined with the oxygen in the atmosphere. The materials in the ancient stromatolites are not oxidized, indicating that the Earth's atmosphere 3 billion years ago had little oxygen.

(a)   (b)

FIGURE 3.2
Fossil stromatolites 1.8 billion years old from Great Slave Lake, Canada (a), appear much like recent deposits formed in Shark Bay, western Australia (b). The ancient deposits, however, do not show signs of oxidation, suggesting that the ancient Earth's atmosphere had little oxygen. (Photos courtesy of S. Awramik.)

There are other rocks that tell the same story. Economically valuable iron-rich rocks—the rocks that form the major iron ranges where iron is mined—were laid down more than 2 billion years ago. It is believed that the iron had been dissolved in the ancient oceans, and formed sediments when it combined with the oxygen produced by photosynthetic organisms. These sediments then became the rocks which provide our iron ore. When most of the iron dissolved in the ancient oceans had been oxidized and deposited, then the free oxygen produced by organisms began to build up in the atmosphere. Thus the atmosphere's concentration of free oxygen began to increase about 2 billion years ago. Our major iron ore deposits as well as free oxygen in the atmosphere are thus a result of life's influence on the biosphere.

On that ancient Earth, blue-green algae were among the primary oxygen-releasing photosynthetic plants. They, like all such organisms, took carbon dioxide out of the air and gave off oxygen. Some of the oxygen was taken up again in respiration, and some of the carbon dioxide returned to the air in that same process. When some of these organisms or their products were buried without being completely decomposed, there was a slight imbalance in the uptake and release of oxygen, and oxygen was slowly added to the atmosphere. Thus early life in the form of these lowly, seemingly insignificant mats of bacteria began to change the entire planet. *Free oxygen in the Earth's atmosphere is the result of 3 billion years of photosynthesis and is therefore a product of life.*

(a)

(b)

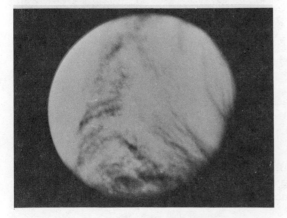

FIGURE 3.3
Photographs of the Earth, Mars, and Venus from spacecraft (a) and data as to their size, distance from the sun, and atmospheric composition (b). (Photos courtesy of NASA.)

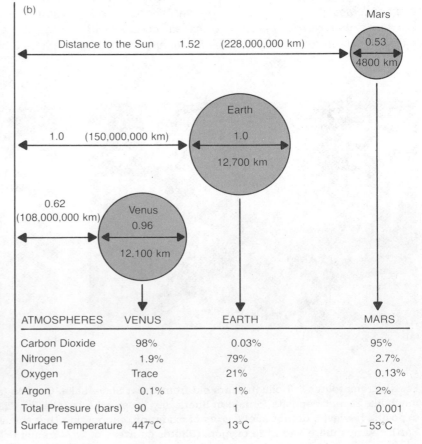

ATMOSPHERES	VENUS	EARTH	MARS
Carbon Dioxide	98%	0.03%	95%
Nitrogen	1.9%	79%	2.7%
Oxygen	Trace	21%	0.13%
Argon	0.1%	1%	2%
Total Pressure (bars)	90	1	0.001
Surface Temperature	447°C	13°C	−53°C

SOME PLANETARY BIOLOGY

Other evidence of life's effect on the Earth can be found by comparing the Earth with other planets in our solar system. The Earth is a peculiar planet when compared with its nearest neighbors, Mars and Venus. These three planets are similar in many ways. They are within a factor of two in their distance from the sun and in their diameter. Since they are similar in size and distance from the sun, one would expect them to be similar in many other characteristics, like the makeup of their atmospheres. However, scientific observations in the last several decades, both from space probes sent to Mars and Venus and from Earth-based telescopic observations, reveal that Mars and Venus have atmospheres that are very different from the Earth's. Their atmospheres are more like one another than either is like the Earth's. Some of the primary characteristics of the atmospheres of the three planets are given in Figure 3.3. The atmosphere of Venus is very dense and that of Mars is very thin, but the chemical compounds that make up their atmospheres are similar in relative abundance [3]. Their atmospheres are mainly carbon dioxide, with nitrogen as the second most important compound. Oxygen is a rare and minor part of their atmospheres. In contrast, the Earth's atmosphere is 79 percent nitrogen, 21 percent oxygen, and only 0.03 percent carbon dioxide. The difference in the relative abundance of the primary constituents of the atmospheres of these three planets is due to the effects of life on the Earth's surface.

LIFE AND THE BIOSPHERE TODAY

Life continues today to affect the characteristics of the Earth's surface, even those areas that we think are far removed from living things, like the summit of an active volcano or the interior of the Antarctic continent. This was shown to be true by samples of the air taken between 1958 and 1971 near the summit of Mauna Loa, Hawaii, one of the largest active volcanoes in the world [4]. These samples, used to measure the carbon dioxide concentration of the Earth's atmosphere, were taken on Mauna Loa because it was far from any local, direct effects of human or other biological activity. Because carbon dioxide is taken up by green plants during photosynthesis and released in the respiration of all oxygen-breathing organisms, a measure of the carbon dioxide in the Earth's atmosphere is like the measure of the breathings in and out of all life on Earth.

The 15-year Mauna Loa measurements are one of the most remarkable observations in our century in regard to the Earth as a living planet. Two important aspects of this record are clearly evident in Figure 3.4a: an annual cycle and a continual upward trend. The annual cycle is extremely regular. A peak is reached in the winter and a trough in the summer. The curve is indeed a measure of the life activities, including human activities, of the entire Northern Hemisphere. In the summer the green plants are most active, and the total amount of photosynthesis exceeds the total amount of respiration; thus carbon dioxide is removed from the atmosphere, and the concentration, as measured on Mauna Loa, decreases. In the winter photosynthesis decreases greatly and becomes less than the total respiration, so the carbon dioxide concentration of the atmosphere increases.

Similar observations have been made in Antarctica for the same 15-year period. (Because Antarctica is so inaccessible, measurements are much more sporadic.) The same trends are observed: an annual cycle and a strong upward trend (Fig. 3.4b). The annual cycle is smaller in amplitude than the Mauna Loa cycle because of the relatively smaller land area in the Southern Hemisphere and the smaller amount of woody vegetation which stores carbon and exchanges carbon dioxide with the atmosphere.

The upward trend in both the Mauna Loa and Antarctica curves for carbon dioxide is thought to be due to the addition of carbon dioxide from the burning of fossil fuels and other human activities, such as the cutting of forests and the burning of wood. These curves show that life touches the entire Earth and that human activities have begun to affect the Earth's entire atmosphere.

If life on the Earth were in a steady state so that the total amount of organic matter were constant and the total photosynthesis equaled the total respiration, then there would not be a net addition or removal of carbon dioxide from the atmosphere. Over geologic ages the Earth has been almost in a steady state in regard to the productivity of the biota, but there has been a slight excess of photosynthesis over respiration, leading to a high concentration of oxygen and low concentration of carbon dioxide in the atmosphere. This slight imbalance over long geological periods has also resulted in deposition in the Earth's sediments of carbon-bearing minerals—oil, coal, organic-rich shales, and calcium carbonate being among the most important.

As a result of life, the Earth's atmosphere is not in a chemical or thermodynamic steady state with the oceans and sediments and land surfaces. This means that if one were able to close off the present atmosphere of the Earth together with the water and the land and keep these without life for a long period, the atmosphere would change from what it is today. Its composition would come into an equilibrium with the oceans and land surfaces.

FIGURE 3.4
Changes in carbon dioxide concentration in the atmosphere at Mauna Loa, Hawaii (a), and the South Pole (b). The concentration is small, and is measured in parts per million (ppm). (From Ekdahl and Keeling, 1973.)

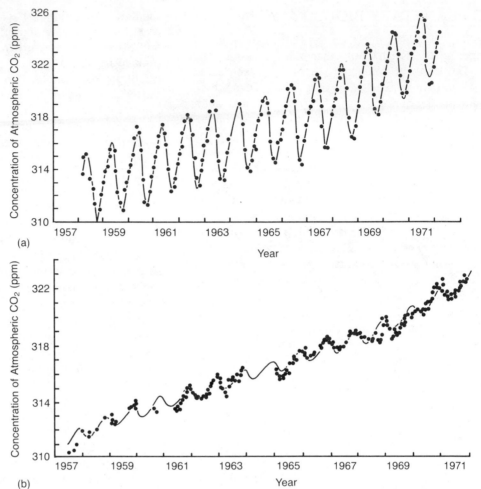

The oxygen would combine with other elements in the waters and the land to form oxides, including carbon dioxide. Eventually, little free oxygen would be found in the atmosphere; however, thousands of years would be required to reach this condition. The atmosphere of this lifeless Earth would resemble that of Mars and Venus. The atmospheres of those planets are much closer to such a chemical or thermodynamic equilibrium than that of the Earth's.

In fact, the Earth's atmosphere is an almost explosive combination of methane and oxygen [5]. Methane is a small organic compound produced by bacterial activity in the absence of oxygen and released from marshes and the intestines of ruminant animals. It is a good fuel and highly explosive. Although its concentration in the Earth's atmosphere is small, approximately 0.00015 percent, only slight increases in the concentration of methane would lead to conditions in which any spark, such as that produced by lightning, would ignite the methane in the present atmospheric concentration of oxygen. This almost explosive combination is totally the result of the activities of microbes. The oxygen concentration in the atmosphere is also in a delicate balance. At present concentrations, lightning strikes can start fires in dry vegetation, but if the oxygen concentration were to increase from 21 to 25 percent even wet twigs and grass could ignite [3]. Thus life provides the means that produce *and* rapidly remove oxygen from the atmosphere.

It has long been believed that life had only a few rather special effects on the Earth. Scientists have long recognized that the oxygen concentration of the Earth's atmosphere is due to biological processes, but it is only recently that they have recognized the extent of life's influence. Life is the principal source of the major gases of the Earth's atmosphere. Living things take up gases from the atmosphere and give off gases to the atmosphere (Table 3.1). The residence time, which is the average time a molecule of a gas spends in the atmosphere, varies greatly from millions or billions of years to a few hours. All the chemical compounds in the atmosphere, except those of the noble gases, are thus strongly affected and controlled by life. The major constituents of the atmosphere are changed, modulated, and controlled by the Earth's entire biota.

TABLE 3.1
Some major gases of the Earth's atmosphere and their sources.

Gas	Principal Source	Residence Time
Nitrogen	Bacteria	Millions to billions of years
Oxygen	Vegetation photosynthesis	Thousands of years
Carbon dioxide	Respiration of living organisms; combustion of fuels	A few years
Carbon monoxide	Bacterial breakdown of certain compounds; incomplete combustion of fuels	A few months
Methane	Bacteria in environments without oxygen	A few years
Nitrous oxide (N_2O)	Bacteria and fungi	10 years
Ammonia	Bacteria and fungi	A few days
NO_x	Reaction of pollutants in sunlight	A few days
Hydrogen sulphide	Bacteria in environments without oxygen	Several hours
Sulphur dioxide	Combustion of fuels; marine algae and green plants	A few days
Hydrogen	Photosynthetic bacteria; oxidation of methane; electrical separation of water	A few years

SOURCE: Modified from Margulis and Lovelock, 1974.

The Mauna Loa and Antarctic observations suggest also that human activities can affect the Earth on a global as well as a local scale. That we are able to influence the Earth as a whole means that the management, wise use, and understanding of our entire planet must now take into account these human activities.

A BRIEF HISTORY OF LIFE'S EFFECT ON THE EARTH

From the study of rocks like the banded-iron formations and the stromatolites, we can reconstruct the history of the Earth's atmosphere, as illustrated in Figures 3.5a and 3.5b. Scientists now believe that the Earth's early atmosphere, before life existed, was devoid of oxygen. Some believe it was primarily an atmosphere of hydrogen, methane, and ammonia. Others believe it was primarily nitrogen, carbon monoxide, and carbon dioxide. Approximately 3.5 billion years ago, about the age of the earliest fossils, there was a shift to an atmosphere with free nitrogen and carbon dioxide. Then, approximately 2 billion years ago, there was a rapid increase in the concentration of oxygen and a decrease in carbon dioxide to approximately present levels. In contrast, on a lifeless Earth the concentration of carbon dioxide and nitrogen in the atmosphere would have remained high. Although oxygen would have increased due to the action of sunlight on water (the sunlight split some water molecules into hydrogen and oxygen), it would not have increased to the same concentration as is found today. In fact, it would have been almost 100 times less. Carbon dioxide would have been 10 times greater; free nitrogen 10 times less.

The Earth's atmosphere differs not only in composition from the atmosphere of a lifeless planet, but also in acidity and temperature. Sunlight and lightning in thunderstorms cause nitrogen and oxygen to form nitrogen oxides. If life disappeared from the Earth, eventually (over a very long time) nitrogen in the atmosphere would be transformed to its chemically stable form as nitrate, the oxide of nitrogen. Nitrate is very soluble in water and forms a strong acid. Enough of this nitrate would then be transferred to the oceans to make the ocean waters more acid [2].

The atmosphere of lifeless planets is affected by sunlight. In particular, the energy from the sun removes hydrogen from the atmosphere, so that planets near to the sun tend to become devoid of hydrogen. Thus the atmosphere of Venus, nearer to the sun than the Earth, has lost much hydrogen, whereas the atmosphere of Mars, further from the sun, has small hydrogen compounds like ammonia.

THE EARTH'S ENERGY BUDGET

Not only did early life change the chemistry of the Earth's surface, it also altered the Earth's heat budget and surface temperature (Fig. 3.6). The temperature of the Earth's surface is a result of energy exchange and of physical characteristics of the surface. Energy is received from the sun, and a very small amount is generated from the Earth's core (produced by radioactive processes that heat up the center of the Earth). Energy is lost by radiation from the Earth's surface to space. The hotter an object is, the more rapidly it radiates heat and the shorter the wavelength of the predominant radiation. That is, a blue flame is hot, and a red flame is cooler. The Earth's surface and the surface of animals, plants, clouds, water, rocks, and so on are so cool that heat is radiated predominantly in the infrared, which is invisible to us (see Fig. 2.2 for comparison) [6].

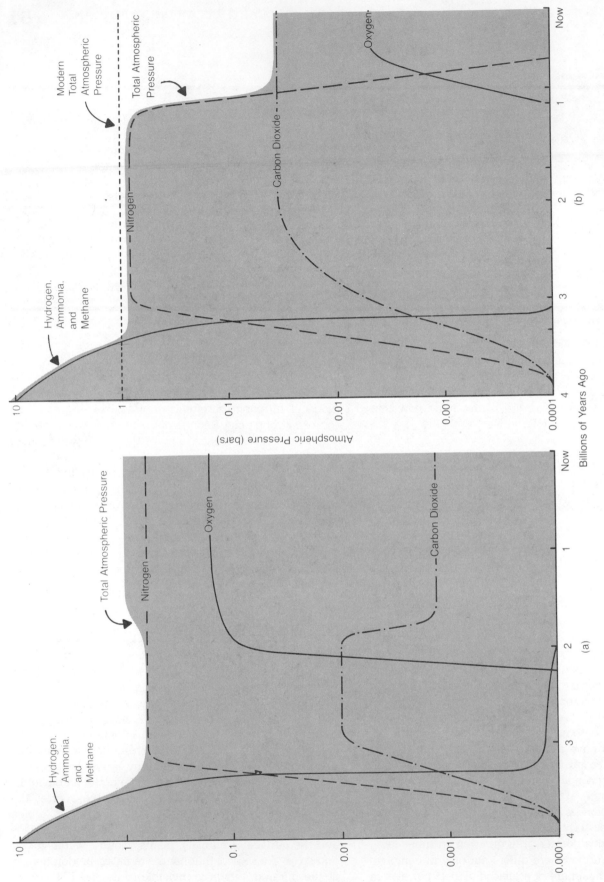

FIGURE 3.5
The history of the Earth's atmosphere as we understand it (a), and a hypothetical history of the atmosphere of a lifeless Earth (b). Without the influence of life, the Earth's atmosphere would be very different from what it is today. (Modified from Margulis and Lovelock, 1974.)

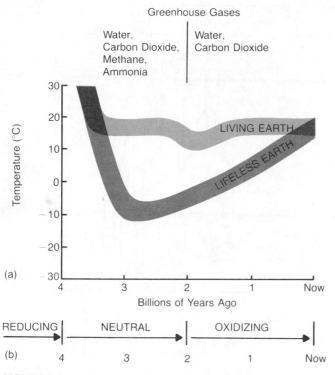

FIGURE 3.6

The history of the Earth's temperature (a). Both the intensity of sunlight and the atmospheric gases affect the Earth's surface temperature. The sun has become hotter, and the Earth's atmosphere has changed as a result of life. A hypothetical lifeless Earth would have undergone much greater temperature changes than has our living Earth. The Earth's chemistry has also been altered by life in three major stages: reducing, neutral, and oxidizing (b). (Modified from Margulis and Lovelock, 1974.)

If an object is cold and gives off less heat than it receives, it will warm up, but as it warms up it will also radiate heat more rapidly. As a result, for any constant input of energy a physical object will eventually reach a temperature that will allow it to radiate heat energy at the same rate that it receives energy.

The rate at which the Earth's surface radiates heat depends on its average "color" and temperature. For example, a white planet radiates heat very differently from a black planet. A perfect emitter of heat is called an *ideal black body,* and black surfaces radiate much more readily than white surfaces. Fluctuations in the amount of ice and the distribution of vegetation over the Earth change the reflecting and emitting characteristics of the Earth's surface. It has been estimated that under present atmospheric conditions a 1 percent change in the amount of sunlight reflected by the Earth would cause approximately a 1.7C° change in the surface temperature.

Changes in the absorption of energy can be due to changes in the cloud cover and in the amount of ice cover on the Earth's land surfaces. Organisms, particularly grasses, trees, and algae, can also change the absorption rate. Such biological changes occur seasonally. Marine algal mats can change from light and highly reflective in one season to nearly black and highly emitting in another. Algae can also produce sediments like calcium carbonate, which when pure are chalky white and have a different reflective characteristic than the sediments produced from a lifeless surface. Because it is the surface characteristics of any physical object that determine its emission of heat energy, a very thin surface—even a single-cell layer of algae—over a large area of water could greatly alter the rate of emission of energy and therefore the temperature of the Earth's surface. Ice reflects 80 to 95 percent of light; a dry grassland 30 to 40 percent; and a conifer forest 10 to 15 percent [6].

The energy exchanged by the Earth's surface is also affected by the chemical composition of the atmosphere. Certain compounds absorb more strongly in some wavelengths of light energy than others (Fig. 3.7). The infrared wavelengths radiated from the Earth's surface are strongly absorbed by water and carbon dioxide. If the water content or the carbon dioxide content of the atmosphere changes, so does the reabsorption of the energy released from the Earth's surface. This leads to what is known as the "greenhouse effect" (Fig. 3.8).

In a greenhouse, visible sunlight passes through the glass and heats up the soil warming the plants. The warm soil emits radiation in longer wavelengths, particularly in the infrared. Because the glass is opaque to these wavelengths, it absorbs and reflects the infrared, which is why a greenhouse is warmer than the outdoors and warmer than a room that does not allow visible light through. (Another reason why a greenhouse is warmer than the outdoors is that a greenhouse prevents heat loss to the flowing air.) The glass in the greenhouse is a one-way window open to visible light and closed to infrared. Carbon dioxide and water vapor act in the same way, although condensed water droplets in clouds are opaque to both infrared and visible light. Thus a change in the atmosphere's carbon dioxide concentration would change the thermal characteristics of the Earth's surface.

If carbon dioxide had been removed from the Earth's early atmosphere, the Earth's surface would have cooled, as if a window in a greenhouse had been opened. According to some scientists, the sun may have become hotter since the ancient times in which life originated, which would have counteracted the cooling process. A lifeless Earth would have a much higher surface temperature than the present Earth or than the ancient Earth when life

FIGURE 3.7
How sunlight changes as it passes through clear and cloudy atmospheres. The Earth's atmosphere absorbs and reflects radiation from the sun. Water vapor, carbon dioxide, and other trace gases absorb the infrared, producing the steep dips in the curve for a clear day. Cloudy skies prevent even more of the sun's energy from reaching the ground. The light under a forest canopy is even less intense, and very little of the infrared reaches the ground. (From Gates, 1965.)

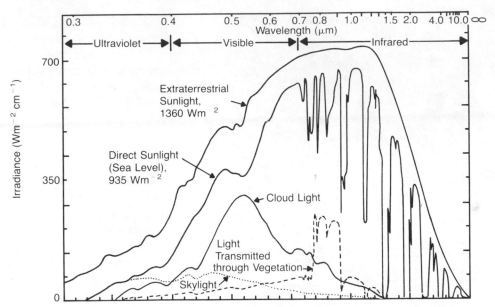

first arose. Venus' surface is much hotter than would be a Venus with life on it because of the high concentration of carbon dioxide in its atmosphere.

To summarize, as life evolved on the Earth and became abundant, it changed the atmosphere's chemistry and the characteristics of the Earth's surface, particularly the reflection and absorption of radiant energy. Both of these changes, physical and chemical, affect the energy budget of the Earth. Any major change in the abundance and distribution of life, today or in the future, on the Earth's surface will affect the reflection and absorption of light and infrared radiant energy and therefore change the Earth's heat budget. Moreover, human activities, particularly those that are the result of a technological civilization, change the Earth's energy budget by introducing

carbon dioxide and pollutants like dust, smoke, haze, or small compounds that absorb in the infrared into the atmosphere. Modern civilization is changing drastically the distribution and abundance of the life forms on the Earth, and this may have concomitant effects on the Earth's energy budget. Those effects will be discussed in later chapters.

GLOBAL BIOGEOCHEMICAL CYCLES

A **biogeochemical cycle** is the cycling of a chemical element through the Earth's atmosphere, oceans, and sediments as it is affected by the geological and biological cycles. It can be described as a series of compartments, or storage reservoirs, and pathways between these reser-

FIGURE 3.8
The "greenhouse effect." (Modified from illustration by Robin Panza.)

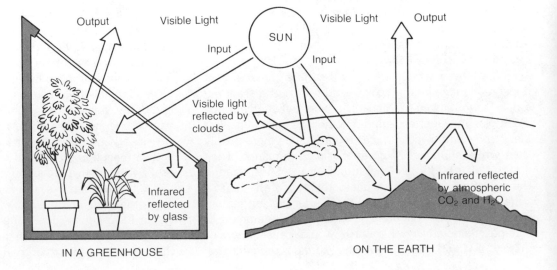

IN A GREENHOUSE   ON THE EARTH

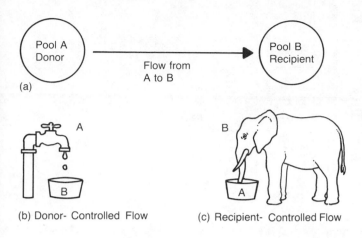

(a)

(b) Donor- Controlled Flow **(c) Recipient- Controlled Flow**

FIGURE 3.9
One unit from a biogeochemical cycle. (Modified from illustration by Robin Panza.)

voirs. An atom or molecule flows from one reservoir to another (Fig. 3.9a). If more molecules flow into a compartment than flow out, the compartment increases in size or storage. If more molecules flow out than flow in, the compartment decreases in size or storage. When the flow in equals the flow out, the compartment is in **steady state.** The material remains within a reservoir for some average time, called the **residence time.** The fraction of the amount stored that is added and lost during a time period is called the **turnover rate.** The exchange between two reservoirs is called the **flow** of the element, and the amount flowing per unit time is called the **flux rate.** The amount in a compartment depends on the inputs and the outputs. The changes in the size of a reservoir depend on the changes in input and output flux rates.

The factors that control the flows between compartments must be understood for science and for managerial purposes. Sometimes flow is controlled solely by one compartment, sometimes by another, and sometimes by processes in between. When the flow is controlled solely by the donor compartment, the flow is called **donor controlled;** when it is controlled by the receiving compartment, it is called **recipient controlled.** Water flowing into a glass from a funnel and flowing out of the glass through a straw to a person drinking the water are examples. The rate at which water flows from the funnel to the glass depends on the size of the opening in the funnel and the pressure exerted by the water in the funnel. This is a donor-controlled flow (Fig. 3.9b). The rate at which water leaves the glass through the straw depends on the pressure exerted by the person sucking on the straw. This is a recipient-controlled flow (Fig. 3.9c). The storage reservoirs in biogeochemical cycles are large units of the

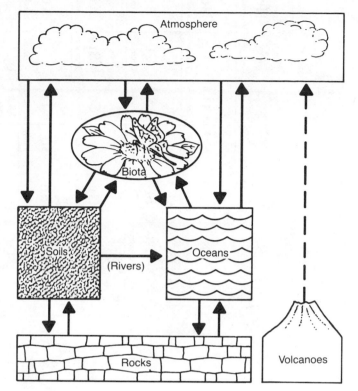

FIGURE 3.10
Generalized biogeochemical cycle. The major parts of the biosphere are connected by the flow of chemical elements and compounds. In many of these cycles, the biota play an important role. Matter from the Earth's interior is released by volcanoes. The atmosphere exchanges some compounds and elements rapidly with the biota and oceans. Rocks exchange material more closely with soils and the oceans.

Earth, such as the atmosphere or all terrestrial vegetation (Fig. 3.10).

Many chemical elements are required for life (Table 3.2). We can divide these into three groups: (1) the "big six" that form the major building blocks of organic compounds: carbon, hydrogen, oxygen, nitrogen, phosphorus, and sulphur; (2) the other **macronutrients** required in large amounts by most forms of life (this group includes potassium, calcium, iron, and magnesium); and (3) **micronutrients,** those required in very small amounts by at least some organisms (this group includes boron, used by green plants; copper, used in some enzymes; and molybdenum, used by nitrogen-fixing bacteria).

There are two major kinds of biogeochemical cycles: those that involve a gaseous phase and those that do not. The cycles that involve a gaseous phase include a residence time in the atmosphere for the chemical element. We will consider several of these cycles as examples.

TABLE 3.2
Periodic table of the elements.

Key

Atomic Number → 20
Element Relatively Abundant in Earth's Crust → *
Environmentally Important Trace Elements → **

Ca
Calcium ← Name
Biological Role

1																	2
H ● Hydrogen																	**He** Helium
3 **Li** ** \ Lithium	4 **Be** × Beryllium											5 **B** Boron	6 **C** ● Carbon	7 **N** ● Nitrogen	8 * **O** ● Oxygen	9 **F** ** \ Fluorine	10 **Ne** Neon
11 **Na** Sodium	12 * **Mg** Magnesium											13 * **Al** Aluminum	14 * ** **Si** Silicon	15 **P** Phosphorus	16 **S** Sulfur	17 **Cl** \ Chlorine	18 **Ar** Argon
19 * **K** ● Potassium	20 ** **Ca** ● Calcium	21 **Sc** Scandium	22 **Ti** Titanium	23 ** **V** \ Vanadium	24 ** **Cr** \ Chromium	25 ** **Mn** ● Manganese	26 * ** **Fe** ● Iron	27 ** **Co** ● Cobalt	28 ** **Ni** × Nickel	29 ** **Cu** \ ● Copper	30 ** **Zn** ● Zinc	31 **Ga** Gallium	32 **Ge** Germanium	33 **As** × Arsenic	34 ** **Se** ● Selenium	35 **Br** × Bromine	36 **Kr** Krypton
37 **Rb** Rubidium	38 **Sr** Strontium	39 **Y** Yttrium	40 **Zr** Zirconium	41 **Nb** Niobium	42 ** **Mo** Molybdenum	43 **Tc** Technetium	44 **Ru** Ruthenium	45 **Rh** Rhodium	46 **Pd** Palladium	47 **Ag** × Silver	48 ** **Cd** × Cadmium	49 **In** Indium	50 ** **Sn** \ Tin	51 **Sb** Antimony	52 **Te** Tellurium	53 ** **I** Iodine	54 **Xe** Xenon
55 **Cs** Cesium	56 **Ba** Barium	57 ● **La** Lanthanum	72 **Hf** Hafnium	73 **Ta** Tantalum	74 **W** Wolfram	75 **Re** Rhenium	76 **Os** \ Osmium	77 **Ir** Iridium	78 **Pt** Platinum	79 **Au** Gold	80 ** **Hg** × Mercury	81 **Tl** \ Thallium	82 ** **Pb** × Lead	83 **Bi** Bismuth	84 **Po** \ Polonium	85 **At** Astatine	86 **Rn** × Radon
87 **Fr** Francium	88 **Ra** Radium	89 ●● **Ac** × Actinium															

58 **Ce** \ Cerium	59 **Pr** Praseodymium	60 **Nd** Neodymium	61 **Pm** Promethium	62 **Sm** Samarium	63 **Eu** Europium	64 **Gd** Gadolinium	65 **Tb** Terbium	66 **Dy** Dysprosium	67 **Ho** Holmium	68 **Er** Erbium	69 **Tm** Thulium	70 **Yb** \ Ytterbium	71 **Lu** × Lutetium
90 **Th** Thorium	91 **Pa** Protactinium	92 **U** × Uranium	93 **Np** × Neptunium	94 **Pu** × Plutonium	95 **Am** × Americium	96 **Cm** × Curium	97 **Bk** × Berkelium	98 **Cf** × Californium	99 **Es** × Einsteinium	100 **Fm** × Fermium	101 **Md** × Mendelevium	102 **No** × Nobelium	103 **Lw** × Lawrencium

Legend

● = Required for all life

◖ = Required for some life forms

\ = Moderately toxic; either slightly toxic to all life or highly toxic to a few forms

× = Highly toxic to all organisms, even in low concentrations

THE CARBON CYCLE

The major aspects of the carbon cycle, which involves a gaseous phase, are shown in Figure 3.11. Carbon exists in the atmosphere in several compounds; the primary one is carbon dioxide, and small organic compounds like methane and ethylene occur in lesser amounts. Certain parts of the carbon cycle are inorganic; that is, they do not depend on biological activities. Carbon dioxide is soluble in water and is exchanged between the atmosphere and ocean or freshwater lakes merely by the physical process of diffusion (Fig. 3.12). Other things being equal and lacking any living systems or source of external energy, the diffusion of carbon dioxide gas will continue in one direction or the other until a steady state is reached between the carbon dioxide in the atmosphere above water and the dissolved carbon dioxide in the water. When carbon dioxide is dissolved in water, a mild acid is formed, which in turn changes the amount of carbon dioxide that can be subsequently dissolved. Dissolved carbon dioxide

forms carbonate and bicarbonate in the water. When a great deal of carbonate is formed, the water becomes saturated, and the carbonate will combine with metallic elements like calcium and be deposited as salt, in which calcium carbonate predominates. These sediments exist in a dynamic steady state with the water and can redissolve when the carbonate concentration in the water decreases. Thus a solution of water over a calcium carbonate rock will eventually come into a steady state just as an air and water solution will come into a steady state.

Carbon dioxide enters the biological cycles through photosynthesis, in which carbon, hydrogen, and oxygen are combined to form organic compounds. The rate of carbon dioxide uptake depends on the amount and activity of photosynthetic plants, on the concentration of carbon dioxide in the atmosphere, and on other environmental conditions. Plants grow faster when the carbon dioxide concentration of the air is increased as long as their other needs are met for light, water, nutrients, and

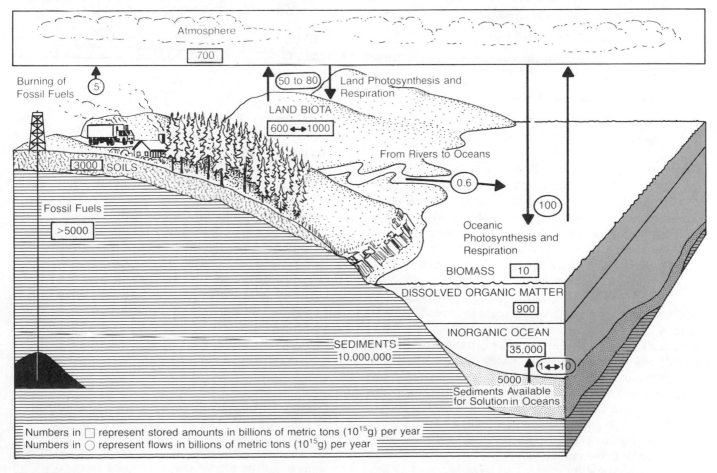

FIGURE 3.11
The global carbon cycle. (From Bolin et al., 1979.)

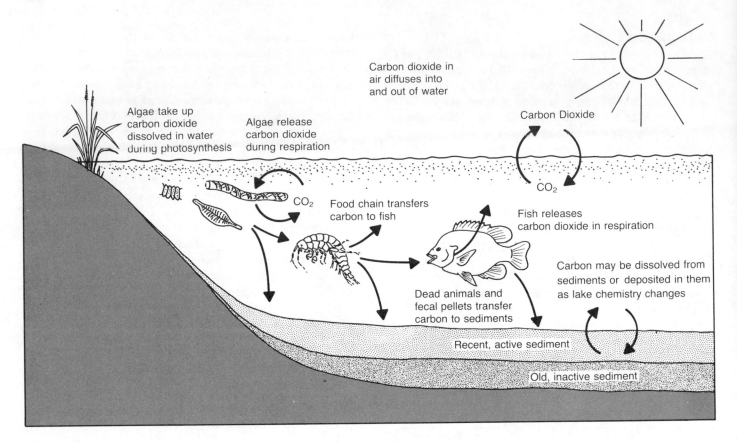

FIGURE 3.12
The carbon cycle in a lake.

so on. Thus the transfer from the atmospheric reservoir to the living organic reservoir is both donor and recipient controlled.

Most of the carbon in living tissue is stored in the woody tissue of vegetation, particularly in forests. The tropical rain forests, the temperate deciduous and evergreen forests, and the boreal forests of the high latitudes represent some of the major storage reservoirs for the Earth's biological carbon. The amount of carbon stored in living tissue is essentially equivalent to the total amount of carbon in the atmosphere.

Carbon leaves the living biota through respiration, which transforms organic compounds to gaseous carbon dioxide. The carbon dioxide is released to the atmosphere or to fresh and marine waters before being emitted to the atmosphere. When the living tissue dies, some of the dead organic matter may be stored (in some cases for a considerable length of time) as soil or as deep-sea sediments. Some of the dead organic matter is food for bacteria and fungi. In this case, carbon returns to the atmosphere as methane and carbon dioxide through the respiration of these organisms.

Terrestrial organic matter can be transported by the geological cycles to the ocean, transported as dissolved or particulate organic matter through the fluvial cycle, or

carried by wind. Some of the organic material is added to deposits which eventually convert to rock, and thus the carbon enters the rock cycle. As we can see in Figure 3.11, most of the carbon in the Earth's surface exists as rock, particularly as carbonates, which are primarily the result of organic deposition in marine ecosystems. Carbonates are formed in the shells of marine organisms and are deposited if the ocean waters become supersaturated with carbonate. The next largest storage reservoirs in the carbon cycle are the oceans, where the dissolved carbon is 50 times the amount in the atmosphere.

The cycle of carbon dioxide between living organisms and the atmosphere represents a large flux, with approximately 10 percent of the total amount of carbon in the atmosphere being taken up by photosynthesis and released as respiration annually.

THE PHOSPHORUS CYCLE

Phosphorus is one of the essential elements for life on the Earth's surface. Although it is in relatively short supply, it is a major nutrient for all living things. Phosphorus is an example of a chemical element without a major gaseous phase in its biogeochemical cycle; it exists in the atmosphere only as small particles of dust. In contrast to the carbon cycle, the phosphorus cycle is slow, and much of it proceeds one way from the lands to the oceans (Fig. 3.13). Phosphorus enters living things from the soil, where it exists in minerals combined with calcium, potassium, magnesium, and iron and as phosphate. Because these minerals are relatively insoluble in water, phosphorus becomes available very slowly through the weathering of rocks or particles of rocks in the soil. In a relatively stable terrestrial ecosystem much of the phosphorus that is taken up by vegetation will be returned to the soil. Some of the phosphorus, however, is inevitably lost to the eco-

Numbers in ☐ represent stored amounts in millions of metric tons (10^{12}g) per year

Numbers in ◯ represent flows in millions of metric tons (10^{12}g) per year

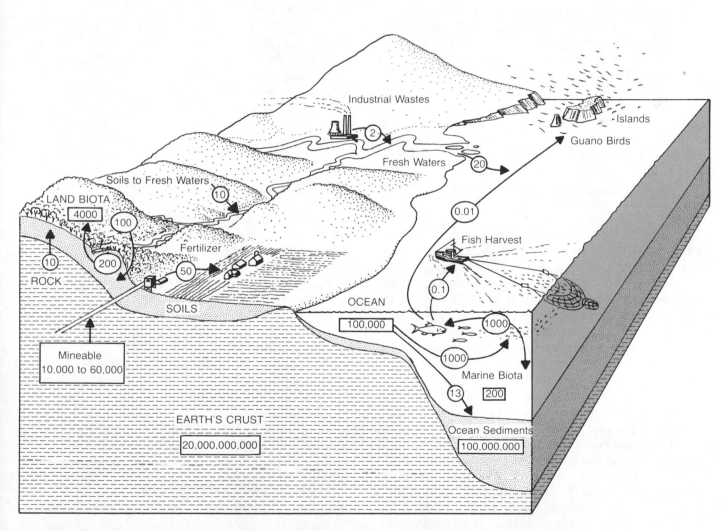

FIGURE 3.13

The global phosphorus cycle. Phosphorus is recycled to soil and land biota by geological uplifting processes, by birds producing guano, and by human beings. While the Earth's crust contains a very large amount of phosphorus, only a small fraction of it is mineable by conventional techniques. Phosphorus is therefore one of our most precious resources. (Values are compiled from various sources. The estimates are approximate and are given here to the approximate order of magnitude. Based primarily on Delwiche and Likens, 1977, and Pierrou, 1976.)

system. It moves out of the soils in a water-soluble form and is transported by waters through rivers and streams to the oceans. Thus the phosphorus cycle is a gradual flow, from the removal of phosphorus from the land to its deposition in the oceans. Because phosphorus is one of the major constituents of agricultural fertilizers, sources of phosphorus such as the guano deposits on oceanic islands are extremely important to us.

The return of phosphorus to the land is also a slow process. We have already learned that ocean-feeding birds provide one pathway to return phosphorus from the ocean to the land. Another major source of phosphorus fertilizers are sedimentary rocks made up of the fossils of marine animals. The richest such mine in the world is "Bone Valley," 40 kilometers east of Tampa, Florida. Between 15 and 10 million years ago, Bone Valley was the bottom of a shallow sea where marine invertebrates lived and died [7]. Through geological processes, Bone Valley was uplifted. In the 1880s and 1890s, the discovery of its phosphate ore led to a "phosphate rush," and now the valley provides more than one third of the world's entire phosphate production and three fourths of U.S. production. Bone Valley represents a very slow return of phosphate to the land, and its redistribution over wide areas occurred only when people used technology to mine, transport, and spread the fertilizer.

Perhaps the most important point about the phosphorus cycle is the slowness of the return from ocean to land. Because it is a crucial macronutrient for all life and because many of its compounds are relatively water-insoluble, phosphorus is likely to be one of the most hard to get, most subject to competition, and most limiting of all the chemical elements necessary for life.

The mining of phosphorus ores poses several environmental issues. Bone Valley, for example, has become an environmental problem. Mined of much of its ore, it is several hundred square kilometers of strip mine with huge slime ponds—a scar on the landscape. Second, what will we do when mines like Bone Valley and the Peruvian islands are exhausted?

Some experts believe that, at current mining costs of $15 per ton, total U.S. reserves of phosphorus are about 2.2 billion metric tons. This quantity is estimated to be only several decades of supply. However, for a higher price per ton, more phosphorus is available. Florida is thought to have 8.1 billion metric tons of phosphorus recoverable with existing methods, and there are large deposits elsewhere [7]. However, the mining processes have negative effects on the landscapes. How to balance the need for phosphorus, the environmental impact of the mining, and the mitigation of these impacts is a major environmental issue.

THE NITROGEN CYCLE

The nitrogen cycle is one of the most important and most complex cycles and is most likely to cause environmental problems for us. One of the "big six" elements, it is required for all living things as an essential part of amino acids that make up proteins. In the first part of this chapter we learned that the biota had greatly changed the nitrogen content of the atmosphere. Life maintains nitrogen in the molecular form (N_2) in the atmosphere, rather than in ammonia (NH_3) or nitrogen oxides, as would occur on a lifeless planet. Nitrogen exists in our atmosphere in seven forms: molecular nitrogen (N_2), oxides of nitrogen (N_2O, NO, and NO_2), and hydrogen-nitrogen compounds (NH, NH_3, and HNO_2).

Nitrogen is comparatively unreactive—in contrast to hydrogen, oxygen, and carbon—and tends to remain in small inorganic compounds. Considerable energy is required to connect molecular nitrogen to some other compounds, and most organisms cannot use molecular nitrogen directly. It must be converted to ammonia (NH_3), nitrate ion (NO_3^-), or amino acids. There are only two major natural pathways for this conversion: lightning and life. By far the greater amount is converted by biological activity (Table 3.3).

The major pathways of the nitrogen cycle are shown in Figure 3.14. One of the more curious and important points we should keep in mind about the nitrogen cycle is that the conversion of molecular nitrogen to ammonia or nitrate, and other chemical transformations of inorganic forms of nitrogen, can be done only by bacteria and blue-green algae—members of the *prokaryotic* forms of life, primitive forms that lack a membrane-bounded cell nucleus. Only these microorganisms carry out certain chemical reactions without which all life would stop.

So important are these conversions of nitrogen that many major higher organisms called *eukaryotes* (with a membrane-bounded cell nucleus) have evolved symbiotic relationships with some nitrogen-transforming microbes.

TABLE 3.3
Crucial nitrogen transformations.

NITROGEN FIXATION
Nitrogen + water combined by bacteria to give ammonia and oxygen.

NITRIFICATION
Ammonia + oxygen combined by bacteria to give nitrite (oxide of nitrogen) and hydrogen and water.

DENITRIFICATION
Nitrate + water combined by bacteria to give molecular nitrogen and oxygen and hydroxide.

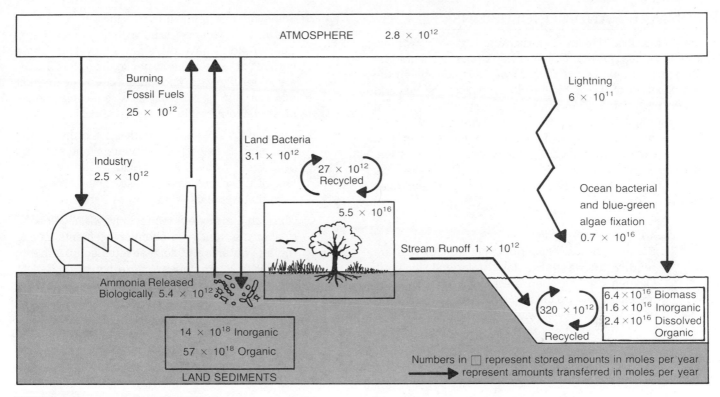

FIGURE 3.14
The nitrogen cycle. (From Garrels et al., 1975.)

These include many flowering plants, particularly the pea family (*Leguminoseae*), which have nodules on the roots that support nitrogen-fixing bacteria (those that convert N_2 to ammonia). The acacia trees, very important in the savannahs of Africa, are also part of such a symbiotic relationship. The trees provide carbohydrates for and an environment conducive to bacterial growth. Nitrogen-fixing bacteria also are symbionts in the stomachs of some animals, particularly the *ruminants*—the cud-chewing, herbivorous mammals with specialized four-chambered stomachs, like cows, deer, moose, and giraffes. The nitrogen fixed by ruminant bacteria provide as much as half of the total nitrogen for the ruminants, with the rest being provided by protein in the green plants they eat. The ruminants provide food for the bacteria, which not only enrich the food with nitrogen (by growing, incorporating nitrogen, and being digested themselves by the ruminants) but also digest plant material that would otherwise not be useable by the mammals.

Ammonia is released into the atmosphere by bacteria decomposition; it is very soluble in water and is returned to the oceans and land in rain as salts of ammonia (ammonium sulphate and ammonium nitrate).

If bacteria only fixed nitrogen, molecular nitrogen would slowly be removed from the atmosphere. But bacteria also carry out **denitrification** (see Table 3.3), releasing the molecular nitrogen as a gas back into the atmosphere. Thus the removal and addition of nitrogen to the atmosphere are primarily controlled by bacterial activity. Nitrogen enters the ocean through the fixation of atmospheric nitrogen by planktonic *Cyanobacteria* which live in the ocean. Nitrogen compounds also flow into the ocean, carried by freshwater runoff from rivers and streams.

In contrast to phosphorus, nitrogen has a major gaseous phase, is highly mobile, and is rapidly recycled. Like phosphorus, nitrogen is a major agricultural fertilizer. Until World War I only natural sources provided nitrogen for fertilizers, but during that war German scientists discovered that electrical discharges could be used to fix nitrogen industrially. Industrial fixation today is a major source of commercial nitrogen fertilizer.

Nitrogen oxides are formed by a high-temperature chemical reaction where nitrogen and oxygen are present. Oxides of nitrogen are a major pollutant from automobiles and are the most difficult to eliminate because they are formed by the high temperature in the cylinders when ordinary air is used.

Nitrogen compounds are thus a bane and a boon for modern technological society. Nitrogen is required for all life, and its compounds are used in many technological processes and devices, including explosives and fuels.

SUMMARY AND CONCLUSIONS

Each chemical element necessary for life is also transformed by life, and its global cycle is greatly affected by the Earth's biota. In this chapter we have considered the complexities of a few of the global cycles, the ways in which life affects them, and the impact of our modern technological society on them. The wise use of our resources and management of our environment depend on our understanding of each of these cycles. Although there are as many biogeochemical cycles as there are chemical elements necessary for life, there are certain general rules for these cycles:

1   Some chemical elements cycle quickly and are readily regenerated for biological activity. Oxygen and nitrogen are among these. Others are easily tied up in relatively immobile forms and are returned only slowly, by geological processes, to locations where they can be reused by the biota. Phosphorus is involved in this kind of cycle.

2   Some biogeochemical cycles include a gaseous phase (a residence time in the atmosphere). These tend to have more rapid recycling. Those without an atmospheric phase, like phosphorus, are more likely to end up as deep-ocean sediments and to be relatively slow to recycle. We are most concerned with this type of cycle.

3   The Earth's biota have greatly altered the cycling of chemicals among the air, waters, and soils and have greatly changed our planet. The continuation of these processes is essential to the long-term maintenance of life on the Earth.

4   Through our modern technology, we have begun to alter the transfer of chemical elements among the air, waters, and soils at rates comparable to natural biological ones. These activities benefit society, but also pose great dangers. To wisely manage our environment, we must recognize both the positive and negative consequences of these activities, attempting to accentuate the first and minimize the second.

REFERENCES

1   HUTCHINSON, G. E. 1950. Survey of contemporary knowledge of biochemistry. 3. The biogeochemistry of vertebrate excretion. *Bulletin of the American Museum of Natural History* 96: 481–82.

2   LOVELOCK, J. E., and MARGULIS, L. 1974. Homeostatic tendencies of the Earth's atmosphere. *Origins of Life* 5: 93–103.

3   LOVELOCK, J. E. 1979. *Gaia, a new look at life on Earth.* New York: Oxford University Press.

4   EKDAHL, C. A., and KEELING, C. D. 1973. Atmospheric carbon dioxide and radiocarbon in the natural carbon cycle: 1. Quantitative deductions from records at Mauna Loa Observatory and at the South Pole. In *Carbon and the biosphere,* eds. G. M. Woodwell and E. V. Pecan, pp. 51–85. Brookhaven National Laboratory Symposium No. 24. Oak Ridge, Tenn.: Technical Information Service.

5   LOVELOCK, J. E., and MARGULIS, L. 1973. Atmospheric homeostasis by and for the biosphere: The Gaia hypothesis. *Tellus* 26: 1–9.

6   GATES, D. M. 1980. *Biophysical ecology.* New York: Springer-Verlag.

7   ANONYMOUS. 1980. Phosphate: Debate over an essential resource. *Science* 209: 372.

FURTHER READING

1   BOLIN, B.; DEGENS, E. T.; KEMPE, S.; and KETNER, P., eds. 1979. *The global carbon cycle.* New York: Wiley.

2   BOWEN, H. J. M. 1979. *Environmental chemistry of the elements.* New York: Academic Press.

3 FORTESCUE, J. A. C. 1980. *Environmental geochemistry: A holistic approach.* New York: Springer-Verlag.

4 GARRELS, R. M.; MACKENZIE, F. T.; and HUNT, C. 1975. *Chemical cycles and the global environment.* Los Altos, Calif.: William Kaufmann.

5 HOLLAND, H. D. 1978. *The chemistry of the atmosphere and oceans.* New York: Wiley-Interscience.

6 HUTCHINSON, G. E. 1954. The biochemistry of the terrestrial atmosphere. In *The Earth as a planet,* ed. G. P. Kuiper, pp. 371–433. Chicago: The University of Chicago Press.

7 LOVELOCK, J. E. 1979. *Gaia, a new look at life on Earth.* New York: Oxford University Press.

8 SHORT, N. M.; LOWMAN, P. D., Jr; FREEMAN, S. C.; and FINCH, W. A., Jr. 1976. *Mission to Earth: LandSat views the world.* Washington, D. C.: U.S. Government Printing Office.

9 WALKER, J. C. G. 1977. *Evolution of the atmosphere.* New York: Macmillan.

STUDY QUESTIONS

1 What characteristics of the planet Mercury, observable by remote sensing, indicate that it is a lifeless planet?

2 It is the year 2500, and you are sent to Venus and told that your job is to convert that planet to a livable habitat for people. Do you believe this is possible? What actions would you take?

3 Describe how the Earth's energy exchange is influenced by life.

4 In what ways do living organisms influence the ozone layers in the atmosphere?

5 What is the evidence that the early Earth had an atmosphere with little oxygen?

6 How do living things influence global chemical cycles?

7 Explain what is meant by the "greenhouse effect."

8 List the major effects life has on the Earth.

4

Birth, Growth,
and Death:
The Ecology of
Populations

INTRODUCTION

In 1911 the U.S. government introduced small groups of reindeer *(Rangifer tarandus)* on two of the Pribilof Islands that lie in the cold Bering Sea between Alaska and Siberia: 4 bucks and 21 does on St. Paul, the largest of the islands, covering 12,000 hectares; 3 bucks and 12 does on St. George, the second largest island (Fig. 4.1) [1]. The reindeer were introduced to provide a much needed source of food for the islands' inhabitants, a group of Aleuts who had been settled there in 1787 by Gerassium Pribilof, the Russian explorer for whom the islands are named.

When the United States purchased Alaska from Russia, the Pribilof Islands and their inhabitants came under U.S. jurisdiction. The Aleuts had survived primarily on what the sea offered, and some additional source of protein seemed necessary to those concerned with management of the islands. The islands had an abundance of the vegetation which the reindeer ate, and there were no wolves or other predators large enough to affect the herbi-vores. In 1922 G. D. Hanna, a wildlife expert, wrote in an article in *Scientific Monthly*, "It would seem that here is the place to maintain model reindeer herds and to determine many of the needed facts for the propagation of these animals on a large scale. At no other place are conditions so favorable" [2].

In spite of favorable conditions, something went very wrong with the Pribilof Islands reindeer. At first, the introductions seemed a success; in the spring of 1912, 17 fawns were born on St. Paul and 11 on St. George (Fig. 4.2). But, ironically, in the year that Hanna wrote his enthusiastic report, the population on St. George had reached a peak of 222 individuals, from which it would decline and never return again. By the 1940s the reindeer herd on St. George numbered between 40 and 60, and in the 1950s the herd became extinct. On St. Paul the herd grew more rapidly, reaching a peak of 2000 animals in 1938, when there was 1 deer for every 6 hectares of the island and 1 for every 5 hectares of rangeland. Soon after the reindeer herd rapidly declined and the St. Paul population numbered only 8 in 1950 and 2 in 1951.

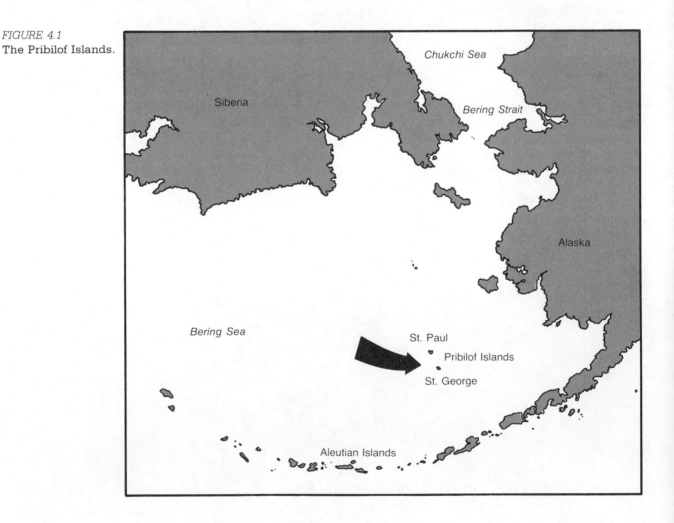

FIGURE 4.1
The Pribilof Islands.

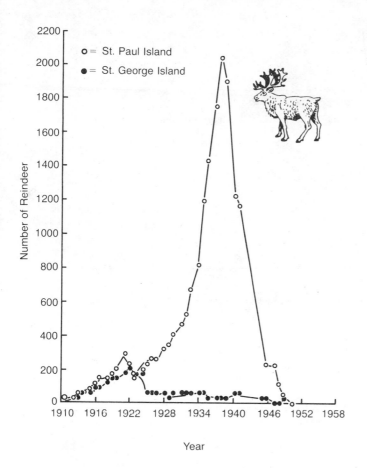

FIGURE 4.2

The population growth of the Pribilof Island reindeer. (Modified from Sheffer, 1951.)

While the reindeer occupied St. Paul and St. George, the herds were monitored and attempts were made to manage them so that they would provide a sustained food supply for the inhabitants. Why did this management go wrong? What happened to the reindeer herd that, in spite of attempts to the contrary, brought it to decline and essentially to extinction? Was its fate inevitable, the result of unassailable laws of nature? And, if so, if one could only discover them, could we use these laws in the wise management of our living resources?

BASIC CONCEPTS OF POPULATION DYNAMICS

People have long been interested in what controls the growth of populations of living things. A **population** is a group of individuals of the same species which interbreed frequently. The initial rapid growth of the reindeer on both Pribilof Islands illustrates the often discussed and well-known capacity for biological populations to grow rapidly. This capacity was recognized as early as Aristotle, whose writings contain some of the earliest discussions in Western culture of population characteristics. In his writings, Aristotle recounts the story of a pregnant mouse who was shut up in a jar filled with millet seed; "after a short while," when the jar was opened, "120 mice came to light" [3].

During the first decade after the introduction, reindeer populations on both St. George and St. Paul islands increased rapidly and, as we have seen, continued to increase rapidly on St. Paul until 1938. The initial growth on both islands is called **exponential growth.** Exponential growth is what Thomas Malthus described in his famous discussion of the inevitability of human populations increasing until they exceed their resource supply. It is commonly recognized that all biological populations have the capacity for exponential growth but rarely achieve it. It was Charles Darwin's recognition that exponential growth was rarely achieved in nature that led him to perceive that there would be competition for survival and to propose his theory of biological evolution.

Since the term *exponential growth* often is used in everyday life, it is important that we understand exactly what is meant when it is said that a population is growing exponentially. The *exponential growth curve* assumes that the *percentage* increase in a population is constant. This means that the larger the population, the larger is the actual number added during any unit of time, so the population increases in absolute number faster and faster (Fig. 4.3). When the percentage increase is constant, the rate of increase does not vary with time, with any characteristics of the population like the health or vigor or proportion of adults in the population, nor with any environmental conditions like the weather or the quality of food.

Clearly, no real population can sustain an exponential rate of growth indefinitely; otherwise, it would eventually require all the matter in the universe. For example, some bacteria cells can divide rapidly. Suppose a cell divided every 2 hours. If there were 2 cells at the beginning, there would be 4 cells after 2 hours, 8 cells after 4, 16 after 6, and 4096 after 1 day. In a matter of weeks the number of cells would require all the matter in the universe.

As another example, suppose a pair of mice were put on an island and that the average population growth was 30 percent per year. There would be 1 million mice in 44 years, 1 billion in 66 years, and 10 trillion in a century. In approximately 800 years the mice would reach a number requiring all the matter in the universe. In many countries the human population is growing at 5 percent per year, which means the population doubles in 14 years. A population of 100 increasing at 5 percent per year would grow

FIGURE 4.3
Exponential and logistic growth. An exponentially growing population increases by a constant percentage every time period. A logistic population initially grows almost at an exponential rate, then slows down and stops growing at its carrying capacity.

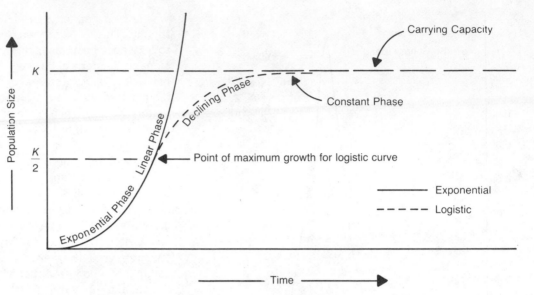

AN OLD EXAMPLE OF THE CALCULATION OF EXPONENTIAL POPULATION GROWTH

Many people have recognized this impossible future for an exponentially growing population. One of the first calculations about the capacity for biological populations to grow was made in the eighteenth century by Count de Buffon in *Natural History, General and Particular.* In that book Count de Buffon observed that the "most ordinary and familiar operation of Nature was the production of organized bodies," and that "here her powers know no limitation." He used as an example the potential contained in the single seed of an elm tree which, he noted, weighed less than one hundredth of an ounce, but could in 100 years grow into a tree whose volume, in Count de Buffon's units, would "amount to ten cubic fathoms."

What would happen if a tree matured in 10 years and produced 1000 new seeds, and each of these after 10 years began to produce 1000 seeds a year? Assuming that each tree had a volume of 10 cubic fathoms after 100 years, Count de Buffon calculated that the elm trees would soon occupy 10,000,000,000,000 cubic fathoms, or 1000 cubic leagues. Ten years after that the elms would occupy "1,000,000,000,000 cubic leagues," and after 150 years the whole globe would be "converted into organized matter" as the result of a single elm seed weighing less than one hundredth of an ounce.*

*From LeClerc, 1812.

to 1 billion in 350 years. If the human population had increased at this rate since the beginning of recorded history, it too would now exceed the matter available in the universe.

If no real population can grow exponentially, what kind of growth can we expect? The study of the growth and regulation of populations is called the study of **population dynamics.** To understand population dynamics, we must first know the primary characteristics that describe a population. Any population has a size, which is its total number of individuals. A population is also characterized by three rates: the rates of birth, death, and growth. When we are only interested in the total number of individuals in a population, these three rates can be related to each other simply: the growth rate equals the birth rate minus the death rate. These factors are related to each other as shown by the equations presented in the box on page 76.

Sometimes we are interested not only in the total number of individuals but in the amount of organic matter, known as the **biomass,** in the population (Fig. 4.4). Whether we want to measure population growth by the number of individuals or by the biomass depends on the question to be answered. If we are trying to determine whether a species is endangered, then the number of individuals is the measure of interest. When the question concerns one population that is a food source for another, then the appropriate measure is often the biomass.

THE LOGISTIC GROWTH CURVE

In 1838 a European scientist, P. F. Verhulst, suggested that a real population would grow according to an S-shaped curve, called the **logistic** (Fig. 4.3). The logistic adds to an exponential growth curve the concept that any real population must be limited eventually by some resource in its environment. This limitation is represented in the simplest possible way by the population's **carrying capacity.** The carrying capacity is defined as the maximum population size that can exist in a habitat or ecosystem over a long period of time without detrimental effects to either population, habitat, or ecosystem. Here the term *detrimental effects* means any population effects that would result in a decrease in the carrying capacity.

In the logistic growth curve it is assumed that each individual has some negative effect, however small, on another, either by decreasing the population's reproductive rates or increasing mortality rates. When a population growing according to the logistic is small, its growth is very close to exponential. However, the logistic growth

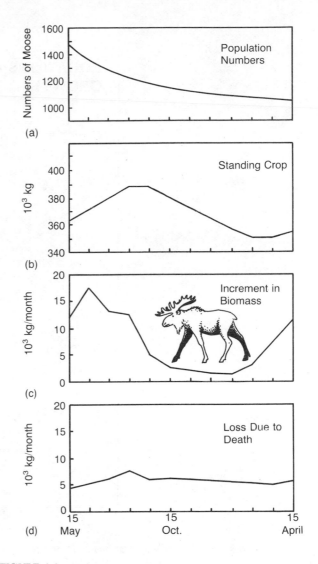

FIGURE 4.4
Population growth measured by changes in population size and by changes in biomass for the moose herd at Isle Royale National Park. The moose is not hunted by human beings, but preyed upon by wolves. The four graphs show four measures of population change over one year. Calves are born in the spring, so the year starts in mid-April. The number of moose (a) declines after the births, as moose are killed and eaten by wolves. However, the biomass (b) of the herd increases rapidly throughout the summer as the moose eat fresh green vegetation. The total biomass decreases in the winter, when the moose have little to eat except twigs of trees and needles of conifers. The net increase in live moose (c) increases rapidly in the spring, then decreases rapidly and is almost zero in the winter. The wolves must have a rather steady diet and kill moose throughout the year at a fairly constant rate (d).

EQUATIONS FOR SIMPLE POPULATION GROWTH

Growth in the number of individuals:

$$\text{Growth rate} = \text{Birth rate} - \text{Death rate}$$

or

$$r = b - d \tag{1}$$

Exponential growth:

$$N_t = N_0 e^{rt} \tag{2}$$

N_t = the number of individuals at time t
N_0 = the original number at the starting time (time zero)
e = the base of natural logarithms
t = time
r = the growth rate as defined in equation (1)

Logistic growth:

$$G = \frac{\Delta N}{\Delta t} = \frac{r N_t (K - N_t)}{K} \tag{3}$$

$\dfrac{\Delta N}{\Delta t}$ = the change in the population per unit time
r = the intrinsic rate of growth as defined in equation (1)
K = the carrying capacity
N_t = the population at the present time t

equation assumes that at every population level the competition for some resource diminishes the growth of the population. *The growth limitation is proportional to the population size*, so the rate of growth decreases as the population size increases. After an initial, almost exponential rate of growth, a logistic population passes into a second phase, where the limitations due to resources cause the population to increase almost along a straight line. That is, the *absolute* numbers added to the population in each time period are constant, and the growth *rate* is declining. This is called the *linear phase* of population growth. Finally, as resources become more limiting, the growth slows down more and more until the population reaches a final fixed size, where births equal deaths, and the population is said to be at its carrying capacity.

A constant carrying capacity is a mathematical convenience that allows a simple description of hypothetical population growth. Although carrying capacity is often considered to be constant in the management of fisheries and wildlife and in theoretical ecology, we will see that it is incorrect to treat it so. Carrying capacities change over time.

The logistic curve has been one of the most important equations in the study of population dynamics, and it is sometimes argued that real populations do in fact grow according to this curve. The logistic growth curve has also been important in the management of biological resources such as oceanic fisheries, endangered species such as the great whales, and game populations such as the white-tailed deer of eastern North America. As we shall see, the application of the logistic to wildlife and fisheries has often led to a management failure because managers have tried to manage a population as if it could be kept at a single population size indefinitely.

One of the appealing aspects of the logistic growth equation is that it has in theory a point of **stable equilibrium**, the point when the population is at its carrying capacity. This point is an *equilibrium* because a logistic population, if undisturbed, would remain at this size indefinitely, since births continue to equal deaths in every time period. This point is also *stable* because a logistic population that is less than or greater than the carrying capacity will return to the carrying capacity. This can be seen most clearly when we graph the rate of change in the population against population size (Fig. 4.5). If the population size exceeds the carrying capacity, then deaths exceed births, the growth rate is negative, and the population returns to the carrying capacity level. On the other hand, if the population falls below this carrying capacity, births exceed deaths, the population growth is positive, and the population size returns once more to the carrying capacity.

The idea that a population has a fixed size to which it will return after it is disturbed fits in nicely with the prescientific beliefs that have pervaded human culture about a balance of nature. A logistic population is constant and stable, suggesting that it is balanced and in harmony with its environment. Although the logistic growth curve is appealing in this way, in general real populations do not follow this curve. Only a few scientific studies suggest that any real populations under natural conditions follow this curve.

In his book on the mathematics of biology, *Elements of Mathematical Biology,* Alfred Lotka describes some hypothetical conditions under which a fixed carrying capacity might be achieved [4]. He describes the growth of flies in a screened laboratory enclosure into which a fixed supply of food was placed each day. As long as there was more food than was required by the population of flies, births would exceed deaths and the population would grow. Eventually a point would be reached when the food supply just met the needs of the population. Any increase in population would result in an excess of deaths over births, and the population would decline again to the carrying capacity made possible by the food supply.

The logistic growth curve has, in fact, been shown to fit the growth of small organisms like microbes or fruit flies kept in laboratory conditions and provided with a constant amount of food at fixed intervals (Fig. 4.6). If logistic growth can be made to occur in a laboratory, why did it not occur for the reindeer on the Pribilof Islands? If it had occurred, the reindeer population would have remained on each island at the maximum level observed rather than undergoing a rapid and extreme decline to extinction.

A review of the history of Pribilof Islands reindeer herds by Victor Sheffer sheds some light on what might have caused the decline [1]. The main foods of the reindeer are grasses and small flowering plants and shrubs found in the interior of the islands. These provide the food for the reindeer during the spring, summer, and fall. The crash of the reindeer population is puzzling because these plants remained abundant on the islands throughout the entire period of the rise and fall of the reindeer population. Thus the decline of the reindeer was not caused by a decrease in the grasses and herbs that provided the bulk of their diet.

The key to the population decline appears to lie in the winter, the time of greatest stress for the reindeer, when they feed on lichens called reindeer moss (of the genus *Cladonia*) obtained by pawing through the snow. Because lichens are very slow growing, they were rapidly depleted by the reindeer. A particularly cold winter in 1940 worsened matters. Island records indicate that in that year a crust of glare ice remained on the snow for weeks. Although the reindeer can paw through as much as 1 meter of soft snow, they had difficulty in digging through this crust. In early spring 150 carcasses, primarily of females, were found on St. Paul Island.

The reindeer ran out of the food that was crucial to

FIGURE 4.5

Population change versus population size for logistic growth.

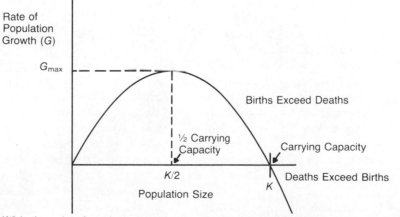

K/2 is the point of maximum growth rate, also known as the point of maximum yield

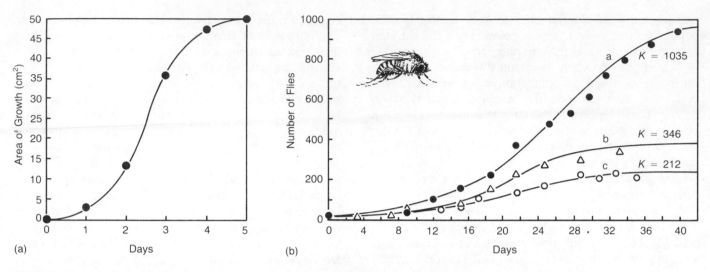

FIGURE 4.6

Examples of logistic growth. Populations of small organisms grown in laboratories can increase along a logistic growth curve because they have a constant supply of food and constant environmental conditions. Part (a) shows the increase in area occupied by a culture of bacteria. The points that measure the area occupied fall along a logistic growth curve. The exponential phase lasts for more than one day; the linear phase lasts from about day 2 to 3. Part (b) shows the growth of the fruit fly *Drosophila melanogaster.* The values for K are the number of flies at the carrying capacity. Curve (a) shows the growth of a population in a pint bottle; curve (c), the growth in a half-pint bottle. The carrying capacity changes with the size available. Curve (b) is for a different genetic strain, and indicates that carrying capacity is affected by genetic characteristics as well as by environmental ones. [Part (a) from Lotka, 1956; part (b) from Hutchinson, 1978; original data from Pearl, 1932. Part (b) reprinted by permission of Yale University Press from *An introduction to population ecology,* copyright © 1978.]

them during the most stressful time of year, and more females died because they were carrying calves and required additional nutrition. In contrast to a hypothetical logistic population, the real reindeer did not adjust instantaneously to changes in their food supply.

Reindeer live a relatively long time as far as animals are concerned, and they starve to death slowly—slowly enough that a large population can have a great effect on future food supplies. The Pribilof Islands reindeer grew rapidly when all their food was in great abundance, and the population rapidly outstripped the capacity of the reindeer moss to sustain the reindeer over a long time. However, the decline of the reindeer took a number of years; during these years the supply of slow-growing lichens was reduced more and more so that the supply of winter food continued to be less than that required by the reindeer population, even though that population grew smaller and smaller.

The history of the Pribilof Island reindeer illustrates several concepts: (1) There was a lag effect in the popula-

tion's response to changes in its habitat (the number of reindeer continued for several years to exceed that which could be supported by the lichens in winter); (2) the death rate was greater in certain parts of the population (after a particularly hard winter in 1940, deaths occurred mainly among females); and (3) the population was controlled by an aspect of its life that occupied a crucial but comparatively short period of time (the food available late in winter).

The logistic equation does not take these concepts into account; it does not include time lags nor does it distinguish among the kinds of individuals in a population or kinds of food. Both the exponential equation and the logistic equation assume that all individuals in a population are equivalent, because in both equations the population is described only by its total number and its three population rates. However, we know that many populations have complicated life histories and are made up of individuals in many different stages in their lives. It is only in the simplest kinds of populations, like bacteria and some of the other single-celled organisms, that all individuals might be considered equivalent, like marbles in a glass, and that an individual might be able to reproduce soon after it itself was produced.

AGE, SIZE, AND SEX:
THE STRUCTURE OF POPULATIONS

In the life of most higher organisms there is a stage of immaturity which can make up a large fraction of the life cycle. The life cycles of some organisms are very complex, with the young having different habits from the adults.

Our view of a life cycle is affected very much by our own cycle and those of domestic mammals. For us and our cats, dogs, horses, and cows there is clearly a differ-

ence between young, mature, and old, but kittens look like little cats and babies are clearly *Homo sapiens*. For much of the plant and animal kingdom, however, the stages in the life cycle differ much more from one another.

For example, parasites have some of the most complex life cycles, which we must understand if we are to control some human diseases. Schistosomiasis, one of the world's major health problems (Fig. 4.7), is caused by a water-borne parasite. The parasite enters a person's body in a free-swimming stage which exists in freshwater ponds and rice paddies. The free-swimming stage penetrates human skin, often through the feet of rice farmers when they wade in the water, and finds its way to the veins and to organs, particularly the liver and bladder, but including many tissues. People severely affected suffer

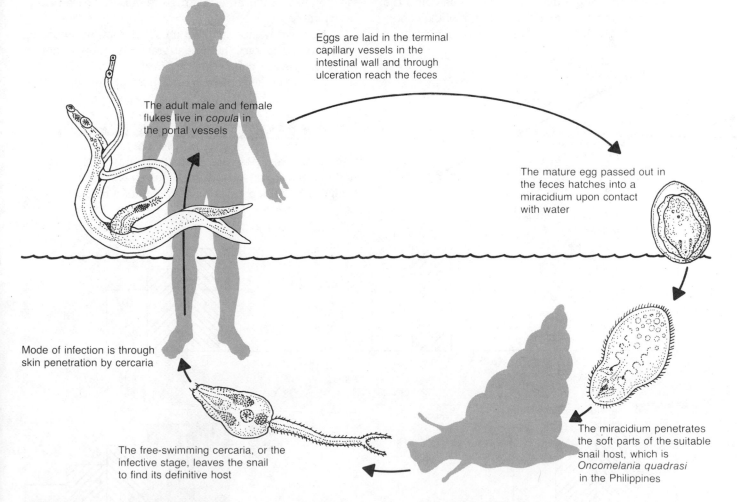

The adult male and female flukes live in *copula* in the portal vessels

Eggs are laid in the terminal capillary vessels in the intestinal wall and through ulceration reach the feces

The mature egg passed out in the feces hatches into a miracidium upon contact with water

Mode of infection is through skin penetration by cercaria

The free-swimming cercaria, or the infective stage, leaves the snail to find its definitive host

The miracidium penetrates the soft parts of the suitable snail host, which is *Oncomelania quadrasi* in the Philippines

FIGURE 4.7
The complex life cycle of *Schistosoma japonicum*, the parasite that causes schistosomiasis, a sometimes fatal disease in humans. Many organisms have complex life histories, which make their management, control, or protection a difficult task. (From Noble and Noble, 1976.)

cough, fever, and enlarged liver and spleen. Severe cases can cause death. The eggs are laid in the human host and eliminated with feces. If the feces contaminate the fresh water, the eggs produce another stage in the life cycle—a parasite of a freshwater snail. The snail parasite then produces the free-swimming form, which penetrates human skin and repeats the cycle. It would be difficult to represent the complex natural history of this parasite with a simple logistic curve.

Cohorts, Survival, and Age Structure   How a population is divided into groups of individuals at each stage in the life cycle is called its **age structure.** In a certain part of a stream 5 trout eggs hatch. Because they are born in the same time period, they are a **cohort** of trout. During the first year 1 dies, and after 1 year, another is caught by a fisherman, leaving 3 alive. This continues until the last one dies in the fifth year. The number of the cohort alive each year can be plotted on a *cohort survivorship curve* (Fig. 4.8).

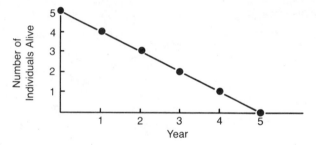

FIGURE 4.8
Cohort survivorship curve.

Imagine that each year in the same part of the stream, 5 eggs hatch and 1 of each cohort dies each year. Figure 4.9 shows the number alive of each cohort as the cohort ages and time passes. Although Figure 4.8 looks identical to Figure 4.9e, the first graph shows the history of 1 cohort and the second shows the *age structure* of the entire population from the part of the stream at year 5. If birth and mortality continue to be exactly the same for each cohort, the age structure will always look the same. This is called a stable or **stationary age structure.**

Suppose births vary, so that in the first year 5 eggs hatch, 3 in the second, and 6 in the third, but 1 individual of each cohort dies every year. Then the curve for the age structure no longer smoothly decreases to the right, but has a sawtoothed shape (Fig. 4.10). Variation in deaths for different ages of that trout would also make the curve sawtoothed. Such an age structure curve is called **nonstationary.**

Now we realize that we must add age structure to population size, birth rate, death rate, and growth rate as another primary characteristic of a population. Consider its importance, for example, in a herd of elephants (Fig. 4.11). A young elephant weighs 225 kilograms, less than one tenth of an adult. Its food requirements and thus its effects on the food supply are very different from that of an adult. So too is its ability to resist predators and disease and to survive on its own. With elephants, the three population rates (birth, growth, and death) will vary with the proportion of the population that is immature and mature;

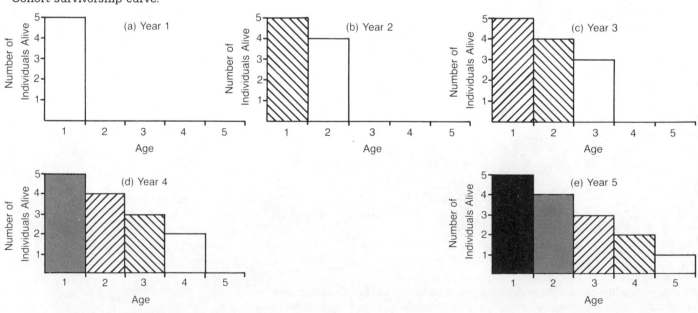

FIGURE 4.9
Stable or stationary age structure.

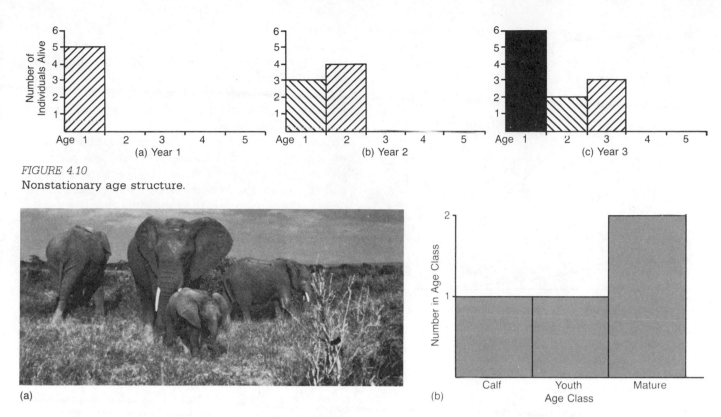

FIGURE 4.10
Nonstationary age structure.

FIGURE 4.11
Age structure in an elephant herd. Elephant herds vary greatly in size, from as few as 4 to more than 60 animals. In this small herd of 4, the age structure, in simple terms, consists of 2 adults, a teenaged youth, and a calf (a). The herd's age structure is represented by the graph shown in (b). Changes in the age structure change the food requirements and the impact elephants have on their environment. (Photo by D. B. Botkin.)

that is, these rates will vary with the population's age structure.

The introduction of ten young, immature elephants into a small park habitat would have very different effects from the introduction of five pregnant females and five mature bulls. The growth rate of the population of ten immature animals would be zero for a number of years, but the mortality rate would be greater than zero. In the population of immature animals one would expect an initial decrease in the number of living elephants, with the possibility that all might die before they reproduce. In contrast, the population of five bulls and five pregnant females would increase in the following year. Moreover, the natural rate of mortality would be smaller among these mature individuals than among the ten immature, so the population as a whole would grow much faster. In addition, because the food requirements of the large animals would be much greater than those of the small, immature elephants, the habitat would be changed more rapidly in the second case than in the first. Thus, for any population that has a complex life cycle and individuals that have a

long lifetime, it is important that we consider how the individuals are divided among ages, or at least among the important stages in the life cycle. Because the logistic and exponential curves do not take age structure into account, they are inadequate to predict accurately the fate of a real population.

As can be seen in Figure 4.12, the number of individuals in each age decreases from the youngest to the oldest in a population with a stationary age structure. The four basic kinds of stationary age structures are also shown in this figure. In the first type, mortality is very low in the young individuals and very high in the old ones, so the population is composed primarily of young individuals. In the second type, the number of individuals dying is constant for each age each year, so the numbers alive at every age declines in a straight line. In the third type, the mortality rate is a constant fraction of the population, so the number remaining alive falls off exponentially with age. Some bird populations approximate this type of survivorship. In the fourth type, the mortality rate is extremely high in the young individuals but decreases as

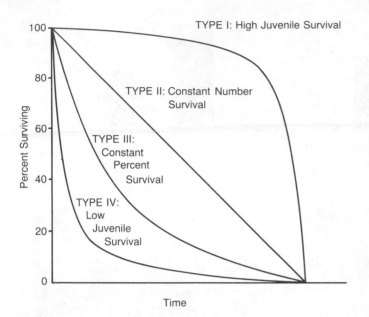

FIGURE 4.12
Stationary age structure graphs. (From Slobodkin, 1980.)

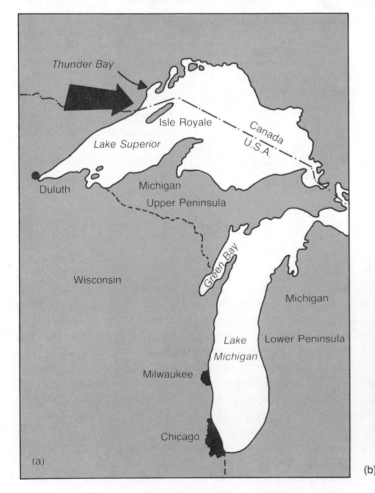

maturity is reached; once an individual has survived to some crucial age, its chances of living a long time are high. Shellfish, like mussels and clams, are examples of organisms with this type of survivorship.

These four types of stationary age structures are useful concepts, even though they are rarely found in nature. If a population like an elephant group in Africa or the Pribilof Island reindeer grew exactly according to a logistic and remained at a constant steady-state population size for a long time, eventually the age structure would become stationary. Because a population's age structure greatly affects the fate of the population and the impact of the population on the habitat, it is important for us to consider what kinds of age structures actually occur in nature.

It is very difficult to obtain data for age structures of wild animal populations—the kind of censuses held for people, with questionnaires, is obviously impossible. Fortunately, with many large mammals, ages can be determined from teeth. In a study of moose at Isle Royale National Park (Fig. 4.13), the age structure of the population was determined by collecting the jaws of the animals at death [5]. The teeth of moose have layers laid down in them, one for each year, and a careful cutting, polishing, and observation of a cross section of a tooth can be used to age these animals. Isle Royale is an undisturbed wilderness area in which there is no hunting or other human activities, and the primary cause of mortality is predation by

FIGURE 4.13
Isle Royale, in northern Lake Superior, is part of the state of Michigan. It is a wilderness of approximately 550 square kilometers (a). Isle Royale's woods and many ponds and lakes, seen from the air, provide a good habitat for moose and wolves (b). (Photo by D. B. Botkin.)

wolves. Thus the age structure of the moose population at Isle Royale provides one of the few examples of such a structure in a true wilderness area.

As we can see in Figure 4.14, the age structure of the Isle Royale moose population does not fit neatly with any of the four patterns we described before. Instead, this age structure has a more complex shape, with three main sections. The number of individuals falls off rapidly in the early ages, indicating that mortality is high among the young. Then there is a small decrease in the numbers until age 15, suggesting that the mature adults have a very low death rate. Finally, as the animals grow old, their death rate increases. This three-stage survivorship curve is characteristic of many mammals, including many human populations, and it is an important curve for us to remember in our discussion of the management of biological resources.

We have seen that a real population of large animals in a natural habitat has a survivorship or age structure curve that is more complex than the idealized ones. Do we ever find stationary age distributions in nature? That is, can a population persist over time with little change? Again, there are few records for such age structures. Studies of the African elephant are a good example. The

elephants, like the moose, can be aged from their teeth [6]. In one program to control the population of elephants in several African parks, a number of animals were harvested and later studied by a team of scientists. The age structures (Fig. 4.15) are sawtoothed and do not have the smooth decrease with age that is required for a stationary population. These real age structures suggest that the population is not in steady state but is responding to some temporal changes in the environment, possibly to changes in annual rainfall, as shown in the last graph (Fig. 4.15d).

In summary, populations of many species are characterized by an age structure. A population in steady state has a stationary age structure, meaning that the propor-

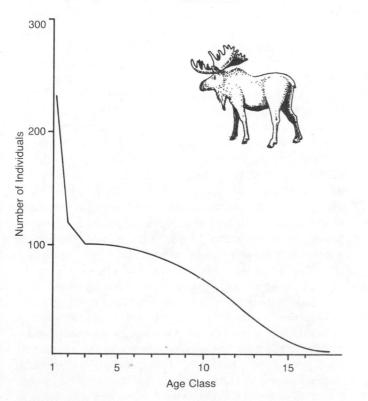

FIGURE 4.14
Age structure of the moose herd at Isle Royale National Park. (From Jordan, Botkin, and Wolf, 1971.)

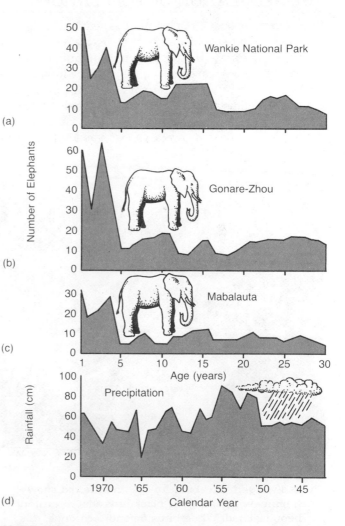

FIGURE 4.15
Age structures of elephant populations in three parks in Zimbabwe (a, b, c). The patterns are similar and seem to follow changes in annual rainfall (d). An age class is large in a year of high rainfall and low when there is little rainfall. (From Wu and Botkin, 1980.)

tion of individuals in each class remains constant in time, and the change from one age class to another is the same in each unit of time. There are four types of idealized age structures, but real populations of many species have more complex age structures that appear to be changing, rather than constant, in time.

The age structure of a population affects its reproduction, mortality, and its effects on other species and on its habitat. When we manage a population of wild creatures, we must remember that our management affects not only total numbers but the distribution of these numbers among ages.

THE REGULATION OF POPULATIONS

In the north woods of Michigan, it is spring and the red-winged blackbird (*Agelaius phoeniceus*) male sings to establish his territory. He sits showing the bright reddish-orange band on his shoulder (Fig. 4.16). In the early morning another male flies nearby and sits on an exposed branch. The first male continues to call and the second approaches. There is a flurry in the air and the males separate. This is repeated several times, when at last the in-

FIGURE 4.16
The red-winged blackbird. Its call and the red showy patch on its wing tell other blackbirds that a territory has been claimed. When birds defend territories, the number that can live in an area is limited—an example of density-dependent population regulation. (Photo courtesy of Paul Lehman.)

truder, threatened with a real attack, flees. The red-winged blackbird has defended his territory. He has limited the density of the population of his own species within a specific area and, since such a territory is required for breeding, affected the reproduction of the population. The population of blackbirds in this breeding territory is therefore controlled by the activities of the individuals themselves.

If a population cannot grow forever, some thing or process must limit the growth. The limitation might be nonbiological (like a windstorm), or biological but external to a population (like a predator), or due to a process that occurs within the population itself. In the last case, a population might be self-regulating, like a population of red-winged blackbirds. Most populations are partially limited by many factors at the same time.

A population that is self-regulating is said to have **density-dependent population regulation.** Under density-dependent population regulation, the rate of population growth is inversely related to population size. Density-dependent population regulation implies that there is a *feedback;* that is, the population in some way responds to its own size or density, bringing about changes in its birth and death rates.

A population that is not self-regulating may be subject to **density-independent regulation.** In density-independent population control, the mechanism of control has no relationship to population size. An example of this would be the death of insects in a forest subject to hurricanes. The population size of the insects has no effect on the frequency or path of hurricanes. Moreover, in a case where all the insects were killed when all the trees were knocked down by the hurricane, the mortality rate would be 100 percent regardless of population density.

Both density-independent and density-dependent population regulation seem possible in nature. Which is more important in nature? What causes either kind? There has been a long-standing controversy in the study of populations as to the relative importance of these regulatory mechanisms. Density-dependent population regulation has been shown to occur for a few species. In some cases this has been demonstrated by experimentation, in others by observation.

An experimental study was done with the freshwater snail *Lymnaea* (Fig. 4.17). Three fenced areas were set up in a pond, and the densities of the snails were changed by moving snails from one area to another. If the population sizes converged (that is, if each population eventually reached almost the same abundance), one could infer that a density-dependent mechanism must exist. In the first

Density   Fenced Areas with Snails Inside

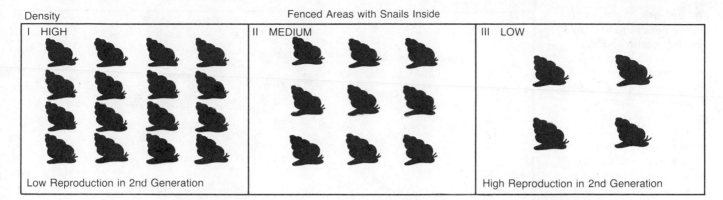

| I HIGH | II MEDIUM | III LOW |

Low Reproduction in 2nd Generation   High Reproduction in 2nd Generation

FIGURE 4.17
Do populations regulate themselves? An experiment with freshwater snails suggests that some species do. See text for further explanation. (From Eisenberg, 1966.)

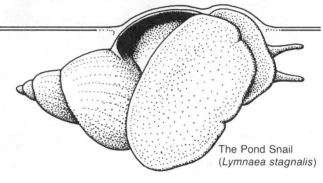

The Pond Snail
(*Lymnaea stagnalis*)

generation the snails survived equally well in all areas, but in the next generation the number of offspring per adult was inversely proportional to the population density. In the high-density area the snails had the lowest number of offspring per adult; in the low-density area the snails had the highest [7]. Density-dependent population regulation occurred, but it took two generations to be expressed.

Some observational studies show that density-dependent population regulation takes place for some large mammals. One of the best studies is of a population of wild Soay sheep *(Ovis aries)* (Fig. 4.18) living on St. Kilda Island north of Scotland [8]. The survival of lambs and adults was studied over a number of years as the total population of sheep changed. Death rates increased when population size increased. Graphs of lamb survival against sheep population size, and adult survival against sheep population size, show that survival decreases as population size increases. Stated more formally, there is a strong negative correlation which is statistically significant [9]. Why did this occur? The sheep lost weight in the winter. Those individuals who lost the most weight tended to have the highest mortality. The situation is reminiscent of the Pribilof Islands reindeer. When ewes were undernourished, they gave less milk and the lambs lost weight and were more likely to die. The less winter food, the higher was the death rate. Thus population size affects the food available per individual, and this leads to a density-dependent control.

It is generally believed that density-dependent population regulation does occur for many populations, but it is very difficult to prove that such regulation exists for any specific population. A population that is not self-regulating would seem to be more prone to great fluctuations and have a greater chance of extinction than a self-regulating one. Although the idea is appealing—and density-dependent population regulation is often said to occur—few data demonstrate clearly that this kind of regulation exists in nature. Either observations must be carried out over many years, as with the Soay sheep, or clever experiments must be conducted over several generations, as with the freshwater snails. More evidence is accumulating that suggests the existence of such regulation, long argued by all who have thought carefully about the control of wild populations.

In the management of wild animals and plants, it is often important to know whether there are natural density-dependent population mechanisms. If they exist, managers may be able to use them to achieve their goals. The stronger the natural density-dependent regulation, the less intense must be the control mechanisms by the manager. In the past wildlife has been managed as if density-dependent population regulation existed and could be relied on, often with poor results.

FIGURE 4.18
Do animals regulate their own abundance? Soay sheep on St. Kilda Island off the coast of Scotland in the north Atlantic appear to do so (a). The sheep graze widely on the island's rugged terrain. The survival rates of lambs and adults decrease as the population increases (b). However, the effect is more pronounced with the lambs. Their survival is nearly 100 percent at low population levels, but falls off to near zero at higher levels. In contrast, some adults die even at low densities, but at high densities adult survival is higher than that of the lambs. [Part (b) modified from Fowler et al., 1980.]

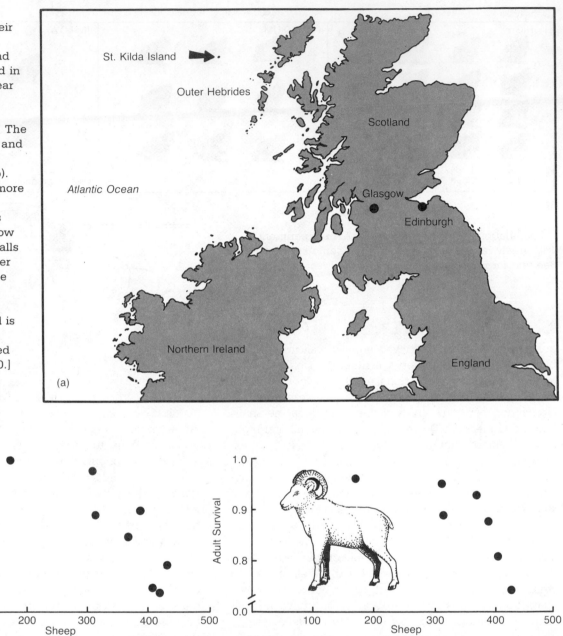

POPULATION INTERACTIONS

It is also possible that density-dependent regulation might occur when populations of two different species interact. In particular, it has often been suggested that predators and prey together form a self-regulating group. There are six major kinds of interactions among species. To understand them, let us consider once more the reindeer on the Pribilof Islands.

A reindeer on those northern wastes appears alone but it carries with it many companions. Like domestic cattle, the giraffe, and the impala of Africa, the reindeer is a ruminant, with a four-chambered stomach (Fig. 4.19).

Its stomach is teeming with microbes—a billion per cubic centimeter—that digest cellulose, take nitrogen from the air in the stomach, and make proteins. The bacteria that digest the parts of the vegetation that the reindeer cannot digest itself require a peculiar environment. They can survive only in an environment without oxygen, and there are few places on the Earth's surface where such an environment exists. Inside the rumen are other species of bacteria, some neither helping nor hurting the reindeer. There are also single-celled, ciliated microbes that feed on the bacteria. Elsewhere in the body of the reindeer, perhaps in its lungs or liver or even in the heart muscle, are parasitic worms. These also require the reindeer to

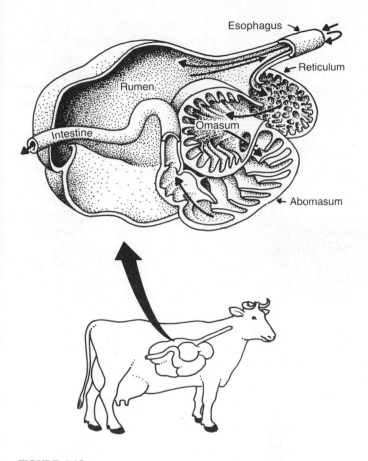

FIGURE 4.19
The microbes in the stomach of a ruminant (cow, deer, moose, giraffe) illustrate many kinds of population interactions. Microbes in the rumen digest plant tissue which the mammal cannot digest alone; both benefit from this relationship. The elaborate structures of the ruminant's stomach benefit the animal and its microbes.

survive, but slowly weaken it, making it easy prey to large predators like wolves or people.

The example of the reindeer shows that populations interact in a variety of ways. Each one of the major kinds of interactions is represented by the reindeer and its companions: *competition, inhibition, predation/parasitism, symbiosis* or *mutualism, protocooperation,* and *commensalism* (Table 4.1). The reindeer and the anaerobic bacteria in the rumen are an example of **mutualism** or **symbiosis**. Each depends on the other; neither could survive without the other. The reindeer provides a highly specialized environment required for the cellulose-digesting bacteria: an oxygen-free, nitrogen-rich atmosphere and a warm, constant-temperature, watery medium with vegetation that the bacteria can digest. The bacteria provide fatty acids and proteins which are an essential source of nutrition for the reindeer.

The single-celled microbes that feed on the bacteria are predators as far as the bacteria are concerned, but they are **commensals** of the reindeer—benefitting from the reindeer but not helping or hurting the reindeer. Other bacteria in the rumen may be cohabiting with the reindeer, but neither they nor the reindeer gain or lose from the relationship. This relationship is neutral. Still others may benefit and aid the reindeer, but each could survive without the other. This is a **protocooperative** relationship.

There are several species of bacteria in the rumen, and these **compete** with one another for the resources—the cellulose in the reindeer's food—and have a negative effect on one another. Some organisms, like the fungi *Penicillium* that live on bread mold, produce antibiotics that inhibit the growth of other organisms, an interaction called **simple inhibition**.

TABLE 4.1
Kinds of interactions among species.

Interaction	Effect[a]		Result
	Species A	Species B	
Neutralism	0	0	Neither affects the other.
Competition	−	−	A and B compete for the same resource; each has a negative effect on the other.
Inhibition	0	−	A inhibits B; A is unaffected.
Parasitism/Predation	+	−	A, the parasite or predator, benefits. A feeds on B, the host or prey, who thereby suffers a direct negative effect.
Mutualism (obligatory symbiosis)	+	+	A and B require each other to survive.
Protocooperation (nonobligatory symbiosis)	+	+	A and B benefit, but can survive separately.
Commensalism	+	0	A requires B to survive; B is not affected significantly.

[a]0 = no effect
+ = positive effect on birth, growth, or survival
− = negative effect on birth, growth, or survival

No species exists alone; every species interacts with others through one of these six ways.

PREDATORS AND PARASITES

Predation and **parasitism** are particularly important in the management of renewable biological resources, including forests, crops, wildlife, fisheries, and endangered species. Some parasites kill desired species. Can the two species interact so that neither becomes extinct, or so that the population sizes are regulated to some degree? To control insect pests of crop plants, we use parasites that live on these pests. Will the parasites and their prey be self-regulating so that neither gets out of hand? In the wild, predators like lions and leopards kill herbivores like impala. Can the lions kill so many impala that both species become extinct? These questions have intrigued people for thousands of years. Classical Greek and Roman authors wrote that predators were necessary to control the growth of their prey. The relationship between predator and prey or parasite and prey has been often studied, but many questions still remain unanswered.

It is possible to imagine a predator and prey which interact in such a way that neither becomes extinct and both remain more abundant than they would alone. For example, imagine a predator that lives on only a single prey, so that without that prey the predator would die. Without food to eat, a constant percentage of the predators dies in each time period. Now imagine that the prey increases exponentially without the predator, and that the death rate of the prey increases as a simple product of the numbers of the predator and prey. Let us add one more simple assumption: The life cycles of the two species are simple, so individuals of the same species can be considered equal to one another. The two hypothetical populations are known as *Lotka-Volterra predator-prey* (named after the two scientists who devised them). Lotka-Volterra predator-prey populations have an interesting possible pattern of abundance. They can follow regular variations in abundance, called *oscillations*, with the predator reaching its peak at the time the prey reaches its minimum. The abundances are out of phase and oscillate indefinitely. Other patterns are possible, but the out-of-phase oscillations have fascinated biologists. The story is told that Volterra became interested in this pattern because two species of fish varied in harvest in just this way: When one was abundant, the other was rare. Furthermore, one was the predator of the other.

If real predators and prey showed such relationships, then we could say that the predator regulated the population of the prey, which, without the predator, would increase beyond its own food supply. However, a hypothet-ical Lotka-Volterra population has simple characteristics. While some species might fit its description, many other species have much more complex life histories and more complex predator-prey interactions.

The actual effect of predation in the regulation of the abundance of prey is not well understood and may vary considerably from case to case. For example, in contrast to the theoretical predictions of the Lotka-Volterra populations, the population of the moose at Isle Royale oscillated prior to the arrival of the wolves, and the oscillations appeared to die away after the wolves arrived. The wolves have an elaborate hunting behavior, the moose feed on many species of plants, and other animals and plants are affected by a harsh and variable climate. These factors may greatly change the patterns of abundance.

If a predator does tend to stabilize the abundance of a prey, then the predator acts as a regulator on the prey population. It is possible that groups of organisms—populations of different species interacting together—may regulate themselves as a group. This would be a more complex kind of density-dependent population regulation. Such a regulation would be an attribute of an ecological community, which we will discuss in Chapter 5.

THE COMPETITIVE EXCLUSION PRINCIPLE

Populations that require the same resources will compete for these resources. Different species of trees in a forest compete for light, for water, for the chemical elements in the soil like nitrogen and phosphorus, and for a space to grow and to set seeds. An important idea in the study of competition is the **competitive exclusion principle.** According to this principle, two species that have exactly the same requirements cannot coexist in exactly the same habitat. Instead, one species will always win out over the other. This species will be the one that is somewhat more efficient in the use of any or all of the resources.

If the competitive exclusion principle is correct, how do species coexist? Some classic studies of the competitive exclusion principle used flour beetles (*Tribolium*) which can live on wheat flour. They make good experimental subjects because the experiments require only containers of wheat flour and the beetles. Experiments show that two species of *Tribolium*, both feeding on wheat, do not coexist. Instead, one species always wins out over the other. Which species wins depends on environmental conditions such as temperature and the moisture content of the flour (Fig. 4.20). The *Tribolium* beetles ''divide up'' a habitat, with one species using the

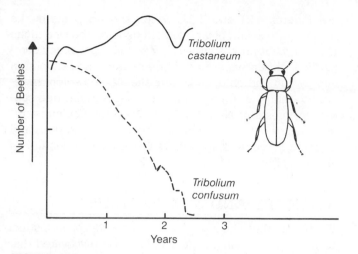

FIGURE 4.20

The competitive exclusion principle. When two competing species of beetles are introduced into the same jar of flour, one persists and the other dies out. Which species wins depends on the flour's temperature and moisture content. In this case, the temperature was 29.5°C, and *Tribolium castaneum* eliminates *Tribolium confusum*. (From Park, 1948.)

warmer, drier locations and the other using the wetter, cooler locations.

This example shows that, although two species are unlikely to coexist indefinitely in a uniform habitat, they

coexist by utilizing different parts of a variable or heterogeneous habitat. Sometimes a habitat can be divided in time rather than in space. For example, in a forest some tree species are adapted to the conditions that occur soon after a catastrophic clearing. At this time the resources are all available in great abundance; there is much light, the nutrients tend to be available in highly soluble form, and trees can grow rapidly. Other tree species seem adapted to old-age forest conditions, when resources are tied up in living and dead trees and are relatively unavailable. In eastern North America, birch trees are characteristic of early stages in forest development and spruce are characteristic of late stages. This partitioning of time following a disturbance will be explored further in our discussion of communities and ecosystems in Chapter 5.

THE ECOLOGICAL NICHE

From the previous discussions, we can see that each species has a unique role. It uses the same resources as its competitors, but it uses them best under certain sets of environmental conditions. We can think of the individuals of a species as having a unique job to do or a "profession." This "profession" is called the **ecological niche**. Where an organism lives is its *habitat;* what it does is its *niche.* Suppose you have a neighbor, Mr. Jones, who is a bus driver. Where he lives and works is your town; what

FIGURE 4.21

The niche concept. Spruce, sugar maple, and pin cherry all grow in the northeastern United States. Spruce grows well in cooler temperatures than either the maple or cherry can tolerate; and the cherry requires bright light, while the spruce and maple can persist in the shade (a). These qualities determine where we will find the different species. Spruce is found in cool areas like the tops of mountains or along cool lakeshores (b). Cherry and sugar maple are found in warmer places, like the valleys below mountains (b) or in the protected interiors of islands (c).

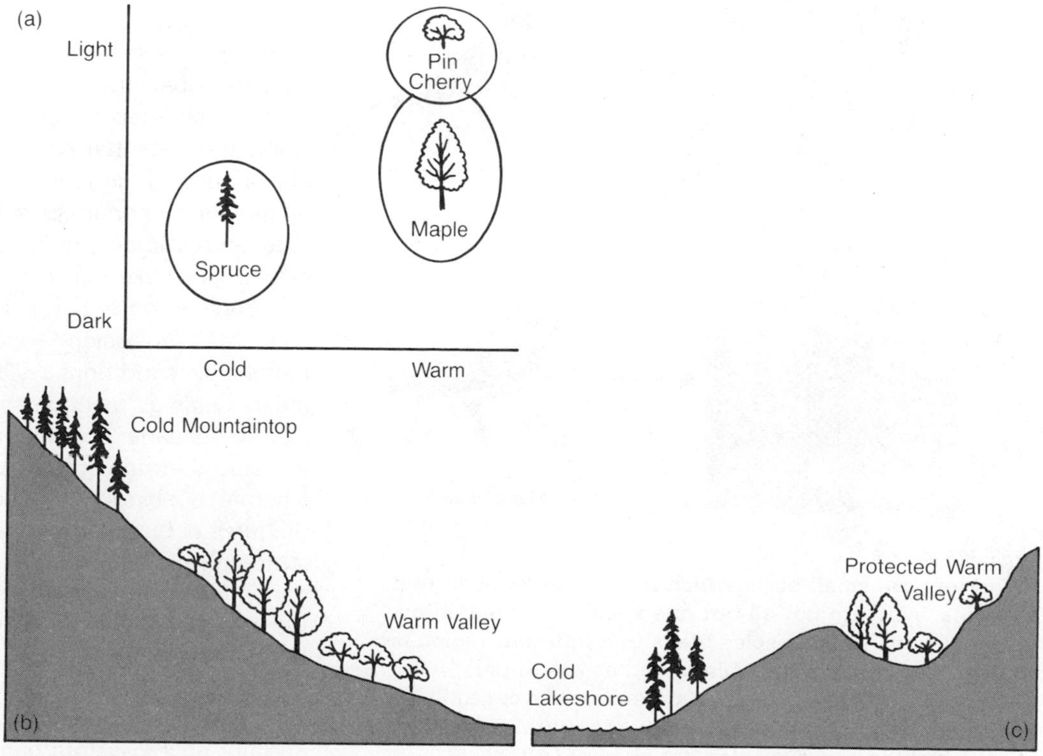

he does is drive a bus. In the same way, if someone says, ''Here comes a wolf,'' you think not only of a creature who inhabits the northern forests but of a predator who feeds on large mammals, possibly you.

Understanding the niche of a species is useful in assessing the impact of development or changes in land use. Will the change remove an essential requirement for some species' niche? A new highway which makes car travel easier might eliminate Mr. Jones' bus route and his niche. In the same way, cutting a forest may drive away prey and eliminate the niche requirements of the wolf.

We can determine the niche of a species by finding out all of the environmental conditions under which it persists. The conditions under which it can persist *without* competitors from other species are called the **fundamental niche**. The conditions under which it persists in the presence of natural competitors are called the **realized niche**. We can picture parts of a niche in diagrams like Figures 4.21 and 4.22, which show aspects of the niches of trees and of warblers. In general, the tree niche is to carry out photosynthesis, grow tall above many competitors, obtain water and nutrients from the soil, obtain energy and carbon dioxide from the air, and so on. In regard to temperature and light, we can see that tree species in forests of New England have different niche requirements. White birch grows under cool conditions and bright light; sugar maple grows under warmer conditions and less light.

SUMMARY AND CONCLUSIONS

We began this chapter with a discussion of individual populations of a single species. We have discovered that every population has certain characteristics that describe its current condition and that may be used to project its future. Among the most important attributes of a population are its total size; its rates of birth, growth, and death; and its structure, in terms of age, size, and sex. Because nothing can achieve an infinite size, every real population must have some upper limit. Biological populations have a great capacity to grow under the appropriate environmental conditions, thus, there must be factors that limit their size.

Populations may be regulated by density-dependent and density-independent mechanisms. Some species seem to have their own density-dependent mechanisms, like the territorial behavior of red-winged blackbirds. Other species may be subjected to a density-dependent regulation because of their interactions with other species. Density-dependent regulation has, at least theoretically, the potential to lead to greater stability and the appearance of harmony and constancy in populations or groups of populations of different species. Density-independent mechanisms include the effects of climatic catastrophes like hurricanes, tornadoes, and fire.

Each species has a habitat (where it lives) and a niche (its ''profession''—what it does under certain environmental conditions). Understanding niche requirements helps us manage species.

To manage populations we must understand and measure their important attributes. A careful manager of a natural population will keep track of the population size, the rates of birth, growth, and death, and the population structure and will understand its habitat and niche requirements. Although these attributes are important in the management of populations, few real populations can be understood or managed if they are considered in isolation from their environment or from other species. A more complete view of the dynamics of populations takes into account an ecosystem perspective, which is the topic of our next chapter.

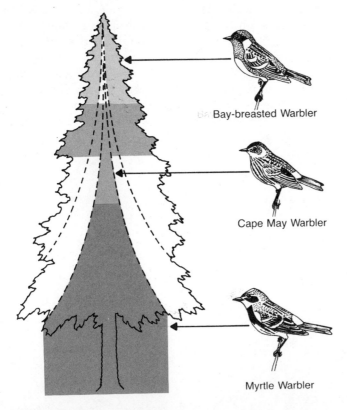

FIGURE 4.22
Warblers are small birds which feed in flocks of several species. Why are not all but one species eliminated by competition? Each species feeds in a different region of a tree: bay-breasted warblers feed at the upper, outermost branches; Cape May feed in the central branches; and myrtle feed on the lower branches. The flocks include other species not shown. (After MacArthur, 1958.)

Bay-breasted Warbler

Cape May Warbler

Myrtle Warbler

REFERENCES

1 SHEFFER, V. B. 1951. The rise and fall of a reindeer herd. *Scientific Monthly* 73: 356–62.

2 HANNA, G. D. 1922. The reindeer herds of the Pribilof Islands. *Scientific Monthly* 15: 181–86.

3 EGERTON, F. N. 1975. Aristotle's population biology. *Arethusa* 8: 307–30.

4 LOTKA, A. J. 1956. *Elements of mathematical biology.* New York: Dover.

5 JORDAN, P. A.; BOTKIN, D. B.; and WOLF, M. I. 1971. Biomass dynamics in a moose population. *Ecology* 52: 147–52.

6 WU, L. S.; and BOTKIN, D. B. 1980. Of elephants and men. *American Naturalist* 116: 831-849.

7 EISENBERG, R. M. 1966. The regulation of density in a natural population of the pond snail *Lymnaea elodes. Ecology* 47: 889–906.

8 GRUBB, P. 1974. Population dynamics of the Soay sheep. In *Island survivors: The ecology of Soay sheep of St. Kilda,* eds. P. A. Jewel, C. Milner, and J. M. Boyd. London: Athlone Press.

9 FOWLER, C. W.; BUNDERSON, W. T.; CHERRY, M. R.; RYEL, R. J.; and STEELE, B. B. 1980. *Comparative population dynamics of large mammals: A search for management criteria.* National Technical Information Service Publication No. PB80–178627. Washington, D.C.: U.S. Department of Commerce.

FURTHER READING

1 DARWIN, C. A. 1859. *The origin of species by means of natural selection or the preservation of proved races in the struggle for life.* London: Murray. (Reprinted variously.)

2 ELTON, C. S. 1927. *Animal ecology.* London: Sedgwick and Jackson.

3 HUTCHINSON, G. E. 1978. *An introduction to population ecology.* New Haven, Conn.: Yale University Press.

4 KREBS, C. 1978. *Ecology: The experimental analysis of distribution and abundance.* 2nd ed. New York: Harper & Row.

5 LACK, D. A. 1954. *The natural regulation of animal abundance.* New York: Oxford University Press.

6 LOTKA, A. J. 1956. *Elements of mathematical biology.* New York: Dover.

7 MAY, R. M. 1973. *Stability and complexity in model ecosystems.* Princeton, N.J.: Princeton University Press.

8 RICKLEFS, R. E. 1973. *Ecology.* Newton, Mass.: Chiron Press.

9 SCHRÖDINGER, E. 1962. *What is life? The physical aspect of the living cell.* Cambridge, England: Cambridge University Press.

10 SLOBODKIN, L. B. 1961. *Growth and regulation of animal populations.* New York: Holt, Rinehart and Winston. (Reprinted 1980 by Dover Press, New York.)

11 WALLACE, R. A.; KING, J. L.; and SANDERS, G. S. 1981. *Biology: The science of life.* Glenview, Ill.: Scott Foresman.

STUDY QUESTIONS

1 Explain the difference between *habitat* and *niche*.

2 What could have been done to improve the management of the Pribilof Island reindeer?

3 You are asked to plan a preserve for the North American wild turkey. What factors would you take into account in planning this preserve? How would you prevent overpopulation?

4 Debate the statement, "Predators are necessary to control the population of prey."

5 It has been said that in nature herbivores are never limited by food supply. What are the arguments for and against this statement?

6 Which of the following is most likely to maintain a constant number over time: (a) silk worms in a terrarium given the same amount of food every day; (b) a European species of snails introduced into southern California, where the climate is similar to the snails' place of origin; (c) goats introduced on an island off the coast of Maine which has a small grassy clearing but is otherwise densely wooded?

7 How can the age structures of a population affect the overall (a) birth rate and (b) death rate? Consider an example of a common and familiar species.

8 It has been said that a "prudent predator" would feed mainly on the very young and very old. Making use of the information you have read on birth rates, death rates, and age structure, explain what advantages this would have for a predator population.

9 Explain how the average age at which women give birth affects the growth rate of the population.

10 It has been suggested that the solution to human population problems is to colonize other planets. Consider the annual increase in the world's population. Do you think this is a practical solution? Explain your answer.

11 Compare and contrast the effect of weather on two of the following: (a) trout in a freshwater stream; (b) tuna in the open ocean; (c) earthworms in the forest near the trout stream in (a); (d) a species of insect which the trout eats (the immature stages of this insect are caterpillars feeding on tree leaves; the mature stage lives only long enough to mate).

5

Cycles and Flows: Ecosystems and Living Communities

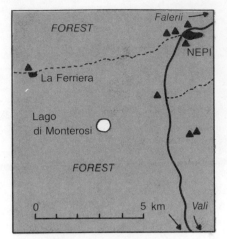

(a) Pre-Roman (Estruscan) Period (before the Roman road)

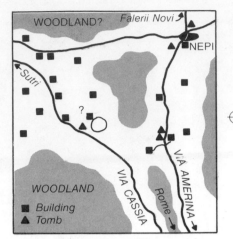

(b) Roman Period (when the Roman road was new)

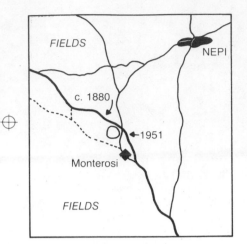

(c) 19th and 20th Centuries (at the time of 20th-century scientific studies)

(d)

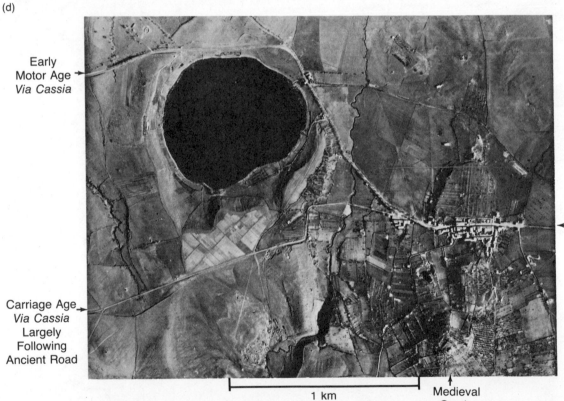

FIGURE 5.1

The history of Lago di Monterosi. Before the Roman period, the lake was surrounded by forests (a). When the Romans built Via Cassia, which passed near the lake, more sediment was deposited in the lake and the lake's organisms changed (b). Scientific studies in the 1960s (c) revealed these effects of Roman road building in the lake sediments. Part (d) shows an aerial view of the lake. (From Hutchinson, 1970.)

94

INTRODUCTION

In Italy in 171 B.C. the Romans constructed a road called *Via Cassia* from Rome north, cutting through what was then an uninhabited forest. About 40 kilometers north of the city the construction passed alongside a small lake named *Lago di Monterosi,* a small body of water roughly circular in shape, no more than 6 meters deep, and with a closed drainage basin.

More than 2000 years later, in the 1960s, the lake lay in the midst of settled and long-used land (Fig. 5.1). Scientists from Italy and the United States removed a long cylinder of deposits from the lake's bottom, called a *core*, which they then studied to reconstruct the history of the lake [1]. This core, a sample of the deposits transported to the lake through the hydrologic cycle (see Chapter 2), was made of mud, sand, shells of small freshwater animals and plants, bits of leaves and twigs, and pollen from the trees around the lake. It can be read like a book under the eyes of modern scientific instruments. Radioactive carbon in the sediments dates the pages of this book, which are sometimes visible as annual cycles in the deposits. The bits and pieces of the organisms tell a biologist what crea-

tures lived in and around the lake. The thickness of a year's deposit, or **varve**, tells a hydrologist the rate of erosion from the surrounding watershed.

The cores show that the lake sediments changed abruptly approximately 2000 years ago, about the time the Romans built Via Cassia. The amount of nitrogen deposited increased suddenly (Fig. 5.2), and there was a great change in the abundances of the microscopic algae (phytoplankton) that floated in the lake. In particular, a blue-green algae called *Aphanizomenon* bloomed. Other changes also took place. The rate of flow of sediments increased, and in these sediments was a greater concentration of alkaline earths (calcium, magnesium, and potassium). Organic production also increased.

The changes in Lago di Monterosi in 171 B.C. were very much like changes that occurred in North American lakes in the 1960s and 1970s. For example, in the summer of 1971 the waters of Medical Lake in the state of Washington became clogged with algae and turned a dark, turbid green; fish and algae died and, blown by the wind, formed masses of stinking algae and fish windrows on the lee shore (Fig. 5.3). Medical Lake, once clear, had become dark and dying [2].

FIGURE 5.2

The effect of Roman road building about 2000 years ago is seen in the lake sediments of Lago di Monterosi. Eroded soil entered the lake, increasing the growth of algae and the rate of sediment deposition. One change which occurred was increased nitrogen deposition in the sediments, shown in this greatly simplified cross-sectional view of the lake (vertical view greatly exaggerated). (From Hutchinson, 1973.)

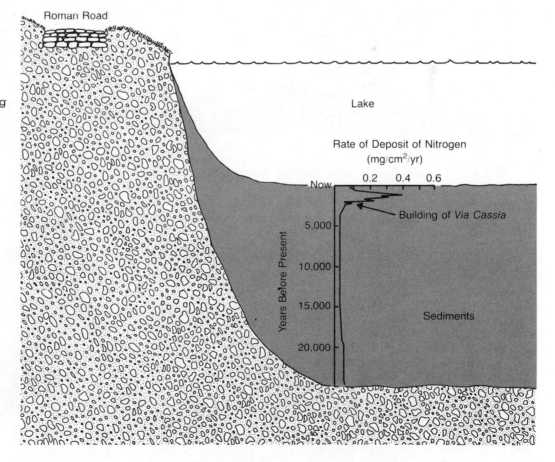

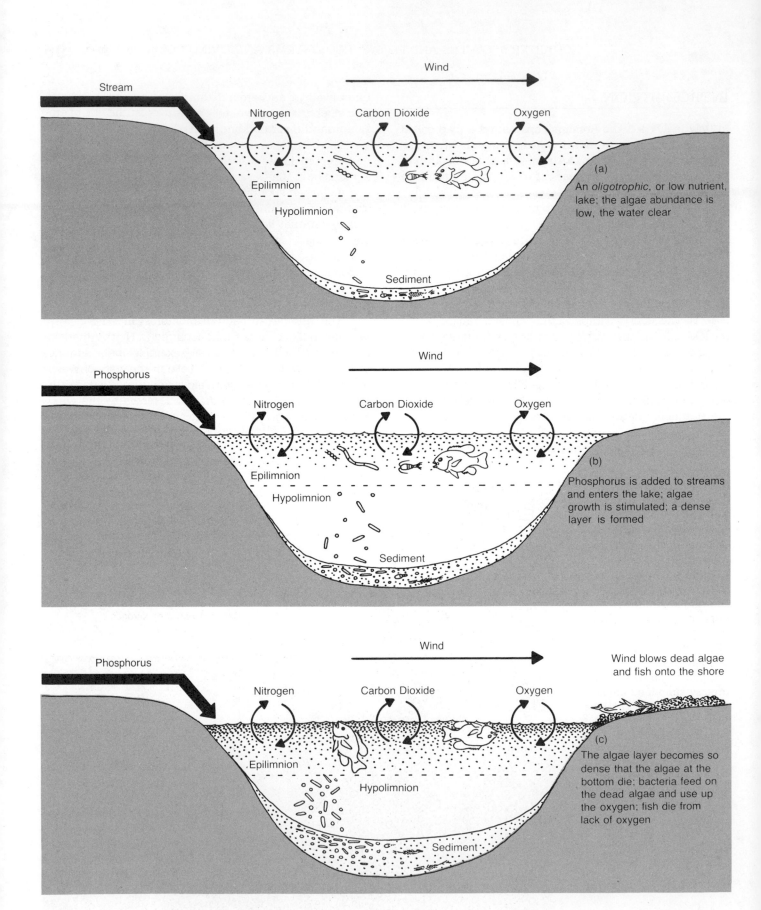

Wind

Stream

Nitrogen   Carbon Dioxide   Oxygen

Epilimnion

Hypolimnion

Sediment

(a)
An *oligotrophic*, or low nutrient, lake; the algae abundance is low, the water clear

Wind

Phosphorus

Nitrogen   Carbon Dioxide   Oxygen

Epilimnion

Hypolimnion

Sediment

(b)
Phosphorus is added to streams and enters the lake; algae growth is stimulated; a dense layer is formed

Wind

Phosphorus

Nitrogen   Carbon Dioxide   Oxygen

Wind blows dead algae and fish onto the shore

Epilimnion

Hypolimnion

Sediment

(c)
The algae layer becomes so dense that the algae at the bottom die; bacteria feed on the dead algae and use up the oxygen; fish die from lack of oxygen

These two lakes, Lago di Monterosi and Medical Lake, were connected across 2000 years and thousands of kilometers. They shared an intricate set of human-induced causes and effects. We can imagine that, soon after the Roman road had been built, Lago di Monterosi underwent changes like those observed in 1971 in Medical Lake. In both, some processes had caused a great increase in blue-green algae—a population explosion reminiscent of that of the reindeer on the Pribilof Islands discussed in Chapter 4.

What are the underlying rules and principles that govern these two events, connecting them over time and space? Those who understand the causes and effects have used their knowledge to clean up lakes so that some lakes that were once dark and dying are now clear and living. The population explosion of blue-green algae in Medical Lake cannot be explained by consideration of that population alone. Nor can it be explained by interactions among a community of different species without reference to the environment. It can only be explained by understanding the interrelationships among the entire community of organisms—all the organisms in the lake—and their local, nonbiological environment. That is, the changes in Medical Lake and Lago di Monterosi can only be explained using the concept of an *ecosystem*.

As we learned in Chapter 4, all biological populations have the potential for exponential growth, but this potential is always limited by resources necessary for growth. Under ordinary circumstances, the growth of any population of algae in a lake is limited on the one hand by the supply of energy and necessary chemical elements. On the other hand, the increase in algae is countered by the death of individuals—through grazing by herbivorous animals, ''starvation'' of those who do not obtain enough resources, and by the sinking of the algae to the bottom of the lake.

The addition of sewage or runoff from fertilized agricultural areas can pollute a lake, leading to the population explosions observed in Medical Lake. Experiments by a group of scientists at the Canadian Freshwater Institute in Manitoba show what causes such increases [3]. They put a plastic sheet down the middle of a lake, dividing it into two ecologically similar parts (Fig. 5.4). In one, fertilizers were added; in the other, nothing was done. Various chemical fertilizers were tried, but only when phosphorus was added to half of the lake did population explosions occur like those observed in Medical Lake. This experiment shows that phosphorus is a primary agent in pollution problems and is a limiting factor for the growth of algae.

The algae in Medical Lake responded to changes in the entire lake ecosystem. The increase in phosphorus led to increased growth of blue-green algae, which became so abundant that they formed dense mats. Algae at the lower part of the mat were shaded by those above and, lacking light, began to die. The dead algae became food for bacteria which, in decomposing the algae, took up oxygen required for their metabolism. The bacteria became so abundant that oxygen was depleted, and the fish began to die from oxygen starvation. The masses of dead plants and animals were blown by the winds to the lake shore. Thus the unpleasant effects resulted from the interactions among different species, the effects of the species on chemical elements in their environment, and the condition of the environment (the lake and the air above it).

The management of Medical Lake required an understanding of the lake as an ecosystem, a perspective that is necessary for anyone tackling environmental problems. In this chapter we shall explain the ecosystem concept, in terms of basic processes that take place within ecosystems, and show that these processes help us to formulate principles of environmental management.

FIGURE 5.3

The eutrophication of a lake. In a lake low in nutrients (called an *oligotrophic* lake), there is little algae, the water is relatively clear, and there is enough dissolved oxygen for the fish (a). When phosphorus is added to a stream, it enters the lake and stimulates the growth of algae (b). The algae become so abundant that a dense layer is formed, cutting off the sunlight and killing the algae on the bottom. The algae are fed upon by bacteria, who use up the oxygen in the water, and the fish die from lack of oxygen. Dead fish and algae are blown to the shore (c). The lake is divided into two major zones. The upper zone, the *epilimnion*, is mixed by the wind and receives much light. The lower zone, the *hypolimnion*, is not mixed by the winds and receives less light. The floating algae grow in the upper layer. Dead algae and other sediments pass through the hypolimnion to the lake bottom. In many lakes, the entire lake mixes, or ''turns over,'' in the spring and fall with changes in the temperature.

FIGURE 5.4
This aerial photograph shows a lake which has been divided into two parts by a large plastic sheet. The Canadian Freshwater Institute added chemical elements to one side only. The results of this experiment demonstrated that phosphorus produced algae blooms and was an important limiting nutrient. (Photo courtesy of D. Schindler.)

THE NATURE OF AN ECOSYSTEM

Sustained life on the Earth is a characteristic of ecosystems, not of individual organisms or populations [4]. Any entity that sustains life over long time periods must have two characteristics: There must be a flow of energy through the living system and a complete cycling of all the chemical elements necessary for life. That is, each element required for growth and reproduction must be made available in a reusable form by that system: wastes converted into food, converted into waste, converted into food. Neither individual cells, populations, nor communities of populations are a sufficient system. No single organism, population, or species produces all of its own food and completely recycles all of its own metabolic products. Green plants in light produce sugar from carbon dioxide and water, and from sugar and inorganic compounds make many other organic compounds, including protein and woody tissue. But no green plant can degrade woody tissue into its original inorganic compounds. Those living things that degrade woody tissue, such as bacteria and fungi, do not produce their own food, but instead obtain their energy and chemical nutrition from the dead tissues they feed on.

From these observations, we know that for complete recycling of chemical elements to take place, there must be several species. But, in addition, the recycling of chemical elements and the flow of energy require a nonliving transport and storage medium: water, air, soil, rock, or a combination of these.

The minimal systems that can allow the required flow of energy and complete chemical cycling are composed of at least several interacting populations of different species and their nonbiological environment. The smallest candidates for such minimal systems are ecosystems. An **ecosystem** is a local community of interacting populations of different species and their local, nonbiological environment.

The term *ecosystem* is applied to areas of the Earth that differ greatly in size, from the smallest puddle of water to a large forest (Fig. 5.5). Ecosystems differ greatly in composition: in the number and kinds of species, in the kinds of and relative proportions of nonbiological constitutents, and in the degree of variation in time and space. Sometimes the borders of the ecosystem are well defined, as in the border between Lago di Monterosi and the surrounding countryside. Sometimes the borders are vague, as in the subtle gradations of forest into prairie in Minnesota and the Dakotas in the United States or in the transition from grasslands to savannahs and forests in East Africa.

Ecosystems can be natural or human-made. A pond that is a part of a waste treatment plant is an example of a human-made ecosystem. Ecosystems can be natural or managed, and the management can vary over a large range of actions. Agriculture can be thought of as partial management of certain kinds of ecosystems.

What is common to all ecosystems is not physical size, but the existence of the processes we have mentioned—the flow of energy and the cycling of chemical elements. Ecosystems also share the pathways for these flows and cycles, which follow along food webs from one trophic level to another.

To summarize, ecosystems have several fundamental characteristics. Ecosystems have *structure*: nonliving and living parts, the living made up of a set of species connected by food webs and trophic levels. Second, eco-

(a)

(b)

(c)

FIGURE 5.5

Some examples of ecosystems. Ecosystems vary in size, but they all have a flow of energy and a cycling of chemical elements. An ecosystem can be a small bog of less than one hectare (a) or a forest watershed of more than 25 or 50 hectares (b). A city waste-treatment pond (c) is a human-produced ecosystem. This pond in Goleta, California, receives abundant nutrients from sewage and thus supports aquatic life and numerous shore birds. Part (a) also illustrates bog succession; the early stages are in the center of the pond, and the latest are at the edges of the photograph. (Photos by D. B. Botkin.)

systems have *processes:* energy flows through them and chemical elements cycle within them. Third, ecosystems *change over time,* through a process called **succession**.

ECOSYSTEM AND COMMUNITY STRUCTURE

An ecosystem has living and nonliving parts. An **ecological community** is the living part of an ecosystem—a set of interacting species. One way in which these individuals interact is by feeding on one another. Energy, chemical elements, and some compounds are thus transferred from creature to creature along **food chains**. Ecologists group organisms in food chains into trophic levels. A **trophic level** consists of all those organisms in a food chain that are the same number of steps away from the original source of energy. Thus green plants are grouped in the first trophic level; herbivores in the second trophic level; carnivores that feed on herbivores in the third; carnivores that feed on carnivores in the third trophic level are in the fourth; and so on. For example, in the oceans (Fig. 5.6) the tiny, single-celled planktonic algae are in the

first trophic level. They are eaten by small invertebrates called zooplankton and by some fish, which are in the second trophic level. Other fish and invertebrates that feed on these herbivores form the third trophic level. The great baleen whales, which filter the seawater for food, feed primarily on small herbivorous crustaceans, so the baleen whales are in the third level. Some fish and marine mammals, like the killer whale, feed on the predatory fish and form higher trophic levels. In the abstract a trophic-level diagram seems simple and neat, but in reality food webs are complex because most creatures feed on several different trophic levels. The harp seal is at the sixth level [5], feeding on flat fish, who feed on sea lances, who feed on pelagic amphipods, who feed on zooplankton, who feed on phytoplankton (Fig. 5.7). In this case, the harp seal is at the highest trophic level [5]. It feeds at several trophic levels from the third through the fifth, and it feeds on predators of some of its prey.

ECOSYSTEM PRODUCTION

Ecological production is the increase in either organic matter or stored energy. Ecologists measure both the pro-

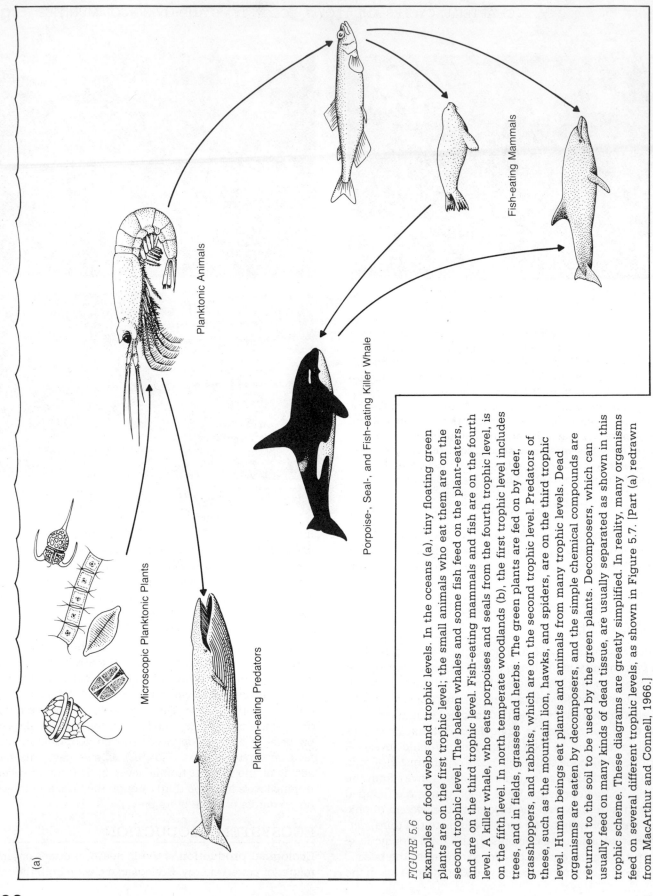

Planktonic Animals

Microscopic Planktonic Plants

Plankton-eating Predators

Porpoise-, Seal-, and Fish-eating Killer Whale

Fish-eating Mammals

(a)

FIGURE 5.6

Examples of food webs and trophic levels. In the oceans (a), tiny floating green plants are on the first trophic level; the small animals who eat them are on the second trophic level. The baleen whales and some fish feed on the plant-eaters, and are on the third trophic level. Fish-eating mammals and fish are on the fourth level. A killer whale, who eats porpoises and seals from the fourth trophic level, is on the fifth level. In north temperate woodlands (b), the first trophic level includes trees, and in fields, grasses and herbs. The green plants are fed on by deer, grasshoppers, and rabbits, which are on the second trophic level. Predators of these, such as the mountain lion, hawks, and spiders, are on the third trophic level. Human beings eat plants and animals from many trophic levels. Dead organisms are eaten by decomposers, and the simple chemical compounds are returned to the soil to be used by the green plants. Decomposers, which can usually feed on many kinds of dead tissue, are usually separated as shown in this trophic scheme. These diagrams are greatly simplified. In reality, many organisms feed on several different trophic levels, as shown in Figure 5.7. [Part (a) redrawn from MacArthur and Connell, 1966.]

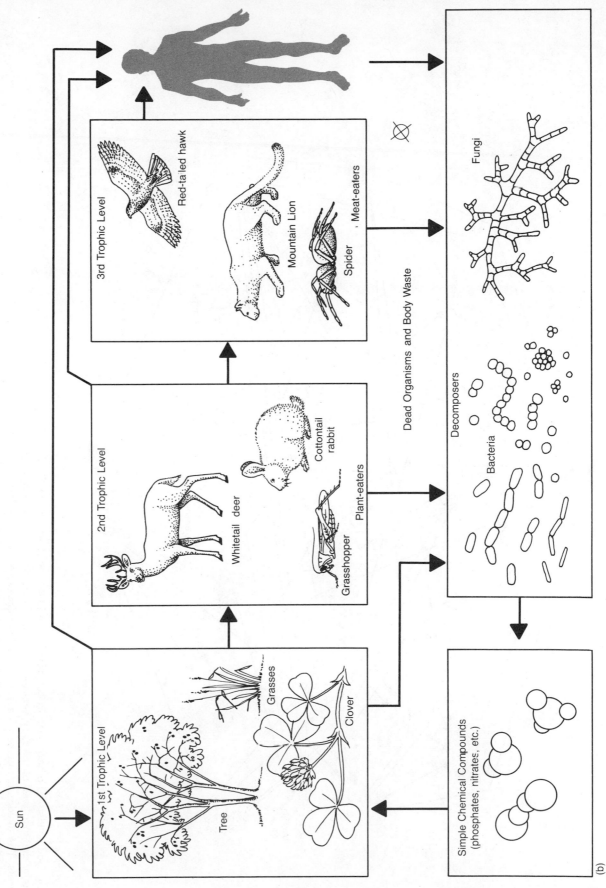

Sun

1st Trophic Level

Tree

Grasses

Clover

2nd Trophic Level

Whitetail deer

Cottontail rabbit

Grasshopper

Plant-eaters

3rd Trophic Level

Red-tailed hawk

Mountain Lion

Spider

Meat-eaters

Dead Organisms and Body Waste

Decomposers

Fungi

Bacteria

Simple Chemical Compounds
(phosphates, nitrates, etc.)

(b)

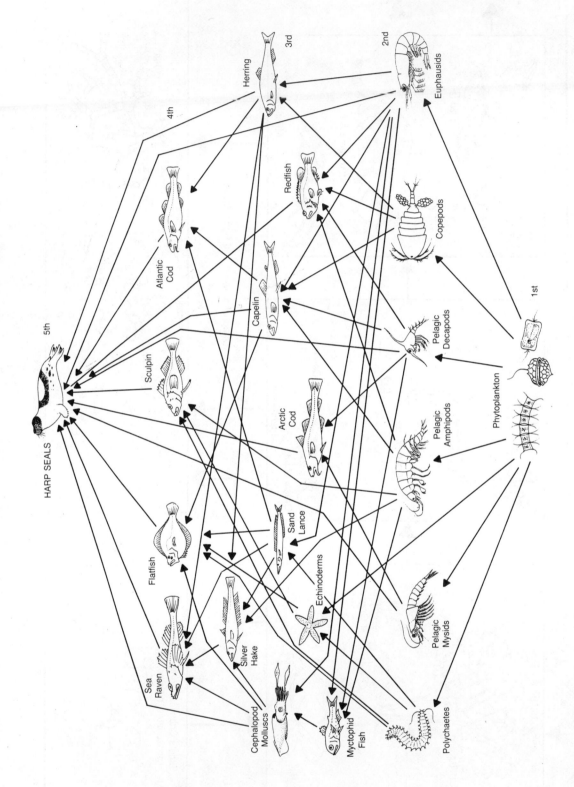

FIGURE 5.7

The food levels of the harp seal. In contrast to the idealized diagrams in Figure 5.6, the actual food web of the harp seal has many connections at several trophic levels. The maximum number of steps in the food chain between the harp seal and sunlight is six, but the harp seal also feeds on intermediate trophic levels. For example, it feeds on amphipods at the third trophic level and on capelin at the fourth level. The harp seal thus competes with its own prey (capelin) for its food (amphipods). Since the harp seal competes with other species, including other marine mammals and human beings, the entire food chain is very complex. (Modified from Lavigne et al., 1976.)

duction of individual trophic levels and of entire ecosystems. Is a field of corn more productive than a forest? Is a fishery that we actively manage more productive than one we leave alone? If we expend energy in fertilizing a forest from a helicopter, do we get more energy in firewood than we would have anyway? Do we get more energy in the added firewood than we used in the helicopter and in making the fertilizer? The answers to these questions are important to our understanding and management of the environment.

The flow of energy and the production of ecosystems are also important in our legal processes, because the concepts of productivity are used in federal laws that regulate the environment. For example, as we shall see in later chapters, the laws governing fisheries have as their goal the maximization or optimization of production.

Thus, to see how our society influences the environment, we must understand the ecological concepts of *energy flow* and *production*. In this context, organisms in an ecosystem are divided into two groups: the **autotrophs**, those who make their own food and are therefore in the first trophic level, and **heterotrophs**, those who must feed on other organisms and make up all the other trophic levels. The production of the first trophic level is called **primary production** (Fig. 5.8). In a forest this is the addition of new biomass of trees, shrubs, and herbs. In the ocean it is the production of algae, including the tiny floating phytoplankton and the large, multicellular algae such as kelp along the shores. In a pond primary production is the production of phytoplankton and the aquatic flowering plants, such as water lilies.

Production of the heterotrophs is called **secondary production**. To understand and measure ecological production, we must make a few other distinctions. First, we must distinguish between gross production and net production. **Gross production** is the increase in biomass or energy content before *any* is used. For example, in a forest it is all the sugars produced by all the green plants before any is used—even by the leaves. **Net production** is what remains stored after energy is used. Living things use energy through the process of respiration for many kinds of functions: growth (making new tissues), moving, repairing, and so forth. The difference between gross and net production is like that between a person's gross and

FIGURE 5.8
Ecological production. See text for further explanation.

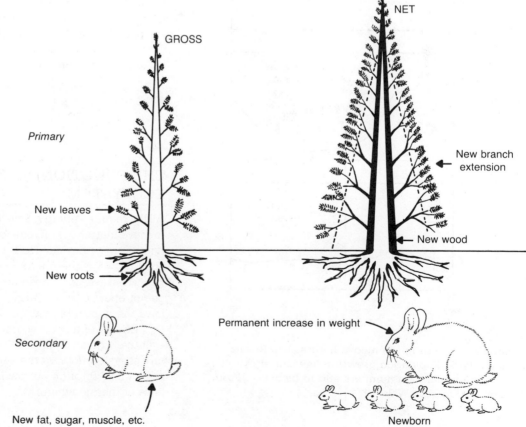

net income. Gross income is the dollars you are paid; net income is what you have left after money is deducted for taxes and other fixed costs. Gross and net ecological production are connected by a simple formula:

Gross production − Respiration = Net production

Net production includes that added in the growth of already living individuals and in the birth of newborn minus the loss in the respiration of living individuals and the loss of individuals through death.

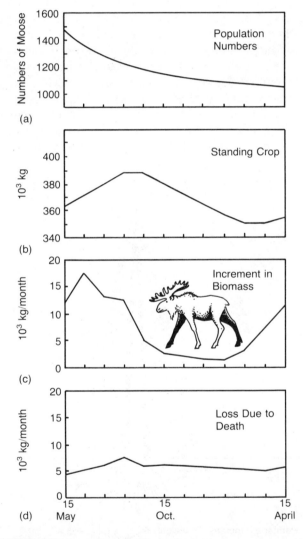

(a)

(b)

(c)

(d)

FIGURE 5.9
Population growth in the moose herd at Isle Royale National Park. Population growth is measured by changes in population numbers and in biomass. (From Jordan, Botkin, and Wolf, 1971.)

Production can be measured in terms of changes in population (i.e., the number of individuals) or in terms of changes in amount of organic matter, or biomass. Biomass and population numbers may change in quite different ways over time. As an example, changes in biomass and numbers were measured for the moose herd at Isle Royale National Park, a wilderness island in Lake Superior. The change in population size was determined during a single year [6]. In May, just after the calves are born, the population number is at its annual maximum, and it declines with deaths through the rest of the year (Fig. 5.9). The total biomass increases during the rest of the spring, summer, and early fall when the forest vegetation is growing and providing abundant food for the moose. The gross increment in live weight of the population remains slightly positive during the winter, when the moose feed on twigs and conifer needles. The loss through death, caused almost completely by wolf predation, occurs at a comparatively steady rate throughout the year. While the number of individuals declines through the year, the biomass rises and then falls, reaching its peak in late summer and its minimum in late winter.

There is one other distinction we must make between *production* and *productivity*. Production is an *amount,* and productivity is a *rate*. For example, the net production of a corn field is the yield in metric tons per hectare. The productivity is the rate of increase—the tons per hectare per day, for example.

With these concepts in mind, we can now analyze the energy flow through ecosystems. The first question we must ask is, What is the function of energy in an ecosystem?

THE FUNCTION OF ENERGY IN AN ECOSYSTEM

All life requires energy. **Energy** is the ability to do work or to move matter. As anyone who has dieted knows, we obtain energy from our food. Our weight is a delicate balance between the energy we take in and the energy we use. What we don't use and don't pass on, we store. Our use of energy, and whether we gain or lose weight, follows the inexorable laws of physics.

Energy enters an ecosystem and is used, stored, and eventually given off (Fig. 5.10). The flow of energy through an ecosystem follows the same laws of physics that govern its flow through our own bodies: the first and second laws of thermodynamics.

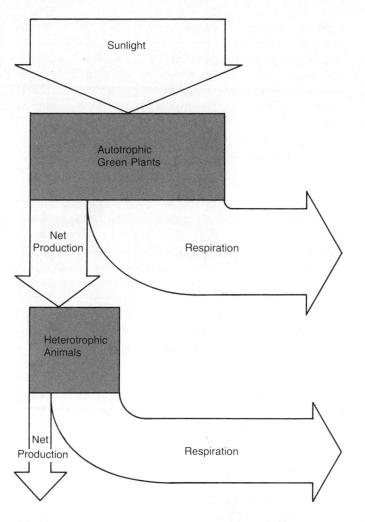

FIGURE 5.10
Diagrammatic view of ecological production and energy flow. Energy flows in from outside the ecosystem, mainly through photosynthesis. The net primary production is available to the heterotrophs.

According to the **first law of thermodynamics,** all energy and matter must be conserved. In other words, the energy (or matter) in a system cannot be created or destroyed. We are often told that we eat to get particular chemical elements—calcium from milk, nitrogen from meat, and so on. But, if all matter must be conserved and one atom of an element is indistinguishable from another, why can't we recycle all the elements in our bodies and only add new elements to replace those lost by accident or required for growth? In a book called *What Is Life?* [7], the twentieth-century physicist Erwin Schrödinger asked this question. Schrödinger maintains that the statement that the *essence* of food is to get us new atoms of chemical elements must be false, because any one atom is the same as any other. We are also told that we eat to obtain energy. But if all energy is conserved, why can't we recycle energy?

Let's imagine a closed system, isolated from the rest of the universe, containing a pile of coal, a tank of water, air, a steam engine, and an engineer (Fig. 5.11). Suppose the engine runs a lathe that makes furniture. The engineer lights a fire to boil the water to run the steam engine. As the engine runs, the heat from the fire gradually warms the entire system. When all the coal is completely burned, the engineer will not be able to boil any more water and the engine will stop. The average temperature of the system now is higher than it was at the beginning. The energy that was in the coal is now dispersed throughout the entire system, much of it as heat in the air. Why can't the engineer collect all that energy, compact it, put it under the boiler, and run the engine? This is where the **second law of thermodynamics** comes in. Physicists have discovered that no real use of energy can ever be 100 percent efficient. Whenever useful work is done, some energy is inevitably converted to heat. If the engineer tried to collect all the energy dispersed in the

FIGURE 5.11
An imaginary closed thermodynamic system, with a steam engine running a lathe. An engineer puts coal and water into the engine, which runs the lathe to make furniture. When the coal is all burned, the energy it contained is dispersed as heat throughout the system.

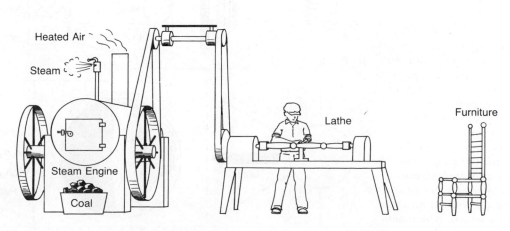

system, she would do more work collecting the energy than would be done by the energy collected.

Our imaginary system began in a highly organized state, with energy compacted in the coal. It ended in a less organized state, with the energy dispersed throughout the system as heat. The energy is said to be *degraded,* and the system is said to have undergone a *decrease in order.* The measure of the decrease in order is called **entropy.** The engineer did produce some furniture, converting a pile of lumber into nicely ordered tables and chairs. In that system, there was a local increase of order (the furniture) at the cost of a general increase in disorder (the state of the entire system). According to the second law of thermodynamics, in any real system any work done results in a net increase in disorder and a decrease in the amount of energy remaining that can be used for useful work.

To answer one of our first questions, our bodies cannot keep reusing the same energy because some of it is converted to heat. When we move, for example, friction in our bodies creates heat. We must get rid of that heat if we are to keep a safe body temperature. We require a continual input of new energy and a continual loss of unusable heat.

What do we get from our food? We get the ability to do useful work and to create order—the order of our bodies. We feed on the negative of entropy, what Schrödinger calls **negentropy** [7]. Negentropy is the ability to create order on a local scale from the flow of energy that distinguishes life and is the essence of what we eat. To summarize, energy must continually be added to an ecosystem in a usable form. It is inevitably degraded into heat and must be released. This is what is meant by the statement that *energy flows one way through an ecosystem.*

From what we have just discussed, it is clear that an ecosystem must lie between a source of usable energy and a sink for degraded (heat) energy. The ecosystem is said to be intermediate between the source and the sink (Fig. 5.12). The energy source, ecosystem, and energy sink form a thermodynamic system. The ecosystem can undergo an increase in order (called a *local* increase) as long as the entire system undergoes a decrease in order.

The physicists' "order" has a special meaning. A disordered system has a great amount of randomness. For example, compare a goldfish in an aquarium with a simple mixture of the same chemical elements, in their inorganic forms, dissolved in a container of water. In the goldfish bowl, certain elements are concentrated in a few specific places—in the fish. This is a highly organized and nonrandom distribution. Also, a sample taken from the goldfish bowl which contains a fish is very different from a sample without fish. But in the container with no life, a sample taken anywhere is likely to be much the same as any other.

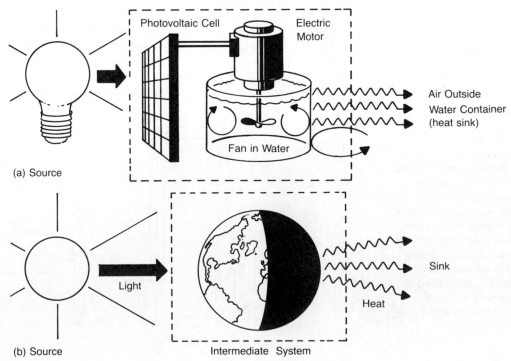

FIGURE 5.12
A system with a source of usable high-quality energy and a sink for degraded (heat) energy is called an intermediate thermodynamic system. In (a), the intermediate system consists of a photocell, electric motor, propeller blades, and water. The energy source is a lightbulb, and the sink is the air outside and around the water container. Energy flows from the lightbulb through the photocell, motor, propeller, and water to the air. The Earth is an intermediate system between the sun (the energy source) and the depths of space, which acts as a sink (b).

(a) Source

Photovoltaic Cell

Electric Motor

Air Outside
Water Container
(heat sink)

Fan in Water

(b) Source

Light

Sink

Heat

Intermediate System

The second law of thermodynamics places a limit on the efficiency with which energy can be used. In terms of energy, no real process can be 100 percent efficient. There are no perpetual motion machines.

Consider a forest in which all life stopped. The forest would lose order, either slowly or quickly. The complex structure of tree stems and roots would break down, and the large, complex organic compounds would oxidize and degrade to smaller, mainly inorganic forms. Perhaps a fire would start and burn the wood, leading to a rapid change. Or perhaps the wood would be buried in a bog and converted to coal, which would last hundreds of millions of years. The rate of decay can vary, but the direction of change is always the same. The physical (nonbiological) processes in the entire universe, on the Earth's surface, or in a single ecosystem tend to lead to decay, disorder, and randomness. Life, however, is a building up—an *aggrading* process.

THE PATHWAYS OF ENERGY IN AN ECOSYSTEM

Energy enters an ecosystem either as heat, when it warms things up, or when it is "fixed" by organisms that can produce their own food from energy and inorganic minerals (Fig. 5.13). In most cases, this fixation of energy is carried out by green plants through photosynthesis, but in some cases, as in warm deep-sea vents, marshes, and the muds of ponds, energy is fixed by chemical processes carried out by bacteria.

When the energy is fixed in photosynthesis, the process is a photochemical one. Simple sugars are produced and then converted to other kinds of compounds: carbohydrates, oils, and proteins. As energy is passed from one organism to another, it is used in many different steps. As an example, consider the flow of energy in a forest, where the energy is fixed by photosynthesis in a tree leaf. Some of the energy is used immediately by the leaf to keep its own life processes going. What is not used there is transported to roots, stems, flowers, and fruits or stored within the leaf cell. Some more energy is lost in the transfer and is used to make other organic compounds—cell walls, protein, and so forth. The feeding of herbivores, like caterpillars, on the leaves, stems, flowers, and fruits requires energy. Their digestive systems use some more energy, as does the process of building animal tissue from the digested food. Herbivores store some of the energy as fat for future growth and movement and, for a caterpillar, transformation into a butterfly.

A carnivore, like a woodland fly-catcher that feeds on grasshoppers or caterpillars, uses some energy in its searching to obtain more energy. As a warm-blooded creature, the bird uses much energy in its metabolic processes just to maintain a constant body temperature.

Energy not used by an organism, and not taken as food by another, remains when the organism dies and is finally used by decomposing organisms. In a forest, dead

FIGURE 5.13
Idealized diagram of energy flow through an ecosystem. Energy enters from outside (mainly as sunlight) and passes through each trophic level. At each level, some is released as heat by respiration (*R*), some is transferred to the next trophic level (*L*), and some is transferred to the decomposers through death and excrement (*D*). (After Bowen, 1966.)

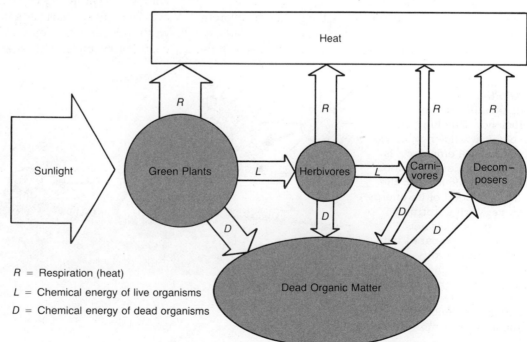

R = Respiration (heat)
L = Chemical energy of live organisms
D = Chemical energy of dead organisms

leaves and twigs are chewed by many animals—earthworms, sowbugs, and termites—and each takes some energy from the dead material. Woody tissues (cellulose and lignin), hair, and feathers, the most difficult materials to digest, can only be digested by certain bacteria and fungi.

THE EFFICIENCY OF ENERGY USE AND ECOLOGICAL ENERGY TRANSFER

The laws of physics and chemistry place certain limits on how efficiently living things obtain and use energy. As energy flows through a food web, it is degraded and less and less of it is usable. Other things being equal, the more energy an individual gets, the more it has for its own use. But organisms also differ in how efficiently they use the energy they obtain. The less an individual uses immediately in metabolism or loses as heat, the more it has for growth or reproduction. A more efficient organism has a selective, or evolutionary, advantage over a less efficient one.

The definition of **efficiency** depends on who uses the term. A farmer thinks of an efficient crop as one that stores a lot of energy and uses relatively little itself. An efficient corn crop is one that fixes a great deal of solar energy as sugar and uses little of that sugar to produce stems, roots, and leaves. In this case, the most efficient crop is the one that has the most harvestable energy left over at the end of the season.

A truck driver, on the other hand, has a different view of efficiency. An efficient truck is one that uses as much energy as possible from its fuel and leaves as little in the product (the exhaust) as possible. If a truck driver measured the efficiency of a truck the way a farmer measures the efficiency of corn, the efficient truck would leave a lot of usable energy in the exhaust gases.

The energy efficiency of organisms is important to us for several reasons. In agriculture, we want to get the greatest production for the least input of energy, so we look for energy-efficient crops. In fisheries, some laws governing the catch of fish require that managers seek a maximum or optimal yield—the most energy-efficient fishery. In conservation, we want to know if a rare, endangered species can use its resources efficiently enough to grow and reproduce and persist. Energy efficiency is also important in biological evolution and competition.

Energy efficiency in the abstract is a straightforward concept. It is a ratio of energy output to energy input. But in practice energy efficiency can mean many different things, and ecologists have defined twenty or more kinds of efficiency. When we discuss energy efficiency, we must be very careful that our comparisons are based on the same definition [8, 9].

A common ecological measure of energy efficiency is called **trophic-level efficiency,** or the ratio of production of one trophic level to the production of the next lower trophic level. This efficiency is never very high. Green plants convert only 1 to 3 percent of the energy that comes in from the sun during the year to new plant tissue.

The efficiency with which herbivores can convert plant energy into herbivorous energy or the efficiency with which carnivores can convert herbivores into carnivorous energy is usually less than 1 percent. It is frequently written in popular literature that the transfer is 10 percent. For example, it is often said that 10 percent of the

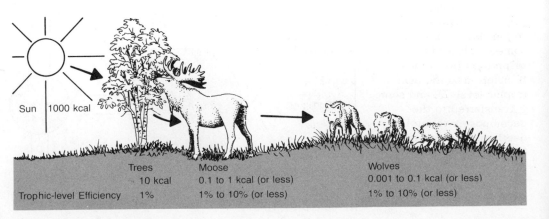

FIGURE 5.14

An example of ecological efficiency. One kind of ecological efficiency, called trophic-level efficiency, is the ratio of net production of one trophic level to the net production of its energy source. With autotrophs, this ratio is their net production compared with the source energy available. Trophic-level efficiency, rarely more than a percent or two, is sometimes much less and probably never more than 10 percent.

	Sun 1000 kcal	Trees	Moose	Wolves
		10 kcal	0.1 to 1 kcal (or less)	0.001 to 0.1 kcal (or less)
Trophic-level Efficiency		1%	1% to 10% (or less)	1% to 10% (or less)

energy in corn can be converted into the energy in a cow. However, in natural ecosystems the organisms in a trophic level tend to take more energy for their own uses than they store for the next trophic level. For example, at Isle Royale National Park the trophic-level efficiency of wolves is about 0.01 of 1 percent [6]. The wolves' trophic efficiency is very low because they use most of the energy they take in for themselves. From the wolves' point of view, they are very efficient, but from the point of view of someone who wants to feed on wolves, the wolves appear to be very inefficient (Fig. 5.14).

The rule of thumb for ecological trophic energy efficiencies is that more than 90 percent of all energy transferred between trophic levels is lost as heat. Less than 10 percent (approximately 1 percent in natural ecosystems) is fixed as new tissue. In highly managed ecosystems such as ranches, however, the efficiencies can be greater, but still no greater than 10 percent.

ECOSYSTEM MINERAL CYCLING

The cycling of chemical elements within an ecosystem and into and out of an ecosystem is called **ecosystem mineral cycling.** Ecosystem mineral cycling connects biological cycles to geological cycles. Every individual requires a number of chemical elements available at the right times, in the right amounts, and in the right ratios. Some chemical elements, such as carbon, hydrogen, oxygen, nitrogen, and phosphorus (Tables 5.1 and 3.2) [10], are required by all living things. Other elements are required by one group of organisms and not by another. For example, sodium is required by vertebrates for nerve impulse transmission and for blood salinity in mammals, but sodium is required by only a few green plants. Each element has a characteristic role: Carbon forms long chains and is the building block for organic compounds; phosphorus is involved in the transfer of energy; and cal-

TABLE 5.1
The macronutrients: chemical elements required by all life forms in large amounts.

Element	Function
Carbon	The primary building block for large organic compounds; forms long chains and combines with many other elements
Calcium	The beam and girder element, important in strong structures—shells, bones, teeth of animals, and cell walls of plants; also involved in nerve impulse transmission and muscle contraction
Hydrogen	The lightest element, a constituent of water and of all organic molecules
Iron	Various functions in enzymes and some respiratory compounds (hemoglobin)
Magnesium	Along with calcium, has structural functions (e.g., in bones and shells); also important in some electrochemical and catalytic roles
Nitrogen	The protein element, occurs in cell proteins, genetic compounds, and all chlorphylls. Nitrogen occurs in very large compounds, but is also important in small compounds, including nitrate (an oxide of nitrogen) and ammonia. Ammonia and nitrate can be taken up by green plants from soils as a nitrogen source.
Oxygen	The respiration element, required in aerobic respiration; one of the major (or big six) elements of organic compounds and a constituent of water
Phosphorus	The energy element, important in energy transfer as the compounds ATP and ADP; also a constituent of nucleic acids; along with calcium and magnesium, important in vertebrate teeth and bones and in some shells of invertebrates
Potassium	Required for certain enzymes; important in nerve cell transmission
Sodium	Essential to animals but not to most plants; important in nerve impulse transmission and in salt balance of vertebrate blood
Sulphur	Required for many proteins; an essential constituent of some enzymes

cium is found in structures that provide form and strength —bones of vertebrates, shells of many ocean invertebrates, and the firm cell walls of plants. As we discussed in Chapter 3, some elements (called macronutrients) are required in large amounts, and others (called micronutrients) are required in small, sometimes trace amounts (Tables 5.1 and Table 3.2).

Chemical elements cycle within an ecosystem from organism to organism through nonliving fluid mediums (water and air) and sediments (soils and rocks). Because some chemical elements have a gaseous phase and enter the atmosphere and others do not (Fig. 5.15), there are great differences in their cycling. Compare the cycling of sulphur (Fig. 5.16), which has a gaseous phase, and the cycling of calcium (Fig. 5.17), which does not [11].

Species that are efficient in using a resource which is present in low concentrations are called **tolerant** of that resource. Other species are efficient in using an element or resource when it is available in large quantities

and inefficient in using the resource when it is rare. Such species are called **intolerant.** We can measure the rate of uptake of a resource—chemical elements in particular—as it changes with the availability and abundance of that element in the environment. From this we can determine whether species are tolerant, intolerant, or intermediate in their response to that chemical resource. As we saw at the beginning of the chapter in the discussion of lake eutrophication, the ecological tolerance to the availability of chemical elements is an important factor in our management of ecosystems.

A population in an ecosystem will be greatly affected by the resource that it needs which is available in the least supply. In fact, some ecologists have argued that the growth and abundance of a population is controlled solely by that resource which is available in the least supply. This is known as **Liebeg's Law of the Minimum.** According to this law, we need only find out which element is available in the least supply to be able to predict which

FIGURE 5.15
Generalized ecosystem mineral cycling. Chemical elements cycle *within* an ecosystem and are exchanged *between* an ecosystem and the biosphere. Organisms exchange elements with the nonliving environment; some elements are taken up from and released to the atmosphere, while others are exchanged with water and soil or sediments. The parts of an ecosystem can be thought of as *storage pools* for an element. The elements move among pools at different *transfer rates* and remain within different pools different average lengths of time called *residence times*. For example, the soil in a forest has an active part, which rapidly exchanges elements with living organisms, and an inactive part, which exchanges elements slowly. Generally, life benefits if elements are kept within the ecosystem and are not lost by erosional processes.

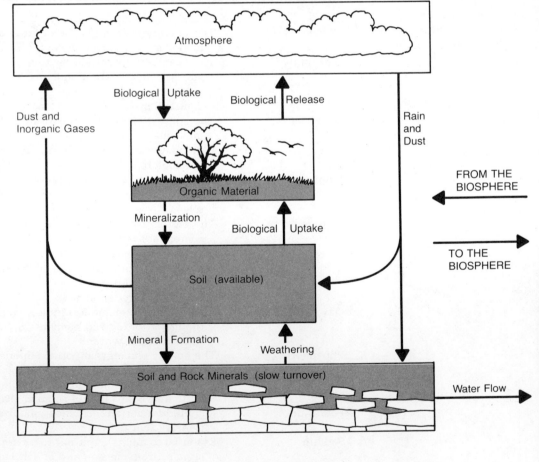

FIGURE 5.16
Annual sulphur cycle in a
forest ecosystem. The
numbers in circles are
amounts transferred (kg/
ha/yr). Other numbers are
amounts stored (kg/ha/yr).
Sulphur has a gaseous
phase as H_2S and SO_2.
(After Likens et al., 1977.)

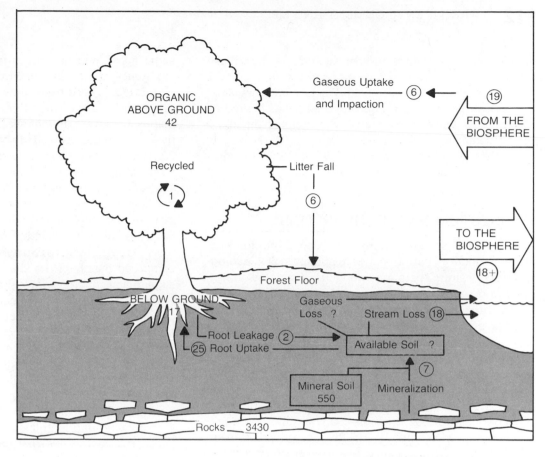

FIGURE 5.16
Annual sulphur cycle in a forest ecosystem. The numbers in circles are amounts transferred (kg/ha/yr). Other numbers are amounts stored (kg/ha/yr). Sulphur has a gaseous phase as H_2S and SO_2. (After Likens et al., 1977.)

FIGURE 5.17
Annual calcium cycle in a
young New England forest
ecosystem. The numbers in
circles are amounts
transferred (kg/ha/yr).
Other numbers are
amounts stored (kg/ha/yr).
Only major transfer
pathways are shown.
(After Likens et al., 1977.)

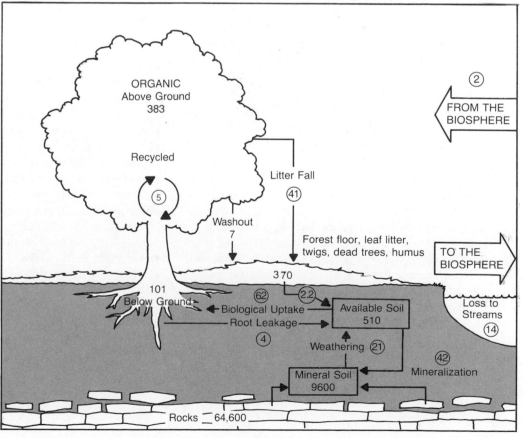

FIGURE 5.17
Annual calcium cycle in a young New England forest ecosystem. The numbers in circles are amounts transferred (kg/ha/yr). Other numbers are amounts stored (kg/ha/yr). Only major transfer pathways are shown. (After Likens et al., 1977.)

111

species will predominate and persist and how the populations in an ecosystem and the ecosystem itself will change over time. The case history of lake eutrophication illustrates this idea. The unpolluted lake was limited by phosphorus. Once phosphorus was supplied in abundance, some organisms (the fish) were limited by oxygen, and others (the algae) were limited by light or some chemical element other than phosphorus.

ECOSYSTEM PATTERNS IN TIME

Ecosystems change over time and space. Over time, ecosystems undergo patterns of development called **ecological succession.** There are two kinds of succession: primary and secondary. **Primary succession** is the initial establishment and development of an ecosystem; **secondary succession** is a reestablishment of an ecosystem. Forests that develop on a new lava flow near a volcano or at the edge of a retreating glacier are examples of primary succession. A forest that develops on an abandoned pasture is an example of a secondary succession, as is a forest that grows after a hurricane, flood, or fire. In secondary succession there are many remnants of a previous biological community, including organic matter and seeds in the soil of a forest, but in primary succession such remnants do not exist or are negligible. We see examples of succession all around us. When a house lot is abandoned in a city, weeds begin to grow, and eventually trees can be found—secondary succession is taking place. Farmers weeding a crop or homeowners weeding their lawns are fighting against the natural processes of secondary succession. Succession, one of the most important ecological processes, has many implications for environmental management.

FOREST SUCCESSION

Succession involves recognizable, repeated patterns of change. One typical pattern of secondary succession is familiar in areas like New England, where a hurricane in 1938 knocked down many trees. In a typical field the year after the hurricane, seeds of short-lived weedy plants and some trees sprouted (Fig. 5.18a). Some time later young trees became established, particularly white pine (Fig. 5.18b), pin cherry, and white and yellow birch. These species have widely distributed seeds and are fast growing, particularly in bright light. After several decades forests of the pioneer species were well established, forming a dense stand of small trees (Fig. 5.18c). Once the forest was established, other species began to grow, including

sugar maple and beech in the lower, warmer areas and red spruce and balsam fir in the higher, cooler locations (Fig. 5.18d). These trees grow more slowly than the ones that came into the forest first, but they have other capabilities. Called *shade-tolerant* by foresters, they grow relatively well in the deep shade of the redeveloped forest. By 30 years, most of the short-lived pin cherry have matured, borne fruit, and died; these trees cannot grow in the shade of the forest and do not regenerate in a forest that has been reestablished. After 50 years, the forest is a rich mixture of birches, maples, beech, and many others. The trees are generally taller, but the forest has many sizes of trees in it. After one or two centuries, these forests will be composed mainly of the shade-tolerant species—beech, sugar maple, fir, and spruce. From this pattern of recovery following the 1938 hurricane, we can abstract several general features of succession.

During succession *the species change* (Fig. 5.19a). Plants that grow rapidly and are short-lived, do well in the bright light and high nutrient conditions that often occur after disturbances, have widely and rapidly dispersed seeds, and tend to dominate the early stages of succession. These are called *pioneer species*. Plants that are slow-growing and long-lived do well in the shade of the forest and have seeds that can persist but may not be widely dispersed. These species dominate later stages and are called *late successional species*.

The biological community changes during succession (Fig. 5.19b). In early stages the amount of living organic matter—the biomass—increases. The diversity of life forms also increases. In the middle stages of succession, there are many species and sizes of trees.

Finally, *the forest ecosystem changes* (Fig. 5.19c). Organic material accumulates in the soil, and the amount of chemical elements stored in the soils and the trees increases. If the forest remains undisturbed for a very long period—hundreds of years or more—it may achieve a temporary steady state. This stage in succession is known as the **climax stage,** or the mature stage. Then the forest will slowly change again. The diversity will decrease (only the most shade-tolerant species will persist), the amount of live organic matter will decrease, and the soil may lose some of its chemical elements. Such very old ecosystems are called **senescent.** This general pattern of succession occurs in many kinds of ecosystems.

In primary succession, the earliest stages involve hardy species which can persist on bare rock or inorganic soil, such as small plants like lichens and mosses (Fig. 5.20). These may aid in retarding erosion and allowing soil to develop on which trees and shrubs may grow.

(a)

(b)

(c)

(d)

FIGURE 5.18

Succession in a New England forest. Sometime after a farm field is abandoned, it is occupied by short-lived weeds and annual and perennial plants. During this time seedlings of pioneer trees sprout (a). After a decade, small trees like the white pine are abundant and are replacing the herbs (b). After 30 years, the pine trees grow into a fairly uniform, dense stand (c). After half a century, many other species of trees have become established. These include late successional species that sprout and grow well in the shade of the forest as well as the older remaining pioneer trees (d). Given enough time (several centuries) without disturbances, the pioneer species will become rare and the forest will be dominated by sugar maples, beech, and other species that grow well in shade. (Photos by D. B. Botkin.)

113

FIGURE 5.19
Species, communities, and the ecosystem change during succession. Vegetation changes from short-lived pioneer weeds (herbaceous plants), to early successional, relatively short-lived trees, and finally to long-lived tree species that can regenerate in the deep forest shade (a). The community changes; biomass and diversity increase, then decrease somewhat (b). The ecosystem changes; soil organic matter and the total storage of chemical elements increase to a peak in late succession, decreasing somewhat afterwards (c). The patterns shown are idealized; real patterns differ depending on local conditions.

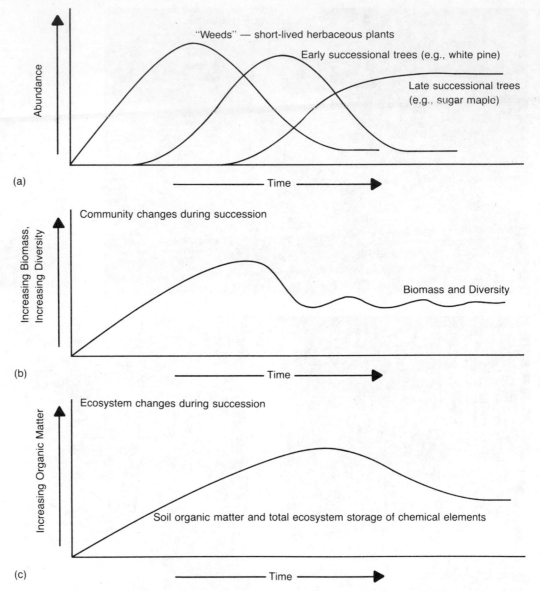

POND AND BOG SUCCESSION

Ponds and bogs also undergo succession. From a geological point of view, a pond is a temporary feature of a landscape, eventually filling in with sediments. Ponds are therefore common in areas with large-scale geological disturbances that shift drainage patterns of streams and create depressions and dams. These phenomena occur in areas subject to glaciation. Minnesota, which lies within the heavily glaciated area of North America, is therefore called the "Land of 10,000 Lakes," whereas Oklahoma, in the southern Great Plains, south of the glacial moraines, has few natural ponds.

When a pond is first created, it tends to have clear water with little sediment, low concentrations of chemical elements, and little organic matter. Over time, streams carry sediments that are deposited in the pond and bring chemical elements to the pond that are suspended in the particles and dissolved in its waters. These enrich the pond, adding the nutrients necessary for life. This increase in chemical elements is called the **eutrophication** of the pond. The young, nutrient-poor pond is called **oligotrophic,** and the old, nutrient-rich pond is called *eutrophic.* The input of sediments and nutrients is sometimes referred to as a *loading* of the pond.

The increase in a pond's chemical elements allows a greater production of plants and animals, leading to an increase in live organic matter and an increase in the organic content of the sediments. In a natural pond this

FIGURE 5.20
Primary succession on the bare rock summit of Mt. Monadnock in New Hampshire. Mosses and lichens grow on the bare rock, and in small depressions where some soil has accumulated, small shrubs and trees are established. Further down the slope, where more soil has accumulated, taller and later successional species are found. (Photo by D. B. Botkin.)

process is usually very gradual, and over a long period the pond may even shift back and forth, from oligotrophic to eutrophic to oligotrophic, as the climate and land vegetation in the watershed supplying the pond's waters change. A study of the sediments accumulated over thousands of years in Berry Pond, a small pond in Massachusetts, reveals these variations. As the forests in the watershed around Berry Pond shifted back and forth among various kinds of trees, the nutrient concentrations in the stream changed from low to high to low and so on, and the pond experienced periods of oligotrophy, eutrophy, oligotrophy, and so on.

When human activities change the rate of sediment and nutrient loading, the pond responds accordingly. Thus, in Lago di Monterosi, the Roman road increased sediment and nutrient loading, eutrophying the lake. The same process occurred in Medical Lake in Washington State. When our activities change the sediment and nu-

trient loading, we can change the pond from eutrophic to oligotrophic, or vice versa. By understanding the natural processes and the effects of our actions on natural ecosystems, we can manage our surroundings more wisely.

A bog is a body of fresh water with inlets, but no surface outlets. The succession of a bog (Fig. 5.21) proceeds from the center of the bog outward. In the quiet waters, plants that form floating mats grow out over the water surface. These are the pioneers; they are short-lived shrubs whose mats of thick, organic matter form a primitive soil, into which other plant species can enter. Then taller and longer-lived forms grow. Meanwhile, sediments including the dead organic material of aquatic animals and plants are carried into the lake by streams and build up on the bog bottom. The bog slowly fills in, and the floating mats and sediments eventually meet to form a firmer base that can support trees. The first trees—cedars and larch—are those adapted to wet ground, such as occurs in the northern forests of Minnesota, Michigan, and New England. But if the process continues undisturbed, the entire bog fills in and a raised, heavily organic soil forms, in which other trees can survive. In some cases, the bog may disappear and be replaced by the tree species characteristic of the mature forests of the well-drained soils of the region.

SAND DUNE SUCCESSION

Plant succession occurs in sand dunes along beaches. The dunes along the shores of Lake Michigan were among the first sites where this phenomenon was studied early in the twentieth century. Sand dunes are geologically unstable and are continually formed, destroyed, and reformed by the action of winds, tides, and storms. Along the east coast of North America, the earliest pioneer plant that survives on a newly formed dune is a species of grass which has long runners that anchor the plant in the sand and soil. The runners have sharp ends that force their way, as they grow, through the sand. This grass helps to stabilize the dune, which in turn allows other plants to establish themselves. Shrubs and then small trees grow on the dune, and eventually a small forest develops, including pines and oaks. A major storm can force the ocean to break through the dune, redistribute the sand, and start the process over again.

In some places, like the shores of Lake Michigan or the coast of Australia, series of dunes extend a considerable distance inland. The interior dunes were deposited earlier and have an older forest.

From these examples we see that on land there is a general pattern of change in the physical structure of plants. Generally, succession begins with small plants

FIGURE 5.21

Bog succession. Ponds and bogs of northern glaciated areas (like Livingston Bog in Figure 5.5a) gradually fill in and disappear. The pond begins as open water with water plants such as the pondweed (*Potamogeton*). A sedge, or floating mat, forms which gradually thickens as it accumulates organic matter and soil particulates (a). Sediments also accumulate in the center of the bog. As the sedge thickens, other plants are able to grow in it, including bog plants like cranberries and young trees like balsam fir and white cedar. The sediments at the bottom of the lake and the floating mat meet at the edges of the pond (b) until finally the entire bog fills in. The soil forms a slightly raised hummock in the center. The floating sedge disappears and is replaced by trees adapted to wet ground, such as the firs and cedars. Eventually, these are replaced by tree species from the surrounding forest that are adapted to drier soils (c).

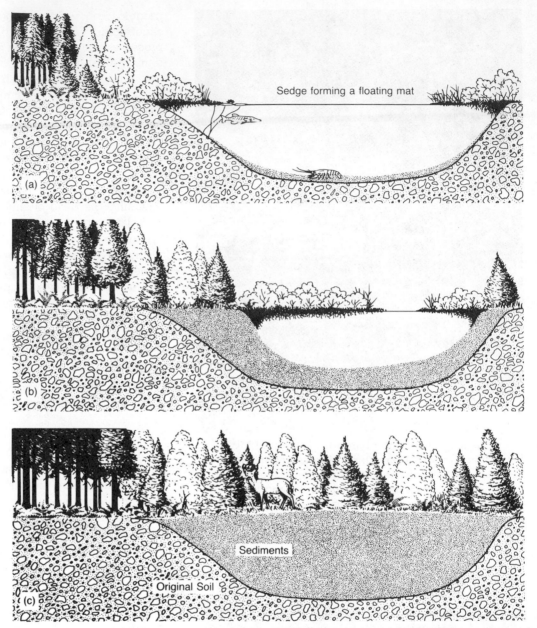

(the dune's grasses and the bog's floating mats) and proceeds to larger and larger plants. The species of animals also change during succession.

SUCCESSION AND WILDERNESS

The succession of dunes and forests makes us wonder about one of the oldest and most intriguing questions in environmental studies: What is nature undisturbed? What is the state of nature when it is unaffected by human beings? In other words, what is a true wilderness? This question is of practical importance in management of our natural resources and the entire planet, as well as important to recreation, aesthetics, and philosophy.

There has long been a common belief that an undisturbed ecosystem has a kind of constancy and permanency. In one of the first books on what we now call ecology, George Perkins Marsh wrote in his nineteenth-century book *Man and Nature* [12] that "nature left undisturbed" achieves "a permanency of form, outline and proportion" and that when disturbed, natural processes repair the "superficial damage" and restore the "former aspect." These natural processes, he wrote, lead to "a condition of equilibrium" which remains "with little fluctuations, for countless ages."

In the twentieth century, ecological theory developed around this same idea. It was believed that succession leads inevitably to an essentially constant state, the "climax state." Evidence developed that the climax forest ecosystem reproduced itself over time, because it was composed of species of trees whose seeds could germinate and seedlings grow and mature in the shade of their parent trees. Evidence accumulated that all the desirable characteristics of a forest—the amount of timber, the total biomass, the fertility of the soil (both in total organic matter and content of chemical elements), and the diversity of species—increased during succession to a maximum at the climax stage [13].

Recent evidence from studies of very long successional patterns casts doubt on these ideas, however. For example, a sequence of dunes in eastern Australia provides information for more than 100,000 years of succession. The patterns observed are surprising. At first, the pattern of succession follows the classic idea; as one walks inland to older and older dunes, the stature of the forest first increases, as does the total soil organic matter and the richness of the species. But then these factors decrease, and the very oldest dunes are poor in vegetation. Gone are the large trees and the great diversity of species. On the oldest dunes one finds shrubs of low stature and a comparatively barren landscape. The soil has been slowly leached of its nutrients, which have been washed so far below the surface that tree roots can no longer reach them. The ecosystem has become *senescent* [14].

Other evidence suggests that mature forests in North America are also "leaky"; that is, they lose chemical elements. A forest in steady state by definition must have a zero net production—no organic matter can be added. Such forests can lose nutrients through geological processes, but have little means to reaccumulate them. In contrast, an earlier successional forest, in which organic matter is accumulating, also increases its storage of chemical elements. Thus a forest that has never been disturbed will ultimately go downhill biologically [15].

It has been suggested that natural areas are subject to disturbances of many kinds, regardless of whether people are involved. Furthermore, such disturbances have existed as part of the environment for so long that animals and plants have adapted to them.

One of the best examples of this point is the role of fire in the North Woods of North America. The Boundary Waters Canoe Area, a million-acre international recreational area lying in northern Minnesota and southern Ontario, is one of the best examples of nature relatively undisturbed by people. The Indians who lived there had relatively little impact, and the French *voyageurs* traveled through this region hunting and trading for furs. Although

logging and farming were done in some places in the nineteenth and early twentieth centuries, for the most part the land has been relatively untouched. And yet the forests show a history of persistent fire. Fires occur somewhere in this forest almost every year, and on the average the entire area is burned at least once in a century. In areas that have been set aside as preserves and have been prevented from burning, the main species change, sometimes from desirable to less desirable ones. In addition, unburned forests appear more subject to insect outbreaks and disease [16]. Thus recent ecological research suggests that wilderness depends on change and that succession and disturbance are continual processes.

Because species have adapted to each stage in succession, it is not always desirable to manage an ecosystem so that it progresses totally to the mature state. A good case in point is the history of an endangered species, a small bird of the North Woods called the Kirtland's warbler (Fig. 5.22a). This bird lives in jack pine forests of Michigan (Fig. 5.22b). The jack pine is an early successional species which regenerates after a fire. When fires were suppressed in Michigan during the first part of the twentieth century, jack pine began to disappear, as did the Kirtland's warbler. Ornithologists wanting to preserve this species recognized that fire and fire-generated succession were needed to save the warbler, so now management practices have changed to promote light fires.

Similarly, the redwood forests of California also appear to require fires. Redwood trees do not regenerate—seeds do not sprout and survive—without the disturbances of light fires. Although not damaged by moderate fires, they may be by severe ones (Fig. 5.22c). The importance of fire is even more striking in the savannahs and grasslands of eastern and southern Africa (Fig. 5.22d), where fires are almost an annual event. These fires are very important in maintaining the production of early successional vegetation, which provides most of the food for the vast herds of animals. Because the fires in Africa appear to have been human-induced for a very long period, some ecologists speculate that the persistence of the African savannahs and grasslands and their great abundance of wildlife are in fact due to a partial management of these areas by primitive people.

Until recently it was generally believed that succession was an inevitable one-way process—that is, each stage had to occur in a fixed sequence, and succession could only go in a single direction, progressing from small plants to large or from low organic matter to high. Recent studies, however, suggest that patterns can be much more complicated. For example, a study of bogs in Great Britain shows that the patterns of succession tend to proceed along a certain vegetation sequence, but can un-

(a)

(b)

(c)

(d)

FIGURE 5.22
Fire and ecosystems. Fire is important in maintaining many ecosystems. The
Kirtland's warbler (a), an endangered bird, lives in jack pine forests. Jack pine (b)
regenerates only after fires and is short-lived; without fire at intervals of 50 years
or less, jack pine is replaced by other species and the warbler's habitat is lost.
Only through intentionally managed fires has this species been protected.
Although redwoods are extremely long-lived, they require fire and other
disturbances (floods) for regeneration and to enable them to dominate over other
species (c). The savannahs and grasslands of Africa burn frequently, and animals
like the impala use the new vegetation growth following fire as an important food
source (d). [Photo (a) courtesy of Michigan Department of Natural Resources;
photos (b), (c), and (d) by D. B. Botkin.]

118

dergo variations of this pattern and even move "backwards" [17].

From our modern perspective, we see succession as part of a dynamic change on the landscape. Succession is the biological process of building up life's effects on the landscape. It is continually countered by physical processes that tend to degrade, such as the slow process of fluvial erosion or the faster processes of hurricanes, tornadoes, or fires. Without disturbances, the aggrading process eventually slows down, but degradation continues. With too much disturbance, the aggrading biological processes do not lead to a significant accumulation of organic material.

The kind of ecosystem that develops in any area depends on this interaction of biological and physical processes. In deserts the rates of growth are so slow that there is no appearance of succession. The kinds of plants that grow on a recently disturbed area are the ones that can be found there long afterward. In an area heavily disturbed by people, like the edge of a highway or an abandoned city lot, the rate of disturbance exceeds the ability of the ecological processes to develop any but the earliest stages in succession.

In the ocean succession occurs where there is relative constancy, such as along the shore in rocky areas between the tides along the coast of the state of Washington or in a coral reef. In the open waters, continually stirred by winds, waves, and currents, there is no perceptible succession.

These case histories provide important lessons for our management of the landscape. First, there is rarely just one desirable state for an ecosystem; rather, there is a series of desirable conditions, which vary with particular social goals. Second, change is sometimes desirable or necessary (as with fire and the Kirtland's warbler). Third, a natural ecosystem is a mosaic of patches, each subject to a slightly different history. These provide a pattern in space that results from the temporal patterns of succession.

ECOSYSTEM STABILITY

Ecosystems are characterized by change, not constancy. Much that we have discussed in this chapter suggests that change is an integral part of ecosystems and ecological communities, but this was not always believed to be so. It was believed that an ecosystem had an equilibrium condition—its climax condition—and that when it was disturbed from this condition it would return to it, like the stability of a clock's pendulum. A pendulum has a rest

point (when it is vertical); when pushed, the pendulum will move back and forth until friction slows it down. Eventually, the pendulum will stop at its original rest point, called a point of *stable equilibrium* (Fig. 5.23). In contrast, an ecosystem is an entity that maintains certain processes (the flow of energy and cycling of chemical elements), but whose species composition can change over time. In some cases the more we attempt to maintain an ecosystem in a static, constant condition, as in the attempt to suppress fires in wilderness areas, the less likely we are to achieve what we want. The only way to manage ecosystems in which disturbance has been a long-term characteristic is to allow natural disturbances to occur.

Thus the pendulum concept is inadequate for the management and wise use of ecological communities and ecosystems. We need to think in terms of the persistence of ecosystems, the resistance to change, the resilience

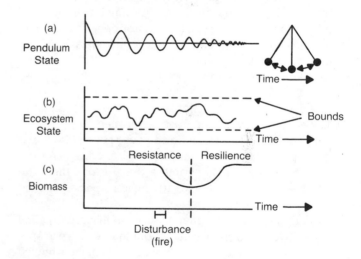

FIGURE 5.23

Ecological stability. A pendulum has a *stable equilibrium.* When disturbed from that rest point, it moves back and forth until friction causes it to slow down (a). It stops at its original rest point. The idea that an undisturbed wilderness will remain in a constant state and will return to it when disturbed is a similar concept of ecological stability. However, ecosystems rarely show the stability of a pendulum. Ecosystem characteristics, such as biomass, the stored amount of an element or compound, or the abundances of species, change over time within some bounds (b). The bounds vary with many factors, including human influence. Ecosystems have a certain resistance to change during a disturbance and a certain ability for recovery afterwards, sometimes referred to as *resilience* (c).

following a change, and the recurrence of an ecosystem's desirable conditions [18,19]. Many of our actions can affect whether an ecosystem will persist, how resistant it will be to undesirable or desirable changes, how resilient it may be in returning to a desirable state following a disturbance, and under what conditions desirable states of the ecosystem can be made to recur or can be observed to recur without management. Later in this book we will find that we must use these concepts to discover a proper relationship between ourselves and our environment.

SPECIES DIVERSITY

How many species are necessary to maintain the flow of energy and cycling of chemical elements along food chains? The discussion of the number of species in ecological communities and ecosystems is called the study of **species diversity.** We have seen that no single species is capable of carrying out all the activities required to sustain life—from the production of new organic compounds to their return to inorganic ones. The simplest imaginable number of species would be 2: an autotroph, which would produce organic compounds, and a decomposer, which would change those compounds back into inorganic ones. Natural ecosystems, however, have many more species. Ecosystems differ greatly from one another in the number of species. The simplest, like the hot springs in Yellowstone National Park, have fewer than 10; and a forest in Great Britain was found to have thousands of species. Tropical rain forests and coral reefs are among the most diverse. Biologists estimate that *in total* there are somewhere between 3 and 10 million species on the Earth.

Adding species is somewhat like an insurance policy for an ecosystem. Suppose each trophic level has one species. We can imagine a farm which grows alfalfa, which is fed to cows, and the cows are eaten by people. If a disease wipes out the alfalfa, the cows and the farmer will die. A farm which grows alfalfa mixed with a number of grasses, however, will still be able to keep its cows alive if all the alfalfa dies out, and the farmer too will survive. If some of the grasses can survive severe drought and others can survive flooding, the farmer will be protected against climatic and weather changes. The more kinds of changes that might occur in the environment, the more kinds of species one would want for insurance.

Does this mean that the more species, the better? Do more species make an ecosystem more stable—that is, do they help insure the persistence of life? Ecologists have argued these points for decades. It remains an open question whether natural ecosystems are always more stable if they have greater species diversity. Species diversity is an important concept in our management of renewable resources, and we will return to this issue in Chapter 12.

SUMMARY AND CONCLUSIONS

Today Lago di Monterosi is a small, peaceful, almost unknown body of water, appearing constant and changed only by people. But like all else on the Earth's surface, it has experienced many changes over many scales of time. Disrupted by glaciers and by the action of Roman road builders, it has responded not as merely a collection of individual species and water, sediments, and air, but as a unit. Only by understanding this unit—this ecosystem—can we understand its responses and predict and manage it and other bodies of water or other areas of the Earth's surface.

That small lake is a unit of *processes:* a flow of energy and the cycling of chemical elements. It has a community of organisms related to each other by a food web made up of trophic levels. The flux of energy and the cycling of elements affect which species dominate the waters at any one time as well as the total biomass and production. A lake undergoes changes in time, called succession, and given enough time may disappear from the landscape as it finally becomes filled in with sediments. For shorter periods of time, that lake will appear stable.

Terrestrial, freshwater, and marine ecosystems have the major characteristics we have just discussed: (1) a set of species which interact as a living community and change over time; (2) an amount of biomass; (3) production of new biomass and loss of existing biomass; (4) a flow of energy; and (5) cycling of chemical elements. Ecosystems are linked through the biosphere.

Some ecosystems exist in regions of frequent variation and disturbances (like fire and storms). Such ecosystems usually require these disturbances to retain the characteristics we desire in them. In ecosystems biological processes tend to build, and physical processes of erosion tend to tear down; the state of an ecosystem is the result of the interplay between these. Without disturbances at least some ecosystems seem to degrade—that is, they slowly lose nutrients, become less productive, and have fewer species and smaller biomass. With too frequent disturbances or when the crucial requirements for life are in short supply, the ecosystem develops slowly. With moderate rates of environmental disturbance and abundant resources, ecosystems seem usually to be most productive and to have the greatest number of species and biomass.

The ecosystem concept is useful in managing our natural resources and in understanding our relationship with the environment.

REFERENCES

1 HUTCHINSON, G. E., ed. 1970. Ianula: An account of the history and development of the Lago di Monterosi, Latium, Italy. *Transactions of the American Philosophical Society* 60: 1–178.

2 MAUGH, T. H. 1979. Restoring damaged lakes. *Science* 203: 425–27.

3 SCHINDLER, D. W.; KLING, H.; SCHMIDT, R. V.; PROKOPOWICH, J.; FROST, V. E.; REID, R. A.; and CAPEL, M. 1973. Eutrophication of lake 227 by addition of phosphate and nitrate: The second, third, and fourth years of enrichment, 1970, 1971, and 1972. *Journal of the Fisheries Research Board of Canada* 30: 1415–40.

4 MOROWITZ, H. J. 1979. *Energy flow in biology.* Woodbridge, Conn.: Oxbow Press.

5 LAVIGNE, D. M.; BARCHARD, W.; INNES, S.; and ORITSLAND, N. A. 1976. *Pinniped bioenergetics.* ACMRR/MM/SC/112. Rome: United Nations Food and Agriculture Organization.

6 JORDAN, P. A.; BOTKIN, D. B.; and WOLF, M. I. 1971. Biomass dynamics in a moose population. *Ecology* 52: 147–52.

7 SCHRÖDINGER, E. 1962. *What is life? The physical aspect of the living cell.* Cambridge, Mass.: Cambridge University Press.

8 SLOBODKIN, L. B. 1960. Ecological energy relations at the population level. *American Naturalist* 95: 213–36.

9 ODUM, H. T. 1957. Trophic structure and productivity of Silver Springs, Florida. *Ecological Monographs* 27: 55–112.

10 BOWEN, H. J. M. 1966. *Trace elements in biochemistry.* New York: Academic Press.

11 LIKENS, G. E.; BORMANN, F. H.; PIERCE, R. S.; EATON, J. S.; and JOHNSON, N. M. 1977. *The biogeochemistry of a forested ecosystem.* New York: Springer-Verlag.

12 MARSH, G. P. 1967. *Man and nature.* (Originally published in 1864, reprinted and edited by D. Lowenthal.) Cambridge, Mass.: Belknap Press.

13 ODUM, E. P. 1969. The strategy of ecosystem development. *Science* 164: 262–70.

14 WALKER, J.; THOMPSON, C. H.; FERGUS, I. F.; and TUNSTALL, B. R. 1981. Plant succession and soil development in coastal sand dunes of subtropical eastern Australia. In *Forest succession: Concepts and applications,* eds. D. West; H. H. Shugart; and D. B. Botkin, pp. 107–31. New York: Springer-Verlag.

15 GORHAM, E.; VITOUSEK, P. M.; and REINERS, W. A. 1979. The regulation of chemical budgets over the course of terrestrial ecosystem succession. *Annual Review of Ecology and Systematics* 10: 53–84.

16 HEINSELMAN, M. L. 1970. Landscape evolution, peatland types, and the environment in the Lake Agassiz Peatlands Natural Area, Minnesota. *Ecological Monographs* 40: 235–61.

17 WALKER, D. 1970. Direction and rate in some British post-glacial hydrospheres. In *Studies in the vegetational history of the British Isles,* eds. D. Walker and R. G. West. Cambridge, England: Cambridge University Press.

18 HOLLING, C. S. 1978. *Adaptive environmental assessment and management.* New York: Wiley-Interscience.

19 BOTKIN, D. B., and SOBEL, M. J. 1975. Stability in time varying ecosystems. *American Naturalist* 109: 625–46.

FURTHER READING

1 BROECKER, W. S. 1974. *Chemical oceanography.* San Francisco: Harcourt, Brace and Jovanovich.

2 EHRLICH, P. R.; EHRLICH, A. H.; and HOLDREN, J. P. 1977. *Ecoscience: Population, resources, environment.* 3rd ed. San Francisco: W. H. Freeman.

3 HUTCHINSON, G. E. 1957. *A treatise on limnology.* New York: Wiley.

4 LIKENS, G. E.; BORMANN, F. H.; PIERCE, R. S.; EATON, J. S.; and JOHNSON, N. M. 1977. *The biogeochemistry of a forested ecosystem.* New York: Springer-Verlag.

5 MCNAUGHTON, S. J., and WOLF, L. L. 1979. *General ecology.* New York: Holt, Rinehart and Winston.

6 MOROWITZ, H. J. 1979. *Energy flow in biology.* Woodbridge, Conn.: Oxbow Press.

7 SCHRODINGER, E. 1962. *What is life? The physical aspect of the living cell.* Cambridge, Mass.: Cambridge University Press.

8 SINCLAIR, A. R. E. 1965. *The African buffalo.* Chicago: The University of Chicago Press.

9 WEST, D. A.; SHUGART, H. H.; and BOTKIN, D. B., eds. 1981. *Forest succession: Concepts and applications.* New York: Springer-Verlag.

STUDY QUESTIONS

1 What are the major actions that can be taken to reduce the undesirable effects of *eutrophication?*

2 Why is the pollution of a lake likely to be an *ecosystem* problem?

3 Distinguish among *ecosystem, community,* and *habitat.*

4 Long space voyages would require the production of food—a biological life support system. Describe the major features necessary for such a life support system.

5 Some experimental sewage treatment facilities make use of ponds in which aquatic animals and plants are allowed to grow. What processes would take place to change raw sewage into drinkable water?

6 You are asked to be the manager of one of two preserves for the Indian lion, an endangered species. In the first preserve there are 200 lions on a heavily eroded landscape. Vegetation and animals are generally sparse. In the second, there are 10 lions, the soil is fertile, and the vegetation is abundant. Which would you choose to manage? Why?

7 It has been said that in conserving endangered species it is better to have a few individuals in a healthy ecosystem than a large number in a poor ecosystem. Explain this statement.

8 Farming has been described as "managing land to keep it in an early stage of succession." Explain this statement.

9 As a general rule, would pollution tend to favor early or late successional species? Why?

10 Distinguish between *trophic level* and *food chain.*

11 Why is it unlikely that there would be a species on the twentieth tropic level? (Consider the net production of various trophic levels.)

12 Describe a forest that is efficient from the point of view of (a) a forester; (b) an earthworm; (c) a bird that feeds on leaf-eating caterpillars.

13 What is a limiting factor? Can there be more than one in the same ecosystem?

14 It has been said that phosphorus is our most precious agricultural fertilizer. Explain why this might be true. In answering this question, consider the global and ecosystem cycles of carbon, nitrogen, and phosphorus.

15 How do (a) net primary production and (b) total biomass change in an ecosystem during succession?

16 Redwood trees reproduce successfully only after disturbances (including fire and floods), yet individual redwood trees may live more than 1000 years. Is redwood an early or late successional species?

6 Who Has Been Where, When? The Geography of Life

INTRODUCTION

American chestnut (*Castanea dentata*) was once a major tree of the Atlantic coastal states. Forests of chestnuts, oaks, and hickories stretched for hundreds of kilometers from Connecticut to Georgia. In 1900, chestnut was important economically as a source of tannin and as a decay-resistant wood used for fence posts, telephone poles, and railroad ties. The nuts were not only a major food for wildlife, but were a commercial feed for hogs.

About the turn of the century a shipload of wood from Europe brought a fungus (*Endothia parasitica*) that was a parasite of the European chestnut, a relative of the American species. The European chestnut and the fungus had existed together for a long time, and the tree was resistant to its disease. Never having been exposed to the fungus, the American chestnut had no resistance to it and succumbed rapidly. The disease was first noticed in 1904 in the New York City Zoological Park. It spread quickly, reaching Connecticut, Massachusetts, Vermont, New Jersey, and Pennsylvania in 4 years, covering all of New England in 20 years, and then crossing Virginia at the rate of 40 kilometers a year [1].

Large sums were spent to prevent the spread of the disease. Between 1911 and 1913 the state of Pennsylvania alone spent more than $250,000 in an attempt to provide a barrier to the fungus by removing chestnut trees from a northeast-southwest line in the middle of the state [1]. The efforts were ineffective because the fungus could survive on other tree species to which it was not fatal. By the 1930s the chestnut had disappeared as a major tree of the vast forests—a victim of **inadvertent introduction** and of previous **biological isolation**. Today chestnut sprouts continue to appear in the New England and Atlantic coastal states, but they die before they reach a height of 10 meters, killed by the fungus which persists in oaks and other trees.

The story of the demise of the American chestnut is a story of **biological geography,** involving geographical barriers, isolations, and reintroductions. American and European chestnuts evolved from a common ancestor, which at one time spread to both Europe and North America. Isolated from one another, these populations evolved along slightly different lines. One evolved in the presence of the fungus and had developed resistance to it, whereas the other did not and had no chance to develop immunity.

Just as geographical factors were important to the chestnut and those who used it, so is the geography of life important to many issues in environmental studies. First, *the geography of life is important to the biosphere.* In fact, life affects the global energy budget. Any major change in the abundance and distribution of life would change the reflection and absorption of light and infrared radiant energy by the Earth, and therefore change the Earth's heat budget. Such major changes in the biota would also affect the global cycling of chemical elements. Human activities, particularly technological ones, have global effects similar in magnitude to the effects of nonhuman life forms on the biosphere. For these reasons, knowledge of the geography of life is necessary for an understanding of the global effects of the biota.

The geography of life is important to ecosystems, species, and individuals. The abundances of individual populations and species are affected by geography; that is, the species that make up a particular ecosystem are in part the consequences of where they came from, who was there before, who came after, and so forth. Biogeography in turn is affected by the Earth's geological and climatological processes. Thus the geography of life integrates much of what is discussed in Chapters 2 through 5.

The geography of life is important to many applications of environmental studies. In studying the geography of life, we discover rules that tell us what kinds of organisms can survive and prevail in each of the Earth's climatic zones. Knowing these rules helps us to plan for the future and to evaluate the impact our activities will have on the landscape.

The principles that we will discuss in this chapter include

1 What lives where depends first on *climate,* second on *substrate,* and third on a species' *ecological attributes* and *biological interactions,* and fourth on *history.* History determines what areas of the Earth have become available to a species. Since biological interactions affect the local substrate (as with plants leaching soils and adding organic matter to soil), and since the Earth's climate and chemistry have been greatly altered by life, the interrelationships among climate, substrate, life, and history are remarkably complex.

2 Similar environments, given enough time, tend to lead to similar ecosystems.

3 Similar environments favor certain morphological and physiological characteristics of individuals. Over enough time and geographical isolation, this leads to *convergent evolution*—the evolution of geographically isolated species that look and function alike, but have different origins and genetic characteristics.

4 The current distribution and abundance of the biota greatly affect the Earth's energy exchange and its biogeochemical cycles.

5 The geography of life affects the biological diversity of any area, the resistance of the local ecosystems to disturbances, and the ability of the ecosystems to recover from disturbances.

6 During our history on the Earth, we have greatly altered the geography of life. This alteration has accelerated since the sixteenth century, with the beginning of Western civilization's exploration of the Earth, and continues to accelerate as our technology increases the ease and amount of transportation of goods and people.

7 Human alterations of the geography of life have brought people great benefits and great suffering.

8 Knowing what is ecologically "natural" to an area requires an understanding of the principles of biogeography.

9 Geographical isolation leads to the global diversification of species and to local extinctions.

10 The number of species on biogeographical "islands" decreases with the distance of the island from the mainland and as the size of the island becomes smaller. This principle was learned from studies of real islands, but it has important implications for management of landscapes. From the point of view of a geographer of life, a natural preserve is a biological island, as is a forest stand managed for timber, a lawn in a suburban yard, or a city park.

11 Given enough time, environmental diversity within a habitat tends to lead to an increase in species diversity. Similarly, a landscape with a diversity of habitats can support more kinds of species than a uniform landscape.

12 Geographical isolation has genetic effects, and the isolation of a small population can have detrimental genetic effects, including *genetic drift* and the *fixation of lethal genes,* which may lead to extinction.

13 A species long isolated in a biogeographical island tends to be highly vulnerable to exotic competitors, parasites, and predators. In contrast, a species that has evolved on a large continental area with many competitors, parasites, and predators tends to be hardier and more resistant to geographical introductions. Such introductions, however, always carry a risk of exposing any population to a competitor, parasite, or predator to which they have little resistance.

AN OVERVIEW OF BIOGEOGRAPHY

The biosphere is where life exists. A thin layer compared with the Earth's volume, the biosphere extends from the depths of the ocean to the summits of mountains, but is mostly contained within a few meters above and below the surface of the land and ocean. More specifically, life

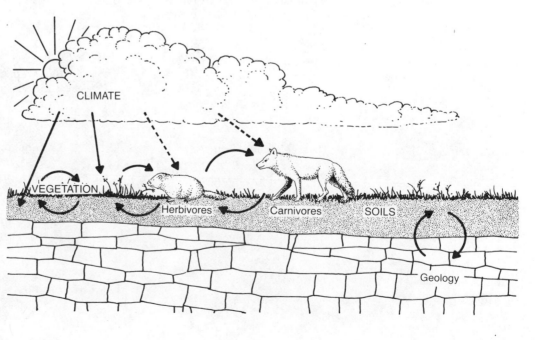

FIGURE 6.1
Interrelationships among climate, geology, soil, vegetation, and animals. Who lives where depends on many factors. Climate, geological features (bedrock type, topographic features), and soils influence vegetation. Vegetation in turn influences soils. The kind of vegetation influences the kinds of animals that will be present. Animals in turn affect the vegetation. Arrows represent a causal relationship; the direction is from cause to effect. A dashed arrow indicates a weak influence and a solid arrow a strong influence. (From Miller et al., 1977.)

TABLE 6.1
The Earth's major biomes and their dominant biota, climate, and substrate.

Biome	Dominant Growth Form	Representative Plants (mostly Northern Hemisphere)
Aquatic Systems		
Open oceans	Plankton, floating algae	Diatoms, plankton (dinoflagellates, etc.)
Estuaries and shores	Multicellular algae, grasses	Seaweeds, eelgrass, marsh grass
Lakes and streams	Algae, mosses, higher plants	Plankton algae, filamentous algae, duckweed, water lilies, pondweed, water hyacinth
Swamps, marshes, bogs	Algae, rushes, etc.	Cattails, water plantains, pipeworts, rushes, sedges, sphagnum moss, tamarack, baldcypress, mangrove
Forests		
Tropical rain forests	Trees, broadleaved evergreen	Many species of evergreen, broad-leaved trees, vines, epiphytes (orchids, bromeliads, ferns)
Tropical seasonal forests	Trees, both evergreen and deciduous	Mahogany, rubber tree, papaya, coconut palm
Temperate rain forests	Trees, evergreen	Large coniferous species (Douglas fir, Sitka spruce, coast redwood, western hemlock, white cedar)
Temperate deciduous	Trees, broadleaved deciduous	Maples, beech, oak, hickory, basswood, chestnut, elm, sycamore, ash
Temperate evergreen	Trees, needleleaved	Pines, Douglas fir, spruce, fir
Boreal coniferous (taiga)	Trees, needleleaved	Evergreen conifers (spruce, fir, pine), blueberry, oxalis
Reduced Forests — Scrubland		
Chaparral, magius, etc.	Shrubs, sclerophyll evergreen	Live oak, deerbrush, manzanita, buckbrush, chamise
Thorn woodlands	Spinose trees and large shrubs	Acacia, large shrubs
Temperate woodlands	Small evergreen or deciduous trees, grass or shrubs	Pinyon pine, juniper, evergreen oak
Grasslands		
Tropical savannah	Grass (and trees)	Tall grasses, thorny trees, sedges
Temperate grasslands	Grass	Bluestem, Indian grass, grama grass, buffalo grass, bluebunch wheat grass
Tundras		
Arctic	Diverse small plants	Lichens, mosses, dwarf shrubs, grass, sedges, forbs
Alpine	Small herbs (grasslike)	Sedges, grasses, forbs, lichens
Deserts		
Tropical warm	Shrubs, succulents	Spinose shrubs, tall cacti, euphorbias
Temperate warm	Shrubs, succulents	Creosote bush, ocotillo, cacti, Joshua tree, century plant, bur sage (in USA)
Temperate cold	Shrubs	Sagebrush, saltbush, shadscale, winterfat, greasewood (in USA)

126

TABLE 6.1 *(continued)*

Biome	General Climate[a]	Most Typical Soils or Substrate[a]
Aquatic Systems		
Open oceans	Wide range of temperatures	Saltwater
Estuaries and shores	Wide range of temperatures from arctic	Salt or freshwater
Lakes and streams	and antarctic cold to tropical warm (with hot springs at one extreme)	
Swamp, Marsh, and Bog	Various, from tropics to arctic	Mud or peat, wet with salt or freshwater
Forests		
Tropical rain forests	125 to 1250 cm annual rain, no dry period (18 to 35°C)	Mainly reddish laterites
Tropical seasonal forests	Marked dry season, generally lower precipitation	
Temperate rain forests	125 to 900 cm precipitation, nearly even throughout year, some snow (-4 to 21°C)	Podzolic, deep humus
Temperate deciduous	60 to 225 cm ppt., droughts rare, some snow (-30 to 38°C)	Podzolic: gray-brown, red, and yellow
Temperate evergreen	35 to 250 cm ppt., evenly distributed or summer dry season, possibly deep snow (-48 to 27°C)	Various podzolic, often shallow and rocky
Boreal coniferous (taiga)	35 to 600 cm ppt., evenly distributed, much snow (-54 to 21°C)	True podzols
Reduced Forests		
Mediterranean type, broad sclerophyll (chaparral, magius)	25 to 90 cm ppt., nearly all during cool season (2 to 40°C)	Variable, but fairly low carbonate
Thorn woodlands and scrubs	Dry tropical climates, between seasonal forest and desert	
Temperate woodlands, including pigmy conifers, oak woodlands	Temperate climates between forest and grassland or desert	
Grasslands		
Tropical savannah	25 to 90 cm rain, warm season thunderstorms, dry during cool season (13 to 40°C)	Variable, lateritic
Temperate grasslands	30 to 200 cm ppt., evenly distributed or high in summer, snow (-46 to 60°C)	Chernozems and related, lime layer
Tundras (arctic and alpine)	10 to 50 (arctic) and 75 to 200 (alpine) cm ppt., snowdrifts and areas blown free of snow (-57 to 16°C—arctic, -52 to 22°C—alpine)	Rocky, patterned ground, permafrost (especially in arctic)
Deserts		
Warm (tropical or temperate)	0 to 25 cm rain, very irregular, long dry seasons (2 to 57°C with high diurnal fluctuations)	Sandy or rocky, sometimes reddish or saline
Cold (temperate or arctic)	5 to 20 cm ppt., most in winter, some snow, long dry season (-40 to 42°C with diurnal fluctuation)	Gray desert; often sandy, rocky, saline

SOURCE: Jensen and Salisbury, 1972.
[a]Data modified and condensed from Billings, 1970.

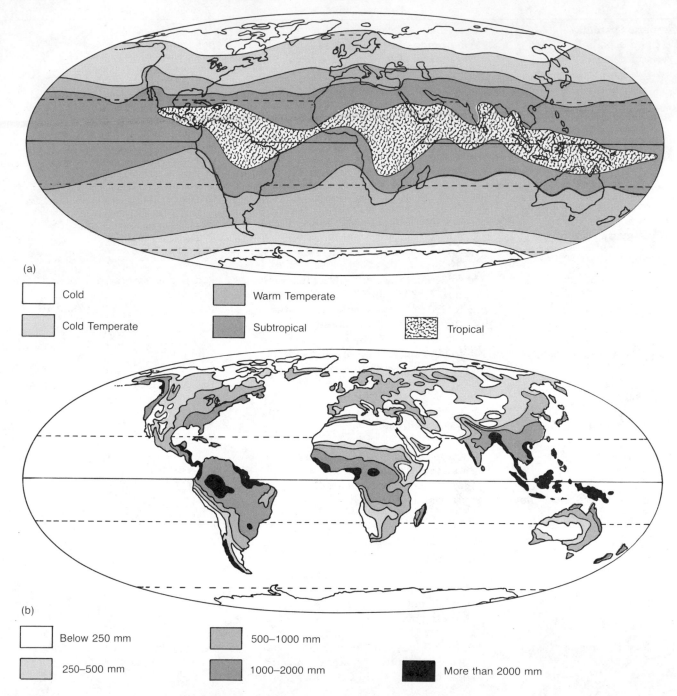

(a)

Cold

Cold Temperate

Warm Temperate

Subtropical

Tropical

(b)

Below 250 mm

250–500 mm

500–1000 mm

1000–2000 mm

More than 2000 mm

FIGURE 6.2

The Earth's major terrestrial biological zones (biomes). The distribution of biomes (c) reflects temperature (a) and rainfall (b) patterns. Climate is usually considered the primary determinant of the geography of the biomes, which are also influenced by history (e.g., who was where in the past) and geology (e.g., land bridges and barriers; the chemical fertility of the bedrock and soils). These maps are greatly simplified and omit the effects of human activities. [Part (a) from Cox et al., 1973, and parts (b) and (c) from Walter, 1973.]

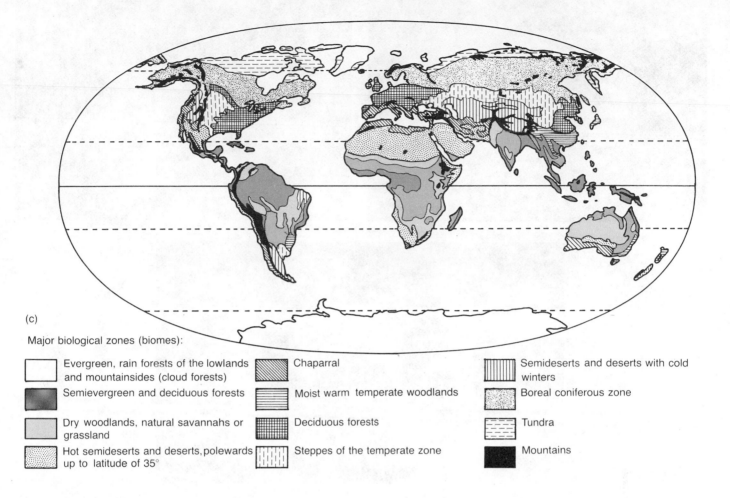

(c)

Major biological zones (biomes):

Evergreen, rain forests of the lowlands and mountainsides (cloud forests)	Chaparral	Semideserts and deserts with cold winters
Semievergreen and deciduous forests	Moist warm temperate woodlands	Boreal coniferous zone
Dry woodlands, natural savannahs or grassland	Deciduous forests	Tundra
Hot semideserts and deserts, polewards up to latitude of 35°	Steppes of the temperate zone	Mountains

occurs in **biomes**, which are the largest recognizable biogeographical units; they are kinds of ecosystems. An example of a biome is the tropical rain forests of the world. Other examples are the alpine tundra, deserts, coniferous forests, and intertidal marine areas (Table 6.1).

The geographical distribution of living things is controlled in large part by climate and soils, but is also influenced by biological interactions (Chapters 2 and 3), the species, and the Earth's history [2,3] (Fig. 6.1). The study of biological geography follows the principle of *uniformitarianism,* which we reviewed in Chapter 1. That is, we assume that processes operating in the present also operated in the past.

As stated in the introduction to this chapter, similar environments lead to similar ecosystems. A comparison of world maps of average temperature (Fig. 6.2a), average rainfall (Fig. 6.2b), and biomes (Fig. 6.2c) makes this clear. Cold, moist areas with rainfall between 35 and 60 centimeters and temperatures ranging from −54° to 21°C that

occur in the higher latitudes of North America, Europe, and Asia are occupied by *boreal forests*—forests dominated by conifers like spruce, fir, and cedar (Fig. 6.3). Mediterranean climates, with rainfall only in the cool seasons, relatively low amounts of annual rainfall (25 to 90 centimeters), and relatively mild temperatures (2° to 40°C), result in *chaparral*—dense, low woodlands of thick-leaved, shrubby vegetation. Chaparral occurs in relatively small, isolated areas of the Earth, primarily the coast of southern California, coastal regions of Chile, and parts of South Africa. These three regions have been isolated from one another for a very long time, and the principal species are not closely related but look similar. The woody plants of these areas are examples of *convergent evolution* and of the second rule of biological geography, the rule of climatic similarity. Given similar climates in two different areas, over enough time similar kinds of biota and biological communities will tend to occur (Fig. 6.4). This is a tendency, not a certainty.

(a)

(b)

(c)

(d)

(e)

FIGURE 6.3
The major biomes are distinguished from one another by the dominant kinds of vegetation as well as certain effects of climate and substrate: (a) boreal forests; (b) temperate deciduous forests; (c) northern prairies; (d) chaparral (California); and (e) tropical forests. Other types are illustrated in Figures 6.4 and 6.5. [Photos: a, Paul Nesbit; b, S. H. Anderson; c, Sharon M. Kurgis; d, Steve Davis, Stanford University/Biological Photo Service; e, courtesy of World Wildlife Fund.]

FIGURE 6.4
Similar climates lead to similar vegetation communities. Plants of low stature with many small stems, such as grasses and sedges, do well under the conditions found in the tundra in the Rocky Mountains of Colorado (a) and in the mountains of the South Island of New Zealand (b). (Photos by D. B. Botkin.)

(a)

(b)

As another example, the deserts of California and Arizona have large succulent plants like the sahuaro cactus and the Joshua tree. Their tiny leaves, sometimes reduced to sharp spines, their thick, green, photosynthetically active stems, and other adaptations store water and reduce water loss (Fig. 6.5a). East and Southern Africa have plants that look similar—such as *Euphorbia candelabra* (Fig. 6.5b) and species of aloe, which also have few or no leaves—and depend on thick, green stems for photosynthesis. Although the plants of the American and African deserts look alike, they belong to different biological families. The sahuaro is a member of the American cactus family; the Joshua tree is a member of the agave fam-

ily; and the *Euphorbia* is a member of the *Euphorbiaceae* (or spurge) family. These plants have been geographically isolated for millions of years, during which they have been subjected to similar climates. The similar climates imposed similar stresses and afforded similar potentials to green plants. The plants evolved to adapt to these stresses and potentials and have come to look alike and prevail in like habitats—a process known as **convergent evolution.** The ancestral differences between these plants which generally look alike can be found in their flowers, fruits, and seeds. Because these organs are *evolutionarily* the most conservative (least changing), they provide the best clues to the genetic history of species.

(a)

(b)

FIGURE 6.5

Examples of convergent evolution. Plants of similar environments have similar shapes. The Joshua tree (*Yucca brevifolia*) and sahuaro cactus (*Cereus giganteus*) (a) of North America look like the giant *Euphorbia* (*Euphorbia candelabra*) of East Africa (b). All three are tall, have green succulent stems which replace the leaves as the major sites of photosynthesis, and have spiny projections, but they are not closely related. The Joshua tree is a member of the agave family, the sahuaro is a member of the cactus family, and the *Euphorbia* is a member of the spurge family. Their similar shapes are a result of evolution under similar climates; all live in dry desert or semidesert areas. (Photos by D. B. Botkin.)

Convergent evolution can occur without such strict isolation. For example, both the baleen whale and the nurse shark feed on small, planktonic marine species and obtain their food by filtering large volumes of sea water. Both are large and have certain external similarities. They have different ancestry, however, because whales are warm-blooded mammals and sharks are not, and the whales have lungs and must breathe air, while sharks have gills and must obtain oxygen from the water.

Convergent evolution is found among many life forms. Large, flightless running birds that feed on seeds and insects have evolved separately in Africa (ostrich), South America (rhea), and Australia (emu) (Fig. 6.6). These birds illustrate that, given enough time, organisms evolve to take advantage of opportunities. In open savannahs and grasslands, a large bird that can run quickly but feed efficiently on small seeds and insects has certain advantages over other organisms seeking the same food. Other examples of convergent evolution are marsupials in Australia and rodents in South America, which have evolved to fill roles similar to those of hoofed mammals in Europe, Asia, and Africa.

Convergent evolution is both a boon and a bane for us. Plants long isolated by oceanic barriers but adapted to similar environments have been transplanted around the world. During the last three centuries, people have transported the most decorative and useful of these plants so that, for example, cities around the world that lie in Mediterranean climates share the same decorative plants. Bougainvillea, a spectacular bright flowering shrub originally native to Southeast Asia, decorates cities as far away as Los Angeles, California, and Salisbury, Zimbabwe. In Mediterranean climates, cities share perhaps a great diversity of transported species. For example, the county courthouse in Santa Barbara, California, is landscaped with trees from Africa, Asia, Australia, and North and South America (Fig. 6.7). But such introductions have also led to problems. An Australian tree, the Wattle tree, was introduced into parts of Africa. In areas where the environment is similar to its Australian one, the Wattle tree has been so successful that it is crowding out native plants and threatening their survival.

Sometimes the benefits of convergent evolution are economic. For example, at the time of European settle-

(a)

(b)

(c)

FIGURE 6.6
Convergent evolution in animals. Large flightless birds have evolved in widely separated regions of the Earth—the ostrich in Africa (a), the rhea in South America (b), and the emu in Australia (c). [Photos (a) and (b) courtesy of Los Angeles Zoo; photo (c) courtesy of Australian Information Service.]

ment of California, Monterey pine existed only in a small area along the California coast, where it appeared as small, twisted trees. Apparently, as climate changed in North America following the last glacial age, the range of this species became restricted; that is, its geographical "island" had shrunk. In the twentieth century this tree has been widely transplanted to Australia, Africa, and parts of Europe where, under climates more like those to which it had adapted over long periods of time, it has become a major timber tree, growing rapidly and forming tall, straight stems of great economic value (Fig. 6.8).

A species introduced into a new geographical area is called an **exotic** species. A species that is native to a particular area and not native elsewhere is called **endemic.** For example, Monterey pine is endemic to the California coast and exotic in New Zealand. A species with a broad distribution, occurring all over the world wherever the environment is appropriate, is called a **cosmopolitan** spe-

cies. The moose is an example of a cosmopolitan member of the northern boreal forests. The house mouse (*Mus musculus*) can be thought of as cosmopolitan, since it occurs in many places where people provide a habitat for it. Species that are found almost anywhere are called

FIGURE 6.7
A positive effect of introductions of exotic species. Trees from around the world add beauty to many cities. For example, the county courthouse in Santa Barbara, California, is landscaped with trees from five continents. In this photograph, the two tall trees in left center, Norfolk Island pine (left) and bunya bunya trees (center), are native to Australia. The tall palms on the right are native to North America. Smaller trees represent Asia, Africa, and South America. (Photo by D. B. Botkin.)

ubiquitous. The bacteria known as *E. coli,* associated with the human intestinal tract, is ubiquitous.

On a local scale, the kinds of species and ecosystems that occur change with soils and with changes in the topographic characteristics of slope, aspect (the direction the slope faces), elevation, and relation to a drainage basin. Change in the relative abundance of a species is referred to as an **ecological gradient** [4].

In mountainous areas, the change in elevation leads to the same biogeographic changes that occur with changes in latitude. A striking example of this can be found in the Grand Canyon in Arizona (Fig. 6.9). The vegetation at the base of the canyon, where it is hot and dry, is that of the Sonoran deserts of the dry parts of that state. On the rim of the canyon grow pinyon pine and juniper, characteristic of the lower mountain regions. At the higher elevations on the nearby San Francisco Mountain there are yellow pine, and above these are Douglas fir and white fir, which grow in abundance in Oregon, Washington, and British Columbia. Above these are Englemann

FIGURE 6.8
Monterey pine in its native habitat is a poorly shaped tree with a twisted and stunted form. Planted elsewhere, it has become a major timber tree and produces a large, straight trunk. Locations such as New Zealand and parts of Great Britain provide an environment perhaps more like the pine's original one than the present climate of the California areas, where it manages to survive. The Monterey pine in California is shown here. (Photo courtesy of Gerald Loomis, Point Lobos State Reserve.)

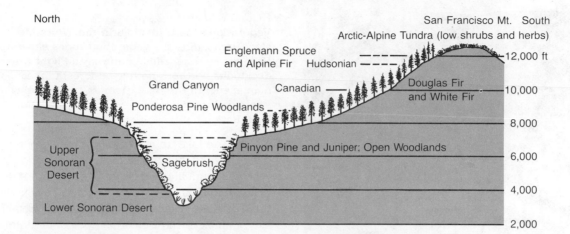

FIGURE 6.9

Altitudinal zones of vegetation in the Grand Canyon and nearby San Francisco Mountain in Arizona. The climate changes with elevation; hottest and driest at the bottom, coldest at the top. These climatic changes due to changes in elevation in the Grand Canyon are approximately equivalent to climatic changes which occur as one moves north through latitudinal zones. Thus, in response to the climate, the vegetation changes with elevation as it would with latitude. In the bottom of the Grand Canyon, one finds plants of the Sonoran desert, including creosote bush and mesquite. The next zone has vegetation from the upper Sonoran, including sagebrush (S) and pinyon and juniper (PJ). The next zone includes ponderosa pine; the following zone (Canadian zone) contains Douglas fir and white fir; and finally, the Hudsonian zone is characterized by Englemann spruce and alpine fir. (From *Natural regions of the United States and Canada* by Charles B. Hunt, W. H. Freeman and Company. Copyright© 1974.)

spruce and Alpine fir, which grow in central Canada and Alaska [5]. This example illustrates a general pattern: *Changes in elevation are equivalent to changes in latitude*. The changes on a mountainside are quite dramatic as one walks up and down and around the mountain from east to south to west to north (Fig. 6.10). However, similar changes can occur in much more subtle but just as definite ways over a relatively flat landscape.

A generalized cross section of the United States (Fig. 6.11) shows the relationships among weather patterns, topography, and biota. Off the West Coast in the Pacific basin occur the pelagic ecosystems: "euphotic" zones with sufficient light for photosynthesis, occupied by small, mainly single-celled algae; and other zones with too little light for photosynthesis, occupied by animals that depend for food on dead organisms that sink from above. Near the shore, particularly in areas of upwelling such as occurs in the Santa Barbara channel, there is an abundance of algae, fish, birds, shellfish, and marine mammals. Where the tides and waves alternately cover and uncover the shore, a thin, long line of intertidal ecosystems is found, dominated by kelp and other large algae that are attached to the ocean bottom; by shellfish like mussels, barnacles, and abalone; by crabs and other invertebrates; and by shore birds like the sandpiper.

Weather—storms, low and high pressure systems—moves from west to east in the Northern Hemisphere. As air masses are forced up over the coastal and Rocky Mountains, they are cooled, and the moisture in them condenses. The West Coast is an area of moderate temperature because water has a high capacity to store heat and the Pacific Ocean modulates the temperature. Annual amounts of rainfall increase with elevation on the west slopes of the mountains. In the south, along the southern California coast, rainfall remains low until the mountains force the air high enough to condense much of its moisture. In general, the colder, wetter heights of the mountains support coniferous forests.

In the north, along the coasts of Washington and Oregon, cool temperatures year-round lead to heavy rains near the shore, producing an unusual temperate climate rain forest. The most famous of these forests occurs in Olympia National Park on the northwestern edge of the state of Washington.

The eastern slopes of the coastal ranges form a *rain shadow*, which occurs in the following way. First, the air that passes over these eastern slopes has given up most of its moisture to the mountains; that is, it is dry as it passes to the east. Then, as the air sinks to lower elevations, it is warmed and can hold more moisture. This dry

← FIGURE 6.10
Vegetation zone transition. In the White Mountains of New Hampshire, the vegetation zones show up clearly. Here we see the transition from the light green of deciduous northern hardwood forests (beech, sugar maple, etc.) to the dark green of the boreal coniferous forest (balsam fir, spruce) at higher elevations. (Photo by D. B. Botkin.)

air tends to take up moisture from the ground, producing the deserts of Utah, California, Arizona, and New Mexico. Whereas annual rainfall in the Olympic peninsula of Washington reaches 375 centimeters (150 inches), east of the Cascades it falls to 20 centimeters (8 inches). Similarly, rainfall in the California coastal mountains reaches 200 centimeters (80 inches) per year, falling to 30 centimeters (12 inches) per year in Reno, Nevada. As the air masses move eastward, they accumulate moisture, and annual rainfall increases.

The same effect occurs in the Rockies. Less than 160 kilometers west of Denver, in the Rockies, the annual rainfall is 100 centimeters (40 inches). One hundred and sixty kilometers east of that city in the Great Plains the rainfall is 30 to 40 centimeters (12 to 16 inches) per year. Average annual rainfall increases steadily as one moves eastward; 50 centimeters (20 inches) at Dodge City, Kansas; 70 centimeters (28 inches) near Lincoln, Nebraska; and 90 centimeters (36 inches) near Kansas City, Missouri [6].

The biomes reflect these changes in rainfall. Just east of Denver are short grass prairies, which are replaced by mixed grass and then tall grass prairies as one moves

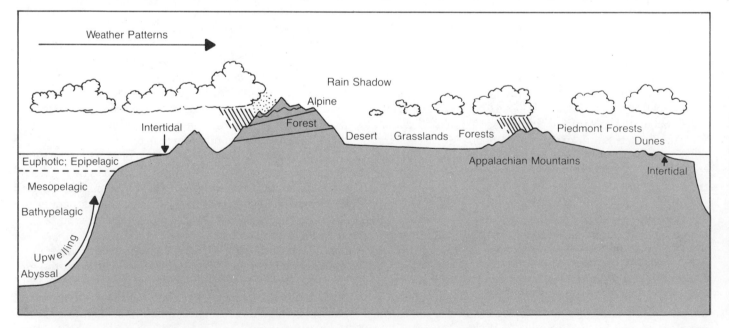

FIGURE 6.11
Generalized cross section of North America showing weather, landforms, and the geography of life. The weather patterns move from west to east.

eastward. Farther east, rainfall reaches levels sufficient to support forests. This occurs near the South Dakota–Minnesota border in the north [where annual rainfall reaches 50 to 64 centimeters (20 to 25 inches)] and in Texas to the south [along a line where annual rainfall reaches 75 to 100 centimeters (30 to 40 inches)]. From there to the East Coast occurs the deciduous and boreal forest of eastern North America [6].

Sometimes these ecological borders, as between the short and tall grass prairies, are said to be subtle, but they were quite striking and visible to some of the early travelers in the West. One such traveler, Josiah Gregg, wrote in his journal in 1831 that to the west of Council Grove, Missouri, at the border of the tall and short grass prairies, the "vegetation of every kind is more stinted—the gay flowers more scarce, and the scant timber of a very inferior quality" while to the east of that place he found the prairies to have "a fine and productive appearance. . . truly rich and beautiful."

The patterns described for the United States occur worldwide. Changes with elevation from warm, dry adapted woodlands to moist, cool adapted woodlands (Fig. 6.11) are found in Spain (Fig. 6.12), where beech and birch, characteristic of mid- and northern Europe (Germany, Scandinavia, etc.) are found at high elevations and alpine tundra is found at the summits [2]. Similar patterns are shown in Figure 6.12b for Venezuela, where a change in elevation from sea level to 5000 meters at the summits of the Andes is equivalent to a latitudinal change from the Amazon basin to the southern tip of the South American continent. The seasonality of rainfall as well as the total amount often determine which ecosystems occur in an area.

BIOMASS AND THE GEOGRAPHY OF NET PRIMARY PRODUCTION

In this section we shall examine the geographic patterns of biomass and net primary production. In general, on the land both biomass and net primary production increase with increasing rainfall and temperature (Table 6.2). The biomass per unit area stored on the land is much greater than is stored in the oceans, because woody plants maintain a large amount of woody tissue. Net primary production is lowest for deserts and tundra and largest for tropical rain forests. Per unit area, the open ocean has a net

TABLE 6.2
Productivity and biomass of the major biomes. Values are total organic matter. (Values are approximate. No accurate global ecological survey has been made.)

Biome	Area (millions of km)	Mean Net Primary Production (g/m²/yr)	Total Net Primary Production (billions of metric tons/yr)	Average Biomass (kg/m²)	Total Biomass (billions of metric tons)
Tropical rain forest	17	2000	34	44	748
Tropical seasonal forest	7.5	1500	11	36	270
Temperate evergreen forest	5	1300	6.4	36	180
Temperate deciduous forest	7	1200	8.4	30	210
Boreal forest	12	800	9.6	20	240
Woodland and shrubland	8	600	4.9	6	48
Savannah	15	700	10.4	4	60
Temperate grassland	9	500	4.4	1.6	14
Tundra and alpine meadow	8	140	1.1	0.7	5.6
Desert	18	70	1.3	0.7	13
Rock, ice, and sand	24	3	0.09	0.02	0.48
Cultivated land	14	640	9.1	1.1	15.4
Swamp and marsh	2	2500	4.9	15	30
Lake and stream	2.5	500	1.3	0.02	0.05
Total continental	149	720[a]	107	12[a]	1835
Open ocean	332	130	42	0.003	1
Upwelling zones	0.4	500	0.2	0.02	0.008
Continental shelf	27	360	9.6	0.01	0.27
Algal bed and reef	0.6	2000	1.1	2	1.2
Estuaries	1.4	1800	2.4	1	1.4
Total marine	361	153[a]	55	0.02[a]	3.9
Total biosphere	510	320[a]	162	3.6[a]	1839

SOURCE: Modified from Whittaker and Likens, 1973.
[a]Average value for total area.

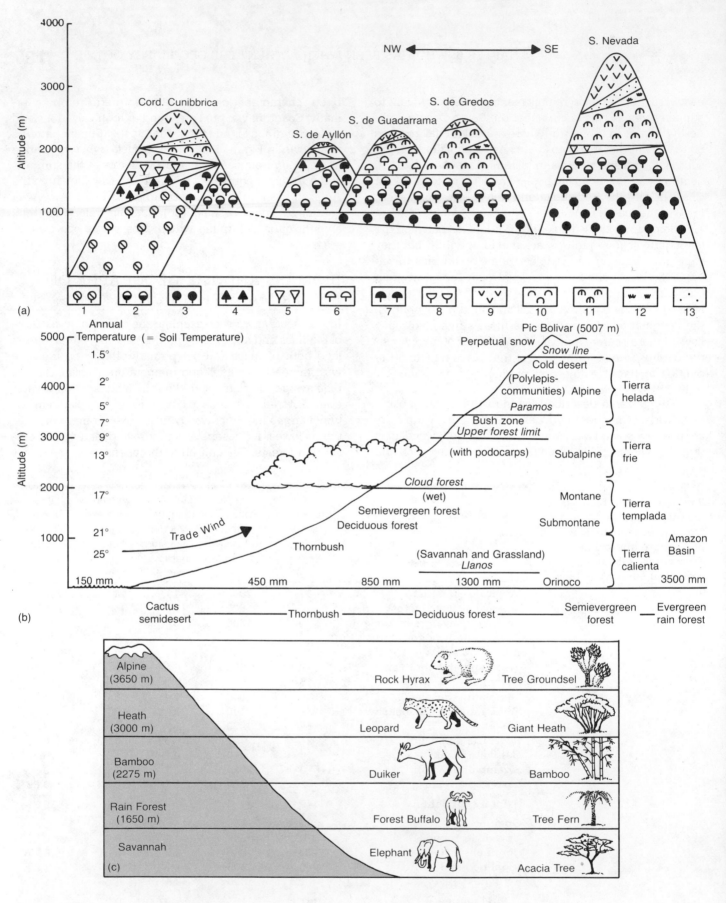

(a)

Annual
Temperature (= Soil Temperature)

1.5°	
2°	
5°	
7°	
9°	
13°	
17°	
21°	
25°	

Pic Bolivar (5007 m)
Perpetual snow
Snow line
Cold desert
(Polylepis-
communities) Alpine   Tierra
                           helada
Paramos
Bush zone
Upper forest limit
(with podocarps)   Subalpine   Tierra
                                frie
Cloud forest
(wet)   Montane
Semievergreen forest   Tierra
Deciduous forest   Submontane   templada
Trade Wind
Thornbush   Amazon
(Savannah and Grassland)   Basin
Llanos   Tierra
                  calienta
150 mm   450 mm   850 mm   1300 mm   Orinoco   3500 mm

(b)

Cactus
semidesert ——— Thornbush ——— Deciduous forest ——— Semievergreen   Evergreen
                                                    forest   rain forest

Alpine
(3650 m)   Rock Hyrax   Tree Groundsel

Heath
(3000 m)   Leopard   Giant Heath

Bamboo
(2275 m)   Duiker   Bamboo

Rain Forest
(1650 m)   Forest Buffalo   Tree Fern

Savannah   Elephant   Acacia Tree

(c)

138

◄*FIGURE 6.12*

Changes in plants and animals with elevation. In Spain (a), the vegetation changes with climate and elevation. Cold climate occurs at high elevation, where vegetation characteristic of northern high-latitude regions is found. An open oak woodland typical of warm, rather dry regions of Spain occurs at the base of the mountains (zone 1). As rainfall increases and temperatures cool, different species of oak dominate (zones 1 through 3); then beeches, typical of much higher latitudes, dominate the forest, followed by birches (which occur in central and northern Europe) and pines. At the summit occurs tundra (zones 9 through 13). Species zones are 1. Deciduous oak forest (*Quercus robur, Qu. petraea*); 2. *Qu. pyrenaica* forest; 3. *Qu. ilex* forest; 4. Beech forest (*Fagus sylvatica*); 5. Birch forest (*Betula verrucosa*); 6. Pine forest (*Pinus sylvestris*); 7. Mixed deciduous forest of oak, basswood, and maple (*Quercus, Tilia, Acer*); 8. High altitude forest of the Sierra Nevada of mountain ash and cherries (*Sorbus, Prunus*, etc.); 9. High alpine grass- and herbaceous-vegetation; 10. Dwarf-shrub heath (*Calluna, Vaccinium, Juniperus*); 11. Broom heath (*Cytisus, Genista, Erica*); 12. Thorn cushion belt; and 13. *Festuca indigesta*—dry sward. Similar changes occur in the mountains of Venezuela (b), from desert cactus at the lowest elevations through wet forests on midslope to dry tundra near the summits. Altitudinal zonations occur for animals on Mt. Kenya in Kenya (c). [Parts (a) and (b) from Walter, 1973; part (c) from Cox et al., 1973.]

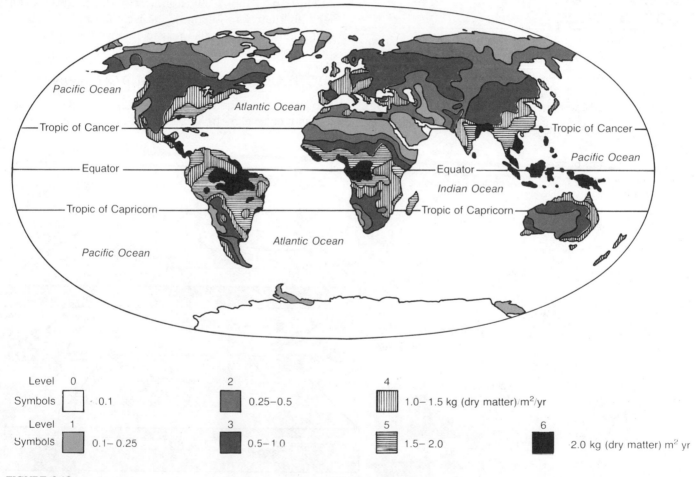

Level	0		2		4	
Symbols		0.1		0.25–0.5		1.0– 1.5 kg (dry matter) m²/yr
Level	1		3		5	6
Symbols		0.1– 0.25		0.5– 1.0	1.5– 2.0	2.0 kg (dry matter) m² yr

FIGURE 6.13

The geography of vegetation production. Vegetation production on the land varies with rainfall and temperature. The greatest productivity occurs in areas of highest rainfall and warm temperatures. Here the potential vegetation production is predicted from average rainfall and temperatures. (From Whittaker and Likens, 1973. UNC biosphere model by H. Lieth. "Miami" model by Elgene Box, H. Lieth, and T. Wolaver.)

primary production similar to that of deserts. The open ocean is considered a desert because of the scarcity of some chemical elements necessary for life. The areas of the oceans that have higher nutrient concentrations—the upwelling zones, continental shelves, reefs, and estuaries —have a net primary production per unit area that is similar to the more productive land biomes. The total net primary production on the land, however, is larger than the total for oceans.

Using correlations between net primary production and temperature and rainfall, it is possible to project average net primary production for the major land areas (Fig. 6.13).

PLATE TECTONICS AND BIOGEOGRAPHY

In Chapter 2 we discussed plate tectonics and the history of the continents. The separation and uniting of continents have repeatedly isolated and reconnected the biota of large regions. During the Earth's history, each major group of living things has originated at some period. When a large-scale separation of landmasses occurred

after the rise of some major group, divergent and convergent evolution took place, leading to the wide dispersal of certain life forms and, through divergent evolution, the rise of many species (Fig. 6.14).

Naturalists recognized these geographic patterns as a consequence of *history* before plate tectonics was understood. In 1876 A. R. Wallace recognized five geographical *realms* for mammals. Each realm was populated primarily by genetically related animals, the marsupials in Australia being a prime example [7]. In some cases land "bridges" have allowed these realms to mix. Thus the large mammals of India appear to have origins in three regions: Europe, India, and Africa. We now understand that Wallace's realms (Fig. 6.15a) and similar realms for vegetation (Fig. 6.15b) result from geological processes that lead alternately to geographical isolation and geographical "bridges" that make possible convergent and divergent evolution.

BIOGEOGRAPHY AND GLACIATION

Ecosystems change over long periods of time in response to changes in climate, including the periods of advances

FIGURE 6.14
Schematic diagram of the develoment of biogeographic realms.

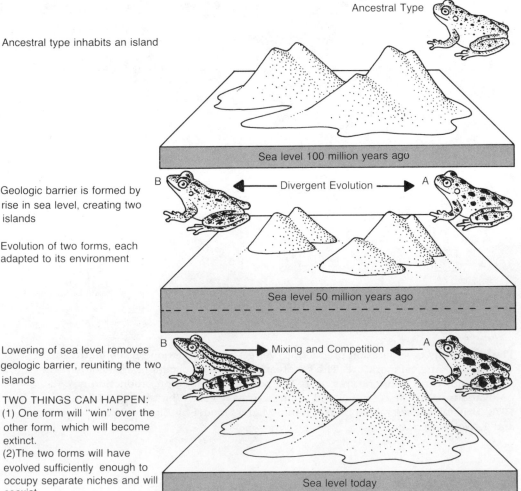

Ancestral type inhabits an island

Geologic barrier is formed by rise in sea level, creating two islands

Evolution of two forms, each adapted to its environment

Lowering of sea level removes geologic barrier, reuniting the two islands

TWO THINGS CAN HAPPEN:
(1) One form will "win" over the other form, which will become extinct.
(2)The two forms will have evolved sufficiently enough to occupy separate niches and will coexist.

Ancestral Type

Sea level 100 million years ago

B   Divergent Evolution   A

Sea level 50 million years ago

B   Mixing and Competition   A

Sea level today

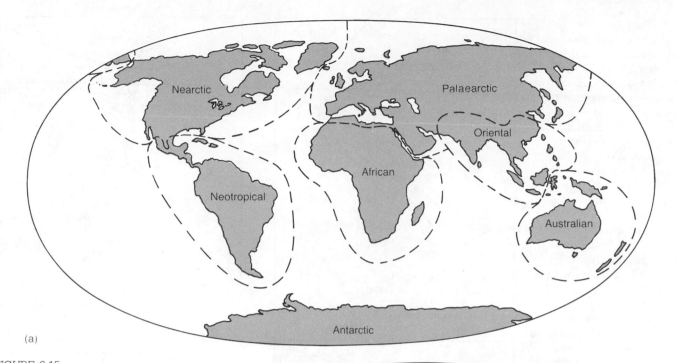

(a)

(b)

FIGURE 6.15
The main biogeographic realms for animals (a) are based on genetic factors. Within each realm the mammals are in general more closely related to each other than to mammals of other realms. Within each realm there has been divergent evolution to meet the needs and opportunities of the environments. The major vegetation realms (b) are also based on genetic factors. Flowering plants within a realm are more closely related to each other than they are to flowering plants of other realms. See Figure 6.4. [Part (a) from Wallace, 1876; part (b) from Cox et al., 1973.]

and retreats of glaciers which have occurred during the last 2 million years. Such changes in ecosystems can be traced by studies of deposits in the bottoms of lakes. Pollen, leaves, and twigs from trees fall into lakes and ponds and become buried in the sediment. As we learned in Chapter 5, these sediments were used to reconstruct the history of Lago di Monterosi in Italy. Studies of many lakes provide a history of the migration of species [8] and

tell us how ecosystems change over long periods of time.

During the glacial episodes, the climate became colder in northern areas and glaciers extended further south. As this happened, individual animals traveled southward. Individual trees could not migrate, but their seeds were carried by the wind and by animals. Thus species of trees migrated through several generations.

The glaciers covered the northern part of North America until about 10,000 years ago, when they began a retreat which has continued to the present with minor fluctuations. As the climate warmed, bare ground was exposed where ice had been, and vegetation was able to colonize the new ground and migrate northward.

Different tree species migrated northward at very different rates. Chestnut, which we discussed at the begin-

FIGURE 6.16
Changes in distribution of tree species during the last 10,000 years in North America. At the end of the last glacial period, as the climate warmed, tree species began to migrate northward. The history of their migration can be traced by studying pollen deposits in lakes. These graphs show the time (in thousands of years) that the pollen of chestnut (a), oak (b), maple (c), and white pine (d) first appeared at different areas of North America. From this information, one can reconstruct the general path of the migration (shown by the arrows). Chestnut, oak, and maple migrated from a refuge in the south central United States; white pine migrated from the southeast, from a refuge now offshore and under water. From a geological perspective, these species have been together in the mid-Atlantic states for a relatively short time. Chestnut reached the vicinity of what is now New York City about 2000 years ago, but oak arrived there 8000 years earlier. The forests first seen by the early European explorers (and thought to be relatively permanent associations of species) were ones that had changed in the preceding several thousand years. (From Davis, 1981.)

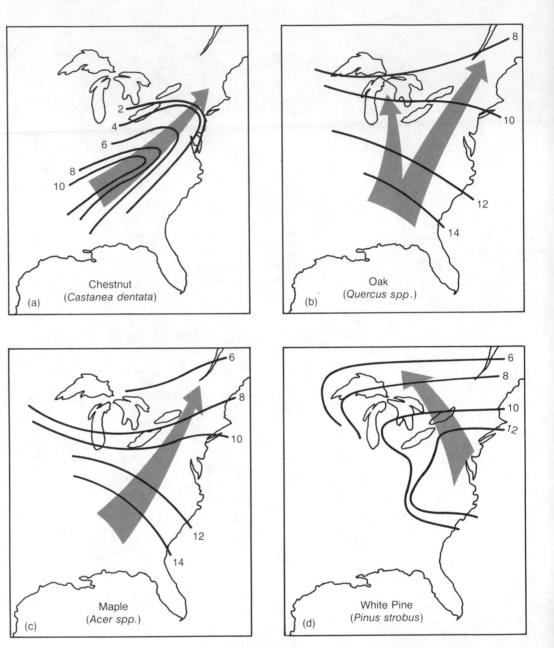

ning of this chapter, moved very slowly. Deposits of its pollen in lakes and ponds indicate that chestnut only reached the area of Connecticut and New Jersey about 2000 years ago (Fig. 6.16a). Oaks returned much faster, reaching this area 8000 years earlier, about 10,000 years before the present (Fig. 6.16b). Thus the oaks and chestnuts which grew in the primeval forests seen by the first

European settlers in the seventeenth century had, from a geological time scale, only recently come together [9].

Further north in New England, the early settlers found maple and white pine growing in the same general areas. However, these species occurred in places hundreds of kilometers apart at the height of the glaciations. Maple grew somewhere in the south, in what is now Louisiana and Mississippi, and had returned northeastward from this southwestern refuge (Fig. 6.16c). White pine had moved northwest, from a southeastern refuge, which was dry land at the height of the glacial period but is now under water offshore (Fig. 6.16d).

These records of pollen in lakes and ponds give us an important insight into the true character of wilderness uninfluenced by human beings. Chestnuts and oaks were thought to be members of the same climax forest, a forest which would persist indefinitely in an area as long as the climate remained the same. Before the pollen records became available, it was commonly believed that the species had migrated as a unit and that a chestnut-oak forest could have been found, although much further south, at anytime during the glacial periods. Similarly, northern forests which included maples and white pine were commonly thought to have moved together across the landscape to the same refuges. Now we know that the cooccurrences of such groups of species depend in part upon historical events. Wilderness has changed greatly over the last several thousand years even without human influence. Ecological communities, at least as far as the tree species are concerned, seem to be relatively loosely coupled groupings.

ISLAND BIOGEOGRAPHY

Islands have a special fascination, especially for naturalists. Darwin's visit to the Galapagos Islands gave him his most powerful insight into biological evolution. There he found many species of finches which were related to a single species found elsewhere. On the Galapagos, each species was adapted to a different *niche,* or way of making a living. Darwin suggested that the finches, isolated from other species that filled these niches on the continents, were able to exploit unused resources. Over a long time the population of finches separated into a number of groups, each adapted to a more specialized role [10]. A similar **adaptive radiation** occurred on the Hawaiian Islands, where a finchlike ancestor evolved into several species, including fruit and seed eaters, insect eaters, and nectar eaters, each with a beak adapted for its food (Fig. 6.17) [8].

Islands have fewer species than continents. The smaller the island, the fewer are the species, as can be seen for the number of reptiles and amphibians in West Indian islands (Fig. 6.18) [11]. The farther the island from the mainland, the fewer also are the species [12].

There are two sources of new species on an island: migrants from the mainland and the evolution of new species (as with Darwin's finches on the Galapagos). Every

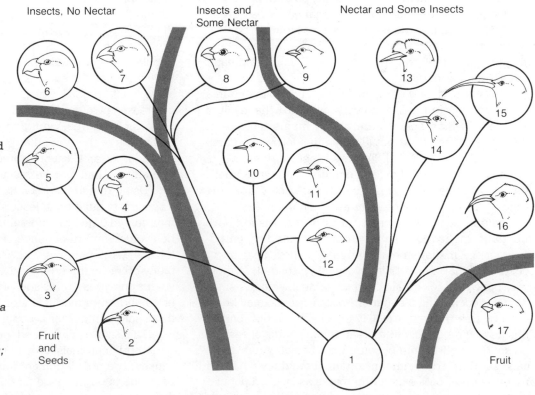

Insects, No Nectar   Insects and Some Nectar   Nectar and Some Insects

FIGURE 6.17
Evolutionary divergence among honeycreepers in Hawaii. Sixteen species of birds, each with a beak specialized for its food, evolved from a single ancestor. The birds evolved to fit ecological niches on the islands. On the North American continent, these niches were previously filled by other species not closely related to the ancestor. The sixteen species, plus the ancestor, are 1. Unknown finchlike colonist from North America; 2–5. *Psittacirostra psittacea, P. kona, P. bailleui, P. cantans*; 6. *Pseudonestor xanthophrys*; 7–9. *Hemignathus wilsoni, H. lucidus, H. procerus*; 10–12. *Loxops parva, L. virens, L. coccinea*; 13. *Drepanidis pacifica*; 14. *Vestiaria coccinea*; 15. *Himatione sanguinea*; 16. *Palmeria dolei*; and 17. *Ciridops anna.* (From Cox et al., 1973.)

Fruit and Seeds

Fruit

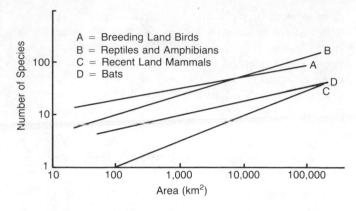

FIGURE 6.18

Islands: their size and number of species. The larger the island, the greater is the number of species. This general rule is demonstrated by a graph of the number of species of birds, reptiles and amphibians, land mammals, and bats for islands in the Caribbean. (Modified from Wilcox, 1980.)

species is subject to some risk of extinction. Extinctions can be caused by *random fluctuations*, *predation*, disease (*parasitism*), *competition*, *climatic change*, or *habitat alteration*. The smaller the island, the smaller is the population of a particular species that can be supported. Other things being equal, the smaller a population, the greater is its risk of extinction.

Thus, over a long time, an island tends to maintain a rather constant number of species, which is the result of the rate at which species are added minus the rate at which they become extinct. These rates follow the curve shown in Figure 6.19 [12]. For any island, the number of species of a particular life form (such as birds, mammals, herbivorous insects, or trees) can be predicted from the island's size and distance from a mainland [3].

Insights gathered from studies of real islands have important implications for environmental studies, particularly for the management of any population, species, or ecosystem that is partially or wholly isolated. Almost every park is a biological island for some species. A small city park, occupying a square between streets, may be an island for trees and squirrels. Even a large national park is an island. For example, the great wildlife parks of eastern and southern Africa are becoming islands of natural landscape surrounded by human settlement and towns. Lions and other great cats exist in the parks as isolated populations, no longer able to roam freely and mix over large areas. Other examples are islands of uncut forests created by logging operations and oceanic islands, where intense fishing has isolated parts of fish populations.

Our management of endangered species and of forests, fisheries, and wildlife can benefit by the study of is-

land biogeography. The knowledge gained in such study is being put to use in a Brazilian project originated by the U.S. World Wildlife Fund. Logging operations in the Brazilian Amazon are being carried out so as to leave uncut islands of many sizes, from a few hectares to hundreds of square kilometers. The intent of the experiment is to learn the minimum size for an island of each major type of life form. From this we can learn what size of forest island must be used in management to protect the forest's species.

What is a sufficiently large island to guarantee the survival of a species? The size varies with the species, but can be estimated. Some islands that seem large to us are too small for species we wish to preserve. For example, a preserve was set aside in India in an attempt to reintroduce the Indian lion into an area where it had been eliminated by hunting and changing patterns of land use. In 1957 a male and 2 females were introduced into a 95 square-kilometer preserve in the Chakia forest called the Chandraprabha Sanctuary [13]. The introduction was carried out carefully and the population counted annually. There were 4 in 1958, 5 in 1960, 7 in 1962, and 11 in 1965, after which they disappeared and were never seen again. Although 95 square kilometers seems large to us, male Indian lions have territories of 128 square kilometers. Within such a territory females and young also live. A population that would be viable for a long time would have a number of such territories, and an adequate preserve would require 640 to 1280 square kilometers. Various reasons were suggested for the disappearance of the lions, including poisoning and shooting by villagers, but regardless of the immediate cause, it was learned that a much larger area was required for long-term persistence of the lions than was set aside.

Using genetic factors, it has been estimated that for a specific population at least 50 breeding adults should be maintained if genetic problems of small populations are to be avoided and the chance of extinction kept acceptably low [14]. A population with 50 breeding adults will have many other nonbreeding individuals, including newborn, young and premature, and old post-breeding. Thus several hundred individuals are, as a rule of thumb, a minimum safe number for a population.

The size of parks and preserves for rare predatory birds and mammals is a particularly difficult issue because these populations require large areas. Can we ever set aside enough space for our endangered species? We will return to this question when we discuss natural resource management in Chapter 12.

A species long isolated on a biogeographical island is highly vulnerable to exotic competitors, parasites, and predators. The dodo existed on Mauritius Island in the Indian Ocean where there were no mammals or other ani-

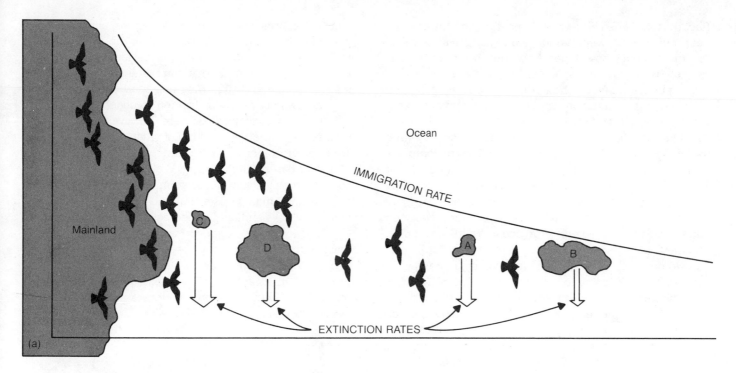

(a)

FIGURE 6.19

Idealized relation between an island's size, its distance from the mainland, and its number of species (a). The nearer to the mainland, the more likely an individual will find an island. Thus, the nearer an island to the mainland, the higher is the rate of immigration. The larger the island, the larger the population it can support, and the chance of persistence of a species increases. Small islands have a higher rate of extinction. The average number of species, therefore, depends on the rate of immigration and the rate of extinction. Thus, a small island near the mainland may have the same number of species as a large island far from the mainland. In (a), the thickness of the arrow represents the magnitude of the rate. In (b), the same relations are presented as a graph. The downward-sloping curves are rates of immigration, which decrease with the distance of the island from the mainland. The upward-sloping curves are rates of extinction, which increase as the island size decreases. Where the lines cross, the rates are equal and the number of species remains the same. A small island far from the mainland has the smallest number of species; a large island near to the mainland has the largest number. (Modified after MacArthur and Wilson, 1963.)

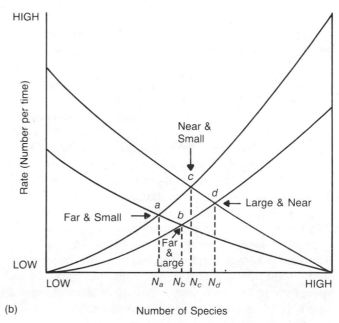

(b)

mals that were predators on the birds' eggs. The dodo nested on the ground. After the island was discovered, dogs, hogs, and rats were introduced there and rapidly destroyed the eggs of the dodo, which had little means of defending them. People easily caught the birds, which were not accustomed to human predation, and ate them. Thus the dodo became extinct.

DIVERSITY AND BIOGEOGRAPHY

There are millions of species on the Earth; estimates range from 3 to 10 million. The range of estimates is large because there are parts of the Earth relatively unexplored for many kinds of species. The number expected to be found in unexplored areas is estimated from the number found in well-studied areas. Many of the unnamed and yet undiscovered species are insects in tropical forests.

Recent discoveries of new ecological communities in the deep ocean near vents which emit gases illustrate how incomplete our knowledge of the number of species is. Why are there so many species? Why is there not just one green plant species on the land, one species of algae in the oceans, one herbivore species on the land and one in the oceans, and so forth?

Our discussion of biogeography in this chapter gives us *some* answers. Geographical barriers and geographical islands lead to convergent and divergent evolution. In different parts of the Earth, different species have similar roles; that is, they fill similar ecological niches. If the Earth's landmasses were all merged into one continent and the waters formed one large ocean, there would probably be fewer species than exist now on the Earth.

Some areas of the Earth have many more species than others. For example, tropical forests have hundreds of species of trees, while forests at high latitudes and high elevations in mountains where the environment is severe have very few species. As a general rule, locations with severe environmental conditions—very limited amounts of the chemical elements necessary for life, little sunlight for plant growth, or temperature extremes—have few species. Locations with abundant resources of many kinds often have many kinds of species. As with so many biological phenomena, this is a general trend, to which there are exceptions. For example, certain semidesert areas which have great local variation in topography may have many kinds of species.

Biologists have long speculated about the reasons why there are so many kinds of species. One of the central issues has concerned the relationship between environmental variation and species diversity. Some suggest that constant environments seem to lead to an increase in diversity. Their argument is that when the environment varies greatly, all species must be generalists—that is, they must be able to survive under a variety of conditions —and there are relatively few ecological niches. Under relatively constant conditions, however, there are many subtle variations in local environments and species can specialize. In fact, the great diversity of tropical rain forests is used as an example of this tendency.

Others argue that variations in the environment provide different kinds of opportunities and allow more kinds of species. Variations in the environment which lead to disturbance and ultimately to ecological succession illustrate how varying climate promotes an increase in diversity. The seasonal changes in stream flow into ponds, which change the total amount of water as well as the abundance of crucial chemical elements, are examples of variations which lead to greater diversity in the plankton.

For there to be a great diversity of species, there must be an opportunity for species to specialize. Certain kinds of spatial and temporal variations can make this possible. Too rapid changes in time or too small changes in space may not provide the potential for high species diversity. On the other hand, a completely uniform environment in space and time may not allow enough environmental diversity for specialization to occur.

The matter cannot rest here, however. Living things themselves can provide a kind of environmental diversity. Trees provide many kinds of microhabitats; their upper branches can be dry and hot, while their lower ones are cool and moist. Even in a uniform climate, trees may produce microclimates of several kinds. In the oceans, coral reefs provide a similarly varied set of habitats.

Biogeography suggests some explanations for the reasons there are so many species. Ecological communities and ecosystems provide other answers, which were discussed in Chapter 5. The great diversity of species provides us with renewable biological resources, which we will discuss in Chapter 12.

SUMMARY AND CONCLUSIONS

The geography of life—the distribution of living things on the Earth—has an important role in the global cycling of chemical elements and in the Earth's pattern of energy exchange with the solar system. The geography of life is a consequence of several factors, including the history of the origin of species, the creation and distinction of ''bridges'' between geographically isolated areas, local climate, topography and substrate, and ecological interactions among species such as competition, predation, and parasitism. The geography of life has long fascinated naturalists and has provided the basis for many important biological ideas, including Darwin's theory of evolution.

The study of the geography of life can help us to decide what kinds of transplants of species will be beneficial and which may be disastrous, although this knowledge has been little used in the past in planning introductions of exotic species.

Knowledge of the biological geography of islands is particularly useful in natural resource management because all preserves, parks, and managed forest stands and rangelands are biological islands. The smaller the island, the greater the amount of active management that is required and the more important a knowledge of island biogeography becomes.

The geography of edible and otherwise useful plants and animals has affected every human culture and society. Ever since the domestication of plants and animals, human beings have been primarily responsible for transporting and introducing exotic species into new areas.

REFERENCES

1 BOTKIN, D. B., and MILLER, R. S. 1974. Complex ecosystems: Models and predictions. *American Scientist* 62: 448–53.

2 WALTER, H. 1973. *Vegetation of the Earth.* New York: Springer-Verlag.

3 MACARTHUR, R. H. 1972. *Geographical ecology.* New York: Harper & Row.

4 WHITTAKER, R. H. 1970. *Communities and ecosystems.* New York: Macmillan.

5 HUNT, C. B. 1967. *The physiography of the United States.* San Francisco: W. H. Freeman.

6 BOTKIN, D. B. 1977. The vegetation of the West. In *The reader's encyclopedia of the American West,* ed. H. R. Lamar, pp. 1216–24. New York: Thomas Y. Crowell.

7 WALLACE, A. R. 1876. *The geographical distribution of animals.* Vol. 1. New York: Hafner. (Reissued in 1962 in two volumes.)

8 COX, C. B.; HEALEY, I. N.; and MOORE, P. D. 1973. *Biogeography.* New York: Halsted Press.

9 DAVIS, M. B. 1981. Quaternary history and the stability of forest communities. In *Forest succession: Concepts and applications,* eds. D. C. West, H. H. Shugart, and D. B. Botkin, pp. 132–53. New York: Springer-Verlag.

10 DARWIN, C. A. 1859. *The origin of species by means of natural selection or the preservation of proved races in the struggle for life.* London: Murray.

11 MACARTHUR, R. H., and WILSON, E. O. 1967. *The theory of island biogeography.* Princeton, N. J.: Princeton University Press.

12 _____. 1963. An equilibrium theory of insular zoogeography. *Evolution* 17: 373–87.

13 NEGI, S. S. 1969. Transplanting of Indian lion in Uttar Pradesh. *Cheetal* 12: 98–101.

14 SOULÉ, M. E. 1980. Thresholds for survival: Maintaining fitness and evolutionary potential. In *Conservation biology,* eds. M. E. Soulé and B. A. Wilcox, pp. 151–69. Sunderland, Mass.: Sinauer.

FURTHER READING

1 DARWIN, C. A. 1909. *The voyage of the Beagle.* New York: Collier & Sons.

2 ELTON, C. S. 1958. *The ecology of invasion by animals and plants.* London: Methuen.

3 HUMBOLDT, A. von. 1896. *Views of nature.* London: George Bell.

4 HUNT, C. B. 1967. *The physiography of the United States.* San Francisco: W. H. Freeman.

5 LIETH, H., and WHITTAKER, R. H. 1975. *Primary productivity of the biosphere.* New York: Springer-Verlag.

6 MACARTHUR, R. H. 1972. *Geographical ecology.* New York: Harper & Row.

7 MACARTHUR, R. H., and WILSON, E. O. 1967. *The theory of island biogeography.* Princeton, N.J.: Princeton University Press.

8 NEILL, W. T. 1969. *The geography of life.* New York: Columbia University Press.

9 PIELOU, E. C. 1979. *Biogeography.* New York: Wiley.

10 WALLACE, A. R. 1876. *The geographical distribution of animals.* Vol. 1. New York: Hafner. (Reissued in 1962 in two volumes.)

11 WALTER, H. 1973. *The vegetation of the Earth.* New York: Springer-Verlag.

STUDY
QUESTIONS

1 Why are introductions of species so often unsuccessful?

2 What is a geological barrier, and why is this concept important in the geography of living things?

3 Other things being equal, on which kind of planet would you expect a greater diversity of species: (a) a planet with intense tectonic activities or (b) a tectonically dead planet? (Remember that *tectonics* refers to the geological processes which involve the movement of tectonic plates and continents, processes that lead to mountain building, etc.)

4 You conduct a survey of city parks. What relationship would you expect to find between the number of species of trees and the size of the parks?

5 What is meant by the statement "Every nature preserve must be managed as if it were an island"?

6 Why are no land mammals native to New Zealand?

7 What are the major factors which determine which species live in a particular location on a continent?

8 Riding in a balloon, you become lost in the clouds and eventually land on an island. The island is characterized by rolling hills with low, dense vegetation. The plants have thick leaves which give off strong, pleasant smells. Are you sorry that you forgot to take your umbrella?

9 What are the consequences of geographical isolation?

10 In Jules Verne's classic novel *The Mysterious Island*, a group of Americans find themselves on an isolated volcanic island inhabited by kangaroos and large rodents closely related to the agoutis of South America. Why is this situation unrealistic? What would make this co-occurrence possible?

11 Why does desert occur *west* of the Andes in Chile when the desert is *east* of the Rocky Mountains in North America?

12 Why is tundra found both in the far north and in mountaintops? What differences, if any, would you expect to find between these two kinds of tundra, *arctic* and *alpine*?

13 Distinguish among *biosphere, biome,* and *ecosystem*.

7

An Ecological
View of Human
Beings: Human
Population
Dynamics

INTRODUCTION

When John Eli Miller died on his farm in Middlefield, Ohio, in 1961, he was the head of the largest family in the United States. He was survived by 5 of his 7 children, 61 grandchildren, 338 great-grandchildren, and 6 great-great grandchildren. Within his lifetime John Miller had witnessed a population explosion. Glenn D. Everett noted in the *Population Bulletin* that it was remarkable that the explosion started with a family of just 7—not all that unusual for nineteenth-century America. Two of his own children died during his lifetime. During most of his long life, therefore, John Miller's family was not unusually large. It is just that he lived long enough to find out what simple multiplication can do, wrote Everett [1]. While the number of children born to John Miller or each of his descendants was not unusually large, the death rate among infants, children, and young adults was very small compared with the history of most human populations. Of 63 grandchildren born to John Miller, 61 survived him, and of 341 great-grandchildren (born to 55 married grandchildren—a little more than 6 children per parent), only 3 had died. All 6 great-great grandchildren were healthy (Fig. 7.1).

John Miller's family illustrates a major factor in our modern population explosion: Modern technology and its medical practices and the supply of food, clothing, and shelter have decreased death rates and accelerated the net rate of growth [1].

Malthus had written 200 years before the following key phrases. He wrote, "I think I may fairly make two postulates. . . .First, that food is necessary to the existence of man. . . .Second, that the passion between the sexes is necessary and will remain nearly in its present state" [2]. The problem as Malthus saw it was that, assuming his two postulates, "the power of population is indefinitely greater than the power in the earth to produce subsistence for man. Population, when unchecked, increases in a geometric ratio. Subsistence increases only in an arithmetical ratio." Malthus' *geometric* growth is the same as *exponential* growth (Chapter 4), which is a constant percentage increase in the population.

Malthus' projections of the ultimate fate of humankind were dire—as black a picture as the most extreme environmentalists of our own time. The power of population growth is so great, he wrote, that "premature death must in some shape or other visit the human race. The vices of mankind are active and able ministers of depopulation . . . But should they fail . . . sickly seasons, epidemics, pestilence and plague, advance in terrific array, and sweep off their thousands and ten thousands." And worst of all, should these fail, "gigantic famine stalks in the rear, and with one mighty blow, levels the population with the food of the world" [2].

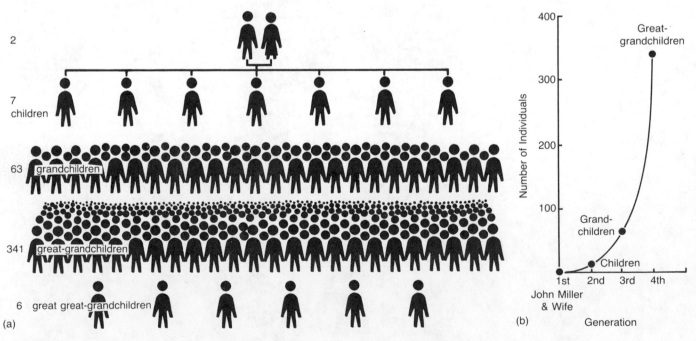

FIGURE 7.1

A simplified family tree of the John Eli Miller family (a). The population explosion of the John Eli Miller family shown in graphic form (b).

Critics of Malthus have continued to point out that his predictions have yet to come true; whenever things have looked bleak, our technological society has found a way out. They argue that in the future our technologies will continue to save us from a Malthusian fate. Who is correct? The question is undoubtedly one of the most important that faces—or has ever faced—humankind. To find the answers, we must apply the basic principles of environmental studies discussed in the first chapter of this book. Like every other biological species, *Homo sapiens* can be characterized as a group of populations, each with rates of birth, death, and growth and with a population structure, which is the relative proportion or the number of individuals of each age and sex. Furthermore, *Homo sapiens* as a single species depends on the biosphere and on ecosystems for its existence. We cannot escape these constraints; we are part of the biosphere and must live within its laws. "Nature," as Paul Sears, a twentieth-century conservationist has written, "is not to be conquered save on her own terms" [3].

Indeed, the story of John Eli Miller's family is *our* story. In the third quarter of the twentieth century the most dramatic increase in the history of the human population has occurred. In merely 25 years the human population of the world increased from 2.5 billion to over 4.4 billion [4]. Although the average rate of growth was only 2 percent per year, when applied to a very large number it leads to huge increases in a short time. It has not always been this way for our species, however. Although new fossil finds continue to push the dates for the origin of *Homo sapiens* further and further back, it is clear that for most of human history the total population was small compared with today's population and had an extremely low, long-term rate of increase.

A BRIEF HISTORY OF THE HUMAN POPULATION

How many people have lived on the Earth? Of course, before written history there were no censuses. The first esti-

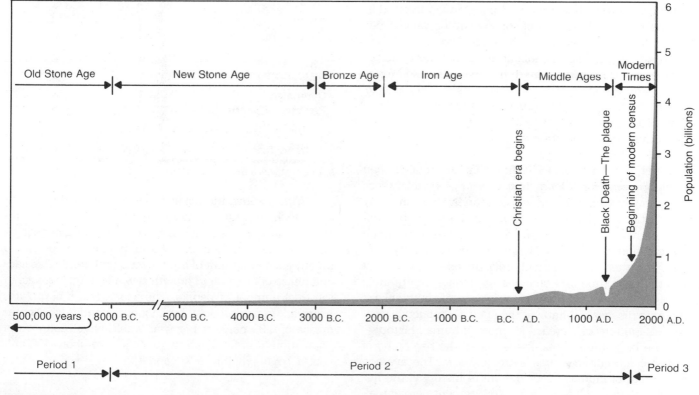

FIGURE 7.2

The history of the human population. For most of *Homo sapiens'* time on the Earth, the population was probably less than 1 million and perhaps only a quarter of a million. It has taken hundreds of thousands of years for the number of people to exceed 1 billion; now, the number will reach 6 billion if current growth rates continue. Note the left-hand side of the graph is not to scale. If the Old Stone Age were to scale, it would stretch 12 meters to the left. (From Desmond, 1962, and Van der Tak et al., 1979.)

mates of population in Western civilization were attempted in the Roman era. During the Middle Ages and the Renaissance, occasionally scholars estimated the number of people. The first modern census, however, was made in 1655 in the Canadian colonies by the French and the British [5]. Sweden began the first series of regular census taking in 1750. The first census in the United States was made in 1790 and has continued every decade since. But most countries did not begin taking censuses until much later. For example, the first Russian census was made in 1870. Even today, many countries do not take censuses or do not do so regularly. And the population of China, the most populous country in the world, remains the greatest unknown. The last census there was made in 1953 and was subject to large, and largely unestimable, errors. However, by studying modern primitive peoples and applying principles of ecology, we can gain a rough idea of the number of people that might have lived on the Earth during our species' history [5] (Fig. 7.2). Edward Deevy, a well-known twentieth-century ecologist, was one of the first to do this [6]. Early human beings were hunters and gatherers, and people who follow this way of life live at a density of somewhere between 1 per 130 to 260 square kilometers in arctic and semiarid regions and 1 per 2.6 square kilometers in the most habitable areas [7]. Deevey estimated the area of the Old World, where early Stone Age people might have been able to live, and calculated that the average early Stone Age population could have been as low as one quarter of a million —less than the population of modern small cities like Hartford, Connecticut.

Such estimates show clearly that the total human population was for a long time very low—certainly less than a few million, which is fewer than live now in many of our cities. We know nothing about the fluctuations in these early populations. Many authorities have assumed that the number of early *Homo sapiens* was fairly constant, but there were undoubtedly periods of boom and bust that accompanied changes in climate, the gradual discovery of new habitable areas, or sporadic outbreaks of epidemic diseases. Whatever the short-term fluctuations, the average rate of increase for most of human history— from the beginning of our species until the beginning of domestic agriculture—was extremely low. The average annual rate of increase over the entire history of the human population is less than 0.00011 percent per year, although the rate in recent decades approaches 2 percent [7]. Beginning with a few individuals, an average annual rate of increase of 0.0015 percent over 1 or 2 million years would result in approximately 5 million by 10,000 B.C. [8], which many experts believe is a reasonable estimate of the number alive then.

TABLE 7.1

Estimated population of the Roman Empire at the beginning of the modern era, about 14 A.D. Notice that the density is generally less than 50 individuals per square kilometer.

	Number	Individuals per Square Kilometer
Europe		
Italy	6,000,000	24
Sicily	600,000	23
Sardinia-Corsica	500,000	15
Iberia	6,000,000	10
Narbonensis	1,500,000	15
Gaul	3,400,000	6
Danube	2,000,000	5
Greece	3,000,000	11
European Total	23,000,000	10
Asia		
Asia (Province)	6,000,000	44
Asia Minor	7,000,000	17
Syria	6,000,000	55
Cyprus	500,000	52
Asian Total	19,500,000	30
Africa		
Egypt	5,000,000	180
Cyrenaica	500,000	33
Province of Africa	6,000,000	15
African Total	11,500,000	26
Grand Total	54,000,000	16

SOURCE: Modified from Borrie, 1970.

With the invention of agriculture and the domestication of plants and animals, the population *density* increased greatly. We know this because primitive peoples who practice agriculture have population densities greatly exceeding that of hunters and gatherers. This second phase of more rapid growth saw a general increase in the total number of people. The average rate of growth, however, remained low by modern standards. An increase of 0.03 percent per year would have brought the population from 5 million in 10,000 B.C. to about 100 million in 1 A.D. [9] (Tables 7.1 and 7.2).

AGE STRUCTURE AND ITS IMPORTANCE FOR PEOPLE

One of the most amazing features of the human population's great rate of increase is that it has occurred in spite of the fact that many aspects of an individual's biological

TABLE 7.2
Estimates of world population by regions from 14 A.D. to 1981 (millions).

Year	World Total	Africa	North America	Latin America	Asia[a]	Europe[a]	Oceania
14	256	23		3	184	44.5	1
350	254	30		5	185	32.6	1
600	237	37		7	168	24.3	1
800	261	43		10	173	34.2	1
1000	280	50		13	172	44.2	1
1200	384	61		23	242	57.5	1
1340	378	70		29	186	90.5	2
1500	427	85	1	40	225	73.8	2
1600	498	95	1	14	305	95	2
1650	470	100	1	7	257	103	2
1750	694	100	1	10	437	144	2
1800	919	100	6	23	595	193	2
1850	1,091	100	26	33	656	274	2
1900	1,571	141	81	63	857	423	6
1920	1,811	141	117	91	966	487	9
1930	2,070	164	134	108	1,120	534	10
1940	2,295	191	144	130	1,244	575	11
1960	3,005	278	199	213	1,660	639	16
1970	3,632	344	228	283	2,056	705	19
1981[b]	4,497	486	254	366	2,608	753	30

SOURCES: Figures for 1400 to 1600 A.D. are from C. Clark (1967, table III); for 1650 to 1910 from the United Nations Department of Economic and Social Affairs, Population Division (1953, chapt. II, table 2); for 1920, from the U.N. *Demographic Yearbook 1962* (1963, table 2); and for 1930 to 1969 from the U.N. *Demographic Yearbook 1969* (1970, table 1). From Matras, 1973, 1981; and from Haub, 1981.
[a]Estimates for Asia exclude Asiatic USSR. Estimates for Europe include Asiatic USSR.
[b]Estimated.

life history have remained unchanged. We began this chapter by pointing out that John Eli Miller's original family of seven children was not especially large in the history of human beings. Nor has the longevity of individuals changed very much. In fact, the chances of a person who is 75 years old living to 90 were considerably greater in ancient Rome than they are today (Fig. 7.3).

The third era of human population growth began in the Middle Ages with the Renaissance in Europe. The population appears to have begun to increase rapidly about 1000 A.D., and then to have decreased with the great plague of the fourteenth century. Another rapid increase occurred with the discovery of the causes of diseases, the invention of vaccines, improvements in sanitation, and other advances in medicine and health, and with the great increase in the production of food, shelter, and clothing due to modern technology [5]. By the seventeenth century the rate of increase was on the order of 0.1 percent and increased about one tenth of a percent every 50 years until 1950 [7] (Table 7.2).

From the information in this section, it is possible to estimate the total number of people who have ever lived on the Earth. This number is about 50 billion [10]. The more than 4 billion people alive today represent about 9 percent of the people who have ever lived!

How do we know what the life expectancy was in ancient Rome or in medieval Europe? Life expectancy can be estimated from studying tombstones that give the age at death, and from these ages the chance of dying can be reconstructed (Fig. 7.4a). The age at death can also be used to infer the survivorship curve of the population (Fig. 7.4b). These reconstructions suggest that death rates were much higher among young people in Rome and medieval Europe than they are now. The modern decrease in the death rate of young people greatly affects population growth because it increases the number of individuals who live long enough to have children.

Another important factor in the rate of population growth is the average age of marriage and therefore the average age that women have children. Societies in which marriage occurs late grow much more slowly than societies in which marriage occurs early. Thus death rates, birth rates, and the age of first childbearing interplay to affect the total growth rate of a human population.

Modern medicine, agriculture, and technology have decreased early death rates. The tendency in modern societies to delay marriage and to have smaller families can offset the effects of decreased early death rates, but the net result is to change the shape of the population's age structure. Figure 7.5a shows the age structure in a rapidly

FIGURE 7.3
Mortality in Rome, medieval England, and the modern United States. This graph shows the chance of dying during the next year for various ages. The chance of dying for York and Rome were reconstructed from the age of death given on tombstones. The chance of dying is lowest in young people in the United States; in contrast, for people older than 70 the chance of dying is *higher* in the United States than in Rome. In other words, someone who lived to be 75 in Rome had a better chance of surviving one year more than today in the United States. However, considerably smaller fractions of the population reached 75 in Rome. In technical terms, the chance of dying given here is called the *age-specific mortality rate.* (Modified from Hutchinson, 1978. Reprinted by permission of Yale University Press from *An introduction to population ecology,* copyright © 1978.)

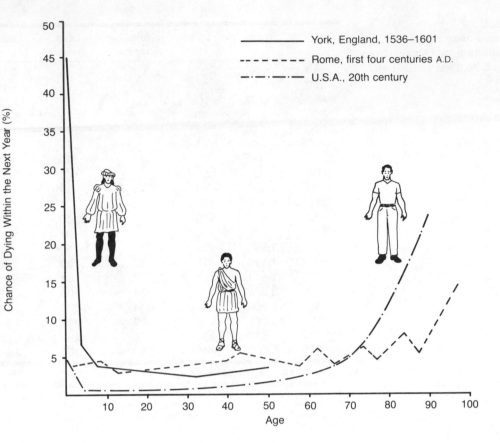

growing population heavily weighted toward youth; Figure 7.5b shows a stable population in which the average age is much older. Population (b) has the advantage that it is tending toward a no-growth condition and may avoid a Malthusian fate. However, population (a) may seem to some a pleasanter population from an individual's point of view. That is, in (a) there is a high proportion of young people, and the average family size is larger, which is desirable in many societies, particularly nontechnological, agrarian ones where the physical strength of young people is especially important. In population (b) there is a greater proportion of older people, who in many societies would no longer work and would be dependent on a much smaller percentage of active younger people for food, clothing, and shelter. Also, in a stable society like (b) there are relatively few opportunities for a young person. Many of the desirable jobs are filled, and

when the population is in a steady state the rate of replacement for these jobs is small. No modern society has achieved a true steady state, but some nations that have begun to approach it by promoting birth control have found the decreasing pool of young people to be an extremely serious problem. This appears to be at least part of the reason that some countries have stopped promoting birth control and have even begun to advocate an increase in births.

Thus one lesson we can learn from the study of the history of human population is the following: What is good for a population may not be pleasant (or good) for the individual. The opposite also holds true: What is pleasant for an individual—you or I—may be detrimental to the population. No modern society has yet figured out how to deal successfully with the problems of a steady-state population or how to maintain a satisfying style of life for individuals while maintaining a constant population. These problems are the most important facing the next few generations.

It is easy enough to show that human societies cannot increase indefinitely. One of the more popular calculations to demonstrate this states that if the human population continues to increase at its present rate, in 6500

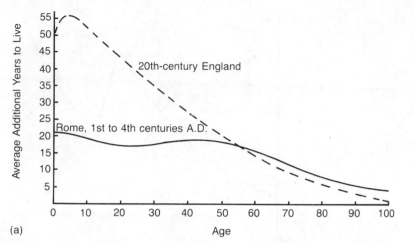

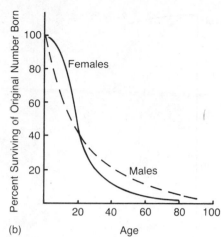

FIGURE 7.4

Life expectancy in ancient Rome and modern England (a). This graph shows the average additional number of years one can expect to live if he has reached a given age. For example, a 10 year old in England could expect to live about 55 more years; a 10 year old in Rome could expect to live about 20 more years. Among the young, life expectancy is greater in modern England than it was in ancient Rome. However, the graphs cross at about age 60. An 80-year-old Roman could expect to live longer than an 80-year-old modern Englishman. The graph for Romans is reconstructed from ages given on tombstones. Part (b) shows the approximate survivorship curve for Rome from the first through the fourth centuries A.D. The percent surviving decreases rapidly in the early years, reflecting the high mortality rates in children in ancient Rome. Females have a slightly higher survivorship until age 20, after which males have a slightly higher rate. (Modified from Hutchinson, 1978. Reprinted by permission of Yale University Press from *An introduction to population ecology*, copyright © 1978.)

years the mass of humanity will form a sphere expanding through the universe at the speed of light. This, as one author has pointed out, would rule out migration to other planets as a solution to population growth [7].

Clearly, at some time in the future the numbers of *Homo sapiens* must stop increasing. The question is not *whether* but *when* the population will cease growing and under *what conditions* people will live. We cannot escape the laws of population growth. With the best of intentions, we continue to act to increase the survival of our young without seeking a sufficient compensating factor in the rates of birth, age of first childbearing, or the mortality of older people. None of the possible solutions is pleasant for us as individuals; which are least objectionable is an ethical, moral, and aesthetic judgment beyond the realm of, but built from, science.

One possible future for humanity is a continued series of booms and busts: short, horrible periods of catastrophe followed by pleasant periods of increase. Since the time of Christ, Western society has experienced one such great disaster, the Black Death. This epidemic disease, caused by the bacteria *Bacillus pestis* and spread

by fleas that live on rodents, was not known before the seventh century in Western recorded history as a major human problem until it occurred in the Roman Empire and in North Africa. The plague probably first appeared in India in the seventh century [11] and spread rapidly northward and westward. However, it did not cause a major widespread epidemic after that until the fourteenth century. Like a true epidemic, it spread rapidly, reaching Italy in 1348 and Spain, France, Scandinavia, and central Europe within two years (Fig. 7.6). In England one third to one fourth of the population died within a single decade. Entire towns were abandoned, and the production of food for the remaining population was jeopardized [12]. Historians attribute the reduction in the work force and the concomitant increased value of labor as one of the factors involved in the growth of the power of the masses and the rise of modern democracy.

Why the Black Death did not become a major epidemic before or after the thirteenth and fourteenth centuries remains somewhat of a mystery. Some attribute the rapid spread of the epidemic to the densely crowded towns and cities with poor sanitation—a consequence,

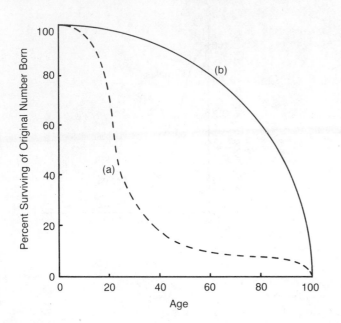

FIGURE 7.5
Two hypothetical human age structures. Curve (a) represents a population weighted toward young individuals. Curve (b) shows a comparatively stable age distribution; a greater fraction survives longer and a higher percentage of the population consists of older individuals.

that is, of the large number of people. As such, the Black Death acted as a *density-dependent* population regulating mechanism (see Chapter 4). However, climatologists point out that the climate deteriorated in the early fourteenth century. In the early part of the century France had many cold and wet years when crops failed. The year 1313, curiously enough, was one of the particularly bad years. In 1315 in Rouen and Chartres in France one observer wrote, "We saw a large number of both sexes . . . from places as much as five leagues away, barefooted, and many even, except the women, in a completely nude condition" [11]. Repeated crop failures and generally poor health in a poor climate may have augmented the spread and impact of the disease.

The Black Death still exists in many parts of the world. Occasional cases occur in North America, and the disease seems to be endemic now among the desert rodents of the southwestern United States. Modern medicine and sanitation contribute greatly to the restriction of this disease, but do not explain entirely the lack of epidemics after the great medieval episode.

An **epidemic,** or an **acute disease,** is a disease that occurs sporadically and causes very high mortality for a short while. There are other diseases, like certain cancers, that are **chronic**—that is, they affect a small but constant fraction of a population. One of the first to recog-

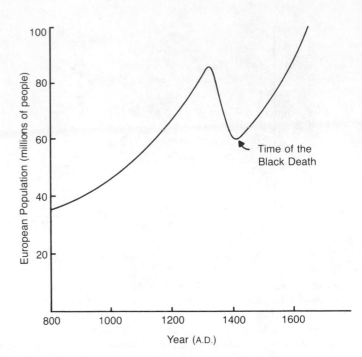

FIGURE 7.6
The effect of the Black Death on European populations. European population in general increased exponentially from the ninth through the seventeenth century, but total population dropped drastically during the fourteenth century as a result of the Black Death and remained low for 200 years. (Data from Langer, 1964, and Matras, 1973.)

nize the difference between acute epidemics and chronic disease was John Graunt, a British merchant who made the first modern English study of the causes of death. In his work titled *Natural and Political Observations Made Upon the Bill of Mortality* (1662), Graunt examined the death records for more than 20 years in several parishes of London. He observed that of the 229,250 deaths, "there died of acute diseases (the plague excepted) about 50,000 or 1/9 parts . . . about 70,000 died of chronical diseases" [13] (Fig. 7.7).

Although modern medicine has greatly reduced epidemic diseases, there has been a much smaller reduction in chronic disease. In fact, modern technology and its toxic pollutants may have increased the number and kinds of chronic disease. The cumulative effects of technology may explain the smaller life expectancy among our older people in comparison with ancient Rome.

POPULATION PROJECTIONS

It is possible to make some population projections from extrapolations from current trends as well as from some

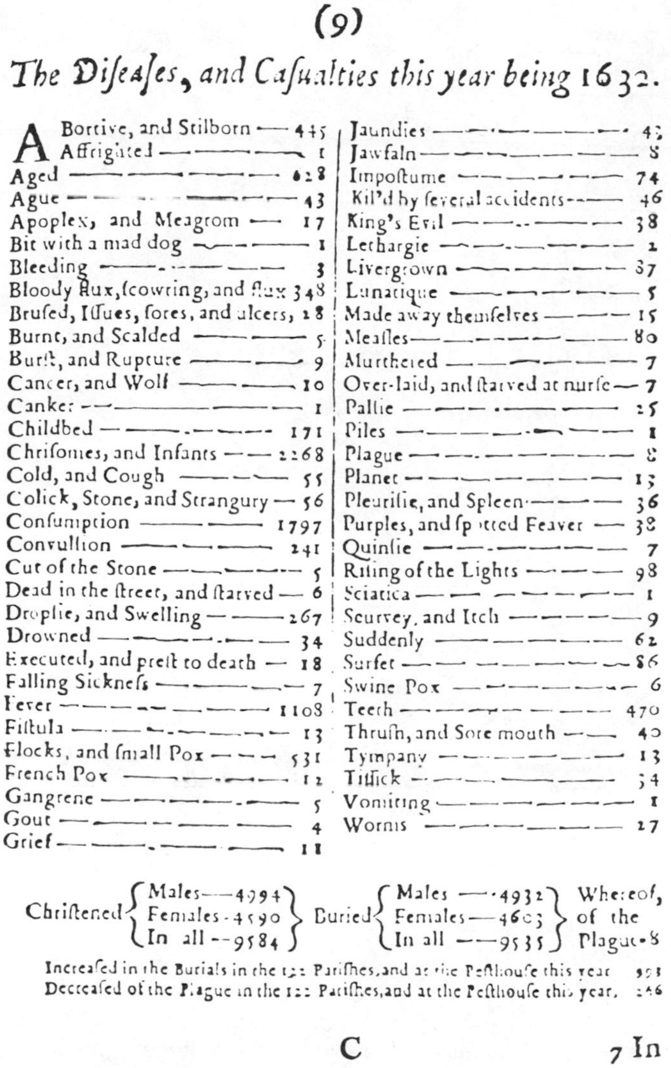

(9)

The Diseases, and Casualties this year being 1632.

Abortive, and Stilborn	445	Jaundies	43
Affrighted	1	Jawfaln	8
Aged	628	Impostume	74
Ague	43	Kil'd by several accidents	46
Apoplex, and Meagrom	17	King's Evil	38
Bit with a mad dog	1	Lethargie	2
Bleeding	3	Livergrown	87
Bloody Flux, scowring, and flux	348	Lunatique	5
Brused, Issues, sores, and ulcers	28	Made away themselves	15
Burnt, and Scalded	5	Measles	80
Burst, and Rupture	9	Murthered	7
Cancer, and Wolf	10	Over-laid, and starved at nurse	7
Canker	1	Palsie	25
Childbed	171	Piles	1
Chrisomes, and Infants	2268	Plague	8
Cold, and Cough	55	Planet	13
Colick, Stone, and Strangury	56	Pleurisie, and Spleen	36
Consumption	1797	Purples, and spotted Feaver	38
Convulsion	241	Quinsie	7
Cut of the Stone	5	Rising of the Lights	98
Dead in the street, and starved	6	Sciatica	1
Dropsie, and Swelling	267	Scurvey, and Itch	9
Drowned	34	Suddenly	62
Executed, and prest to death	18	Surfet	86
Falling Sickness	7	Swine Pox	6
Fever	1108	Teeth	470
Fistula	13	Thrush, and Sore mouth	40
Flocks, and small Pox	531	Tympany	13
French Pox	12	Tissick	34
Gangrene	5	Vomiting	1
Gout	4	Worms	27
Grief	11		

Christened { Males — 4994, Females — 4590, In all — 9584 } Buried { Males — 4932, Females — 4603, In all — 9535 } Whereof, of the Plague 8

Increased in the Burials in the 122 Parishes, and at the Pesthouse this year 993
Decreased of the Plague in the 122 Parishes, and at the Pesthouse this year 266

C 7 In

FIGURE 7.7
John Graunt's summary of the number of people dying of various causes in parishes near London in 1632. He studied similar records for a 20-year period. (From Graunt, 1662.)

insight into changes in resources, standards of living, and social conditions [14]. The Population Reference Bureau projects that the total human population will grow from 4.4 billion in 1981 to 6.1 billion in the year 2000, with an average annual increase of 1.7 percent [4]. The age structures in developed countries will move toward a greater percentage of older people (Fig. 7.8a). The less developed countries, however, will continue to have a higher percentage of young people and a higher growth rate than the more developed countries (Fig. 7.8b). Thus the greater percentage of the world population will live in the less developed countries (Fig. 7.9).

SOME GENERALIZATIONS ABOUT HUMAN POPULATIONS

Human beings cannot escape the laws of population growth. Here we should remind ourselves of the fifth fundamental concept in environmental studies (Chapter 1): *Individual populations are capable of rapid exponential growth, but this is rarely achieved in nature; control of populations is the norm.* The degree of control and the causes of control determine many features of ecosystems and the life around us. Modern medicine, sanitation, and technology have decreased death rates, particularly among the young. There has been little to compensate for this—either in the reduction of the number of offspring or an increase in death rates among the old. Although the overall rate of growth is a small percentage (approximately 2 percent per year), the number of individuals added to the world population is huge and beyond our ability to imagine.

Recent advances in medical technology suggest that soon we may also increase longevity and the survivorship of individuals with genetic or chronic ailments. Such technologies as recombinant DNA—essentially altering our own genetic makeup—may truly lead to a "fountain of youth," that is, to a marked increase in the total length of life and in the length of active, productive life. While these developments are pleasant prospects for each of us as individuals, they will only increase the social problems we are already facing as a result of current population trends. If these techniques succeed, the future growth rate of the human population may exceed an exponential curve, and the age structure may become weighted more and more to older individuals. The resources available per person will decrease and individual opportunity will likely decrease in turn. While the advances in medicine are to be marveled at and welcomed, they cannot be continued without careful consideration of how human society will deal with the new age structure and population densities. Here we should remind ourselves of the fourth concept used in environmental studies: *Sustained life on the Earth is a characteristic of ecosystems, not of individual organisms or populations.* For life—including human life—to be sustained, there must be a flow of energy and a complete cycling of chemical elements necessary for life.

The remarkable thing about human society is that it has a way of "short-circuiting" the rates of natural processes. This short-circuiting began with the development of agriculture and underwent a second major spurt with the development of cities. The Scientific and Industrial revolutions that began after the European Renaissance provided a third major change; and in our time, the cybernetic, space-age revolution is leading to a fourth period in

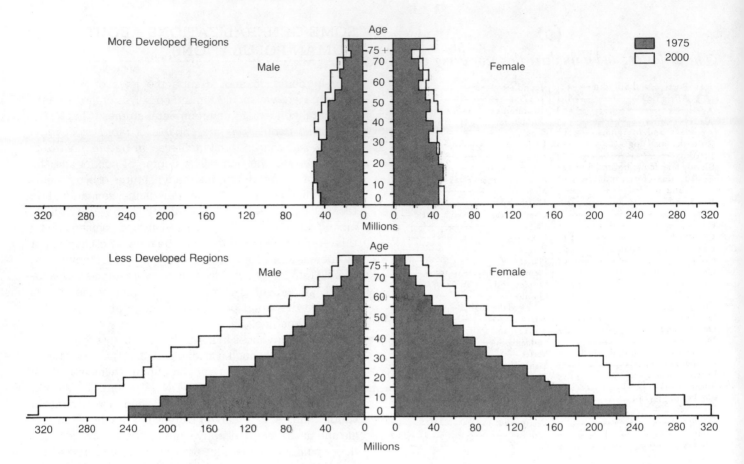

FIGURE 7.8

Age and sex composition of the Earth's human population for 1975 and projected for the year 2000. (From Council on Environmental Quality and the Department of State, 1980.)

human history. All human social entities, however, are subject to the same ecological constraints. The Industrial and Scientific revolutions have led to rapid changes in our environment, but *all technology, from the most primitive to the most advanced, causes some change in the environment.*

Human societies and civilizations have waxed and waned, and the reasons for these rises and falls have fascinated people throughout written history. Recently, some evidence has begun to accumulate that some of

these rises and falls may have been affected significantly, if not caused totally, by changes in environmental conditions. For example, periods of climatic warmings are correlated with periods of increased human exploration of the Earth; and as we saw in the example of the Black Death, some catastrophes seem correlated with periods of bad weather [12]. The climatologist Reid Bryson has found evidence that a period of increased rainfall in the fourteenth century in North America correlates with an expansion of American Indian maize farming in the mid-

FIGURE 7.9

The human population from 1000 to 2000 A.D. In 1000 A.D., 60 percent of the people ➡ lived in Asia, 17 percent in Europe and Russia, 18 percent in Africa, and 4 percent in the Americas. In 1975, 58 percent of the people lived in Asia, 18 percent in Europe, 14 percent in the Americas, Australia, and New Zealand, and 10 percent in Africa. It is projected that by 2000, 59 percent will live in Asia, 13 percent in Europe, 13 percent in Africa, and 15 percent in the Americas plus Australia and New Zealand. (Data for 1975 and 2000 from Council on Environmental Quality and the Department of State, 1980, and Haub, 1981. Remaining data from Desmond, 1962.)

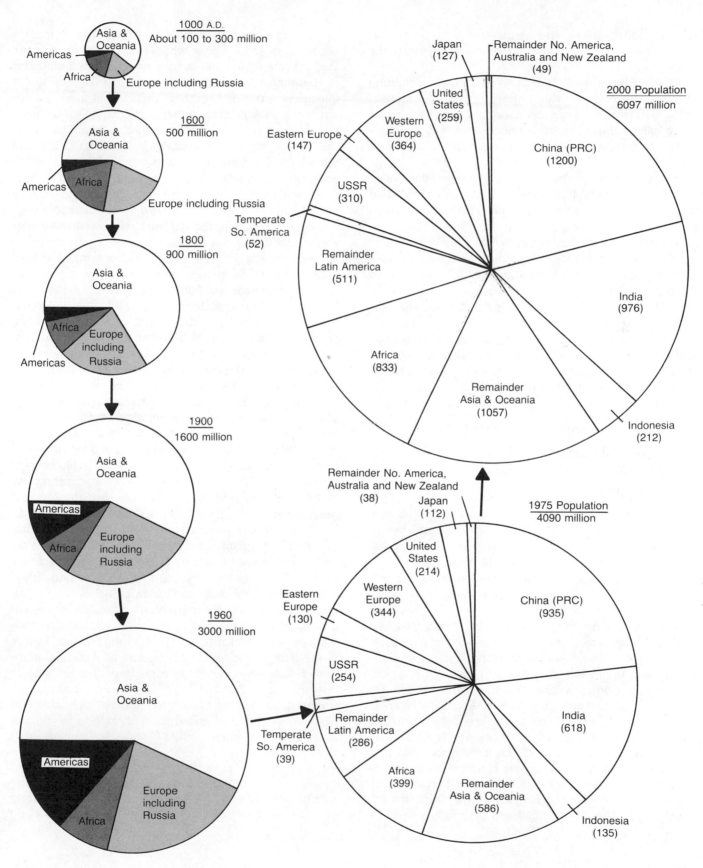

1000 A.D.
About 100 to 300 million

Asia & Oceania

Americas

Africa

Europe including Russia

1600
500 million

Asia & Oceania

Americas

Africa

Europe including Russia

1800
900 million

Asia & Oceania

Africa

Europe including Russia

Americas

1900
1600 million

Asia & Oceania

Americas

Africa

Europe including Russia

1960
3000 million

Asia & Oceania

Americas

Africa

Europe including Russia

2000 Population
6097 million

Japan (127)

Remainder No. America, Australia and New Zealand (49)

United States (259)

Western Europe (364)

Eastern Europe (147)

USSR (310)

Temperate So. America (52)

Remainder Latin America (511)

China (PRC) (1200)

India (976)

Indonesia (212)

Africa (833)

Remainder Asia & Oceania (1057)

1975 Population
4090 million

Remainder No. America, Australia and New Zealand (38)

Japan (112)

United States (214)

Western Europe (344)

Eastern Europe (130)

USSR (254)

Temperate So. America (39)

Remainder Latin America (286)

China (PRC) (935)

India (618)

Indonesia (135)

Africa (399)

Remainder Asia & Oceania (586)

western United States, and a subsequent dry period correlates with a great decrease in Indian settlements in the same area [15].

Others have speculated that one factor contributing to the demise of the Mayan civilization was the overuse of the land in farming maize and the subsequent destruction of the soil. Changes in sedimentation rates in fresh waters appear to correlate with the rise and fall of that civilization.

Indeed, none of these factors alone is sufficient to cause the rise and fall of human civilizations; people and their societies are too complex to be explained by a single factor. On the other hand, we cannot afford to ignore the importance of such influences. Because our own society has the power to affect the entire biosphere, we may need these lessons from history to learn how to avoid violating the principles of environmental studies, particularly the first: *The Earth is the only suitable habitat we know, and its resources are limited.* We must also confront the thirteenth principle: *What is necessary to sustain life is a scientific question; what effects specific actions have on the life around us and our environment are applied scientific questions; but what we choose is ultimately an ethical question beyond the reach of science.*

HUMAN BEINGS AS AN ECOLOGICAL FACTOR IN PREHISTORY

As stated before, the Industrial and Scientific revolutions have led to rapid changes in our environment, but *all technology, from the most primitive to the most advanced, causes some changes in the environment.*

In 1582, the explorer Verrazano sailed along the Carolina coast of North America and saw ''verie great fiers.'' In 1610, Henry Hudson's crew also saw ''a great fire'' somewhere in what is now New Jersey, along the coast south of Sandy Hook. Throughout the colonial period, reports of fires by explorers are common, and the fires are usually said to have been intentionally set by the American Indians to improve the habitat in various ways. For example, William Wood wrote in *New England's Prospect* in 1634 that it was ''the custome of the Indians to burne the wood in November, when the grass is withered, and leaves dryed. It consumes all the underwood, and rubbish, which would over grow the country, making it impassable, and spoil their much affected hunting'' [16]. Where fires were not lit, the woods were so thick that riding through them was difficult and getting lost easy.

It was suggested that the fires were set by Indians because ''in some places where the Indians dyed of the Plague some foureteene years agoe, is much underwood . . . because it hath not been burned'' [16, 17]. One might question the reliability of these reports, but there is modern evidence that corroborates it. In 1701 a Dutch family, the Mettlers, obtained rights to a large area of land near the present city of New Brunswick, New Jersey. Much of the land was cleared for farming, but some was left as woodlots. One small stand of this woodlot, about 30 hectares, was never burned or cut after 1701 and is the only known virgin stand along the New Jersey piedmont (Fig. 7.10). Known today as the William L. Hutcheson Memorial Forest, this stand is a nature preserve administered by Rutgers University and is one of the few remaining links with the forests Verrazano and Hudson saw.

Today Hutcheson Forest has a dense underbrush that is almost impossible to walk through and not penetrable on horseback. As late as 1749, however, this same area was described by the Swedish naturalist Peter Kalm as so open a woods that one could ride a horse and carriage through it [18]. Recent studies of the forest link fire to these changes [19]. When a fire burns through the bark of a tree, a scar is left that can be seen in the cut stump. The tree continues to grow outward and seals in the scar. Because a tree ring is laid down each year, we can determine how many years ago the fire occurred as well as the number of years between fires. A study of the oldest trees in Hutchinson Forest reveals that fire scars occur approximately every 10 years until 1701, the year the Mettlers moved onto the land, and no fire scars occur afterward.

The burning not only opened up the forests, but changed the species, favoring those trees that better withstand fires. At the time of European settlement, the eastern Piedmont (literally, the foothills) was primarily a forest of chestnuts, oaks, and hickories, believed by ecologists in the first half of the twentieth century to be the climax species. However, after 250 years without fires, the great oaks and hickories are slowly being replaced by less fire-resistant but more shade-tolerant maples. Thus primitive people changed many aspects of the ecosystems they inhabited. Fires set by human beings, a fungus introduced by human beings (which, as we learned in Chapter 6, killed the chestnut), and long-term climatic changes have all altered the forests of the eastern piedmont.

Fire was one of the first major ecological tools used by human beings to change the environment for their own benefit. Indeed, fire has been used around the world by early peoples to clear the land for improved travel and

FIGURE 7.10
William L. Hutcheson Memorial Forest in New Jersey is one of the few remaining uncut stands of oak-hickory forests in the Atlantic coastal states. This photograph shows the forest from the outside. (Photo by D. B. Botkin.)

hunting or for farming. In Africa the persistence of the great open plains and savannahs is believed to depend in large part on annual or almost annual fires. It is also believed that many of these fires were set by people for perhaps tens of thousands of years.

Thus the species *Homo sapiens* has been a major ecological factor in vast areas of the Earth's surface for a very long time, although the history of the human population might suggest otherwise. As we reviewed in the first section of this chapter, the total number of human beings was very small for a very long time, which would seem to imply that there were too few people to have a great effect. But, as the story of fire in New England and the Atlantic coastal states illustrates, there is evidence that our ancestors greatly affected the environment throughout history. Prehistoric people changed the environment by burning; by decreasing the abundance of some animals through hunting; by increasing the abundance of others when they altered habitats and made them more favorable to those species; by domesticating plant and animals; by changing erosion rates by agricultural and other land-clearing practices and thus altering soils as well as vegetation; and by transporting organisms into new areas from which they had been isolated by geographical boundaries.

All of these effects could have been very great. Paul Martin, an American anthropologist, has proposed that the great Pleistocene extinctions, which were discovered from the fossil record, could have been caused by the arrival of the ancestors of the American Indians when they first arrived in North America via a land bridge from Asia. Martin suggested that the Indians could have acted like a newly introduced predator confronting a prey with no prior experience or adaptation to that predator. He argued that a small number of people, concentrated along a thin but densely populated line and migrating south and eastward from Alaska down through North America, could have extinguished entire species with available techniques, including fire [20]. Others believe that climatic changes must have played an important or causal role. Most likely a number of factors operated together to cause the extinctions. The important point for our study is that aboriginal *Homo sapiens* could have been one of these major factors (Fig. 7.11).

A similar, better documented series of extinctions occurred on New Zealand, whose islands were separated from continents prior to the evolution of mammals and were too distant to allow their introduction. Birds, however, did migrate there and evolved to fill many niches, including those of large herbivores usually filled by mammals. When the Polynesian Maoris arrived there, they quickly extinguished the largest of these flightless birds.

In Scotland many species of animals, including the lynx (*Lynx lynx*), brown bear (*Ursus arctos*), wolf (*Canis lupes*), wildcat (*Felis silvestris*), beaver (*Castor fiber*), and reindeer (*Rangifer tarandus*), lived in the Neolithic Age,

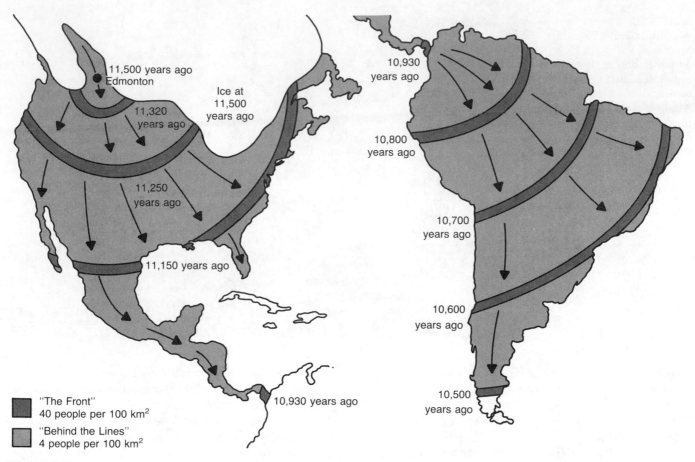

FIGURE 7.11

Paul Martin's map showing a hypothetical narrow band of migrating people extending down through North America. These people were able to act as predators on the North American fauna. (From Martin, The discovery of America, *Science* 179:969–74. Copyright 1973 by the American Association for the Advancement of Science.)

but then disappeared. Hunting and replacement by domestic animals are believed to have been responsible for their disappearance [21].

This great impact of early people on the landscape further alters our ideas about wilderness. Indeed, by the time the wilderness concept arose as a part of conservation or even became a part of "nature to be admired," as in the early nineteenth century, there were few areas of the Earth that had not been strongly changed by people for a long time. True wilderness—nature unaffected by human beings—is a rarity.

In the conterminous United States, the fraction of land that meets the criterion of true wilderness (unaffected now or in the recent past by human beings) is very small. Parts of the Rocky Mountains, some of the more inhospitable parts of our national parks, and parts of the deserts would seem to be true wilderness. East of the

Rockies there are even fewer wilderness areas. Isle Royale may indeed be the only true wilderness in the North Woods of the United States. In New Hampshire, a U.S. Forest Service study revealed only two small forested areas in the entire state that had not been logged, cleared, farmed, or settled.

Antarctica and the deep sea are true wilderness, as are parts of the higher Himalayas. But most of Africa, India (outside of the highest mountains), Europe (except for the snow-covered peaks of the Alps), and North and South America (except for the Amazon basin and parts of Tierra del Fuego) have all been modified by human beings. In Central and South America the practice of shifting, or *milpa* agriculture, has caused changes in forest composition over large areas [21].

In Europe, Neolithic people cleared forests in Denmark and other areas of northern Europe. Pollen deposits in lakes in Denmark show that tree species were removed between 2500 and 2300 B.C. and replaced by pollen from herbaceous plants, which were then replaced by species like birch that regenerate after fire. Replicas of Neolithic axes with flint blades and ash handles were used by two archeologists, Jorgen Traelssmith and Sven Jorgensen, to show that forests could indeed be cleared effectively with them [22]. The evidence suggests that cutting and burning were used together to first clear and then maintain open land.

The following is an ecological picture of *Homo sapiens* in Europe since the end of the last ice age. As the ice retreated northward, tundralike, open areas appeared where Neolithic people hunted; as the climate warmed further, forests developed, and for a while people moved away or became less abundant. Then around 4000 to 5000 years ago, Neolithic people began to clear the land and move into large areas of former forests, where they grazed cattle and grew crops. Thus the impact of *Homo sapiens* on northern Europe was widespread long before the beginning of the modern era [23].

HUMAN BEINGS AS AN ECOLOGICAL FACTOR FROM THE BEGINNING OF CIVILIZATION TO THE MODERN ERA

The story of the rise of civilization and its many direct influences on the landscape is well known from history books and need not be repeated here. Recent studies suggest that these influences may have been greater than was formerly believed, strongly affecting population growth and the development—the rises and falls—of human cultures and civilizations.

For example, the medieval period in Europe was formerly pictured as comparatively static, with little expansion in the development of the land. The environment, however, underwent great change through human activities. In 900 A.D. a large part of what is now modern Germany was forested; settlement of these forests began about that time, and the land was steadily cleared. By 1900 only a small fraction of the original forested area remained (Fig. 7.12).

Technological advances during the medieval period, including the invention of the modern horse collar, accelerated land clearing. Prior to this, horses were collared with modified oxbows which, because horses lacked oxlike shoulders, choked them and prevented them from pulling with their entire strength. The new bridle allowed plowing of rougher land and increased the rate of land clearing [24]. Much land that was abandoned during the

Black Death was repopulated. By the time of the great era of exploration, much of England's forests had been cleared. Some historians suggest that the need for timber, particularly for the British navy, was an important factor in the British colonization of North America [24].

Thus, through history, our species has greatly modified the Earth's surface. In succeeding chapters we will examine current environmental issues so that we learn to manage our environment wisely.

SUMMARY AND CONCLUSIONS

Throughout most of our history, the human population and its average growth rate have been small. It has been estimated that approximately 50 billion people have lived on the Earth, and the present population of more than 4 billion represents about 9 percent of those who have ever lived on our planet. The growth of the human population can be divided into three major phases. In the early, hunter-gatherer phase, the average growth rate was very small. The second period, beginning about 10,000 years ago, experienced more rapid growth. The third era, which began in the Middle Ages, continues to the present day. Except for the great plagues of the fourteenth century, the population has grown rapidly and its rate of increase has become greater. This third phase demonstrates that, like other biological populations, *Homo sapiens* is capable of a geometric increase.

This rapid increase has occurred without much of an increase in longevity, but through a great increase in early survival. In fact, once one reached 75 or older in ancient Rome, one's chance of living another year was greater than in the modern United States.

Projections of the growth of the human population make use of age structures. A population's age structure has important environmental and social effects. A human population with zero growth will have a greater proportion of its population in older people; this in itself is an adjustment our societies will have to make in the future.

Although today we have a great impact on our environment, people have always affected their environment to some degree. Indeed, for the last 10,000 years human impact has been quite large. North American Indians and other peoples used fire in a way that markedly changed the landscape, and forests have been cleared for many centuries.

Our population is increasing so rapidly that we must confront the problems caused by decreasing the growth rate, deal with changes in age structure, and live within our resources. An understanding of human population processes is essential for wise solutions to these problems.

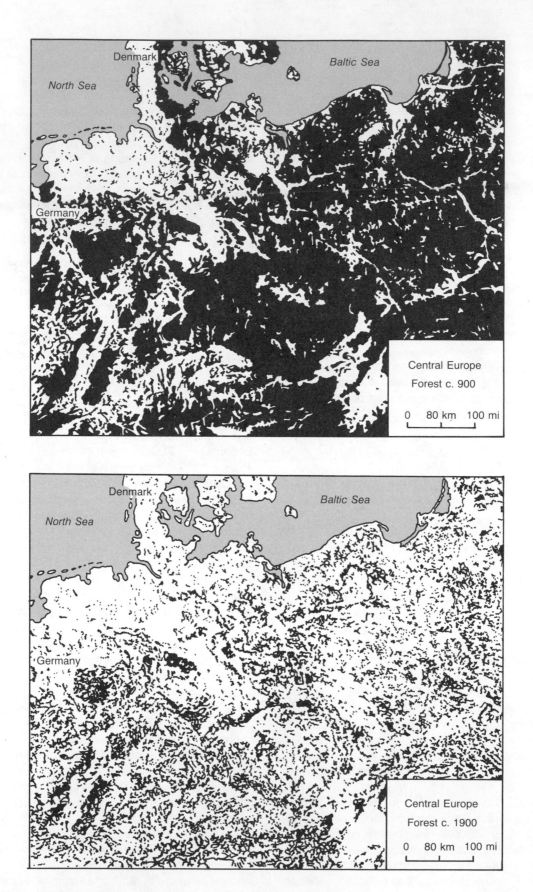

FIGURE 7.12
About 900 A.D. most of central Europe was in forest (a). By 1900, due to the
activities of human beings, all but a small fraction of this area has been cleared (b).
(From Thomas, 1974. After Schlüter, 1952.)

REFERENCES

1 EVERETT, G. D. 1961. One man's family. *Population Bulletin* 17: 153–69.

2 MALTHUS, T. R. 1817. *An essay on the principle of population.* London: Murray.

3 SEARS, P. 1935. *Deserts on the march.* Norman: University of Oklahoma Press.

4 HAUB, C. 1981. *1981 world population data sheet.* Washington, D.C.: Population Reference Bureau.

5 DESMOND, A. 1962. How many people have ever lived on Earth? *Population Bulletin* 18: 1–19.

6 DEEVY, E. S., Jr. 1960. The human population. *Scientific American.* 203: 194–204.

7 THOMLINSON, R. 1965. *Population dynamics: Causes and consequences of world demographic change.* New York: Random House.

8 DUMOND, D. E. 1975. The limitations of human population: A natural history. *Science* 187: 713–21.

9 BORRIE, W. D. 1970. *The growth and control of the world's population.* London: Weidenfield and Nicolson.

10 WESTING, A. H. 1981. A note on how many people that have ever lived. *Bioscience* 31: 523–24.

11 MATRAS, J. 1973. *Populations and societies.* Englewood Cliffs, N.J.: Prentice-Hall.

12 LE ROY LADURIE, E. 1971. *Times of feast, times of famine: A history of climate since the year 1000.* Garden City, N.Y.: Doubleday.

13 GRAUNT, J. 1662. *Natural and political observations made upon the bill of mortality.* London: Roycraft. (Reprinted in 1973 by Gregg International Publishers, Ltd., Germany.)

14 COUNCIL ON ENVIRONMENTAL QUALITY and THE DEPARTMENT OF STATE. 1980. *The global 2000 report to the President: Entering the twenty-first century.* Vols. I and II.

15 BRYSON, R. A. 1980. Ancient climes on the Great Plains. *Natural History* 89: 65–73.

16 WOOD, W. 1634. *New England's Prospect.* Prince Society (eds.), Boston, 1865.

17 DAY, G. M. 1953. The Indian as an ecological factor in the northeastern forest. *Ecology* 34: 329–46.

18 KALM, P. 1750. *Travels in North America.* Reprinted in 1963 by Dover Books, New York, from the English version, A. B. Benson (translator), 1770 (2 volumes).

19 BARD, G. E. 1954. Mettler's woods. *The Garden Journal,* July-August.

20 MARTIN, P. S. 1973. The discovery of North America. *Science* 179: 969–74.

21 HEIZER, R. F. 1955. Primitive man as an ecological factor. *Anthropological Society Papers* 13: 1–31.

22 IVERSON, J. 1956. Forest clearing in the Stone Age. *Scientific American* 194: 36–41.

23 COLLINS, D. 1969. Cultural traditions and environment of early man. *Current Anthropology* 10: 267–316.

24 CARROLL, C. F. 1970. The forest civilization of New England. Ph.D dissertation, Brown University.

FURTHER READING

1 BORRIE, W. D. 1970. *The growth and control of the world population.* London: Weidenfield and Nicolson.

2 CHAPMAN, W. B., Jr. 1981. *Human ecosystems.* New York: Macmillan.

3 KEYFITZ, N. 1968. *Introduction to the mathematics of population.* Reading, Mass.: Addison-Wesley.

4 MALTHUS, T. R. 1817. *An essay on the principle of population.* London: Murray.

5 MATRAS, J. 1973. *Populations and societies.* Englewood Cliffs, N.J.: Prentice-Hall.

6 MAY, J. M. 1958. *The ecology of human disease.* New York: M. D. Publications.

7 THOMAS, W. L., Jr. 1974. *Man's role in changing the face of the Earth.* 9th ed. (2 vols.) Chicago: The University of Chicago Press.

8 UNITED NATIONS. 1980. *Population bulletin of the United Nations.* No. 12. ST/ESA/SER. New York: United Nations.

STUDY QUESTIONS

1 What are the principal reasons that the human population has grown so rapidly in the twentieth century?

2 What is meant by the statement "Technology continues to prove Malthus wrong"?

3 Why is the density of people who live by hunting and gathering lower than the density of people who practice agriculture?

4 Why is it important to consider the age structure of a human population?

5 Three characteristics of a population are the birth rate, growth rate, and death rate. How have each of the following been affected by (a) modern medicine; (b) modern agriculture; (c) modern industry?

6 What is meant by the statement "What is good for an individual is not always good for a population"?

7 Strictly from a biological point of view, why is it difficult for a human population to achieve a constant size?

8 What environmental factors are likely to increase the chance of epidemic disease outbreaks?

9 Why is it so difficult to predict the growth of the Earth's human population?

10 Before the beginning of the Scientific and Industrial revolutions, what factors tended to decrease the size of the human population?

11 What is meant by the statement "Human beings have always been an important ecological factor"?

12 What are the major ways that early human beings changed their environment?

13 How might changes in climate during the Ice Ages have offered new opportunities to early human beings?

14 Would you rather start out to seek your fortune with a bachelor's degree in a rapidly expanding population or in a steady-state population?

In Part One of this book, we came to see the Earth's surface as the biosphere—a large heterogeneous system greatly modified by, and sustaining, life. The biosphere is a large system within which materials and energy flow are temporarily stored. The solid surface (the geological sediments, structures, and formations) provides localized, aggregated deposits of minerals and energy—our non-renewable resources. These geological features also provide local regions which vary in their ability to serve as sinks for our wastes.

The Earth's waters and atmosphere are its fluid mediums through which energy and matter flow rapidly. They, too, are resources. They constantly renew and refresh us with oxygen, rapidly transport our wastes, and perform many kinds of "public service" functions.

Just as dirt is soil out of place, so all wastes can be

Part II

Resources For Human Society

viewed as resources that have been used, moved, or transformed so that they are in the wrong place and either are now no longer useful or are toxic and dangerous.

Thus we see that some resources are local or regional (such as deposits of iron ore, or the timber in the tropical rain forests), while others are global (the air's oxygen). Resources are also divided into those that are renewable and those that are not. Renewable resources are those that can be regenerated; air, water, forests, fisheries, and wildlife are renewable resources. Nonrenewable resources cannot be regenerated in any ordinary sense. Once a nonrenewable resource has been found, refined, and dispersed, it is gone as a local rich deposit forever.

To understand how we can best use our resources and manage our wastes, we must see the Earth's biosphere in its largest sense—as a vast system that life has altered for more than 3.5 billion years. From this point of view, resources are things that living organisms need or human beings desire; wastes are things that have been used and are no longer useful and may be toxic or dangerous if left close to home or not disposed of properly.

Before the rise of modern civilizations, waste disposal was a local problem. In Stone Age culture, people threw used, unwanted materials into their backyards. Such backyard garbage dumps, called *middens,* are a primary source of information for archeologists who search for information to reconstruct the history of primitive peoples.

Even during the early phases of the Industrial Revolution, wastes were generally dumped into rivers, lakes, oceans, or the atmosphere. As long as human production of wastes was small compared with the capacity of the air and waters, this method worked, and the atmosphere, rivers, lakes, and oceans provided a "public service" by moving what we didn't want to where we didn't care what it did.

Similarly, our problems with resources are problems of scale—the relative scale of people's needs and desires compared with the abundance and rate of production of resources. When the human population is small, technology limited, and resources abundant, there is no perceived resource problem, as occurred in the United States in the eighteenth and nineteenth centuries. When human population is high, needs and desires great, and technology highly developed, then resources become limiting.

In Part Two we will explore the biosphere's local and global resources. We will begin with a global view and examine the two fluid mediums, air and water, that are our constant source of renewal and constant sink for many of our wastes. In Chapters 8 and 9 we will explore the problems generated by waste disposal and the management of our air and water resources. In Chapters 10 and 11 we will examine our nonbiological and nonrenewable mineral and energy resources. In Chapter 12 we will examine renewable resources. Throughout this entire section, we should remind ourselves of the first fundamental principle of environmental studies: *The Earth is the only suitable habitat we have, and its resources are limited.* We will also become more aware of the last principle: *All technology, from the most primitive farming to the most recent inventions, cause some change in the environment.*

8 The Atmosphere

INTRODUCTION

In London, during the first week of December, 1952, the air became stagnant and the cloud cover reflected much of the incoming solar radiation. The humidity climbed to 80 percent, and the temperature dropped rapidly until the noontime temperature was about −1°C. A very thick fog developed, and the cold and dampness increased the demand for home heating. Because the primary fuel used in homes was coal, emissions of ash, sulphur oxides, and soot increased rapidly. The stagnant air became filled with pollutants, not only from home heating fuels but from automobile exhaust. At the height of the crisis, visibility was greatly reduced and automobiles had to use their headlights during midday. Between December 4 and 10, an estimated 4000 people died from the pollution. Figure 8.1 shows the increase in sulphur dioxide and smoke and the accompanying deaths during this period. The siege of smog finally ended when the weather changed and the air pollution was dispersed; the environment, not human activities, finally solved this problem. Since the beginning of the Industrial Revolution and before, people had survived in London and other major cities in spite of the weather and of pollution. What had finally gone wrong?

Our atmosphere is able to absorb great amounts of pollution. Three things can happen to pollutants: The pollutants can be transported away by winds, washed out by precipitation, or chemically transformed within the atmosphere to harmless compounds.

During the 1952 London smog crisis, the average residence time for sulphur dioxide was only about five hours, and sulphur dioxide was being rapidly transformed by the atmosphere. The stagnant weather conditions combined with the number of homes burning coal and cars burning gasoline, however, exceeded the atmosphere's ability to remove or transform the pollutants; even the usually rapid natural mechanisms for removing sulphur dioxide were saturated. As a result, sulphur dioxide remained in the air and the fog became acid, adversely affecting people and other organisms, particularly green vegetation.

The 1952 London smog crisis was a landmark event. Finally, human activities had exceeded the natural abilities of the atmosphere to serve as a sink for the removal of wastes. The crisis was due in part to a positive, or reinforcing, feedback situation. Burning fossil fuels added particulates to the air, which increased the formation of fog and decreased visibility and light transmission; the dense, smoggy layer increased the dampness and cold and accelerated the use of home heating fuels. The worse the weather and the pollution, the more people acted so as to further worsen the weather and the pollution (see Fig. 8.1).

Prior to December, 1952, London was well known for its fogs; what was relatively little known was the role of coal burning in the accentuation of fog conditions. Since 1952, London fogs have been greatly reduced because coal has been replaced to a large extent by a much cleaner gas as the primary home heating fuel. A foggy day in London is now not so common and no longer seems quite so romantic.

SOME HISTORY OF AIR POLLUTION

People have long recognized the existence of atmospheric pollutants, both natural and human-induced. Perhaps one of the first laws attempting to control air pollution was enacted in 1273, when the King of England convinced Parliament to pass an act that prohibited the burning of soft coal in London. Enforcement evidently was strict, and one man was reportedly executed for burning forbidden coal. The word *smog* was probably introduced by a physician at a public health conference in 1905 to denote poor air quality resulting from a mixture of smoke and fog. Leonardo da Vinci wrote that in 1550 a blue haze formed from materials emitted into the atmosphere from plants; he had observed a natural photochemical smog whose cause is still not completely understood today.

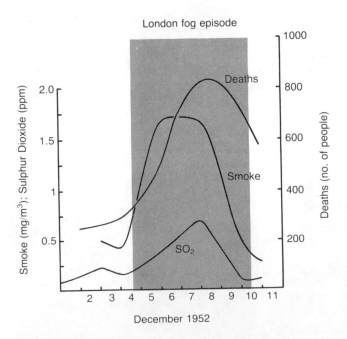

FIGURE 8.1
Graph showing the relationship between the number of deaths and the London fog of 1952. (Modified from Williamson, 1973.)

Acid rain was first described in the seventeenth century, and by the eighteenth century it was known that smog and acid rain damaged plants in London. Beginning with the Industrial Revolution in the eighteenth century, air pollution became more noticeable; and by the middle of the nineteenth century, particularly following the American Civil War, concern with air pollution increased.

Two major pollution events, one in the Meuse Valley in Belgium in 1930 and the other in Donora, Pennsylvania, in 1948, were responsible for raising the level of scientific research about air pollution. The Meuse Valley event lasted approximately one week and caused 60 deaths and numerous illnesses. The Donora, Pennsylvania, event caused 20 deaths and 14,000 illnesses. By the time of the Donora event, people recognized that meteorological conditions were an integral part of the production of dangerous smog events. This view was reinforced by the 1952 London smog crisis, after which regulations to control air quality began to be formulated. Today in the United States legislation resulting from the Clean Air Act of 1977 has reduced emissions, but much more needs to be done to reduce air pollution to satisfactory levels in many areas.

CHARACTERISTICS OF AIR POLLUTION

The atmosphere is one of our great resources. The movement of air across the Earth's surface, as local winds or the vast movement of weather fronts, continually renews the air around us. The atmosphere is a complex chemical factory, with many little understood reactions taking place within it. Many of these reactions are strongly influenced by sunlight and by compounds produced by life.

Chemical pollutants can be thought of as compounds that are in the wrong place at the wrong time in the wrong concentrations. As long as a chemical is transported away or degraded rapidly compared with its rate of production, there is no pollution problem. Pollutants that enter the atmosphere through natural or human-induced emissions may be degraded not only within the atmosphere but also by natural processes in the hydrologic and geochemical cycles; on the other hand, pollutants that leave the atmosphere may become pollutants in the water or geological cycles.

Ever since life began on the Earth, the atmosphere has been an important resource for chemical elements and a medium for the deposition of wastes. The earliest plants that carried out photosynthesis dumped oxygen—the element that was their waste—into the atmosphere. The long-term increase in atmospheric oxygen, in turn, made possible the development and survival of higher life forms that required high rates of metabolism and rapid use of energy. For our biological ancestors and for ourselves, oxygen became a necessary resource for respiration, the process by which we burn our internal biological fuels and provide the energy to sustain our life processes.

The air we breathe is a mixture of nitrogen (78%), oxygen (21%), argon (0.9%), carbon dioxide (0.03%), and other trace elements and compounds, including methane, ozone, hydrogen sulphide, carbon monoxide, oxides of nitrogen and sulphur, hydrocarbons, and various particulates. Except for molecular argon and other noble or inert gases, all the compounds in the Earth's atmosphere are either primarily produced by biological activity, primarily removed by biological activity, or greatly affected by the biota. Although the atmosphere has been greatly modulated by life during the last 3.5 billion years, we consider most of these alterations to be natural; that is, they have produced an atmosphere whose makeup is relatively constant and essential to our own survival.

As the fastest moving fluid medium in the environment, the atmosphere has always been one of *Homo sapiens'* most convenient places to dispose of unwanted materials. Ever since fire was first used by *Homo sapiens,* the atmosphere has all too often been a sink for waste disposal.

There are two main groups of air pollutants: primary and secondary. **Primary pollutants** are those that are emitted directly into the air, including particulates, sulphur oxides, carbon monoxide, nitrogen oxides, and hydrocarbons. **Secondary pollutants** are those produced through reactions among primary pollutants and normal atmospheric compounds. As an example, over urban areas ozone forms through photochemical reactions among primary pollutants and the natural atmospheric gases. Thus ozone becomes a serious pollution problem on bright, sunny days in areas where there is much primary pollution. This has been particularly well documented for southern California cities like Los Angeles, but occurs worldwide under appropriate conditions.

The primary pollutants that account for more than 90 percent of air pollution problems in the United States are particulates, hydrocarbons, carbon monoxide, nitrogen oxides, and sulphur oxides (Table 8.1). Each year approximately 250 to 300 million metric tons of these materials enter the atmosphere above the United States from human-related processes. About half of this is carbon monoxide, and the other four each account for 8 to 15 percent. At first glance, several hundred million tons of pollutants appears to be a very large amount. However, if this amount of material were uniformly distributed in the atmosphere, it would only amount to about 3 parts per million by weight.

TABLE 8.1
Major air pollutants: sources, effects, and abatement strategies.

Air Pollutant	Emissions (% of total)		Major Sources of Anthropogenic Component	Health Effects (Humans)	Abatement Strategy
	Natural	Anthropogenic			
Particulates	89%	11%	Industrial processes 51% Combustion of fuels (stationary sources) 26%	Toxic effects through several mechanisms, including interference with respiratory tract, adsorbed or absorbed toxic gas on or in particles, and intrinsic toxic particles	Capture before they enter the atmosphere by collectors, settling chambers, or precipitates
Sulphur oxides (SO$_x$)	55%	45%	Combustion of fuels: (stationary sources, mostly coal) 78% Industrial processes 18%	Irritation of respiratory system, diminished lung function, aggravates asthmatic people and older people or people with chronic respiratory problems	Minimize sulphur content of coal; remove from exhaust gases before release into the atmosphere
Carbon monoxide (CO)	91%	9%	Transportation (mostly automobiles) 75% Agricultural burning 9%	Toxic, easily enters blood with increasing concentrations; impairment of psychomotor functions; headache, fatigue, drowsiness, coma, death	Control emissions from automobiles
Nitrogen dioxide (NO$_2$)	Nearly all		Transportation (mostly automobiles) 52% Combustion of fuels (stationary sources, mostly natural gas and coal) 44%	Toxic, respiratory tract problems with increasing concentration; nasal irritation; breathing discomfort; acute respiratory problems; accumulation of fluids; death (based on animal studies)	Remove from exhaust of automobiles; modify vehicles to reduce the amount of nitrogen dioxide generated

TABLE 8.1 (continued)
Major air pollutants: sources, effects, and abatement strategies.

Air Pollutant	Emissions (% of total)		Major Sources of Anthropogenic Component	Health Effects (Humans)	Abatement Strategy
	Natural	Anthropogenic			
Ozone (O_3)	Is a secondary pollutant derived from reactions with sunlight, NO_2, and oxygen (O_2)		Concentration that is present depends on reaction in lower atmosphere involving hydrocarbons and thus automobile exhaust	Toxic, with increasing concentration; nose and throat irritation; fatigue; lack of coordination	Control the primary pollutants—NO_2 and hydrocarbons
Hydrocarbons (HC)	84%	16%	Transportation (mostly automobiles) 56% Industrial processes 16% Evaporation of organic solvents 9% Agricultural burning 8%	Aromatic varieties irritate mucous membranes; may cause injury when inhaled at concentrations greatly exceeding standards	Control emissions at stationary sources (power plants); control emissions from automobiles

Unfortunately, pollutants are not uniformly distributed. They are produced locally or regionally. It is useful to distinguish two major kinds of pollution sources: **point** and **nonpoint sources.** An example of a point source is a smokestack from a large power plant, and an example of a nonpoint source is the exhaust from all of the automobiles during the rush hour in Los Angeles, California.

POINT SOURCES OF AIR POLLUTION AND SMOG

A famous example of a point source of air pollution is the smelters for the refining of nickel and copper ores at Sudbury, Ontario. Sudbury contains one of the world's major nickel and copper ore deposits. Within a few kilometers of each other, there are 3 smelters, 13 mines, 1 open-pit mine, 6 concentrators, 2 iron-ore recovery plants, and 1 copper refinery. The ores contain a high percentage of sulphur, and the smelter stacks emit large amounts of sulphur dioxide as well as particulates containing nickel, copper, and other metals toxic to living things (Fig. 8.2). In 1970 one of the smelters, the Cooper Cliff smelter, discharged approximately 2.7 million metric tons of sulphur dioxide into the atmosphere. In addition,

nickel has been found contaminating soils 50 kilometers from the stacks, and the combination of the acid rain from the sulphur dioxide and the particulates containing heavy metals has devastated forests that once surrounded Sudbury over the last 50 years. An area of approximately 250 square kilometers is nearly devoid of vegetation (Fig. 8.3), and damage to forests in the region is visible over an area of approximately 3500 square kilometers.

Attempts to minimize the pollution problem close to the smelting operation by increasing the smokestack height have spread the problem even more widely. Acid rain, once local, is now widespread. For example, in 1977 rainfall several kilometers east of the Sudbury stacks was often highly acid and thus potentially harmful to people and other living things, especially plants. Secondary effects, in addition to loss of vegetation, include soil erosion and drastic changes in the soil chemistry due to the influx of the heavy metals.

ACID RAIN

Combustion of large quantities of fossil fuels as well as the smelting of ores have created more than local problems. The effluents from these point sources have created **acid**

FIGURE 8.2
The smelter stacks at Sudbury, Ontario, are among the tallest in the world. (Photo by D. B. Botkin.)

FIGURE 8.3
The vegetation close to the Sudbury stacks is completely killed, and further away, only short-lived, small plants adapted to the disturbance survive. (Photo by D. B. Botkin.)

rain that spreads regionally. Fossil fuels and many ores contain sulphur, which is converted to sulphur dioxide during the combustion of fuels and smelting of ores. In any high-temperature process involving ordinary air, oxides of nitrogen are also produced. Sulphur oxide and nitrogen oxides combine with water to form sulphuric and nitric acids. In 1977 alone 27.4 million metric tons of sulphur oxide and 23 million metric tons of nitrogen oxide were released into the atmosphere in the United States (Fig. 8.4).

Although the oxides of sulphur and nitrogen are the primary contributors, other acids are also involved in the acid rain problem. For example, hydrochloric acid is emitted from coal-fired power plants, and sulphuric oxide is primarily emitted from stationary sources such as power plants burning coal or oil. Nitrogen oxide, however, is emitted from both stationary and transportation-related sources such as automobiles. In 1977, approximately 56 percent of the nitrogen oxide discharged into the atmosphere resulted from the combustion of fossil fuels at power plants and other stationary sources, whereas 40 percent was released by transportation-related sources. Because it is expected that the burning of fossil fuels will increase dramatically in the next several decades (especially coal), the acid rain problem will probably increase during this period [1].

Acid rain moves downwind from the source and spreads widely. Sulphur oxides produced in Great Britain, Germany, and other central European countries are producing severe acid rain problems in Norway and Sweden. In North America, industrial point sources in the northeastern United States and adjacent Canada have spread acid rain from that area north to Labrador and the Arctic Ocean.

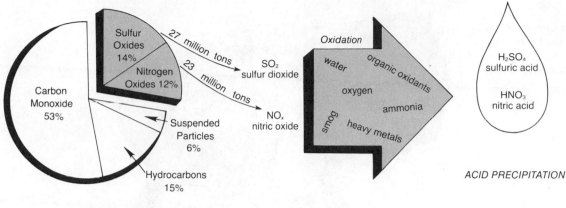

1977 EMISSIONS

FIGURE 8.4

Emission of sulphur and nitrogen oxides and the formation of acid rain. (Modified after U.S. Environmental Protection Agency, 1980b.)

In recent years, attempts to reduce local air pollution near point sources have led to construction of taller and taller stacks. These reduce the local concentrations of air pollutants, but increase the regional effects by spreading the pollution more widely. In this case, dumping waste into someone else's backyard has created more rather than fewer problems.

All rainfall is slightly acidic (Fig. 8.5) because water reacts with atmospheric carbon dioxide to produce a carbonic acid. Thus pure rain has a pH (a numerical value used to describe the strength of an acid) of about 5.6. Acid rain is defined as precipitation in which the pH is below 5.6. Because the pH scale is logarithmic, a pH value of 3 is 10 times more acidic than a pH value of 4 and 100 times more acidic than a pH value of 5. Automobile battery acid has a pH value of 1. It is rather alarming to

learn that in Wheeling, West Virginia, rainfall has been measured with a pH value of 1.5, nearly as acidic as stomach acid and far more acidic than lemon juice or vinegar.

Perhaps more important than isolated cases of very acid rain is the recent growth of the problem. It was not too many years ago that acid rain was believed to be primarily a European problem. However, in recent years acid rainfall has spread from a relatively small area in the northeastern United States to nearly all of eastern North America. Furthermore, the problem is not restricted to the East Coast of the United States; urban centers on the West Coast, such as Seattle, San Francisco, and Los Angeles, are now beginning to record acid rainfall.

Since fossil fuels are a limited resource and may be exhausted within the next century, a large amount of the acid rain problem is, from a geological time scale, a short-term effect. However, the biological effects may be longer lasting. For example, if salmon are eliminated from Norwegian rivers, they may never return. Acid rain can affect life and ecosystems on land and in water, but the effects appear particularly severe in freshwater lakes [2,3].

GEOLOGY, CLIMATE, AND ACID RAIN

Geology and climatic patterns as well as types of vegetation and soil composition all affect potential impacts of acid rain. Figure 8.6, which shows areas of the United States sensitive to acid rain, is based on some of these factors. Particularly sensitive areas are those in which the bedrock cannot buffer the acid rain, including terrain dominated by granitic rocks, as well as areas in which the soils have little buffering action. Areas least likely to suffer damage will be those in which the bedrock contains an abundance of limestone or other carbonate material or in which the soils contain a horizon rich in calcium carbon-

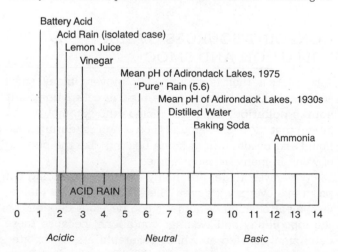

FIGURE 8.5

pH scale. (Modified after U.S. Environmental Protection Agency, 1980b.)

FIGURE 8.6
Areas in the United States sensitive to acid rain. (From U.S. Environmental Protection Agency, 1980b.)

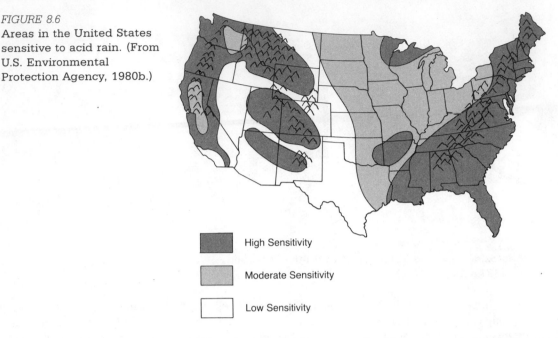

■ High Sensitivity

▨ Moderate Sensitivity

□ Low Sensitivity

ate. Soils in sensitive areas are damaged as nutrients are leached out by the acid, and as soils are depleted of nutrients and other minerals, plant productivity is adversely affected. Acid rain's ability to remove material was dramatically demonstrated recently when a block wall of a greenhouse in Lyme, New Hampshire, actually effervesced as acid rain dissolved holes in the blocks.

Acid rain not only affects forests and lakes, it leaches exposed rock. Since many famous architectural structures are made of soft, easily eroded rocks, acid rain has damaged buildings of great historical value and considerable beauty. For example, classical buildings on the Acropolis in Athens, Greece, have shown considerably more rapid decay in this century than in previous ones. As another example, statues at Herten Castle near Reckinghausen Westfolia, Germany, show the deterioration due to acid rain [4].

In the United States, cities along the eastern seaboard are more susceptible to acid rain today because emissions of sulphur and nitrogen oxides are most abundant there. However, the problem is expected to move westward. Rock types such as marble and limestone, which dissolve in weak acids, are particularly susceptible to damage. Interestingly, the geologic aspect of the problem may also help to predict future effects of acid rainfall. Since 1875 the Veteran's Administration has provided over 2.5 million tombstones to various national cemeteries. Fortunately, these tombstones have come from only three rock quarries and have a standard size and shape. Therefore, these stones located in various parts of the country will be valuable in assessing damages caused by acid rainfall because they provide a variety of dates and

locations to work from. Research will provide valuable data on air pollution and the meteorological patterns that contribute to acid rainfall [1].

Standards to control sulphur oxide emissions from future power plants will begin to be effective after 1995. Unfortunately, this program does not address the continued emissions from existing power plants during the next 15 years. The technology to clean up coal so that it will burn clean is already available, although the cost of removing the sulphur makes the coal more expensive. However, if nothing is done, the consequences of burning sulphur-rich coal during the next 15 years will probably be very expensive to us and to future generations.

NONPOINT SOURCES OF AIR POLLUTION AND SMOG

Air pollution that is produced from very many small sources scattered over a wide area is called **nonpoint source pollution.** Examples include the automobile pollution that develops around Los Angeles during rush hours and pollution from home heating, like the burning of wood in many Vermont homes.

Wherever there are many sources producing air pollutants over a wide area, there is potential for the development of smog. Whether smog develops depends on the topography and weather conditions, because these determine the rate at which pollutants are transported away from their sources and converted to harmless compounds in the air. When the rate of production exceeds the rate of chemical transformations and of transport,

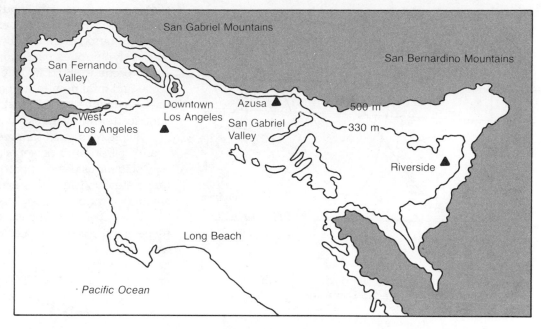

FIGURE 8.7
Part of southern California showing the Los Angeles basin (south coast air basin). (Modified after Williamson, *Fundamentals of air pollution,* © 1973, Figure 10.13. Reprinted with permission of Addison-Wesley, Reading, MA.)

dangerous conditions may develop. Cities that are situated in a geological "bowl" surrounded by mountains are more susceptible to smog problems than cities in open plains. Cities where certain kinds of weather conditions, such as temperature inversions, occur are also particularly susceptible. Both the surrounding mountains and the temperature inversions prevent the pollutants from being transported by the winds and weather systems. The production of smog is particularly well documented for Los Angeles, which has mountains surrounding the urban area and lies within a region that tends to have stagnating air conditions that promote smog (Fig. 8.7).

There are three major types of smog: **photochemical smog,** which is sometimes called LA-type smog or brown air; **sulphurous smog,** which is sometimes referred to as London-type smog or grey air; and the recently identified **particulate smog.** Solar radiation is particularly important in the formation of photochemical smog (Fig. 8.8). The reactions that occur in the development of photochemical smog are complex and involve both nitrogen oxides (NO_x) and organic compounds (hydrocarbons).

The development of photochemical smog is directly related to automobile usage. Figure 8.9 shows a characteristic pattern in terms of how the nitrogen oxides, hydrocarbons, and oxidants (mostly ozone) vary throughout a typical smoggy day in southern California. Early in the morning, when commuter traffic begins to build up, the concentrations of nitrogen oxide (NO) and hydrocarbons

begin to increase. At the same time, the amount of NO_2 may decrease due to the photodisassociation of nitrogen dioxide into nitrogen oxide plus atomic oxygen. The atomic oxygen is then free to combine with molecular oxygen to form ozone, so after sunrise the concentration

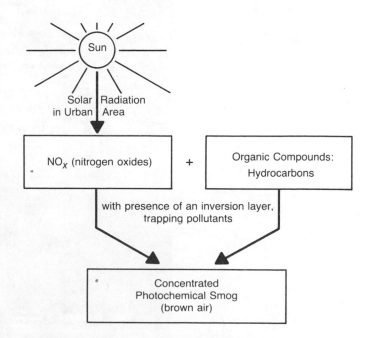

FIGURE 8.8
Idealized diagram showing how photochemical smog may be produced.

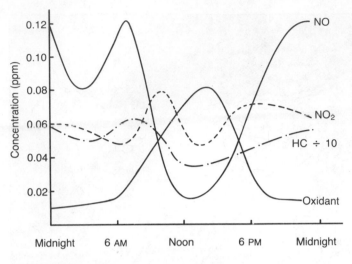

FIGURE 8.9
Idealized diagram showing the development of photochemical smog over the Los Angeles area on a typical warm day.

FIGURE 8.10
Smog in the San Fernando Valley, California. Photograph (a) was taken on a relatively clear day (note the mountains in the background), and photograph (b) was taken on a smoggy day (note restricted visibility). Both of these photographs were taken from the Veteran's Hospital in the San Fernando Valley looking toward the San Gabriel Mountains. (Photos by Jacqueline Keller.)

of ozone also increases. Shortly thereafter oxidized hydrocarbons react with nitrogen oxide to increase the concentration of nitrogen dioxide by midmorning. This reaction causes the nitrogen oxide concentration to decrease and allows ozone to build up, producing the midday peak in ozone and minimum in nitrogen oxide. As the smog matures, visibility may be greatly reduced (Fig. 8.10) due to light scattering by aerosols.

Sulphurous smog is produced primarily by burning coal or oil at large power plants. Figure 8.11 illustrates how sulphur oxides and particulates combine under certain meteorological conditions to produce a concentrated sulphurous smog.

The importance of particulate smog is just beginning to be appreciated. Several cities, including San Francisco, have an air quality problem even though the sunlight necessary for the development of true photochemical smog is at a minimum (i.e., this problem is observed in the winter in San Francisco). Sources that burn fossil fuel, such as automobiles, home gas heaters, and power

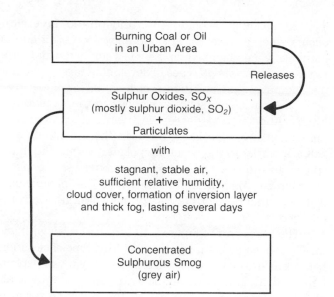

FIGURE 8.11
Idealized diagram showing how concentrated sulphurous smog might develop.

TABLE 8.2
Sizes of particulates, with examples of possible cleaning devices.

plants, produce very small primary particulates. When these particulates are exhausted from their sources, they have a diameter of about 0.1 micron (1 micron = 10^{-6} meter). These particulates can grow under certain situations (such as is found in a photochemical smog atmosphere) and inhibit visibility. As we learn more about this new type of particulate smog, we will have to develop strategies to combat it. One difference between small particulate smog and photochemical smog is the lack of oxidants, primarily ozone, associated with the small particulate smog. Thus strategies worked out in the Los Angeles basin to control ozone may not be applicable to San Francisco and other cities plagued by small particulate smog.

The importance of these very small particles that come from auto exhausts and the burning of oil and coal in power plants has not been recognized until very recently as a source of primary pollutants. Table 8.2 shows the sizes of several aerosols, including auto exhausts, and lists several examples of cleaning devices to remove particulates. Of course, trapping particulates before they

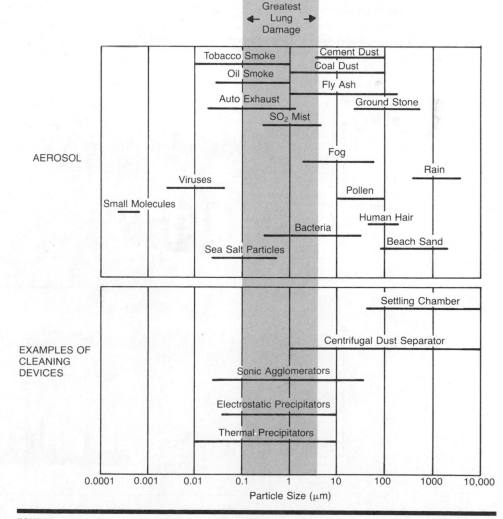

SOURCE: Modified from Giddings, 1973, and Hidy and Brock, 1971.

enter the atmosphere is the only effective way to remove them.

METEOROLOGY AND AIR POLLUTION

Meteorological conditions can determine whether polluted air is only a nuisance or becomes a major health problem. The primary adverse effects of photochemical smog are damage to green plants and aggravation of chronic illnesses in people. Most of these effects are due to relatively low-level concentrations of toxins over a long period of time. Smog periods in the Los Angeles basin or other areas with photochemical smog generally do not cause large numbers of deaths.

Sulphurous smog, on the other hand, has been associated with killer fogs which, over a period of days, can lead to an increase in deaths and illnesses. These fogs often tend to occur in winter rather than in summer, in contrast to photochemical smog.

Although photochemical and sulphurous smog have different causes, certain meteorological conditions can aggravate both of them. Restricted circulation in the lower atmosphere associated with inversion layers may lead to pollution events. An **atmospheric inversion** occurs when warmer air is found above cooler air and is particularly a problem when there is a stagnated air mass. Figure 8.12 shows two types of developing inversions that may aggravate air pollution problems. In the upper diagram, which is somewhat analogous to the situation in the Los Angeles area, descending warm air forms a semipermanent inversion layer. Because the mountains also act as a barrier to the pollution, polluted air moving in response to the sea breeze and other processes tends to move up canyons, where it is trapped. The photochemical smog that develops occurs primarily during the summer and fall.

The lower part of Figure 8.12 shows a valley with relatively cool air overlain by warm air. This type of situation can occur in several ways, one of which we will explain here. When cloud cover associated with a stagnant air mass develops over an urban area, the incoming solar radiation is blocked by the clouds, which absorb some of the energy and thus heat up. On the ground, or near the surface of the Earth, the air cools. If there has been a fair amount of humidity, then as the air cools the dewpoint is reached and a thick fog may form. Because the air is cold, people living in the city burn more fuels to heat their homes and factories, so more pollutants are delivered into the atmosphere. As long as the stagnant conditions exist,

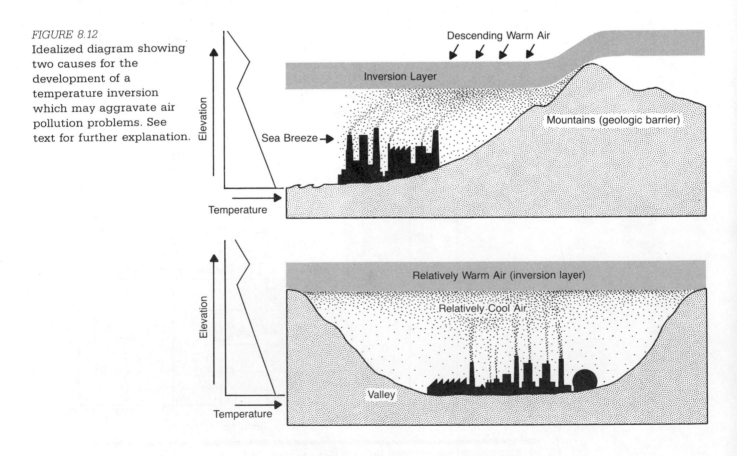

FIGURE 8.12
Idealized diagram showing two causes for the development of a temperature inversion which may aggravate air pollution problems. See text for further explanation.

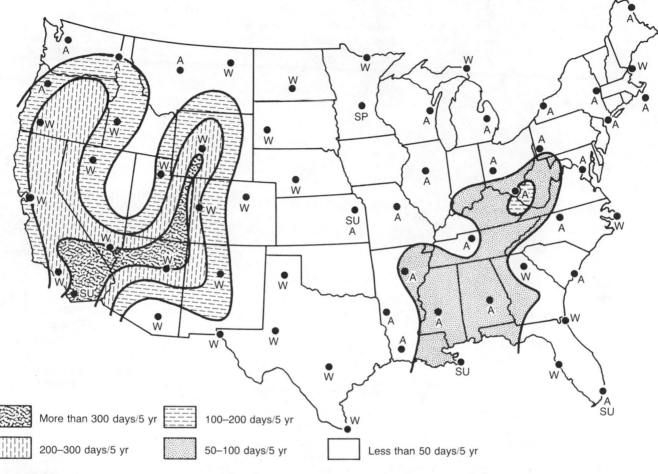

FIGURE 8.13
Number of days in a 5-year period characterized by reduced dispersion of air pollutants which existed for at least a 48-hour (2-day) period. The time of the year when most of these periods occurred are shown by the seasons: SP (spring), A (autumn), W (winter), and SU (summer). (Modified after Holzworth, 1968, as presented in Neiburger, Edinger, and Bonner, 1973.)

the pollutants will build up. This sort of scenario has led to very serious pollution problems, as described in the beginning of this chapter.

Evaluating meteorologic conditions can be extremely helpful in predicting which areas will have potential air pollution problems. Figure 8.13 shows the number of days in a 5-year period for which conditions were favorable for reduced dispersion of air pollution for at least a 48-hour period. This illustration clearly shows that most of the problems should be located in the western United States. Of particular interest is that the San Diego area in southern California may eventually have a very serious pollution problem because the number of days with reduced dispersion of pollutants is greater there than in Los Angeles, for example. The map also shows why areas such as Denver, Colorado, and Phoenix, Arizona, have a pollution problem.

THE NATIONAL AIR QUALITY STANDARDS

Air quality in urban areas is often reported as good, moderate, unhealthy, very unhealthy, or hazardous (Table 8.3). These levels or stages are derived from monitoring the concentration of five major pollutants: total suspended particulates, sulphur dioxide, carbon monoxide, ozone, and nitrogen dioxide. The ozone level is a good indicator of the amount of photochemical smog in the urban air because ozone is produced as oxides of nitrogen combine with hydrocarbons in the presence of sunlight to form the complicated organic particles of smog.

Table 8.4 lists the National Ambient Air Quality Standards (NAAQS) as of 1979. The air quality standards are likely to be revised in the next few years as we learn more about air pollution and how to measure it and study its ef-

TABLE 8.3
Definition of Pollutant Standard Index (PSI) values.

PSI Index Value	Air Quality Level	Pollutant Level					Health Effect	General Health Effects	Cautionary Statements
		TSP (24-hour), $\mu g/m^3$	SO_2 (24-hour), $\mu g/m^3$	CO (8-hour), mg/m^3	O_3 (1-hour), $\mu g/m^3$	NO_2 (1-hour), $\mu g/m^3$			
500	Significant harm	1000	2620	57.5	1200	3750			
400	Emergency	875	2100	46.0	1000	3000		Premature death of ill and elderly. Healthy people will experience adverse symptoms that affect their normal activity.	All persons should remain indoors, keeping windows and doors closed. All persons should minimize physical exertion and avoid traffic.
300	Warning	625	1600	34.0	800	2260	Hazardous	Premature onset of certain diseases in addition to significant aggravation of symptoms and decreased exercise tolerance in healthy persons.	Elderly and persons with existing diseases should stay indoors and avoid physical exertion. General population should avoid outdoor activity.
200	Alert	375	800	17.0	400[c]	1130	Very unhealthful	Significant aggravation of symptoms and decreased exercise tolerance in persons with heart or lung disease, with widespread symptoms in the healthy population.	Elderly and persons with existing heart or lung disease should stay indoors and reduce physical activity.
100	NAAQS	260	365	10.0	240		Unhealthful [a]	Mild aggravation of symptoms in susceptible persons, with irritation symptoms in the healthy population.	Persons with existing heart or respiratory ailments should reduce physical exertion and outdoor activity.
50	50% of NAAQS	75[b]	80[b]	5.0	120		Moderate[a]		
0		0	0	0	0		Good[a]		

SOURCE: U.S. Environmental Protection Agency, 1980a.
[a]No Index values reported at concentration levels below those specified by "Alert Level" criteria.
[b]Annual primary NAAQS.
[c]400 $\mu g/m^3$ was used instead of the O_3 Alert Level of 200 $\mu g/m^3$.

TABLE 8.4
National Ambient Air Quality Standards (NAAQS).

Pollutant	Averaging Time	Primary Standard Levels	Secondary Standard Levels
Particulate matter	Annual (geometric mean)	75 $\mu g/m^3$	60 $\mu g/m^3$
	24 hour[b]	260 $\mu g/m^3$	150 $\mu g/m^3$
Sulphur oxides	Annual (arithmetic mean)	80 $\mu g/m^3$ (0.03 ppm)	—
	24 hour[b]	365 $\mu g/m^3$ (0.14 ppm)	—
	3 hour[b]	—	1300 $\mu g/m^3$ (0.5 ppm)
Carbon monoxide	8 hour[b]	10 mg/m^3 (9 ppm)	10 mg/m^3 (9 ppm)
	1 hour[b]	40 mg/m^3 (35 ppm)	40 mg/m^3 (35 ppm)
Nitrogen dioxide	Annual (arithmetic mean)	100 $\mu g/m^3$ (0.05 ppm)	100 $\mu g/m^3$ (0.05 ppm)
Ozone	1 hour[b]	235 $\mu g/m^3$ (0.12 ppm)	235 $\mu g/m^3$ (0.12 ppm)
Hydrocarbons (nonmethane)[a]	3 hour (6 to 9 A.M.)	160 $\mu g/m^3$ (0.24 ppm)	160 $\mu g/m^3$ (0.24 ppm)

SOURCE: U.S. Environmental Protection Agency, 1980a.
[a]A nonhealth-related standard used as a guide for ozone control.
[b]Not to be exceeded more than once per year.

fects on human population and the biosphere in general. Recent studies have shown that some of the equipment used in the past has yielded fallacious data, resulting in air quality standards that may not reflect actual pollution conditions.

Pollution problems vary in different regions of the United States. For example, in the Los Angeles basin, nitrogen oxides and hydrocarbons are particularly troublesome because they do combine in the presence of sunlight to form photochemical smog. Furthermore, in Los Angeles most of the nitrogen oxides and hydrocarbons are emitted from automobiles, a nonpoint source [5]. In other urban areas, such as in Ohio and the Great Lakes region in general, air quality problems result from emissions of sulphur dioxides and particulates into the air from industry and coal-burning power plants, which are point sources. This is not to say that automobiles are not a problem in areas outside of Los Angeles, but rather to emphasize contrasting conditions (Fig. 8.14). It is also important to keep in mind that automobiles produce many small particulates, whose pollution potential is only beginning to be understood. Thus areas such as San Francisco, with a relatively cool climate, have a serious air pollution problem from automobile exhaust which resembles photochemical smog, but lack the necessary sunlight to produce that type of smog. Of course, the small particulates are also abundant in Los Angeles, where the prob-

lem is more pronounced because of the photochemical smog.

Examination of Figure 8.14 indicates some interesting aspects of the air pollution problem on a national scale. For example, the nitrogen dioxide problem is primarily a southern California problem and is related to automobiles. Oxidants, however, are a problem in both the southwestern and northeastern United States as are particulates. Thus the oxidants and particulates may be related to factors such as general circulation conditions. The pattern of oxidants and particulates in Figure 8.14 correlates well with Figure 8.13, especially for the particulates, which we might expect to be directly related to meteorologic conditions that favor reduced dispersion of pollutants.

Data from 25 major metropolitan areas between 1974 and 1977 suggest that the total number of "unhealthful" and "very unhealthful" days has declined (Fig. 8.15). Although these data do not mean that air pollution has been eliminated, they do indicate that the nation's air quality is improving. However, in 1977 the urban areas of New York and Los Angeles had unhealthful air about two thirds of the time. Only about 40 percent of the urban areas in the United States for which good data were available in 1977 reported unhealthful readings less than 10 percent of the days, and three cities reported an increase in air pollution between 1974 and 1977. The pollutants responsible for

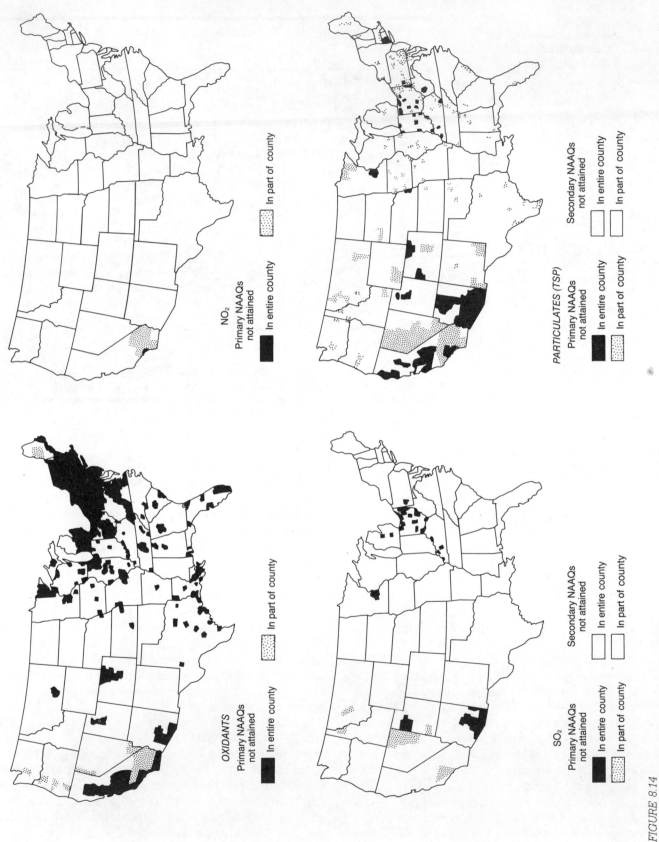

NO₂
Primary NAAQs
not attained
▓ In entire county
░ In part of county

PARTICULATES (TSP)
Primary NAAQs
not attained
▓ In entire county
░ In part of county

Secondary NAAQs
not attained
□ In entire county
□ In part of county

OXIDANTS
Primary NAAQs
not attained
▓ In entire county
░ In part of county

SO₂
Primary NAAQs
not attained
▓ In entire county
░ In part of county

Secondary NAAQs
not attained
□ In entire county
□ In part of county

FIGURE 8.14
Nonattainment of selected air pollutants for August, 1977. (From Council on Environmental Quality, 1978.)

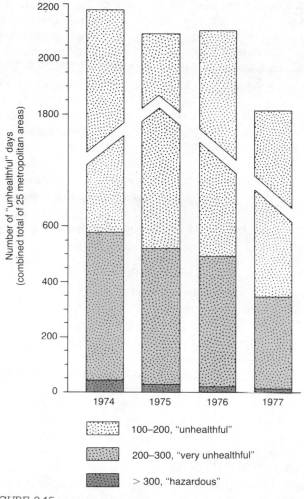

FIGURE 8.15
Data from 25 major metropolitan areas between 1974 and 1977 show that the total number of unhealthful and very unhealthful days declined during that period. See text for further explanation. (From Council on Environmental Quality, 1978.)

Methods used include a variety of techniques, some of which are shown in Table 8.2 and Figure 8.16. The control of small particulate material of either primary or secondary origin from moving sources such as automobiles is much more difficult. As we learn more about these very small particulates, new methods will have to be devised to control them.

Sulphur oxides, most of which are released in combustion of coal or oil, can best be controlled by either minimizing the sulphur content of the fuel itself or removing the gases from the exhaust before it enters into the atmosphere. The obvious solution would be to use coal with a low sulphur content. Unfortunately, most of the low-sulphur coal is located in the western United States, whereas most of the coal is actually burned in the eastern part of the country. Therefore, the use of low-sulphur coal is a solution only in isolated cases where it is economically feasible. Another possibility is cleaning up relatively high-sulphur coal by washing it to remove the sulphur. In the washing process finely ground coal is washed with water, and the iron sulphide (mineral pyrite) settles out

sending the pollution index into the unhealthful range in most urban areas were carbon monoxide and hydrocarbons, both of which are associated with automobile emissions (see Table 8.1). Thus it appears that one way to improve urban air quality is to obtain a reduction in automobile exhaust pollution [6].

EMISSION CONTROL AND ENVIRONMENTAL COST

Each air pollutant requires a different abatement strategy (see Table 8.1). Control of coarse particulates from power plants and industrial sites is best achieved through collection or capture before they enter the atmosphere.

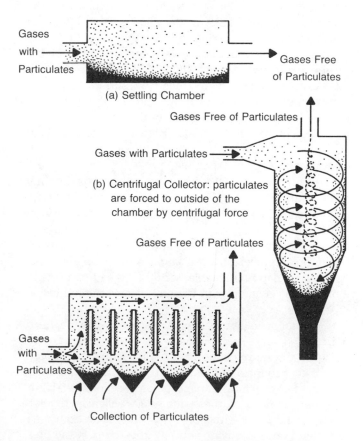

FIGURE 8.16
Idealized diagrams of some of the devices being used to control emissions of particulates before they enter into the atmosphere.

because of its relatively high density compared with coal. The washing process removes some of the sulphur, but it is expensive. Another option is coal gasification, which converts coal that is relatively high in sulphur to a gas in order to remove the sulphur. The gas obtained from coal is quite clean and can be transported relatively easily, augmenting supplies of natural gas. The synthetic gas produced from coal is now fairly expensive, but commercial-scale production on a trial basis may begin in the 1980s [7].

Sulphur oxide emissions from stationary sources such as power plants can also be reduced by removing the oxides from the gases in the stack before they reach the atmosphere. Perhaps the most highly developed technology for the cleaning of gases in tall stacks is *wet scrubbing*. In this method the gases are treated with a slurry of lime (calcium oxide, CaO) or limestone (calcium carbonate, $CaCO_3$). The sulphur oxides react with the calcium to form insoluble calcium sulphides and sulphates, which are collected and then disposed of. A major problem with this method is that the residue containing the calcium sulphides and sulphates is a sludge that must be disposed of at a land disposal site. Because the sludge can cause serious water pollution if it interacts with the hydrologic cycle, it must be treated carefully. Furthermore, scrubbers are very expensive and add significantly to the total cost of electricity produced by burning coal. The capital cost of equipment to accomplish this scrubbing may be as much as 20 percent of the cost of the power plant, adding to the cost of electricity by as much as 30 to 50 percent or even more as scrubbers are added to existing power plant facilities [8].

Another approach to reducing sulphur oxide emissions is to utilize high smokestacks in hopes that the pollutants will be more effectively dispersed further from the ground. This practice, while perhaps alleviating critical problems very close to the site, may spread air pollution problems over an even wider area. This effect has been particularly noticeable with the spread of acid precipitation associated with the emission of sulphur oxides [8].

Control of pollutants such as carbon monoxide, nitrogen oxides, and hydrocarbons in urban areas is best achieved through pollution control measures for automobiles, because most of the anthropogenic portion of these pollutants comes from transportation. (see Table 8.1). Furthermore, control of these materials will also regulate ozone in the lower atmosphere, which forms from reactions with nitrogen oxides and hydrocarbons in the presence of sunlight.

The control of nitrogen oxides from automobile exhausts is accomplished by recirculating exhaust gas, which dilutes the air-to-fuel mixture being burned in the engine. The dilution reduces the temperature of combustion and decreases the oxygen concentration in the burning mixture (that is, it makes a richer fuel). This method produces fewer nitrogen oxides. Unfortunately, the effect is just the opposite on hydrocarbon emissions, which are greater for rich fuels (those with a low air-to-fuel ratio). Nevertheless, exhaust recirculation to reduce nitrogen oxide emissions has been in common practice in the United States since 1975 [7].

The two most common devices used to remove carbon monoxide and hydrocarbon emissions from automobiles are the *catalytic converter* and the *thermal exhaust reactor* [7]. In both devices the carbon monoxide is changed or converted to carbon dioxide and the hydrocarbons to carbon dioxide and water. In the high-temperature or thermal exhaust reactor, the addition of outside air (oxygen) assists in the more complete combustion of exhaust fumes in the high-temperature chamber. The catalytic converter works at a lower temperature and involves a metal catalyst material over which the exhaust gases are circulated with air. Oxidation then occurs, removing the carbon monoxide and hydrocarbons from the exhaust. Catalytic converters have one major problem, namely, the catalyst bed may be rendered ineffective by a number of substances, including lead additives in gasoline. As a result, there has been a tremendous shift to nonleaded gasoline in recent years.

It has been argued that the automobile emission regulation plan in the United States has not been very effective in reducing pollutants. The pollutants may be reduced while the car is relatively new, but many people simply do not take care of their automobile well enough to insure that the emission control devices work over the life of the automobile. In fact, the evidence suggests that some of these devices become less efficient every year following purchase. Because of these adverse aspects of emission control, it has been suggested that effluent fees replace automobile controls as the primary method of regulating air pollution in the United States [9]. Under this scheme, vehicles would be tested each year for emission control, and fees would then be assessed of the owners on the basis of the test results. The fees would provide a positive incentive for the purchase of automobiles that pollute less, and the annual inspections would insure that pollution control devices are properly maintained. Although there is considerable controversy at this time over enforced pollution inspections, such inspections are common in a number of areas and are expected to increase as air pollution abatement becomes imperative.

Damage from air pollution and the costs to clean up the air run into billions of dollars per year in the United

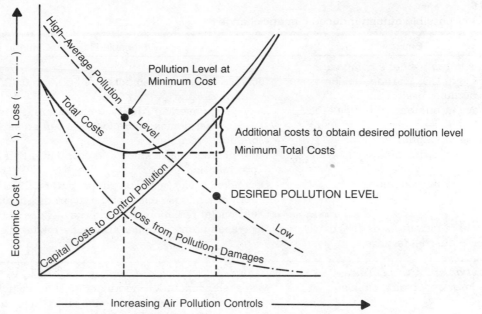

FIGURE 8.17
Idealized diagram of some of the relationships between economic cost and increasing air pollution controls. See text for further explanation. (Modified after Williamson, 1973.)

States. Although a detailed discussion of the cost of air pollution is beyond the scope of this text, some of the variables that must be considered are shown in Figure 8.17. Notice on this idealized graph that with increasing air pollution controls the capital cost to control air pollution increases, and as the controls for air pollution increase the loss from pollution damages decreases. The total cost is thus the sum of these two items, and the minimum is well defined in terms of a particular average pollution level. This graph also shows that if the desired pollution level is lower than that at which the minimum total cost occurs, then additional costs to obtain the desired pollution level will be necessary. This type of diagram, while valuable in looking at some of the major variables, does not consider adequately all of the loss from pollution damages. For example, long-term exposure to air pollution may aggravate or lead to chronic diseases in human beings, with a very high cost. How do we determine what portion of the cost is due to air pollution? In spite of these drawbacks, it seems worthy to reduce the air pollution level below some particular standard. Thus, in the United States the ambient air quality standards have been developed as a minimum acceptable pollution level.

AIR POLLUTION AND THE BIOSPHERE

Air pollutants and other modifications of the land associated with technological civilization affect not only local areas or regions, but the entire biosphere (Table 8.5). The fundamental problem is that the world population takes up and emits materials to the atmosphere at amounts and rates similar to those of natural processes. For example, the annual production of carbon dioxide from the burning of fossil fuels is approximately one tenth the amount emitted by the respiration of all living things.

Pollution of the biosphere has two kinds of effects: climate alterations and chemical changes in the atmosphere that may be hazardous to living things. Particulates and carbon dioxide may be instrumental in climatic change, and chemicals that alter the atmosphere's ozone layer are direct hazards to life.

CARBON DIOXIDE AND CLIMATE

The carbon dioxide concentration of the atmosphere is increasing because we are burning fossil fuels whose primary combustion product is carbon dioxide. Fossil fuels are the result of the burial long ago of unoxidized organic materials—much of which were the products of photosynthesis by green plants.

Today there is a major controversy over the possible effects of increasing carbon dioxide in the Earth's atmosphere. It is generally accepted that around the turn of the century the concentration of carbon dioxide in the atmosphere was about 290 parts per million (0.029%) and that since that time there has been a steady increase to the present level of approximately 330 parts per million (0.033%) (see Fig. 3.4). The projected value for the year 2000 is 380 parts per million (0.038%), almost 20 percent higher than the present concentration. This increase, however, is only half that expected if all of the carbon dioxide from the burning of fossil fuels had remained in the

TABLE 8.5
Examples of possible human-induced climatic changes.

Cause	Probable-Potential Climatic Effects
Changes in atmospheric composition	
Increase in CO_2 (carbon dioxide) by burning fossil fuels	Increase in average temperature of 0.2 to 0.3C° for a 10% increase in carbon dioxide
Increase in number of particulates	Decrease in average surface temperature; increased precipitation; more urban dust domes; more fog in urban areas
Increase in sulphur	More acid precipitation
Increase in hydrocarbons	More photochemical smog and ozone in urban areas; more acid precipitation
Increase in fluorocarbons	Decreased ozone in stratosphere; potential tropospheric warming due to both greenhouse effect of fluorocarbons in atmosphere and increase in ultraviolet radiation reaching the surface of the Earth
Increase in nitrous oxide (N_2O) from industrial fertilizers	Decreased ozone in upper atmosphere (20 km); warmer Earth surface
Modification of Earth's surface	
More large reservoirs, canals	Increased evaporation; more water in atmosphere; more precipitation in some areas; larger greenhouse effect (due to water in atmosphere)
Urbanization	Dust domes; heat islands; local reduction of wind; increased precipitation; lower relative humidity
Deforestation	Increased albedo (reflectivity); decreased evapotranspiration; increased wind
Overgrazing marginal semiarid lands	Desertification; less evapotranspiration; increased wind

SOURCE: Modified after Anthes et al., 1981.

atmosphere. The fate of the rest of the carbon dioxide is the subject of a major scientific debate. Some scientists argue that it is being absorbed by the oceans; others argue that land plants, particularly trees in forests, are growing faster and taking up the additional carbon dioxide.

The extrapolation of the increase in carbon dioxide until the year 2000 is based on several assumptions that may not be entirely correct. First, it is assumed that the ocean will continue to absorb about 50 percent of the carbon dioxide emitted. Second, it is assumed that large-scale and significant atmospheric circulation changes will not take place. Given these assumptions, the temperature increase due to atmospheric carbon dioxide should be approximately 0.5C° by the year 2000. This temperature increase is unlikely to create major problems, but if the carbon dioxide content continues to build for several hundred more years then a temperature rise of 2 to 3C° may be possible and would cause major problems. A change of a few degrees Centigrade in the mean annual temperature might be sufficient to melt the polar ice caps, raising sea levels around the world and flooding major cities. For this reason and others, the carbon dioxide in the atmosphere is being very carefully monitored. Figure 8.18 shows the climate of the Earth in terms of mean annual temperature change for the last 150,000 years. What happens in the next few thousand years may be affected significantly by human-induced changes in the composition of the atmosphere.

The actual mechanism by which carbon dioxide heats the atmosphere is referred to as the "greenhouse effect." When incoming short-wavelength radiation from the sun enters a greenhouse, it is reemitted as long-wave radiation that is trapped by the glass. In the atmosphere, carbon dioxide effectively traps or absorbs long-wave radiation emitted from the Earth, and this trapped radiation heats up the atmosphere. Thus, as the amount of carbon dioxide in the atmosphere increases, the atmosphere has a corresponding increase in temperature (see Fig. 3.8). Actually, a greenhouse heats up because the air inside is retarded from circulating with outside air, but the effect of trapping long-wave radiation is still known as the "greenhouse effect."*

---

*Water vapor and liquid water droplets contribute to the Earth's greenhouse effect. Fluorocarbons in the atmosphere may also cause a greenhouse effect.

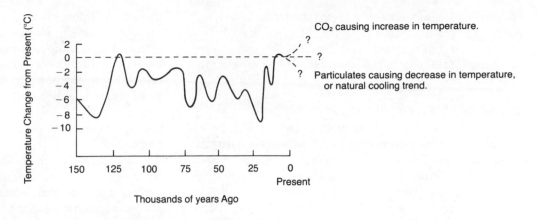

FIGURE 8.18

Changes in the Earth's mean annual temperature for the past 150,000 years. Also shown are possible future changes related to emission of carbon dioxide into the atmosphere. See text for further explanation. (Modified after Committee for the Global Atmospheric Research Program, National Academy of Sciences, 1975.)

Biological and Ecological Effects of an Increase in Carbon Dioxide Concentration

An increase in the atmosphere's carbon dioxide concentration can greatly affect living things at local, regional, and global levels. At the local level, green plants grow better when the air around them is enriched with carbon dioxide. This has been known since the eighteenth century, when Joseph Priestly discovered that a green plant grew better in a bell jar in which a mouse had died (and therefore added carbon dioxide from its own respiration and from the respiration accompanying its decay) than the plant did in another bell jar. Thus a worldwide increase in carbon dioxide might increase annual net vegetation production, if no other environmental factor interfered to limit the growth. On the other hand, the complex interactions among species in a natural ecosystem might tend to stabilize the net production and prevent such increases.

The increasing carbon dioxide concentration in the atmosphere could have other, indirect effects. If the climate were to warm, some kinds of vegetation would increase and others would decrease. The effects of changes in weather patterns, including rainfall, could be complicated and therefore cannot be predicted at this time.

The problem of increasing atmospheric carbon dioxide due to burning of fossil fuels is still unresolved. There is no general agreement on what the effects will be, primarily because we are not certain as to all the storage compartments available in the carbon cycle. That is, the ocean may be able to account for a lesser or greater amount of carbon dioxide in the future, and there may be other sinks for carbon dioxide that we have not considered [10,11].

PARTICULATES IN THE ATMOSPHERE

How particulate material in the atmosphere will affect global changes in the atmosphere's mean annual temperature is uncertain. What is more certain, however, is that more particulates are being added to the atmosphere through human activity. Today only about 20 percent of all the particles are from human sources, but this may increase to approximately 50 percent by the year 2000. Particulates come from primary sources, such as burning coal and other fossil fuels, or from secondary sources, as produced from reactions involving photochemical smog. However, recent evidence suggests that secondary particulate production may not be a major process in photochemical smog. That is, secondary particles may result from primary particulates which grow larger as photochemical smog develops and matures.

Regardless of how they are produced, particulates have two important effects. First, particulates act as condensation nuclei and therefore cause an increase in precipitation or fog. Second, particulates affect the amount of sunlight reaching the Earth. As the total amount of particulates in the atmosphere increases, a larger percentage of incoming solar radiation may be reflected away from the Earth, causing the Earth's mean annual temperature to decrease. On the other hand, some particles may absorb incoming solar radiation, causing an increase in the atmospheric and land surface temperature. This second effect is observed in urban areas where particulates are concentrated. Furthermore, if particles filter out of the atmosphere and are deposited on snow, then a greater portion of the solar radiation will be absorbed, making available more radiation to heat the atmosphere. There seems little doubt that particulates in the atmosphere can interfere with incoming solar radiation.

Catastrophic volcanic eruptions have caused slight global cooling for several years. This was particularly apparent when Krakatoa erupted in 1883 and blasted dust into the stratosphere that circled the Earth for several years, causing a slight cooling [8]. The eruption of Mt. St.

Helens in 1980 undoubtedly affected the stratosphere, but it probably will not change the global surface weather. As with carbon dioxide, there is a considerable controversy surrounding the effects that particulates will have on future climates.

THREATS TO STRATOSPHERIC OZONE

Ozone (O_3) is produced in the stratosphere at altitudes of 16 to 60 kilometers above the Earth when two reactions take place. The first reaction is the splitting of an oxygen molecule into atomic oxygen by sunlight. The second reaction, involving a union of a molecule of oxygen (O_2) with an atom of oxygen (O) to make ozone (O_3), only takes place when a third molecule (a *catalyst*) is present.

Ozone is destroyed naturally by ultraviolet radiation. As the ozone is destroyed, it performs a service function for organisms at the Earth's surface by greatly reducing the amount of ultraviolet radiation that reaches the Earth. In the stratosphere, ozone is constantly being formed and destroyed and therefore is maintained in a rough equilibrium. Any reduction in ozone is potentially dangerous because more ultraviolet light would reach the Earth and possibly cause an increase in skin cancer. The relationship between ultraviolet radiation and skin cancer is estimated to be such that a 1 percent depletion in ozone would result in a 2 percent increase in skin cancer.

Another aspect of the ozone depletion problem involves potential heating or cooling in various parts of the atmosphere. If there were less ozone, more radiation would reach the Earth and would heat up the lower atmosphere. On the other hand, as ozone is depleted in the stratosphere, the upper atmosphere cools and hence less thermal radiation from the stratosphere reaches the Earth's surface. This tends to cool the lower atmosphere. When ozone is destroyed by solar radiation, some of the solar energy is converted into heat, which warms the stratosphere. (Thus we can see why the temperature increases with height in the stratosphere, forming a naturally occurring temperature inversion, and why there is little vertical mixing in the stratosphere.) Because ozone depletion will result in a cooling of the stratosphere, stratospheric gases will radiate less thermal (infrared) radiation to the Earth's surface and therefore contribute to a cooling trend in the lower atmosphere. The protracted effects of stratospheric ozone depletion on the temperature in the atmosphere and at the Earth's surface obviously remain speculative.

Human-induced changes in the amount of ozone in the stratosphere is the subject of considerable controversy. It has been suggested that aerosols, particularly fluorocarbons, will destroy some ozone. Others have suggested that nitrous oxide emitted by industrial fertilizers

through interactions with the biosphere may rise and eventually destroy stratospheric ozone. As is the case with carbon dioxide and particulates in the atmosphere, the potential effects of ozone depletion and the processes involved are not completely understood. Although ten years ago it was believed that the proposed supersonic transport (SST) would release materials into the stratosphere that would deplete ozone, that work is now being overturned.

To summarize, there is little doubt that emission of carbon dioxide, particulates, and other materials through human activity does affect the atmosphere. What is needed is more careful monitoring and modeling of various reactions that take place in the atmosphere so that we can make better predictions. It is an exciting area for scientists interested in environmental matters, atmospheric physics, chemistry, global ecology, and biological and chemical oceanography.

SUMMARY AND CONCLUSIONS

There are two main groups of air pollutants: primary and secondary. Primary pollutants are those emitted directly into the air, such as particulates, sulphur oxides, carbon monoxide, nitrogen oxides, and hydrocarbons. Secondary pollutants are those produced through reactions among primary pollutants and other atmospheric compounds. A good example of a secondary pollutant is ozone, which forms over urban areas through photochemical reactions among primary pollutants and natural atmospheric gases.

Every year approximately 250 to 300 million metric tons of primary pollutants enter the atmosphere above the United States from human-related processes. Considering the enormous volume of the atmosphere, this is a relatively small amount of material. If it were distributed uniformly, there would be little problem with air pollution. Unfortunately, the pollutants are not generally evenly distributed but rather are concentrated in urban areas or other areas where the air naturally lingers.

Two major kinds of pollution sources are point and nonpoint. An example of a point source is a smokestack in a large power plant. Nonpoint sources are dispersed over a larger area, such as the automobiles in the Los Angeles area.

Combustion of large quantities of fossil fuels emits sulphur and nitrogen oxides into the atmosphere, resulting in a problem known as acid rain. Environmental degradations associated with acid rain include loss of fish and other life in lakes, damage to trees and other plants, leaching of nutrients from soils, and damage to stone statues and buildings in urban areas. During the next decade or so, the acid rain problem is likely to remain as one of the most serious environmental problems facing the United States and other industrialized nations.

There are three major types of smog: photochemical, sulphurous, and particulate smog. Each type of smog has particular environmental problems that vary with geographic region, time of year, and local urban conditions.

Meteorological conditions greatly affect whether or not polluted air is a problem in a certain urban area. In particular, restricted circulation in the lower atmosphere associated with temperature inversion layers may lead to pollution events.

Air quality in urban areas is usually reported in terms of whether the quality is good, moderate, unhealthy, very unhealthy, or hazardous. These levels or stages are defined in terms of the Pollution Standard Index (PSI) and National Ambient Air Quality Standards (NAAQS). There is some indication that the nations's air quality has improved in recent years; however, there are still numerous areas where urban air quality is unhealthful a good deal of the year.

The relationships between emission control and environmental cost are complex. The minimum total cost is a compromise between capital costs to control pollution and losses or damages resulting from such pollution. If additional controls are necessary to lower the pollution to a more acceptable level, then additional costs are incurred. These costs can increase quite rapidly beyond a certain level of pollution abatement.

Air pollutants and modification of the land associated with civilization affect local, regional, and even global climates. Problems of particular concern include the increase in the atmosphere's carbon dioxide, which may eventually cause a global warming; the increase in particulates emitted into the atmosphere, which may locally or over a short period affect atmospheric conditions; and the increase in fluorocarbon and nitrous oxide (from fertilizers) emissions, which may eventually reduce stratospheric ozone. There is currently considerable debate over the magnitude and extent of problems resulting from air pollution. We do know, however, that the effects are measurable and may be increasing.

REFERENCES

1 U.S. ENVIRONMENTAL PROTECTION AGENCY. 1980. *Acid rain.*

2 GORHAM, E. 1976. Acid precipitation and its influence upon aquatic ecosystems—An overview. *Water, Air and Soil Pollution* 6: 457–81.

3 WRIGHT, R. F.; DALE, T.; GJESSING, E. G.; HENDRY, G. R.; HENRIKSEN, A.; JOHANNESSEN, M.; and MUNIZ, I. P. 1976. Impact of acid precipitation on freshwater ecosystems in Norway. *Water, Air and Soil Pollution* 6: 483–99.

4 WINKLER, E. M. 1976. Natural dust and acid rain. *Water, Air and Soil Pollution* 6: 295–302.

5 DETWYLER, T. R., and MARCUS, M. G., eds. 1972. *Urbanization and environment.* Belmont, Calif.: Duxbury Press.

6 MARSH, W. M., and DOZIER, J. 1981. *Landscape.* Reading, Mass.: Addison-Wesley.

7 STOKER, H. S., and SEAGER, S. L. 1976. *Environmental chemistry: Air and water pollution.* 2nd ed. Glenview, Ill.: Scott, Foresman.

8 ANTHES, R. A.; CAHIR, J. J.; FRASER, A. B.; and PANOFSKY, H. A. 1981. *The atmosphere.* 3rd ed. Columbus, Ohio: Charles E. Merrill.

9 LYNN, D. A. 1976. *Air pollution—Threat and response.* Reading, Mass.: Addison-Wesley.

10 LOVINS, A. B. 1979. Foreword. In *A golden thread,* eds. K. Butti and J. Perlin. Palo Alto, Calif.: Cheshire Books.

11 EATON, W. W. 1978. Solar energy. In *Perspectives on energy,* 2nd ed., eds. L. C. Ruedisili and M. W. Firebaugh, pp. 418–36. New York: Oxford University Press.

FURTHER READING

1 ANTHES, R. A.; CAHIR, J. J.; FRASER, A. B.; and PANOFSKY, H. A. 1981. *The atmosphere.* 3rd ed. Columbus, Ohio: Charles E. Merrill.

2 BATTAN, L. J. 1974. *Weather.* Englewood Cliffs, N.J.: Prentice-Hall.

3 BRODINE, V. 1973. *Air pollution.* New York: Harcourt, Brace and Jovanovich.

4 CLAIRBORNE, R. 1970. *Man and history.* New York: Norton Press.

5 GATES, D. M. 1972. *Man and his environment: Climate.* New York: Harper & Row.

6 HODGES, L. 1973. *Environmental pollution.* New York: Holt, Rinehart and Winston.

7 NEIBURGER, M.; EDINGER, J. G.; and BONNER, W. D. 1973. *Understanding our atmospheric environment.* San Francisco: W. H. Freeman.

8 TORIBARA, T. Y.; MILLER, M. W.; and MORROW, P. E., eds. 1980. *Polluted rain.* New York: Plenum Press.

9 WILLIAMSON, S. J. 1973. *Fundamentals of air pollution.* Reading, Mass.: Addison-Wesley.

STUDY QUESTIONS

1 Compare and contrast the London 1952 fog event with smog problems in the Los Angeles basin.

2 Why do we have air pollution problems when the amount of pollution emitted into the air is a very small fraction of the total material in the atmosphere?

3 What is the difference between point and nonpoint sources of air pollution? Which is easier to manage?

4 Distinguish between primary and secondary pollutants.

5 We know that natural rainfall is slightly acidic. Why then are we concerned about acid rain?

6 What is particulate smog? How does it differ from other smog? What problems does it pose for the future?

7 Why is it so difficult to establish national air quality standards?

8 In a highly technological society, is it possible to have 100 percent clean air? Is it feasible or likely?

9 Carbon dioxide is a nutrient as well as a pollutant. Explain this paradox.

9 The Waters

INTRODUCTION*

The city of Seattle, Washington, lies between two major bodies of water, saltwater Puget Sound to the west and freshwater Lake Washington to the east. Beginning in the 1930s, the freshwater lake was used for disposal of sewage. By 1954, 10 sewage treatment plants that removed disease organisms and much of the organic matter had been built along the lake. With an additional treatment plant added in 1959, 76,000 cubic meters of effluent flowed per day from the treatment plants into the lake. Smaller streams that fed the lake brought in additional untreated sewage.

The lake's response to these effluents was a major bloom of undesirable algae in 1955 which affected fishing and the general aesthetics of the lake. Public concern increased immediately, and the city's mayor that year appointed an advisory committee to determine what might be done. Some of the committee members were scientists, including those connected with the University of Washington in Seattle, who had studied these changes in the lake for years and knew the lake's biota well.

The committee advised the city to change its sewage treatment methods and to divert the sewage effluents from the lake into Puget Sound. The sound was much larger, and its waters were flushed rapidly and exchanged with the open ocean because of the strong currents and tides that flowed in and out of the sound. A sewage diversion project began in 1963 and was completed in 1968. Sewage that had polluted the lake was taken by pipes to a very deep point far offshore in the sound.

By 1969 changes were noticeable in the lake. The unpleasant algae decreased in abundance, and the surface waters became two and one-half times clearer than they had been five years before. Oxygen concentrations in deep water increased immediately to levels above that observed in the 1930s, favoring an increase in the fish. Phosphorus in the sewage effluent had been the major stimulant of the algal growth and was the major limiting factor for the lake's organisms. With the elimination of sewage flow into the lake, the amount of phosphorus decreased and the undesirable algae decreased in response. Since then the lake has continued to improve and to return to conditions that are considered desirable by the city residents.

The story of the restoration of Lake Washington represents many important aspects of the relationship between human beings, the biosphere, and our water resources. Most importantly, Lake Washington is a success story and shows that public concern, long-term and accurate scientific information, and an understanding of ecosystem processes combined with appropriate policies, laws, and regulations can improve our environment without detriment to other social, economic, or aesthetic factors.

CHARACTERISTICS OF WATER POLLUTION

Pollution of waters occurs when too much of an undesirable or harmful substance flows into a body of water, exceeding the natural ability of that water body to remove the undesirable material or convert it to a harmless form. When Seattle was a small city, the pollution of streams flowing into Lake Washington had little noticeable effect. Natural biological degradation and inorganic processes were sufficient to take care of the effluents. By the 1950s, however, the city had become so large that the rate at which effluents were being dumped into the lake greatly exceeded the lake's capacity to remove them or transform them into harmless forms.

As in the case of Lake Washington, we have three ways to deal with water pollution: reduce the sources; transport the pollutants to where they will not do damage; or convert the pollutants to harmless forms. In the Lake Washington case, the second method was the easiest and most practical because of a unique situation—that is, the nearness of Puget Sound with its rapid rate of flushing with the Pacific Ocean. This method was less expensive than adding improved treatment processes to remove the phosphorus. The effluent flowing from the city sewage to the lake had already been subjected to secondary treatment, meaning that it was free of disease-causing organisms and major organic compounds.

Few other cities are so fortunate as Seattle, but there are other success stories in the treatment of water pollution. One of the most notable is the cleanup of the Thames River in England. For centuries London's sewage had been dumped into that river, and there were few fish to be found downstream in the estuary. In recent decades, however, improvement in water treatment has led to the return of a great number of species of fish—some not seen in the river in centuries.

These case histories show that we can manage our water resources to our benefit, but in the future much still remains to be done. As the world's population and the industrial production of many goods increases, the use of waters will also accelerate. Today world per capita use of water is 710 cubic meters per year, and the total human use of water is 2600 cubic kilometers per year. It is estimated that by the year 2000 world use of water will more

---

*This case history is based on Edmonson, 1975.

TABLE 9.1
Water budgets for the continents.

Continent	Precipitation (cm/yr)	Evaporation (cm/yr)	Runoff (cm/yr)	Runoff (km³/yr)
Africa	69	43	26	7,700
Asia[a]	60	31	29	13,000
Australia	47	42	5	380
Europe	64	39	25	2,200
North America	66	32	34	8,100
South America	163	70	93	16,600
All continents	469	257	212	47,980

SOURCE: Modified after Budyko, 1974.
[a]Includes the USSR.

than double to 6000 cubic kilometers per year—a significant fraction of the naturally available fresh water.

The average total runoff from the Earth's rivers is approximately 48,000 cubic kilometers (Table 9.1), but its distribution is far from uniform. Some occurs in relatively uninhabited regions, such as Antarctica, which produces 2300 cubic kilometers, or about 5 percent of the Earth's total runoff. South America, which includes the relatively uninhabited Amazon basin, provides 16,600 cubic kilometers, or about one third of the Earth's total runoff; and in North America, total runoff is about one-half of that for South America, or 8100 cubic kilometers per year. Unfortunately, much of the North American runoff occurs in sparsely or uninhabited regions, particularly in the northern parts of Canada and Alaska.

On a global scale, total water abundance is not the problem; the problem is water's availability in the right place at the right time in the right form. Water is a heterogeneous resource that can be found in either a liquid, a solid, or a gaseous form at a number of locations at or near the Earth's surface. Depending upon the specific location of water, the residence time may vary from a few days to

many thousands of years (Table 9.2). Furthermore, more than 99 percent of the Earth's water is unavailable or unsuitable for beneficial human use because of its salinity (sea water) or location (ice caps and glaciers). Thus the amount of water for which all the people on Earth compete is much less than 1 percent of the total.

Compared with other resources, water is used in tremendous quantities. In recent years the total amount of water used on Earth per year is approximately 1000 times the world's total production of minerals, including petroleum, coal, metal ores, and nonmetals [1]. Because of its great abundance, water is generally a very inexpensive resource. But, because the quantity and quality of water that is available at any particular time are highly variable, statistical statements about the cost of water on a global basis are not particularly useful. Shortages of water have occurred and will continue to occur with increasing frequency. Such shortages lead to serious economic disruption and suffering by people [2].

The U.S. Water Resources Council has estimated that water use in the United States by the year 2020 will exceed surface water resources by 13 percent. As early as

TABLE 9.2
The world's water supply (selected examples).

Location	Surface Area (km²)	Water Volume (km³)	Percentage of Total Water	Estimated Average Residence Time of Water
Oceans	361,000,000	1,230,000,000	97.2	Thousands of years
Atmosphere	ь10,000,000	12,700	0.001	9 days
Rivers and streams	—	1,200	0.0001	2 weeks
Groundwater: shallow, to depth of 0.8 km	130,000,000	4,000,000	0.31	Hundreds to many thousands of years
Lakes (freshwater)	855,000	123,000	0.009	Tens of years
Ice caps and glaciers	28,200,000	28,600,000	2.15	Up to tens of thousands of years and longer

SOURCE: Data from U.S. Geological Survey.

1965, 100 million people in the United States used water which had already been used once before, and by the end of the century most of us will be using recycled water. How can we manage our water supply, use, and treatment to maintain adequate supplies?

WATER AS A UNIQUE LIQUID

To understand water as a necessity, as a resource, and as part of the pollution problem, we must understand its characteristics, its role in the biosphere, and its role in living things. Water is a unique liquid; without it, life as we know it seems unlikely. Compared with most other common liquids, water has the greatest capacity to absorb or store heat and is a good liquid solvent. Because many natural waters are slightly acidic, they can dissolve a great variety of kinds of compounds, from simple salts to minerals, including sodium chloride (common table salt) and calcium carbonate (calcite) in limestone rock. Water also reacts with complex organic compounds, including many amino acids that are found in the human body. Compared with other common liquids, water has a high surface tension—a property that is extremely important in many physical and biological processes that involve moving water through or storing water in small openings or pore spaces. Among common compounds and molecules, water is the only one whose solid form is lighter than its liquid form, which is why ice floats. If ice were heavier than liquid water and were to sink to the bottom of water bodies, the biosphere would be greatly different from what it is, and life—if it existed at all—would be greatly altered [3].

A peculiar feature of the biosphere is that the temperatures at the Earth's surface are near what is known as the **triple point** of water. The triple point is the temperature and pressure of water vapor in the atmosphere at which the three phases of water—solid (ice), liquid, and gas (water vapor)—can exist together.

These qualities of water led L. J. Henderson to write early in the twentieth century that our planet was peculiarly and amazingly "fit" to support life. For the support of life, he wrote, "there are no other compounds which share more than a small part of the qualities of fitness of water" [3].

THE WATER CYCLE

As with all geochemical cycles, the cycle of water involves its flow from one storage "compartment" to another. In its simplest form, the global water cycle can be viewed as water flowing from the oceans to the atmosphere, falling from the atmosphere as rain onto the ocean, land, or fresh water, and then returning to the oceans as runoff or to the atmosphere by evaporation.

The problems and potentials of water differ with the uses and needs on the land, the fresh waters, and the oceans. In the rest of this chapter, we will discuss water in each of these parts of the biosphere.

SURFACE RUNOFF

Surface runoff is important because of its effect on erosion and the transport of materials. Water moves materials either in a dissolved state or as suspended particles, and surface water can dislodge soil and rock particles on impact (Fig. 9.1). The size of particles and the amount of suspended particles moved by surface waters depend in part on the volume and depth of the water and the velocity of flow. That is, the faster a stream or river flows, the larger the particles it can move and the more material is transported.

The flow of water on the land is divided into watersheds. A **watershed,** or drainage basin, is an area of ground in which any drop of water falling anywhere in it will leave in the same stream or river (assuming, that is, that the drop is not consumed by the biosphere, evaporated, stored, or transported out of the watershed by subsurface flow). The amount of surface-water runoff from a watershed or drainage basin and the amount of sediment carried vary significantly. The variation results from the diverse climatic, geologic, physiographic, and land-use characteristics of the particular drainage basin together with variations of these factors with time. For instance, even the most casual observer is aware of the difference in the amount of sediment carried by streams in flood stage (they are often more muddy) compared with what appears to be carried at low-flow conditions.

The principal geologic factors affecting surface-water runoff and sedimentation include rock and soil type, mineralogy, degree of weathering, and structural characteristics of the soil and rock. Fine-grained, dense, clay soils and exposed rock types with few fractures generally allow little water to move downward and become part of the subsurface flows. The runoff from precipitation falling on such materials is comparatively rapid. Conversely, sandy and gravelly soils, well-fractured rocks, and soluble rocks absorb a larger amount of precipitation and have less surface runoff.

The biota affect the stream flow in several ways. Vegetation decreases runoff by increasing the amount of rainfall intercepted and removed by evaporation. An experimental clear-cutting of an entire forested watershed at Hubbard Brook, New Hampshire, increased the stream flow in midsummer, when the flow was lowest [4]. The clear-cutting reduced evaporation, making more water available for subsurface runoff. Runoff eventually dimin-

FIGURE 9.1

Raindrops strike the earth with enough force to tear apart unprotected soil and separate particles from one another. They wash the soil and generally move soil particles downslope, (Photos courtesy of U.S. Department of Agriculture.)

ished as the forest became reestablished following timber harvesting. Vegetation also retards erosion by roots binding or holding soil particles in place and by intercepting the rain and reducing its impact on the soil surface. In southern California, where fires are common and the soils erodable, areas that burn during a dry season frequently are subject to hazardous mudslides the next rainy season.

Animals affect water flow by removing vegetation; in fact, large mammals create paths that can turn into small channels during intense precipitation. Soil organisms alter the soil's physical structure, sometimes allowing greater percolation into the soil and reducing runoff and erosion. Because it is cohesive (its particles stick together) and tends to hold water, a highly organic soil will tend to erode less rapidly than a sandy soil with low cohesion, high porosity, and high permeability.

SEDIMENT IN RIVERS

The amount of sediment carried by rivers varies with geologic, climatic, topographic, physical, vegetative, and other conditions. Hence, some rivers are consistently and noticeably different in their clarity and appearance, as can be inferred from Table 9.3. This table reflects varying degrees of human influence as well as sizeable differences in the sediment load per unit area in various parts of the world. For instance, on the average, the Lo River of China carries nearly 200 times more suspended load than does

the Nile River of Egypt. Within the United States, the Mississippi is not as "muddy" as the Missouri and the Colorado rivers. Variations in the natural sediment yield

TABLE 9.3

Some major rivers of the world ranked by sediment yield per unit area.

River	Drainage Basin (10^3 km^2)	Sediment Load per Year (tons/km^2)
Amazon	5776	63
Mississippi	3222	97
Nile	2978	37
Yangtze	1942	257
Missouri	1370	159
Indus	969	449
Ganges	956	1518
Mekong	795	214
Yellow	673	2804
Brahmaputra	666	1090
Colorado	637	212
Irrawaddy	430	695
Red	119	1092
Kosi	62	2774
Ching	57	7158
Lo	26	7308

SOURCE: Data from Holman, 1968.

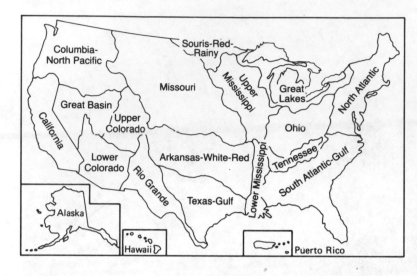

Region	Estimated Sediment Yield (metric tons/km²/yr)		
	High	Low	Average
North Atlantic	4240	110	880
South Atlantic-Gulf	6480	350	2800
Great Lakes	2800	40	350
Ohio	7391	560	2780
Tennessee	5460	1610	2450
Upper Mississippi	13,660	40	2800
Lower Mississippi	28,760	5460	18,220
Souris-Red-Rainy	1650	40	175
Missouri	23,470	40	5250
Arkansas-White-Red	25,760	910	7710
Texas-Gulf	8140	320	6310
Rio Grande	11,700	530	4550
Upper Colorado	11,700	530	6310
Lower Colorado	5670	530	2100
Great Basin	6240	350	1400
Columbia-North Pacific	3850	120	1400
California	19,510	280	4550

FIGURE 9.2

Water resource regions of the United States, and estimated ranges in sediment yield from drainage areas (within these regions) of 260 square kilometers or less. (From Water Resources Council, 1978.)

for different river basins within the United States are summarized in Figure 9.2.

The difference between the sediment loads carried by major rivers is so great that it is part of American folklore. A story is told about a man who claimed that the muddier Mississippi water was more wholesome to drink than the clear waters of the Ohio. He said, "You look at the graveyards; that tells the tale. Trees won't grow worth shucks in a Cincinnati graveyard, but in a St. Louis graveyard they grow upwards of eight hundred foot high. It's all on account of the water the people drunk before they laid up. A Cincinnati corpse don't richen a soil any" [5].

Rivers and lakes have been major resources throughout human history. They provided the first major means for transporting large quantities of goods and people, and rivers, lakes, and ocean harbors are the sites of major cities. This is no accident. Not only do rivers, lakes, and oceans provide transportation, but rivers have provided a major source of energy as water power. Ever since the rise of the Egyptian civilization, the fertile soils deposited on floodplains have been a major agricultural resource.

GROUNDWATER

Groundwater is the water that penetrates the soil and reaches the water table. Some groundwater is derived di-

rectly from precipitation. Other sources of groundwater include water that infiltrates from surface waters, including lakes and rivers; storm-water retention or recharge ponds; and waste-water treatment systems such as cesspools and septic-tank drain fields.

The soil and rocks that groundwater passes through act as natural filters. Under the right conditions, this filtering system cleanses the water, trapping disease-causing microorganisms and particulates that contain toxic compounds. The water actually exchanges materials with the soil and rocks. If the soil or rock surface is already highly contaminated or contains naturally toxic elements, the water may be rendered toxic by these natural processes. For instance, water may become toxic by dissolving sufficient amounts of a toxic element or mineral, such as arsenic. Most often, the groundwaters dissolve a mixture of minerals and some gases which can be nuisances to some human uses. Some examples are iron as ferrous hydroxide, which colors the water brown and leaves a brown discoloration in porcelain; calcium carbonate, which creates the so-called hardness of water; and hydrogen sulphide, which produces a "rotten egg" odor.

People's perceptions about groundwater affect the way they view our water resources. First, people tend to assume that water is available when, where, and in the amounts they want. We turn on a faucet and expect water—it is somebody·else's responsibility to see that we have it. Second, since groundwater is out of sight, it is out of mind and/or mysterious. Third, groundwater is not as easily measured quantitatively as surface water. Therefore, precise quantitative values of groundwater reserves are not available, and we rely on estimates of the probable reserves.

WATER SUPPLY*

The water supply at any particular point on the land surface depends upon several factors in the hydrologic cycle, including the rates of precipitation, evaporation, stream flow, and subsurface flow. The various uses of water by

*Much of the discussion concerning water supply and use is summarized from Water Resources Council, 1978, *The Nation's Water Resources, 1975–2000*, Vol. 1.

people also significantly affect water supply. A concept useful in understanding water supply is the **water budget,** or the inputs and outputs of water in a system. On a continental scale, the water budget for the conterminous United States is shown in Figure 9.3. The amount of water vapor passing over the United States every day is approximately 152,000 million cubic meters (40,000 billion gallons), and of this approximately 10 percent falls as precipitation in the form of rain, snow, hail, or sleet. Approximately two thirds of the precipitation evaporates quickly or is transpired by vegetation. The remaining one third, or about 5510 million cubic meters (1450 billion gallons) per day, enters the surface or groundwater storage systems; flows to the oceans or across the nation's boundaries; is used by consumption; or evaporates from reservoirs. Unfortunately, only a portion of this water can be developed for intensive uses due to natural variations in precipitation that cause either floods or droughts. Thus only about 2565 million cubic meters (675 billion gallons) per day are considered to be available 95 percent of the time [2].

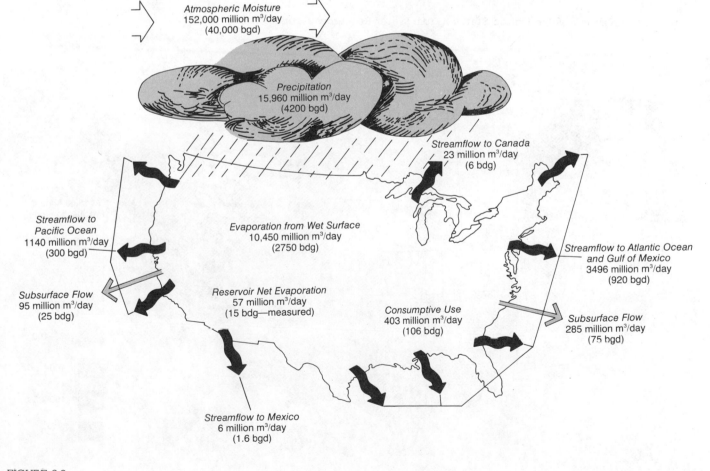

FIGURE 9.3
Water budget for the United States. (From Water Resources Council, 1978.)

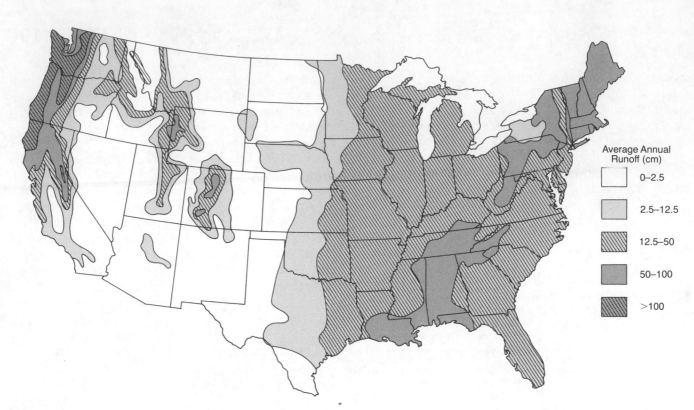

FIGURE 9.5
Average annual runoff of the United States. (From Water Resources Council, 1978.)

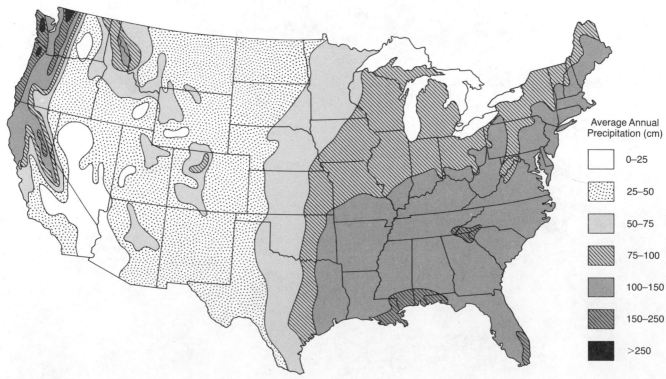

FIGURE 9.4
Average annual precipitation for the United States. (From Water Resources Council, 1978.)

200

On a regional scale, it is critical to consider annual precipitation and runoff patterns in order to develop water budgets. Figures 9.4 and 9.5 illustrate the variability of precipitation and runoff for the conterminous United States. Potential problems with water supply can be predicted in areas where average precipitation and runoff are relatively low, such as in the southwestern and Great Plains regions of the United States as well as in some of the intermontane valleys in the Rocky Mountain area. The theoretical upper limit of surface water supplies is the mean annual runoff, assuming it could be successfully stored. Unfortunately, storage of all the runoff is not possible because of evaporative losses from large reservoirs and a limited number of suitable sites for reservoirs. As a result, there are bound to be shortages in water supply in areas with low precipitation and runoff. Strong conservation practices are necessary to insure an adequate supply [2].

Because there are large annual variations in stream flow, even areas with high precipitation and runoff may periodically suffer from droughts. For example, the dry years of 1961 and 1966 in the northeastern United States and 1976 and 1977 in the western United States produced serious water shortages. Fortunately, in the more humid eastern United States stream flow tends to vary less than in other regions, and drought is less likely [2].

Nearly one half of the U.S. population uses groundwater as a primary source for drinking water. Fortunately, the total amount of groundwater available in the United States is enormous, accounting for approximately 20 percent of all water withdrawn for consumptive uses. Within the conterminous United States the amount of groundwater within 0.8 kilometer of the land surface is estimated to be between 125,000 and 224,000 cubic kilometers. To put this in perspective, the lower estimate is about equal to the total discharge of the Mississippi River during the last 200 years. Unfortunately, due to the cost of pumping and exploration, much less than the total quantity of groundwater is available. Figure 9.6 shows the major aquifers in the conterminous United States capable of yielding more than 0.2 cubic meter (50 gallons) per minute [2].

In many parts of the country groundwater withdrawal from wells exceeds natural inflow. In such cases, water is being mined and can be considered a nonrenewable resource. Groundwater overdraft is a serious problem in the Texas–Oklahoma–High Plains area, California, Arizona, Nevada, New Mexico, and isolated areas of Louisiana, Mississippi, Arkansas, and the south Atlantic–Gulf region. In the Texas–Oklahoma–High Plains area alone, the overdraft amount is approximately equal to the natural flow of the Colorado River [2].

WATER USE

In order to discuss water use, it is important to distinguish between instream and offstream uses. **Offstream uses** remove water from its source, such as water taken for drinking, for washing and sewage, and for agricultural irrigation. **Consumptive use** refers to water that does not return to the stream or groundwater resource immediately after use [2].

Instream water use includes the use of rivers for navigation, hydroelectric power generation, fish and wildlife habitat, and recreation. These multiple uses usually create controversy because each instream use requires different conditions to prevent damage or detrimental effects. Fish and wildlife require certain water levels and flow rates for maximum biological productivity which may differ from the requirements for hydroelectric power generation, which requires large fluctuations in discharges to match power needs. Similarly, both of these may conflict with shipping and boating. The discharge necessary to move the sediment load in a river may require yet another pattern of flow. Figure 9.7 shows some of these conflicting demands in an idealized manner. A major problem concerns how much water may be removed from a stream or river and transported to another location without damaging the stream system. This is a problem in the Pacific Northwest, where certain fish like the steelhead trout and salmon are on the decline partly because human-induced alterations in land use and stream flows have degraded fish habitats.

In California demands are being made on northern rivers for reservoir systems to feed the cities of southern California. In our modern civilization, water is often moved vast distances from areas with abundant rainfall to areas of high usage. In California two thirds of the state's runoff occurs north of San Francisco, where there is a surplus of water, while two thirds of the water use occurs south of San Francisco, where there is a deficit. In recent years canals constructed by the California Water Project and the Central Valley Project have moved tremendous amounts of water from the northern to the southern part of the state.

Many large cities in the country must seek water from areas further and further away. For example, New York City has imported water from nearby areas for over 100 years. The water use and supply in New York City represent a repeating pattern. Originally, local groundwater, streams, and the Hudson River itself were used. However, water needs exceeded local supply, so in 1842 the first large dam was built more than 48 kilometers north of the city. As the city expanded rapidly from Manhattan to Long Island, water needs again increased. The sandy soils of Long Island were at first a source of drink-

FIGURE 9.6
Groundwater resources for the United States. Shown are major aquifers that are capable of yielding more than 0.2 cubic meter per minute of groundwater. (From Water Resources Council, 1978.)

Patterns indicate areas underlain by productive aquifers (capable of yielding 0.2 m³ or more per minute of fresh water to wells). Areas without patterns are underlain by smaller capacity or less extensive aquifers.

EXPLANATION

Watercourse: Productive aquifer adjacent to and capable of replenishment by perennial streams.

Unconsolidated aquifers: Mostly sand and gravel.

Consolidated rock aquifers: Mostly volcanic rocks in the Northwest; mostly sandstone and limestone elsewhere.

Combination aquifers: Sand and gravel aquifers overlying productive rock aquifers.

FIGURE 9.7
Idealized diagram showing instream water uses and the varying discharges for each use. Discharge is the amount of water passing by a particular location and is measured in cubic meters per second.

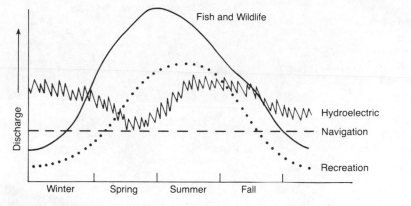

ing water, but this water was removed faster than rainfall replenished it. Local cesspools contaminated the groundwater, and salty ocean water intruded. A larger dam was built at Croton in 1900. Further expansion of the population created the same pattern: initial use of groundwater; pollution, salinification, and exhaustion of this resource; and the subsequent building of new, larger dams further and further upstate in forested areas. The pattern contin-

ues with development of tract housing in eastern Long Island, where there are now problems of pollution, salinification, and exhaustion of the resource. Eventually, the cost of obtaining water from long distances and competition for the available water will place an upper limit on the water supply of the city and its environs.

As more and more water is needed for cities and agriculture, conflicts will increase and intensive argument

FIGURE 9.8
Withdrawals of fresh water in the United States in 1975 and projected to the year 2000. (From Water Resources Council, 1978.)

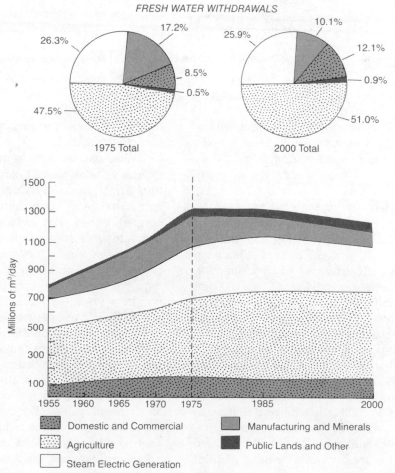

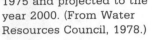

FIGURE 9.9
Freshwater consumption for the United States in 1975 and projected to the year 2000. (From Water Resources Council, 1978.)

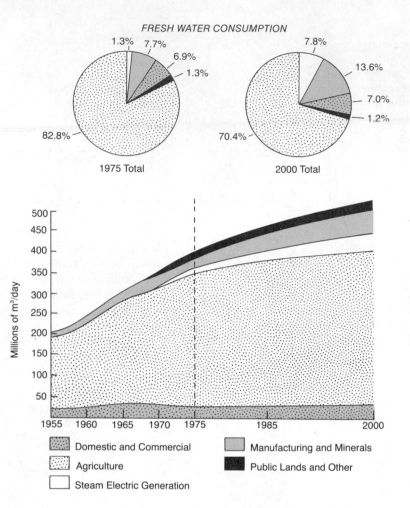

will center on instream water use. An important, fruitful area of research is to evaluate more carefully what flows are necessary to maintain a natural river system.

Offstream water use in the United States today amounts to over 1280 million cubic meters (338 billion gallons) per day, or more than one half the average flow of the Mississippi River, which drains approximately 40 percent of the conterminous United States. Water consumption amounts to about 410 million cubic meters (107 billion gallons) per day. Trends in water use and consumption from 1955 to 1975 and those projected to the year 2000 are shown in Figures 9.8 and 9.9. These diagrams show that agricultural purposes account for nearly half of the total withdrawals and 80 percent of the total water consumed. Water withdrawals for steam generation of electricity are also high, but they are small compared with agricultural use. It is hoped that conservation practices, more water-efficient technology, and recycling of water will reduce withdrawals. By the year 2000 it is expected that withdrawals will decrease, while water consumption will increase slightly. Increases in consumption will be due to population growth and a greater water demand from in-

dustries, manufacturing, agriculture, and steam generation of electricity [2].

What can be done to use water more efficiently and reduce withdrawal and consumption? Improved agricultural irrigation could reduce withdrawals by between 20 and 30 percent. Such improvements include lined and covered canals that reduce seepage and evaporation; computer monitoring and scheduling of water releases from canals; a more integrated use of surface waters and groundwaters; night irrigation; improved irrigation systems (sprinklers, drip irrigation); and better land preparation for water application.

Domestic use of water accounts for only about 6 percent of the total national withdrawals. Because this use is concentrated, it poses major local problems. Withdrawal of water for domestic use may be substantially reduced at a relatively small cost with more efficient bathroom and sink fixtures, night irrigation, and drip irrigation systems for domestic plants. For steam generation of electricity, water removal could be reduced as much as 25 to 30 percent by different types of cooling towers that use less or no water. Manufacturing and industry might curb water withdrawals by increasing in-plant treatment and recycling of water or developing new equipment and processes that require less water [2]. Because the field of water conservation is changing so rapidly, it is expected that a number of innovations will reduce the total withdrawals of water for various purposes even though consumption will continue to increase [2].

SURFACE WATER POLLUTION

As with atmospheric pollutants, water pollutants are categorized as emitted from point or nonpoint sources. Point sources are discrete and confined, such as pipes that empty into streams or rivers from industrial or municipal sites. In general, point-source pollutants from industries are controlled through on-site treatment or disposal and are regulated by permit. Municipal point sources are also regulated by permit. In older cities in the northeastern and Great Lakes areas of the United States, most point sources are outflows from combined sewer systems. These sewer systems combine storm water flow with municipal waste. During heavy rains, urban storm runoff may exceed the capacity of the sewer system, causing it to back up and overflow and delivering pollutants to nearby surface waters.

Nonpoint sources are diffused and intermittent and are influenced by factors such as land use, climate, hydrology, topography, native vegetation, and geology. Common urban nonpoint sources include urban runoff from streets or fields, which contains all sorts of pollutants from heavy metals to chemicals and sediment. Rural sources of nonpoint pollution are generally associated with agriculture, mining, or forestry. Nonpoint sources are difficult to control.

There are many types of materials that may pollute surface (or ground) water. Our discussion will focus on a few selected examples. Sediment pollution, another major problem, will be discussed along with land use and soil erosion.

SOME IMPORTANT KINDS OF WATER POLLUTANTS

Dead organic matter in streams decays. Bacteria carrying out this decay require oxygen. If there is enough bacterial activity, the oxygen in the water can be reduced to levels so low that fish and other organisms die. A stream without oxygen is a dead stream for fish and many organisms we value. The amount of oxygen required for such biological decomposition is called the **biological oxygen demand (BOD)**, a commonly used measure in water quality

management. BOD is measured as milligrams per liter of oxygen consumed over five days at 20°C. Dead organic matter is contributed to streams and rivers from natural sources (such as dead leaves from a forest) as well as from agriculture and urban sewage. Approximately 33 percent of all BOD in streams results from agricultural activities, but urban areas, particularly those with combined sewer systems, also add considerable BOD to streams.

The Council on Environmental Quality defines the threshold for a water pollution alert as a dissolved oxygen content of less than 5 milligrams per liter of water. The idealized diagram in Figure 9.10 illustrates the effect of BOD on dissolved oxygen content for a stream with a sewage input. Three zones are recognized: the pollution zone, with a high BOD and reduced dissolved oxygen content; an active decomposition zone; and a recovery zone in which the dissolved oxygen increases and the BOD is much reduced. A stream has some capability to degrade organic waste after it enters the stream. Problems result when the stream is overloaded with biological-oxygen-demanding waste and its natural cleansing function is overpowered.

Disease-carrying microorganisms are important biological pollutants. Among the major water-borne human diseases are cholera and typhoid. Because it is often difficult to monitor the disease-carrying organisms directly, we use the count of human fecal coliform bacteria as a common measure of biological pollution and a standard measure of microbial pollution. These common, harmless

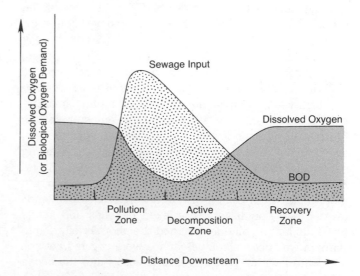

FIGURE 9.10
Idealized diagram of the relationship between dissolved oxygen and biological oxygen demand (BOD) for a stream following the input of sewage. See text for further explanation.

FIGURE 9.11
Relationship between land use and average nitrogen and phosphorus concentration (in milligrams per liter). (From Council on Environmental Quality, 1978.)

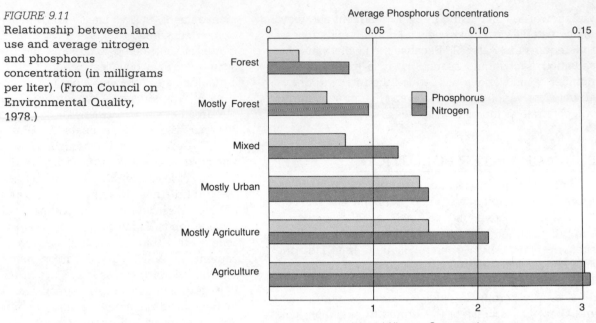

forms of bacteria are normal constitutents of human intestines and are found in all human waste. The threshold used by the Council on Environmental Quality for pollution is 200 cells of fecal coliform bacteria per 100 milliliters of water.

As we saw in the case of Lake Washington, nutrients lead to lake eutrophication and a form of water pollution. The primary pollutant is phosphorus, which is released from agricultural sources such as farm fields and feed lots or urban sewage treatment plants (Fig. 9.11). Most of the phosphorus (66%) released to streams comes from agricultural sources [2].

Oil discharged into surface water (usually the ocean) has caused major pollution problems. Several large oil spills from submarine oil drilling operations have occurred in recent years, such as the 1969 oil spill in the Santa Barbara Channel and the 1979 Yucatán Peninsula spill in Mexico. The latter is the world's largest spill to date, spewing out about 3 million barrels of oil before being capped in 1980. Both spills were caused by an oil well blowing out, and both caused damage to beaches and marine life when the oil drifted ashore. Favorable winds averted a major disaster on Texan beaches and inland wetlands, as the oil only touched the shore briefly. Santa Barbara was not so lucky. Beaches were covered with oil and waterfowl killed. Oil is also released into the ocean from oil tankers. A recent example was the July 29, 1979, collision 80 kilometers northwest of Tabago in the Caribbean of the *Atlantic Empress*, loaded with 276,000 metric

tons of crude oil, with the *Aegean Captain*, carrying 200,000 metric tons of crude. The total amount of oil entering the sea exceeded 216,000 metric tons, making this incident the largest spill from an oil tanker to date. The *Atlantic Empress* sunk, losing or burning its cargo, while the *Aegean Captain* spilled about 20,000 metric tons.

Heavy metals such as mercury, zinc, and cadmium are dangerous pollutants and are often deposited with natural sediment in the bottom of stream channels. If these are deposited on floodplains, then the heavy metals may become incorporated in plants, food crops, and animals. If they are dissolved and the water is withdrawn for agriculture or human use, heavy metal poisoning can result.

The heating of waters, primarily from hot water emission from industrial and power plants, causes thermal pollution. There are several problems with heated water. Even water several degrees warmer than the surrounding water holds less oxygen. Warmer water favors different species than cooler water and may increase growth rates of undesirable organisms, including certain water plants and fish. On the other hand, the warm water may attract and allow better survival of certain desirable fish species, particularly during the winter.

Many synthetic organic and inorganic compounds are hazardous chemical wastes formed as a by-product of industrial processes. Radioactive materials in water are also dangerous pollutants. Of particular concern is the

TABLE 9.4

Thresholds used by the Council on Environmental Quality in analyzing the nation's water quality.

Indicator	Abbreviation	Threshold Level
Fecal coliform bacteria	FC	200 cells/100 ml[a]
Dissolved oxygen	DO	5.0 mg/l[b]
Total phosphorus	TP	0.1 mg/l[c]
Total mercury	Hg	2.0 μg/l[d]
Total lead	Pb	50.0 μg/l[d]
Biochemical oxygen demand	BOD	5.0 mg/l[e]

SOURCE: From Council on Environmental Quality, 1979.
l = liter; ml = milliliter; mg = milligram; μg = microgram.
[a]Criteria level for "bathing waters" from EPA "Redbook."
[b]Criteria level for "good fish populations" from EPA "Redbook."
[c]Value discussed for "prevention of plant nuisances in streams or other flowing waters not discharging directly to lakes or impoundments" in EPA "Redbook."
[d]Criteria level for "domestic water supply (health)" from EPA "Redbook." Criteria level for preservation of aquatic life is much lower.
[e]Value chosen by CEQ.

possible long-term exposure of people to low dosages of radioactivity. Chapter 14 covers chemical and radioactive wastes in more detail.

This list of pollutants certainly is not complete. For some of the pollutants we have discussed, Table 9.4 lists the thresholds used by the Council on Environmental Quality as indicators of water quality.

Since the 1960s there has been a serious attempt in the United States to reduce water pollution and thereby increase water quality. The basic assumption is that people have a real desire for safe water to drink, to swim in, and to use in agriculture and industry. At one time water quality near major urban centers was considerably worse than it is today, and there was at least one instance in which a U.S. river was inadvertently set on fire. In recent years there have been a number of very encouraging success stories; perhaps the best known is the Detroit River. In the 1950s and early 1960s the Detroit River was considered a dead river, having been an open dump for sewage, chemicals, garbage, and urban trash. Tons of phosphorus were discharged each day into the river, and a film of oil up to 0.5 centimeter thick was often present. Aquatic life was damaged considerably, and thousands of ducks and fish were killed. Although today the Detroit River is not a pristine stream, the improvement from industrial and municipal pollution control has been considerable. Oil and grease emissions were reduced by 82 percent between 1963 and 1975, and the shoreline is usually clean. Phosphorus and sewage discharges have also been greatly reduced. Fish once again are found in the Detroit River. Other success stories include New Hampshire's Pemigewasset River, North Carolina's French Broad River, and the Savannah River in the southeastern United States. These examples are evidence that water pollution abatement has positive results [6].

GROUNDWATER POLLUTION

Approximately one-half of all people in the United States today depend on groundwater as their source of drinking water, and we have long believed that groundwater is in general quite pure and safe to drink. Therefore, it may be alarming for some people to learn that groundwater may in fact be quite easily polluted by any one of several sources (Table 9.5) and that the pollutants may be very difficult to recognize. One of the best known examples is the Love Canal near Niagara Falls, New York, where burial of chemical wastes has caused serious water pollution and health problems.

Groundwater pollution differs in several ways from surface water pollution. Groundwater often lacks oxygen, which is helpful in killing aerobic types of microorganisms but which may provide a "happy home" for anaerobic varieties. Bacterial degradation of pollutants, generally confined to the soil or a meter or so below the surface, does not occur readily in groundwater. Furthermore, the channels through which groundwater moves are often very small and variable. Thus, the rate of movement is much reduced (except, perhaps, in large solution channels in rocks such as limestone), and the opportunity for dispersion and dilution is very limited.

Most soils and rocks can filter out solids, including pollution solids. However, this ability varies with different sizes, shapes, and arrangements of filtering particles. Clays and other selected minerals capture and exchange some elements and compounds when they are in solution. Such exchanges are important in the capture of pollutants like heavy metals.

Overpumping of groundwater may lead to infiltration of salt water from the ocean (Fig. 9.12). The intrusion of salt water into freshwater supplies has caused serious

TABLE 9.5
Classification of sources of groundwater pollution and/or contamination.

Wastes		Nonwastes
Sources designed to discharge waste to the land and/or groundwater	Sources that may discharge waste to the land and groundwater unintentionally	Sources that may discharge a contaminant (not a waste) to the land and groundwater
Spray irrigation	Surface impoundments	Buried product storage tanks and pipelines
Septic systems, cesspools, etc.	Landfills	Accidental spills
Land disposal of sludge	Animal feed lots	Highway deicing salt stockpiles
Infiltration or percolation basins	Acid water from mines	Ore stockpiles
Disposal wells	Mine spoil piles and tailings	Application of highway salt
Brine injection wells	Waste disposal sites for hazardous chemicals	Product storage ponds
		Agricultural activities

SOURCE: Modified after U.S. Environmental Protection Agency, 1977, and Lindorff, 1979.

problems in coastal areas of New York, Florida, and California.

Long Island, New York, is a good example of an area with groundwater problems. Two counties on the island, Nassau and Suffolk, with a population of several million people are entirely dependent on groundwater for their water supply. Two major problems associated with groundwater in Nassau County are intrusion of salt water and shallow aquifer contamination [7]. Figure 9.13 shows the general movement of groundwater under natural conditions for Nassau County. Salty groundwater is restricted from inland migration by the large wedge of fresh water moving beneath the island.

In spite of the huge quantities of water in Nassau County's groundwater system, intensive pumping in recent years has caused water levels to decline as much as 15 meters in some areas. As groundwater is removed near coastal areas, the subsurface outflow to the ocean de-

FIGURE 9.12
Idealized drawing showing how saltwater intrusion might occur. The upper drawing shows the groundwater system near the coast under natural conditions, and the lower shows a well with both a cone of depression and a cone of ascension. If pumping is intensive, the cone of ascension may be drawn upward, delivering salt water to the well. The *H* and 40*H* represent the height of the freshwater table above sea level and the depth of salt water below sea level, respectively.

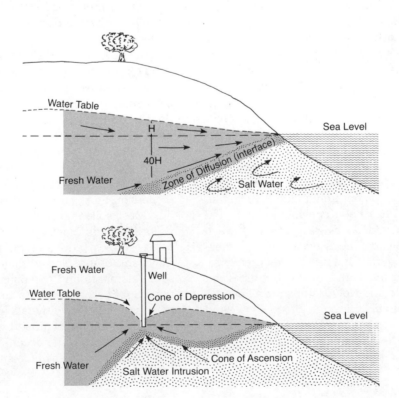

FIGURE 9.13
Idealized drawing of the general movement of fresh groundwater for Nassau County, Long Island, New York. (From Foxworthy, 1978.)

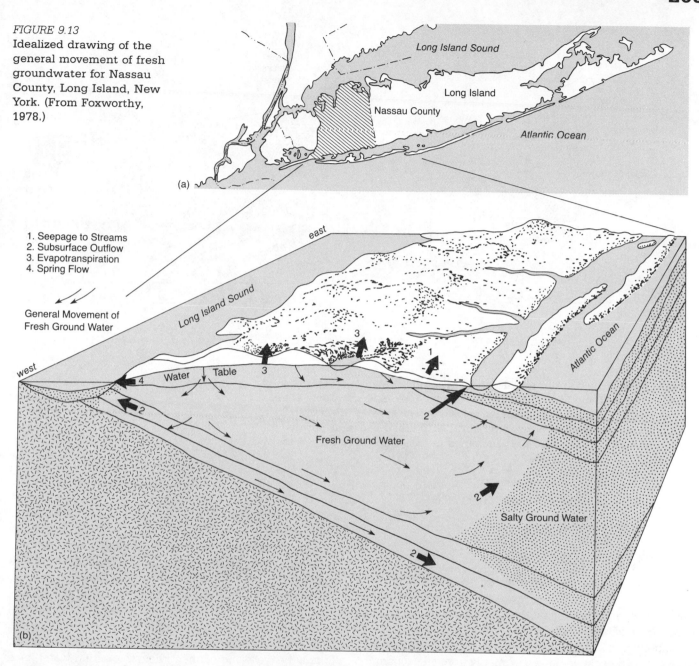

1. Seepage to Streams
2. Subsurface Outflow
3. Evapotranspiration
4. Spring Flow

General Movement of Fresh Ground Water

creases, allowing salt water to migrate inland. Some saltwater intrusion has occurred in Nassau County, but is not yet a serious problem.

The most serious groundwater problem on Long Island is shallow aquifer pollution associated with urbanization. Sources of pollution in Nassau County include urban runoff, household sewage from cesspools and septic tanks, salt used to deice highways, and industrial and solid waste. These pollutants enter surface waters and then migrate downward, especially in areas of intensive pumping and declining groundwater levels. Figure 9.14 shows the extent of high concentration of dissolved ni-

trate in deep groundwater zones. The greatest concentrations are located beneath densely population urban zones, where water levels have dramatically declined and nitrates from sources such as cesspools, septic tanks, and fertilizers are routinely introduced into the hydrologic environment [7].

Drilling of deep wells, such as those for petroleum exploration, also can cause degradation of freshwater aquifers. Some wells allow considerable material to enter freshwater aquifers. This was particularly a problem during the early phases of petroleum and other well drilling (especially for salt), when little was known about the

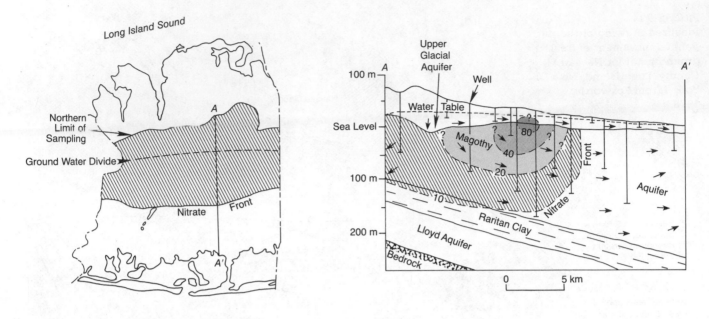

FIGURE 9.14

Extent of high concentration of dissolved nitrate in the groundwater zone of Nassau County, Long Island, New York. The greatest concentrations are located beneath densely populated urban zones, where water levels have dramatically declined and nitrates are more abundant due to urban waste disposal and horticultural practices. Contours are shown in milligrams per liter of dissolved nitrate. (From Foxworthy, 1978.)

depths at which fresh waters could occur in many parts of the United States and throughout the world.

WATER MANAGEMENT

Luna Leopold recently suggested that a new philosophy of water management is needed—one based on geologic, geographic, and climatic factors as well as the traditional economic, social, and political factors. He argues that the management of water resources cannot be successful so long as it is naively perceived primarily from an economic and political standpoint. However, this is how water use is approached. The term *water use* is appropriate because we seldom really *manage* water [8]. The essence of Leopold's water management philosophy is summarized in this section.

Surface water and groundwater are both subject to natural flux with time. In wet years there is plenty of surface water and the near-surface groundwater resources are replenished. During these years we hope that our flood-control structures, bridges, and storm drains will withstand the excess water. Each of these structures is designed to withstand a particular flow (for example, the 20-year flood), which, if exceeded, may cause damage or flooding.

All in all, we are much better prepared to handle floods than water deficiencies. During dry years, which must be expected even though they may not be accurately predicted, we should have specific strategies to minimize hardships. For example, there are subsurface waters in various locations in the western United States that are either too deep to be economically extracted or have marginal water quality. These waters may be isolated from the present hydrologic cycle and therefore are not subject to natural recharge. Such water might be used when the need is great. However, advance planning to drill the wells and connect them to existing water lines is necessary if they are to be ready when the need arises. Another possible emergency plan might involve the treatment of water waste. Re-use of water on a regular basis might be too expensive, but advance planning to re-use treated water during emergencies might be a wise decision [8].

When dealing with groundwater that is naturally replenished in wet years, we should develop plans to use surface water when available and not be afraid to use groundwater in dry years; that is, the groundwater might be pumped out at a rate exceeding the replenishment rate in dry years. During wet years natural recharge and artificial recharge (pumping excess surface water into the ground) will replenish the groundwater resources. This water management plan recognizes that excesses and deficiencies in water are natural and can be planned for.

LAND-USE HYDROLOGY AND SOIL EROSION

Many human activities affect the pattern, amount, and intensity of surface-water runoff, erosion, and sedimentation. Sediment production for possible land-use changes (timber harvesting, urbanization, and conversion to farmland) is idealized in Figure 9.15, and Figure 9.16 summarizes estimated and observed variations in sediment yield under various historical changes in land use.

Figures 9.15 and 9.16 suggest that the effects of land-use change on the drainage basin and its streams may be quite dramatic. Streams in naturally forested or wooded areas may be nearly stable; that is, there is no excessive erosion or deposition. Converting forested land to agriculture generally increases the runoff and sediment yield or erosion of the land. As a result, streams become muddy and may not be able to transport all of the sediment delivered to their channels. Therefore, the channels will partially fill with sediment, possibly increasing the magnitude and frequency of flooding.

The change from agricultural, forested, or rural land to highly urbanized land is even more dramatic. First, during the construction phase, there is a tremendous increase in sediment production, which may be accompanied by a moderate increase in runoff. The response of streams in the area is complex and may include both channel erosion and aggradation, resulting in wide, shallow channels. The combination of increased runoff and shallow channels increases the flood hazard.

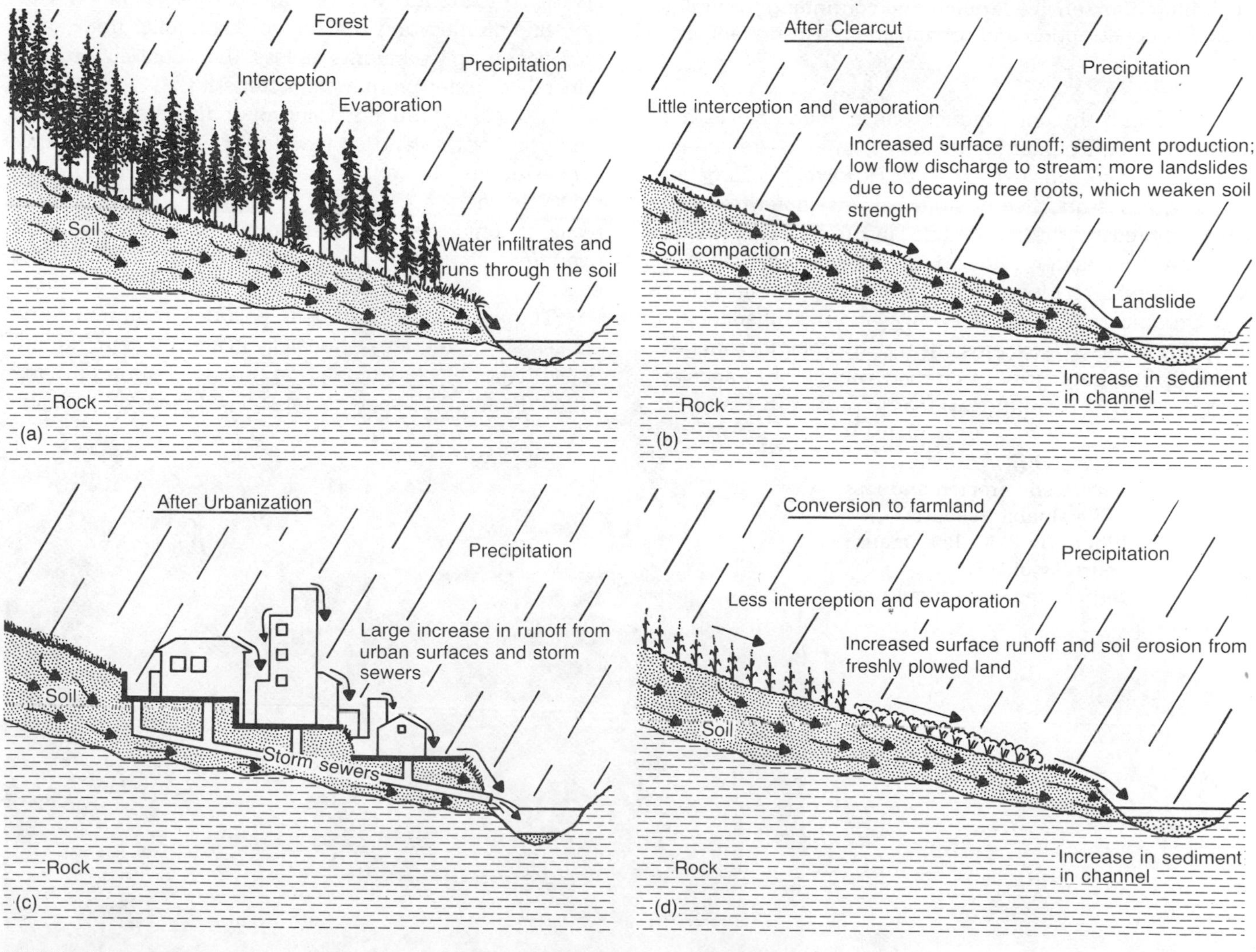

FIGURE 9.15
Idealized diagram showing water relations for a natural forested slope (a) and following several land-use changes: after clearcut (b), after urbanization (c), and after conversion to farmland (d). See text for further explanation.

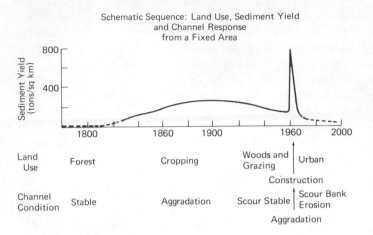

FIGURE 9.16

The effect of land-use change on sediment yield in the Piedmont region of the United States before the beginning of extensive farming and continuing through a period of construction and urbanization. (After Walman, 1967.)

Following the construction phase, much of the land is covered with buildings, parking lots, and streets, and sediment yield drops to a low level. However, the large impervious areas and use of storm sewers increase runoff, which increases the magnitude and frequency of flooding. The streams may respond to the lower sediment yield and higher runoff by eroding (deepening) their channels.

Urbanization is not the only land-use change that causes increased soil erosion and hydrologic changes. Recently the popularity of off-road vehicles has accelerated demand for recreational areas to pursue this interest.

FIGURE 9.17

Severe soil erosion and loss of a valuable resource. (Photo by F. M. Roadman, courtesy of U.S. Department of Agriculture.)

SEDIMENT POLLUTION

Sediment consists of rock and mineral fragments ranging in size from sand particles less than 2 millimeters in diameter to silt, clay, and even finer colloidal particles. By volume, it is our greatest water pollutant. In many areas sediment is choking streams; filling lakes, reservoirs, ponds, canals, drainage ditches, and harbors; burying vegetation; and generally creating a nuisance that is difficult to remove. It is truly a resource out of place. It depletes a land resource (soil) at its site of origin (Fig. 9.17), reduces the quality of the water resource it enters, and may deposit sterile materials on productive croplands or other useful land (Fig. 9.18) [9].

Some polluting sediments are debris from the disposal of industrial, manufacturing, and public wastes. Such sediments include trash directly or indirectly discharged into surface waters. Most of these sediments are very fine-grained and difficult to distinguish from naturally occurring sediments unless they contain unusual minerals or other unique characteristics.

Artificially induced polluting sediments come from disruption of the land surface for construction, farming, deforestation, off-road vehicle use, and channelization works. In short, a great deal of sediment pollution is the result of human use of the environment and civilization's continued change of plans and direction. It cannot be eliminated, only ameliorated.

The solution to sediment pollution requires sound conservation practices, especially in areas where tremendous quantities of sediment are produced during urbanization. Figure 9.19 shows a typical sediment and erosion

FIGURE 9.18
Sediment pollution
damaging productive
farmland. (Photo by R. B.
Brunstead, courtesy of U.S.
Department of Agriculture.)

FIGURE 9.19
Idealized diagram showing
an example of a sediment-
and erosion-control plan for
a commercial development.
(Courtesy of Braxton
Williams, U.S. Soil
Conservation Service.)

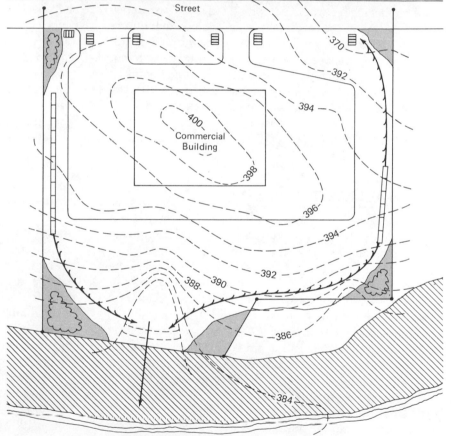

Commercial Sediment and Erosion Control Plan

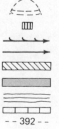

Sediment basin

Storm sewer catch basin

Diversion

Drainage pattern

Buffer zone

Undisturbed area

Perennial stream

Straw bale diversion

Contour showing
elevation   - - 392 - -

213

control plan for a commercial development. The plan calls for diversions to collect runoff and a sediment basin to collect sediment and keep it on the site, thus preventing stream pollution.

That sediment control measures can reduce sediment pollution in an urbanizing area is demonstrated by a study in Maryland [10]. The suspended sediment transported by the northwest branch of the Anacostia River near Colesville, Maryland, with a drainage area of 54.6 square kilometers, was measured over a 10-year period (1962 to 1972). During that time, urban construction within the basin involved about 3 percent of the area each year. The total urban land area in the basin was about 20 percent at the end of the 10-year study. Sediment pollution was a problem because the soils are highly susceptible to erosion and there is sufficient precipitation to insure their erosion when not protected by a vegetative cover. Most of the sediment was transported during spring and summer rainstorms. A sediment control program initiated between 1965 and 1971 reduced sediment yield by an estimated 35 percent. The basic sediment control principles were to tailor the development to the natural topography, expose a minimum amount of land, provide protection for exposed soil, minimize surface runoff from critical areas, and trap eroded sediment on the construction site [10].

CHANNELIZATION

For over 200 years, Americans have lived and worked on floodplains. Of course, building houses, factories, public buildings, and farms on the floodplain invites disaster, but floodplain residents have refused to recognize the natural floodway of the river for what it is: part of the natural river system. As a result, flood control and drainage of wetlands have become prime concerns. Historically, people have attempted to control flooding by constructing dams and levees, or even by rebuilding the entire stream to more efficiently remove the water from the land. Every new project lures more people to the floodplain in the false hope of controlling all floods.

Channelization of streams consists of straightening, deepening, widening, clearing, or lining existing stream channels. Basically, it is an engineering technique to control floods, drain wetlands, control erosion, and improve navigation [11]. Of these four objectives, flood control and drainage improvement are cited most often in channel improvement projects.

Thousands of kilometers of streams in the United States have been modified, and thousands more kilometers of channelization are being planned or constructed. Considering that there are approximately 5.6 million kilo-

meters of streams in the United States, and the average yearly rate of channel modification is only a very small fraction of 1 percent of the total, the impact may appear small. However, many of the alterations occur in flood-sensitive areas, and a change of 1 percent per year means all U.S. waterways will be altered in one century.

Channelization is not necessarily a bad practice, but its adverse environmental effects are not being adequately addressed. In fact, little is known about these effects, and little is being done to evaluate them [11].

Opponents of channelizing natural streams emphasize that the practice inhibits the production of fish and wetland wildlife and degrades the streams' aesthetic qualities. Their argument is as follows: Drainage of wetlands eliminates habitats of certain plant and animal species. Cutting trees along the stream eliminates shading and cover for fish and exposes the stream to the sun, which results in damage to plant life and heat-sensitive aquatic organisms. Cutting of bottom-land (floodplain) hardwood trees destroys the habitats of many animals and birds and also facilitates stream erosion and siltation. Straightening and modifying the stream bed degrades the diversity of flow patterns, particularly the pool environment; changes peak flow; and degrades feeding and breeding areas for aquatic life. Finally, the conversion of wetlands with a meandering stream to a straight, open ditch seriously degrades the aesthetic value of a natural area [11]. Figure 9.20 summarizes some of the differences between natural streams and those modified by people.

Examples of channel work projects that have adversely affected the environment are well known. For example, channelization of the Blackwater River in Missouri enlarged the stream channel, reduced biological productivity, and increased downstream flooding. The natural meandering channel was dredged and shortened in 1910, nearly doubling the natural gradient. This evidently initiated a cycle of channel erosion which is still in progress, and the channel has shown no tendency to resume meandering. The channel's cross-sectional area has increased more than 1000 percent, and as a result, many bridges have collapsed and have had to be replaced. One particular bridge over the Blackwater collapsed due to bank erosion in 1930 and was replaced by a new bridge 27 meters wide. This bridge had to be replaced in 1942 and again in 1947. The 1947 bridge was 70 meters wide, but it too collapsed from bank erosion. Since channelization, erosion has progressed at an average rate of 1 meter in width and 0.16 meter in depth per year [12].

A reduction in the biologic productivity of the Blackwater River occurred because the channel was excavated in easily erodible shale, sandstone, and limestone, producing a smooth stream bed that does not easily sup-

FIGURE 9.20
Comparison of a natural
stream with a channelized
stream. (Modified after
Corning, 1975.)

NATURAL STREAM

Suitable water temperatures:
adequate shading;
good cover for fish life;
minimal temperature variation;
abundant leaf material input.

Pool—Riffle Sequence
Pool—silt, sand,
and fine gravel
Riffle—coarse gravel

Sorted gravels provide diversified habitats
for many steam organisms.

CHANNELIZED STREAM

Increased water temperatures:
no shading; no cover for fish life;
rapid daily and seasonal temperature
fluctuations; reduced leaf material input.

Mostly Riffle

Unsorted gravels;
reduction in habitats; few organisms.

POOL ENVIRONMENT

High Flow

High Flow

Diverse water velocities:
high in pools, lower in riffles. Resting areas
abundant beneath banks, behind large rocks, etc.

May have stream velocity higher than
some aquatic life can stand.
Few or no resting places.

Low Flow

Low Flow

Sufficient water depth to support fish and
other aquatic life during dry season.

Insufficient depth of flow during dry
season to support fish and other aquatic
life. Few if any pools (all riffle).

port bottom fauna after it has been disturbed. As a result, the channelized reaches contain far fewer fish than the natural reaches [12].

Since the channelized portion of the Blackwater has a larger channel, it can carry more water without flooding than the natural, downstream channel. As a result, when heavy rains fall, the runoff that is carried by the upstream channelization section is discharged into the unchannelized downstream section, and frequent flooding results. The downstream end of the Blackwater channelization project terminates where the rock type changes to a more resistant limestone, making dredging financially unfeasible [12].

The channelization of the Blackwater River is a good example of trade-offs and unforeseen degradations caused by channelization. The upper reaches benefitted from the utilization of more floodplain land. However, this benefit must be weighed against the loss of farmland to erosion, the cost of bridge repair, the loss of biological life in the stream, and downstream flooding. A more reasonable trade-off might have been to straighten the channel less, still providing a measure of flood protection while not causing such rapid environmental degradation. The difficult question is, How much straightening can be done before it causes unacceptable damage to the river system?

Not all channelization causes serious environmental degradation, and in some cases drainage projects can even be beneficial. Such benefits are probably best observed in urban areas subject to flooding and in rural areas where previous land use has caused drainage problems. Other examples can be cited to show where channel modification has improved navigation or reduced flooding and has not disrupted the environment.

WASTE-WATER TREATMENT

The quality of water used for industrial and municipal purposes is often degraded by addition of suspended solids, salts, bacteria, and oxygen-demanding material. By law, these waters must be treated before being released back into the environment. Such treatment in the United States now costs approximately $12 billion per year, and the price is expected to double during the next 10 years (Fig. 9.21). Therefore, we can see that waste-water treatment is big business. Conventional methods include disposal and treatment of household waste water by way of septic-tank disposal systems in rural areas and centralized water treatment plants that collect waste water in cities from sewer systems. Recently, innovative approaches to waste-water treatment have been application of waste water to the land, aquaculture, and waste-water renovation and re-use.

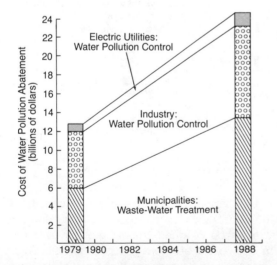

FIGURE 9.21
Cost of water pollution abatement for U.S. municipalities, industry, and electrical utilities from 1979 to 1988. (Data from Council on Environmental Quality and the Department of State, 1980.)

SEPTIC-TANK DISPOSAL SYSTEM

In recent years in the United States people have been moving in great numbers from rural to urban or urbanizing areas. In many instances, a city's sewage system and waste-water treatment facility have not been able to keep pace with the growth. As a result, the individual septic-tank disposal system continues to be an important method of sewage disposal. Unfortunately, all land is not suitable for installation of a septic-tank disposal system, so evaluation of individual sites is usually required by law before a permit can be issued. An alert buyer will check to make sure that the site is satisfactory before purchasing property on the fringe of an urban area requiring a septic-tank disposal system. Failure to do so has made many buyers unhappy, and they may pass the property on to another unsuspecting person.

The basic parts of a septic-tank disposal system are shown in Figure 9.22. The sewer line from the house leads to an underground septic tank in the yard. The tank is designed to separate solids from the liquid, digest and store organic matter through a period of detention, and allow the clarified liquid to discharge into the *seepage bed* or *absorption field,* which is a system of piping through which the treated sewage may seep into the surrounding soil. As the waste water moves through the soil, it is further treated by the natural processes of oxidation and filtering. By the time the water reaches any freshwater supply, it should be safe for other uses.

Sewage absorption fields may fail for several reasons. The most common cause is poor soil drainage (Fig. 9.23), which allows the effluent to rise to the surface in wet weather. Poor drainage can be expected in areas with clay or compacted soil with low permeability, a high water table, impermeable rock near the surface, or frequent flooding.

WASTE-WATER TREATMENT PLANTS

Conventional waste-water treatment falls into two broad classes: **primary treatment,** which involves the mechanical removal of solid material in the water; and **secondary treatment,** which involves biological oxidation of dissolved organic material (Fig. 9.24). Primary treatment lowers the biological oxygen demand (BOD) to some extent, but it is during secondary treatment that the BOD is greatly reduced. A third class of treatment known as **tertiary treatment** removes remaining solids, particularly dissolved minerals, or organic compounds. Probably the most common type of tertiary treatment is chlorination, which removes disease-causing organisms. Other ad-

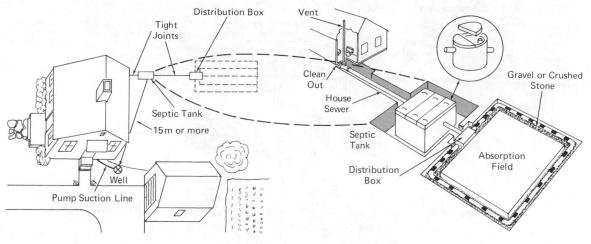

FIGURE 9.22
Generalized diagram showing septic-tank sewage disposal system (right) and location of the absorption field with respect to the house and well. (After Indiana State Board of Health.)

vanced tertiary treatment processes which remove undesirable chemicals (e.g., phosphorus) that promote unpleasant algae growth in streams and lakes or dissolved organic compounds or inorganic minerals are not yet in widespread use. However, tertiary treatment will become more common in the future as more stringent water quality standards are enforced.

LAND APPLICATION OF WASTE WATER

The innovative practice of applying waste water to the land involves the fundamental belief that waste is simply a resource out of place. The idea is sometimes expressed as the **waste-water renovation and conservation cycle,** as shown schematically in Figure 9.25. The major processes in the cycle are as follows: return of treated

FIGURE 9.23
Effluent from a septic-tank sewage disposal system rising to the surface in a backyard. Septic systems in poorly drained soils will not function well during wet weather. (Photo by W. B. Parker, courtesy of U.S. Department of Agriculture.)

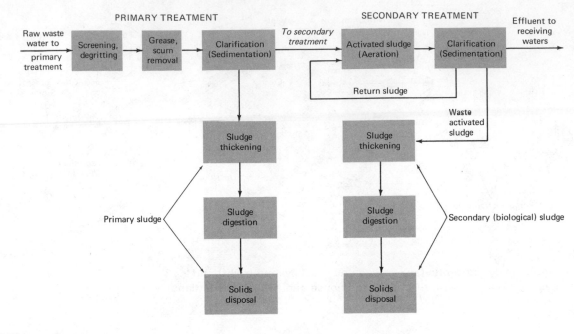

FIGURE 9.24

Flow chart showing various procedures in primary and secondary treatment of waste water. (From American Chemical Society, 1969.)

waste water by a sprinkler or other irrigation system to crops; renovation, or natural purification by slow percolation of the waste water through the soil to eventually recharge the groundwater resource with clean water; and re-use of the water by pumping it out of the ground for municipal, industrial, institutional, or agricultural purposes [13].

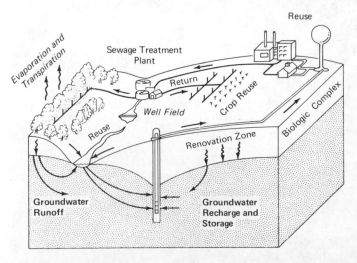

FIGURE 9.25

Idealized diagram showing the waste-water renovation and conservation cycle. (After Parizek and Myers, 1968.)

Recycling of waste water is now being practiced at a number of sites around the United States. In a large-scale waste-water recycling program near Muskegon and Whitehall, Michigan (Fig. 9.26), raw sewage from homes and industry is transported by sewers to the treatment plant, where it receives primary and secondary treatment. The wastes are then chlorinated and pumped into a network that transports the effluent to a series of spray irrigation rigs. After the waste water trickles down through the soil, it is collected in a network of tile drains and transported to the Muskegon River for final disposal. This last step is an indirect tertiary treatment using the natural physical and biological environment as a filter. To date this excellent system removes most of the potential pollutants as well as heavy metals, color, and viruses.

Waste water is also being applied experimentally to freshwater marshes and forest lands. In the future we can expect more tertiary treatment of waste water using biological systems.

AQUACULTURE AND WASTE-WATER TREATMENT

Use of municipal sewage in fish ponds is evidently a very old practice. For example, in China, where aquaculture started, waste disposal in fish ponds was quite common; in fact, it is reported that latrines were even built directly

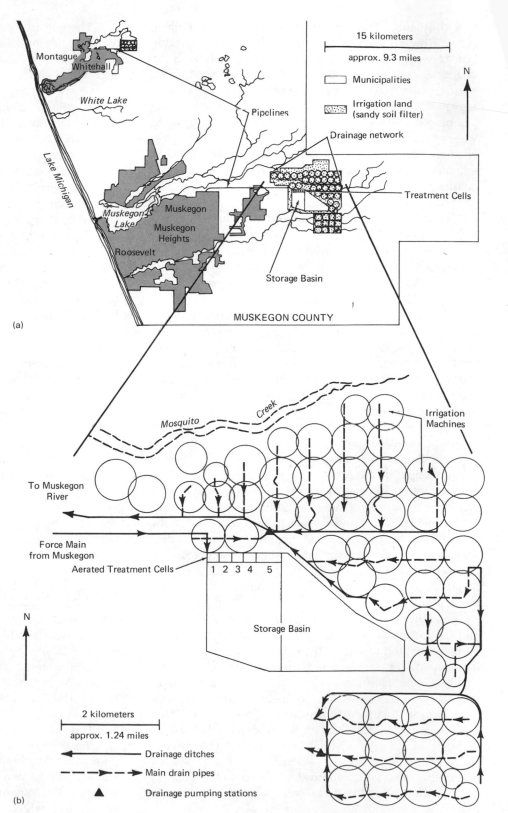

FIGURE 9.26
(a) Two effluent-application areas near Muskegon and Whitehall, Michigan, will eliminate the necessity to discharge inadequately treated industrial and municipal waste into surface waters. The two subsystems are designed to replace four municipal treatment facilities. The effluent-on-land system is designed to handle the 1992 requirements of the county, which has a population of about 170,000. (b) Raw sewage from the sewers of Muskegon is transported by way of a pressure main to the Muskegon–Mona Lake reclamation site. There the effluent flows through three aerated treatment lagoons (1, 2, and 3) and then is diverted either to a large storage basin or into the settling lagoon (4) before going into the outlet lagoon (5). Waste water leaving the outlet lagoon is chlorinated and then pumped through a piping system that carries it to spray irrigation rigs. Circles on the diagram indicate the radius sweep by the spray rigs. After the effluent trickles down through the soil, it is collected by a network of underground pipes and pumped to surface waters. (From Chaiken, Poloncsik, and Wilson, 1973.)

ver the ponds. This method had two benefits: First, it solved a waste disposal problem; and second, the addition of waste caused an increase in the fish yield. Of course, adding waste to the water may have also created a health hazard to the people working in the ponds or eating the fish, but this potential hazard was evidently not recognized [14]. Today aquaculture is teamed up with waste disposal systems at several sites. Two examples, one from northern California and the other from Israel, will emphasize the spectrum of relationships between aquaculture and waste-water treatment.

The waste-water treatment system utilized by the city of Arcata in northern California services approximately 20,000 people. The waste water comes mostly from homes, with minor inputs from lumber and plywood plants, and is treated by standard primary and secondary methods. It is then chlorinated and dechlorinated before being discharged into Humboldt Bay.

Oxidation ponds—part of the secondary treatment— are where aquaculture begins. Pacific salmon fingerlings are raised in water drawn from the oxidation ponds. During the winter the water is aerated so that it is not toxic to the fish, and during the summer the pond water is diluted slightly with sea water from the bay to provide a better habitat for the young fish.

Using waste water for aquaculture comes from the philosophy that our domestic sewage is really water plus fertilizer. That is, waste water is a resource that may actually grow fish. In the Arcata example, the young fish are released into the ocean by way of streams and Humboldt Bay, where presumably they will return a few years later as mature salmon. Thus the waste water is not directly used to feed fish that people consume. Rather, the fish enter the natural environment for some period before returning to local streams to spawn. In some respects, aquaculture may be thought of as an advanced biological tertiary treatment; as the fish and other living things in the pond utilize the waste, there is no reason to chlorinate and dechlorinate the water. Thus no chlorinated hydrocarbons are produced from the chlorination, and no costly dechlorination units need to be installed [15].

The experiments at Arcata, California, certainly indicate that further research to meld aquaculture with waste-water treatment is a worthwhile endeavor. There have been problems with fish being killed either by the water or by predation by birds, but overall the survival rate has improved with experience. The Kohoe salmon and trout have a better survival rate than do the Chinook salmon, suggesting that some fish species adapt better to these specialized ponds.

There is little doubt that organic waste, when added to ponds, increases the supply of food available, and that these ponds produce higher yields of fish. The organic waste from sewage carries nutrients (nitrogen and phosphorus, as well as trace elements) that are important for phytoplankton, which are a significant link in the food chain of fish ponds. In commercial fish ponds supplemental fertilization has increased the biomass of plants and other organisms that are natural food for the fish. Data from a commercial fish farm in Israel clearly illustrate the benefits of applying waste water to fish ponds. The entire sewage from a community of approximately 500 people is treated by primary methods, diluted with fresh water, and then spread among three fish ponds with a combined area of 2.7 hectares. A fourth pond receives the flow of manure and wash water from a dairy serving the community. Data suggest that these ponds produce high-quality fish that have experienced no major health problems. The fat content of the fish in the waste-water ponds is lower than for fish that receive supplemental rations of pelletized food. The lower fat content is probably the result of eating more natural food [14].

The consumption of fish raised in waste-water ponds presents cultural problems. People are not generally willing to buy and eat such fish even though they may be of high quality. Fish produced in the Arcata experiment may be more palatable because they are released into the natural environment and return with the natural stock to local streams, where they are harvested. Thus the systems most likely to succeed will be those which use biological systems for advanced tertiary treatment and include natural ecosystems for at least part of the life cycle of the fish involved.

WATER RE-USE

Water re-use generally refers to the use of waste water following some sort of treatment and is often discussed in terms of an emergency water supply; a long-term solution to a local water shortage; or a fringe benefit to water pollution abatement. Figure 9.27 shows some of the locations in the United States where data has been or is being collected as part of water re-use research or implementation [16].

Water re-use can be inadvertent, indirect, or direct (Fig. 9.28). **Inadvertent re-use** of water results when water is withdrawn, used, treated, and returned to the environment without specific plans for further withdrawals and use, which nevertheless occur (Fig. 9.28a). Such

FIGURE 9.27
Some of the locations in the United States where data have been or are being collected as part of a water re-use research or implementation program. (Modified after Kasperson, 1977.)

▲ Study Proper
△ Pilot Study
○ Consulting Engineering Firms
● Public Health Departments

use patterns occur along many rivers and, in fact, are accepted as a common and necessary procedure for obtaining a water supply. There are several risks associated with inadvertent re-use. Inadequate treatment facilities may deliver contaminated or poor-quality water to users. Because the fate of disease-causing viruses during and after treatment is not completely known, we are uncertain about the environmental health hazards of treated water. In addition, each year many new chemicals are introduced into the environment, some of which may cause birth defects, genetic damage, or cancer in humans. Unfortunately, harmful chemicals are often difficult to detect, and their effects on humans may be hidden if the chemicals are ingested in low concentrations over many years [16]. In spite of these problems, inadvertent re-use of water will by necessity remain a common pattern. If we recognize the potential risks, we can plan to minimize them by using the best possible water treatment available.

Indirect water re-use (Fig. 9.28b) is a planned endeavor, one example of which is the waste-water reclamation cycle shown in Figure 9.25. Similar plans have been used in southern California, where several thousand cubic meters of treated waste water per day have been applied to surface recharge areas. The treated water eventually enters into groundwater storage to be re-used for agricultural and municipal purposes.

Direct water re-use refers to treated water that is piped directly to the next user (Fig. 9.28c). In most cases the "user" will be industrial or agricultural activity. Very little direct use of water is planned (except in emergen-

cies) for human consumption because of cultural attitudes. Figure 9.29 summarizes these attitudes: Direct ingestion is accepted least, whereas uses in which there is no body contact are generally much more acceptable.

DESALINATION

Desalination of sea water which contains about 3.5 percent salt (each cubic meter of sea water contains about 40 kilograms of salt) is an expensive form of water treatment practiced at several hundred plants around the world. The salt content must be reduced to about 0.05 percent. Large desalination plants produce 20,000 to 30,000 cubic meters of water per day at a cost of about 10 times that paid for traditional water supplies in the United States. Desalinated water has a "place value," which means that the price increases quickly with the transport distance and elevation increase from the plant at sea level. Because the various processes that actually remove the salt require energy, the cost of the water is tied to ever-increasing energy costs. For these reasons, desalination will remain an expensive process that will be used only when alternative water sources are not available.

Middle Eastern countries in particular will continue to use desalination. In many arid regions, including the Middle East, there are brackish ground and surface waters with a salinity of about 0.5 percent (one seventh of sea water). Obviously, desalination of this water is less expensive and plants may be located at inland sites.

FIGURE 9.28
Idealized diagrams showing three types of water re-use. (Modified from Symons, 1968, by Kasperson, 1977.)

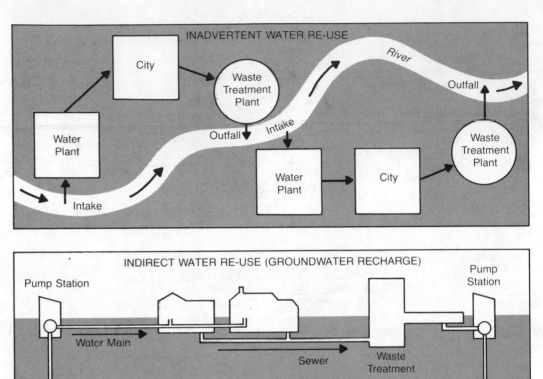

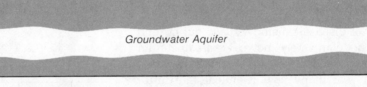

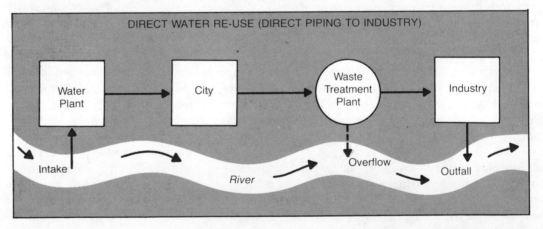

SUMMARY AND CONCLUSIONS

An obvious and well-known detrimental aspect of human use of surface water, groundwater, and atmospheric water is the pollution of rivers and groundwaters. We are beginning to understand that solutions to many hydrologic problems require integrating all aspects of the water cycle.

Water is one of the most abundant and important renewable resources on Earth. However, more than 99 percent of the Earth's water is unavailable or unsuitable for beneficial human use because of its salinity or location. The pattern of water supply and use on Earth at any particular point on the land surface involves interactions between the biological, hydrologic, and rock cycles. To evaluate a region's water sources and use patterns, a water budget is developed to define the natural variability and availability of water.

During the next several decades it is expected that the total water withdrawn from streams and groundwater will decrease slightly, but that the consumptive use will increase because of greater demands from a growing

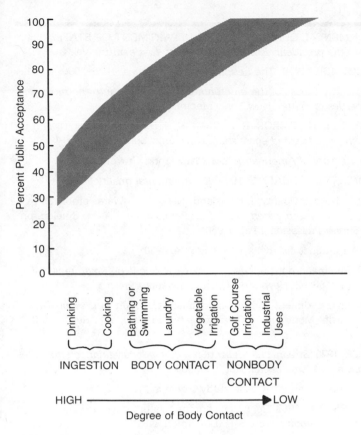

FIGURE 9.29
Diagram showing the public acceptance of use of treated waste water. In general, people have a negative attitude toward direct re-use, that is, drinking of treated waste water. (Modified after Kasperson, 1977.)

population and industry. Water withdrawn from streams competes with other instream needs, such as maintaining fish and wildlife habitats and navigation, and may therefore cause conflicts.

Water pollution specifically refers to degradation of water quality as measured from physical, chemical, or biological criteria. These criteria take into consideration the intended use for the water, departure from the norm, effects on public health, and ecological impacts. Water pollutants have point or nonpoint sources, as do air pollutants. The major water pollutants are oxygen-demanding waste (BOD), pathogens and fecal coliform bacteria, nutrients, synthetic organic and inorganic compounds, oil, heavy metals, radioactive materials, and heat. Since the 1960s there has been a serious attempt to improve water quality in the United States. Although the program seems to have been successful, water quality in many areas is still substandard.

The movement of water down to the water table and through aquifers is an integral part of the rock and hydrologic cycles. In moving through an aquifer, groundwater may improve in quality. However, it may also be rendered unsuitable for human use by natural or artificial contaminants.

In the case of groundwater pollution, the physical, biologic, and geologic environments are considerably different from those of surface water. The ability of many soils and rocks to physically or otherwise degrade pollutants is well known, but not so generally known is the ability of clays and other earth materials to capture and exchange certain elements and compounds.

Pollution of an aquifer may result from disposal of wastes on the land surface or in the ground. It can also result from overpumping of groundwater in coastal areas, which leads to intrusion of salt water into freshwater aquifers.

Principal human influences affecting runoff and sediment production include varied land uses (especially urban) and construction projects to control floods, such as channel improvement and reservoirs.

Sediment pollution, natural or artificial, is certainly one of the most significant pollution problems. A great deal of sediment pollution is a result of human activity, particularly construction, agriculture, and urbanization. Although the problem cannot be completely eliminated, it can be minimized.

Channelization is the straightening, deepening, widening, cleaning, or lining of existing streams. The most commonly cited objectives of channel modification are flood control and drainage improvement. Although many cases of environmental degradation resulting from channelization can be cited, is is not necessarily a bad practice. In fact, there are several cases where channel modification was clearly a beneficial practice. Nevertheless, the present state of channel modification technology does not insure that environmental degradation will not result. Therefore, until new design criteria compatible with natural stream processes are developed and tested, channelization should be considered only as a solution to a particular flood or drainage problem where its impact and alternatives have been carefully studied to minimize environmental disruption.

Waste-water treatment in the United States costs about $12 billion per year, and the cost is expected to double in the next 10 years. Conventional methods of water treatment include large central treatment plants. Application of waste water to farm or forest land, use of waste water in aquaculture, and waste-water renovation for direct and indirect re-use are recent innovations.

Desalination of sea water in specific instances will continue, but large-scale desalination is not likely due to increasing costs of the energy used in the treatment process and in transporting the water to use sites.

Water resource management is in need of a new philosophy that considers geologic, geographic, and climatic factors and utilizes creative alternatives.

REFERENCES

1 COUNCIL ON ENVIRONMENTAL QUALITY and THE DEPARTMENT OF STATE. 1980. *The global 2000 report to the president: Entering the twenty-first century.* Vol. 2.

2 WATER RESOURCES COUNCIL. 1978. *The nation's water resources, 1975–2000.* Vol. 1.

3 HENDERSON, L. J. 1913. *The fitness of the environment: An inquiry into the biological significance of the properties of matter.* New York: Macmillan.

4 LIKENS, G. E.; BORMANN, F. H.; PIERCE, R. S.; EATON, J. S.; and JOHNSON, N. M. 1977. *The biogeochemistry of a forested ecosystem.* New York: Springer-Verlag.

5 BOTKIN, B. A. 1944. *A treasury of American folklore.* New York: Crown.

6 COUNCIL ON ENVIRONMENTAL QUALITY. 1979. *Environmental quality.*

7 FOXWORTHY, G. L. 1978. Nassau County, Long Island, New York—Water problems in humid country. In *Nature to be commanded,* eds. G. D. Robinson and A. M. Spieker, pp. 55–68. U.S. Geological Survey Professional Paper 950.

8 LEOPOLD, L. B. 1977. A reverence for rivers. *Geology* 5: 429–30.

9 ROBINSON, A. R. 1973. Sediment, our greatest pollutant? In *Focus on environmental geology,* ed. R. W. Tank, pp. 186–92. New York: Oxford University Press.

10 YORKE, T. H. 1975. Effects of sediment control on sediment transport in the northwest branch, Anacostia River Basin, Montgomery County, Maryland. *Journal of Research* 3: 487–94.

11 U.S., CONGRESS, HOUSE. 1973. *Stream channelization: What federally financed draglines and bulldozers do to our nation's streams.* H. Rpt. 93–530.

12 EMERSON, J. W. 1971. Channelization: A case study. *Science* 173: 325–26.

13 PARIZEK, R. R., and MYERS, E. A. 1968. Recharge of groundwater from renovated sewage effluent by spray irrigation. *Proceedings of the Fourth American Water Resources Conference,* pp., 425–43.

14 HEPHER, B., and SCHROEDER, G. L. 1977. Waste water utilization in Israel aquaculture. In *Wastewater renovation and reuse,* ed. F. M. D'Itri, pp. 529–59. New York: Marcel Dekker.

15 ALLEN, G. H., and CARPENTER, R. L. 1977. The cultivation of fish with emphasis on salmonids in municipal wastewater lagoons as an available protein source for human beings. In *Wastewater renovation and reuse,* ed. F. M. D'Itri, pp. 479–528. New York: Marcel Dekker.

16 KASPERSON, R. E. 1977. Water re-use: Need prospect. In *Water re-use and the cities,* eds. R. E. Kasperson and J. X. Kasperson, pp. 3–25. Hanover, N.H.: University Press of New England.

FURTHER READING

1 DUNNE, T., and LEOPOLD, L. B. 1978. *Water in environmental planning.* San Francisco: W. H. Freeman.

2 LEOPOLD, L. B. 1974. *Water: A primer.* San Francisco: W. H. Freeman.

3 MILLS, D. H. 1972. *An introduction to freshwater ecology.* Edinburgh: Oliver and Boyd.

4 MORISAWA, M. 1968. *Streams.* New York: McGraw-Hill.

5 SCHUMM, S. A. 1977. *The fluvial system.* New York: Wiley.

6 STOKER, H. S., and SEAGER, S. L. 1976. *Environmental chemistry: Air and water pollution.* 2nd ed. Dallas: Scott, Foresman.

7 UNESCO. 1978. *World water balance and water resources of the Earth.* Paris: UNESCO Press.

8 WATER RESOURCES COUNCIL. 1978. *The nation's water resources, 1975–2000.* Vol. 1.

STUDY QUESTIONS

1 Would the strategy used to deal with water pollution in Seattle, Washington, have worked in Sante Fe, New Mexico? Why or why not?

2 If water is one of our most abundant resources, why are we concerned about is availability in the future?

3 How does clearcut logging affect surface runoff?

4 Why did fish in streams decrease in abundance following logging in New England?

5 Which is more important from a national point of view: conservation of water use in agriculture or in urban areas?

6 Distinguish between instream and offstream uses of water. Why is instream water use so controversial?

7 Compare and contrast surface water pollution with groundwater pollution. Which is easier to treat and why?

8 Why is sediment pollution considered to be one of the major pollution problems?

9 Compare the benefits of stream channelization with its adverse environmental effects.

10 While rafting down a wild river in the wilderness, your guide argues that all channelization is bad. Would you agree or disagree? Why?

11 In the summer you buy a house with a septic system which appears to function properly. In the winter, effluent discharges at the surface. What could be the possible environmental causes of the problem?

12 What are possible socially acceptable, beneficial uses of waste water?

13 In a city along an ocean coast, rare water birds are found to inhabit a pond which is part of a sewage treatment plant. How could this have happened? Is the water in the sewage pond polluted? Consider this question from the birds' and your point of view.

14 When and where is desalination the answer to water supply problems?

10

Mineral Resources: Using, Conserving, and Recycling What Is Not Renewable

INTRODUCTION

It was recently discovered that ash from the incineration of sewage sludge in Palo Alto, California, contains large concentrations of gold (30 parts per million), silver (660 parts per million), copper (8000 parts per million), and phosphorus (6.6 percent). Each metric ton of the ash contains approximately 1 troy ounce of gold and 20 ounces of silver. The gold is concentrated, above natural abundance by a factor of 7500 times, making the "deposit" double the average grade that is mined today. Silver in the ash has a concentration similar to rich ore deposits in Idaho. Copper is concentrated in the ash by a factor of 145, similar to that of a common ore grade. The ash in the Palo Alto dump represents a silver and gold deposit with a 1980 value of about $10 million, and gold and silver worth approximately $2 million are being concentrated and delivered each year [1]. Commercial phosphorus deposits vary from 2 to 16 percent, so the ash is a phosphorus resource of great value.

The most likely source of the metals in the Palo Alto sewage is the large electronics industry as well as the photographic industry located in the area. Gold in significant amounts has only been found in the sewage of one other city, and silver is usually present in much smaller concentrations than at Palo Alto. Thus Palo Alto's unique urban ore offers an unusual opportunity to study and develop methods to recycle valuable materials concentrated in urban waste [1]. The city has now employed a private company to extract the gold and silver.

THE IMPORTANCE OF MINERAL ORES

Metals in mineral form are generally extracted from naturally occurring, anomalously high concentrations of earth materials. When metals are concentrated in anomalously high amounts in ore bodies by geologic processes—or by urban processes, such as in Palo Alto—**ore deposits** are formed. It was such natural deposits that allowed early peoples to exploit copper, tin, gold, silver, and other metals while slowly developing the skills in working with metals.

Availability of mineral resources is one measure of the wealth of a society. Those people who have been successful in the location, extraction, or importation and use of minerals have grown and prospered. Without mineral resources, modern technological civilization as we know it would not be possible.

Many mineral products are found in the typical American home (Table 10.1). In fact, the standard of living in a modern technological society depends to a certain

TABLE 10.1

Selected examples of mineral products in a typical American home.

Building materials: sand, gravel, stone, brick (clay), cement, steel, aluminum, asphalt, glass.

Plumbing and wiring materials: iron and steel, copper, brass, lead, cement, asbestos, glass, tile, plastic.

Insulating materials: rock wool, fiberglass, gypsum (plaster and wallboard).

Paint and wallpaper: mineral pigments (such as iron, zinc, and titanium) and fillers (such as talc and asbestos).

Plastic floor tiles, other plastics: mineral fillers and pigments, petroleum products.

Appliances: iron, copper, and many rare metals.

Furniture: synthetic fibers made from minerals (principally coal and petroleum products); steel springs; wood finished with rotten-stone polish and mineral varnish.

Clothing: natural fibers grown with mineral fertilizers; synthetic fibers made from minerals (principally coal and petroleum products).

Food: grown with mineral fertilizers; processed and packaged by machines made of metals.

Drugs and cosmetics: mineral chemicals.

Other items, such as windows, screens, light bulbs, porcelain fixtures, china, utensils, jewelry: all made from mineral products.

SOURCE: U.S. Geological Survey, 1975.

extent on the per capita availability of mineral resources [2]. Other things being equal, one's standard of living increases with the availability in a useful form of all resources, including metals, nonmetals, energy, soil, air, and water. Our ingenuity allows us to circumvent some problems of resource availability, but it is impossible to maintain an ever-growing population or a constant average standard of living as long as population continues to grow and resources are finite. Thus our resource problem is fundamentally a people problem—too many people chasing after a limited supply of resources.

The mineral resource cycle is idealized in Figure 10.1. Inspection of this diagram reveals that many components of the resource cycle are connected to waste disposal. In fact, the major environmental impacts of mineral resource utilization are related to waste products. Wastes produce pollution that may be toxic to humans, are dangerous to natural ecosystems and the biosphere, and are aesthetically undesirable. They may attack and degrade other resources such as air, water, soil, and living things—in other words, they may destroy entire ecological systems.

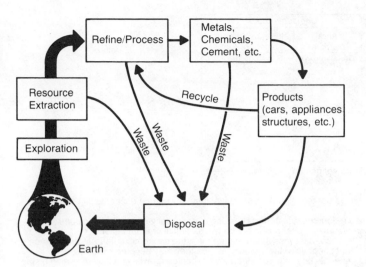

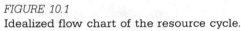

FIGURE 10.1
Idealized flow chart of the resource cycle.

Useful consumption of mineral resources is that which minimizes environmental degradation while allowing for continued use and conservation. If we are careful in mining, transporting, manufacturing, and recycling our mineral resources, we may be able to use natural ecological processes in ecosystems and the biosphere to aid in waste recycling and disposal. These "public service" functions of ecosystems and the biosphere are important components of the total utilization of resources.

MINERAL RESOURCE ABUNDANCE

The Earth's outer layer or crust is silica-rich, made up of mostly rock-forming minerals containing silica, oxygen, and a few other elements. The elements are not evenly distributed in the crust, as eight (oxygen, 46.4%; silicon, 28.2%; aluminum, 8.2%; iron, 5.6%; calcium, 4.2%; sodium, 2.4%; potassium, 2.1%; and titanium, 0.6%) account for over 99 percent by weight. Remaining elements are found (on the average) in trace concentrations. Fortunately, the geologic cycle occasionally concentrates elements in a local environment to a greater degree than would be found on the average.

The ocean, covering nearly 71 percent of the Earth, is another reservoir for many materials. Most elements in the ocean have been weathered from crustal rocks and transported to the oceans by rivers. Other elements are transported to the ocean by wind or glaciers. Ocean water contains about 3.5 percent dissolved solids, most of which is chlorine (55.1% by weight). Each cubic kilometer of ocean water contains about 2.0 metric tons zinc, 2.0 metric tons copper, 0.8 metric ton tin, 0.3 metric ton silver, and 0.01 metric ton gold. These concentrations are low compared with the crust, where corresponding values (in metric tons per cubic kilometer) are zinc, 170,000; copper, 86,000; tin, 5700; silver, 160; and gold, 5. Thus, if we deplete rich crustal ore deposits, we would be more likely to extract metals from lower grade ore deposits or even common rock rather than from ocean water. On the other hand, if mineral extraction technology becomes more efficient, this prognosis could change.

RESOURCES AND RESERVES

To this point we have casually referred to *resources* without a specific definition. **Mineral resources** are broadly defined as elements, chemical compounds, minerals, or rocks that are concentrated in a form that can be extracted to obtain a usable commodity. It is also assumed that a resource can be extracted economically or at least has the potential for economical extraction. A **reserve,** on the other hand, is that portion of a resource which is identified and from which usable materials can be legally and economically extracted at the time of evaluation (Figure 10.2).

The main point about resources and reserves is that *resources are not reserves!* An analogy from a student's personal finances will help clarify this point. A student's reserves are the liquid assets, such as money in the pocket or bank, whereas the student's resources include the total income the student can expect to earn during his or her lifetime. This distinction is often critical to the student in school because resources are "frozen" assets or next year's income and cannot be used to pay this month's bills [3].

Regardless of potential problems, it is very important from a long-range planning perspective to estimate future resources. This requires a continual reassessment of all components of a total resource by considering new technology, probability of geologic discovery, and shifts in economic and political conditions [2].

The example of silver will illustrate some important points about resources and reserves. The Earth's crust (to a depth of 1 kilometer) contains almost 2 million million (2×10^{12}) metric tons of silver—an amount much larger than the annual world use, which is approximately 10,000 metric tons. If this silver existed as pure metal concentrated into one large mine, it would represent a supply sufficient for several hundred million years at current levels of use. Most of this silver, however, exists in extremely low concentrations, too low to be extracted economically with current technology. The known reserves of silver, reflecting the amount we could obtain immediately with known techniques, is about 200,000 metric tons, or a 20-year supply at current use levels.

FIGURE 10.2

Classification of mineral resources used by the U.S. Geological Survey and the U.S. Bureau of Mines. (After Secretary of the Interior, U.S. Bureau of Mines, 1980a.)

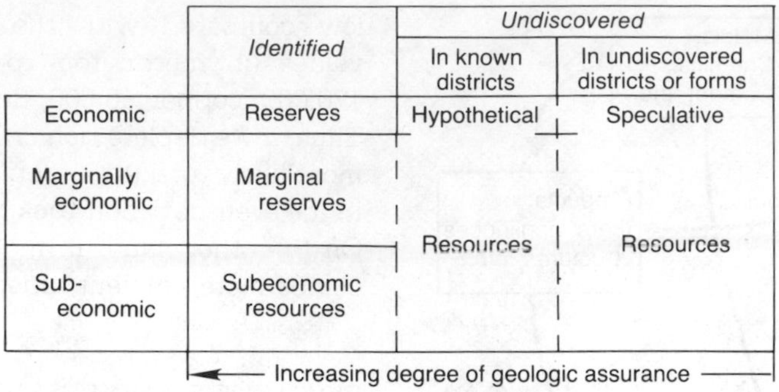

		Undiscovered	
	Identified	In known districts	In undiscovered districts or forms
Economic	Reserves	Hypothetical	Speculative
Marginally economic	Marginal reserves	Resources	Resources
Sub- economic	Subeconomic resources		

⟵ Increasing degree of geologic assurance ⟶

Reserves: Identified resources from which a usable mineral or energy commodity can be economically and legally extracted at the time of determination.
Marginal reserves: That part of identified resources that border on being economically producible. Includes resources that would be producible with postulated changes in economic, technical or legal factors.
Subeconomic resources: That part of identified resources that does not meet the economic criteria of reserves or marginal reserves. However with sufficient economic or technologic change they may become reserves.
Hypothetical resources: Undiscovered materials that may reasonably be expected to exist in a known mining district under known geologic conditions.
Speculative resources: Undiscovered materials that may occur either in known types of deposits in a favorable geologic setting where no discoveries have been made, or in as-yet-unknown types of deposits that remain to be recognized.

The problem with silver, as with all mineral resources, is not with its total abundance but with its concentration and relative ease of extraction. When an atom of silver is used, it is not destroyed, but remains an atom of silver. It is simply dispersed and may become unavailable. In theory, given enough energy, all mineral resources could be recycled, but this is not possible in practice. Consider lead, which is mined and used in gasolines. This lead is now scattered along highways across the world and deposited in low concentration in forests, fields, and salt marshes close to these highways. Recovery of this lead is, for all practical purposes, impossible.

MINERAL RESOURCE AVAILABILITY

The Earth's chemical resources can be divided into several broad categories, depending on our use of them: elements for metal production and technology; building materials; minerals for the chemical industry; and minerals for agriculture. Metallic minerals can be classified according to their abundance. The abundant metals include iron, aluminum, chromium, manganese, titanium, and magnesium. Scarce metals include copper, lead, zinc, tin, gold, silver, platinum, uranium, mercury, and molybdenum.

Some mineral resources, such as salt (sodium chloride), are necessary for life. Primitive peoples traveled long distances to obtain salt when it was not locally available. Other mineral resources are desired or considered necessary to maintain a certain level of technology.

The basic issue with mineral resources is not actual exhaustion or extinction, but the cost of maintaining an adequate stock within an economy through mining and recycling At some point, the costs of mining exceed the worth of the material. When the availability of a particular mineral becomes a limitation, there are four possible solutions: find more sources; recycle what has already been obtained; find a substitute; or do without. Which choice or combination of choices is made depends on social, economic, and environmental factors.

The availability of a mineral resource in a certain form, in a certain concentration, and in a certain total amount at that concentration is determined by the Earth's history and is a geological issue. What is a resource and when it becomes limiting are ultimately social questions. Before metals were discovered, they could not be considered resources. Before smelting was invented, the only metal ores were those in which the metals appeared in their pure form. Originally, gold was obtained as a pure or "native" metal. Now gold mines are deep in the Earth, and the recovery process involves reducing tons of rock to ounces of gold.

In reality, then, mineral resources are limited, which raises important questions. How long will a particular resource last? How much short- or long-term environmental deterioration are we willing to concede to insure that resources are developed in a particular area? How can we make the best use of available resources? These questions have no easy answers. We are now struggling with ways to estimate better the quality and quantity of resources.

We can use a particular mineral resource in several ways: rapid consumption, consumption with conservation, or consumption and conservation with recycling. Which option is selected depends in part on economic, political, and social criteria. Figure 10.3 shows the hypo-

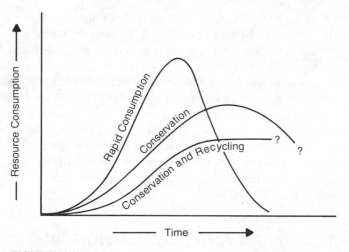

FIGURE 10.3
Idealized diagram showing several hypothetical depletion curves. See text for further explanation.

thetical *depletion curves* corresponding to these three options. Historically, with the exception of precious metals, rapid consumption has dominated most resource utilization. However, as more resources become in short supply, increased conservation and recycling are expected. Certainly the trend toward recycling is well established for metals such as copper, lead, and aluminum.

We usually think of mineral resources as the metals used in structural materials, but in fact (with the exception of iron) the predominant mineral resources are not of this type. Sodium and iron are used at a rate of approximately 0.1 to 1 billion tons per year. Nitrogen, oxygen, sulphur, potassium, and calcium are each consumed at about the rate of 10 to 100 million tons per year. Of these five elements, four are used primarily as soil conditioners

or fertilizers. Elements such as copper, aluminum, and lead have annual world consumption rates of 1 to 10 million tons, and gold and silver have annual consumption rates of 10,000 tons or less. Of the metallic minerals, iron makes up 95 percent of all the metals consumed, and nickel, chromium, cobalt, and manganese are used mainly in alloys of iron (as in stainless steel). Therefore, we can conclude that the nonmetallics, with the exception of iron, are consumed at much greater rates than elements used for their metallic properties.

As the world population and the desire for a higher standard of living increase, the demand for mineral resources expands at a faster and faster rate. From a global viewpoint, our limited mineral resources and reserves threaten our affluence. Approximately 9900 kilograms of new mineral material (excluding energy resources) are required each year for each person in the United States (Fig. 10.4). Predicted world increases in the use of iron, copper, and lead compared with population increases (Fig. 10.5 and Table 10.2) illustrate that the rate of production of these metals would have to increase by several times if the world per capita consumption rate were to rise to the U.S. level. Such an increase is very unlikely; affluent countries will have to find substitutes for some minerals or use a smaller proportion of the world annual production.

Domestic supplies of many mineral resources in the United States and many other affluent nations are insufficient for current use and must be supplemented by imports from other nations. A significant percentage of the total U.S. resource demand is now supplied by imports (Fig. 10.6 and Table 10.3). Of course, the fact that a mineral is imported today into the United States does not mean necessarily that it does not exist in quantities that

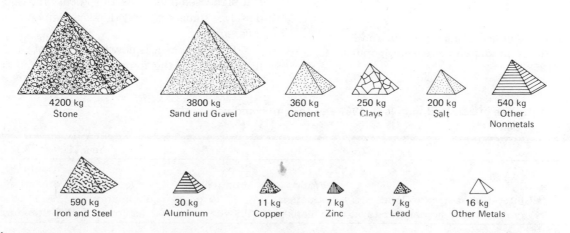

FIGURE 10.4
Amount of new mineral materials required annually by each citizen in the United States. (After Secretary of the Interior, 1975.)

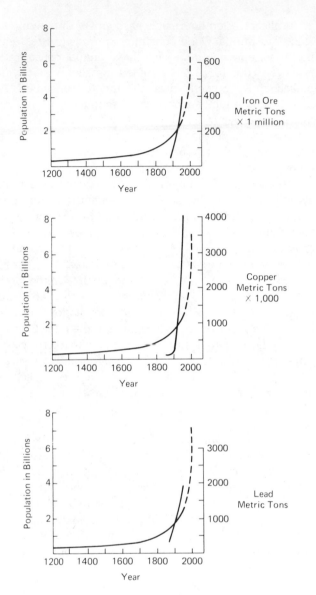

FIGURE 10.5
Predicted increase in use of copper, lead, and iron compared with the predicted increase in population. (From Park, 1968.)

could be mined within the country. Rather, it suggests that there are economic, political, or environmental reasons that make it easier, more practical, or more desirable to import the material.

How do we plan for the future? Because estimates of reserves and resources change with new technology and social, political, and economic conditions, a continual reassessment is necessary for long-range planning. As an aid to planning, the U.S. Geological Survey and the U.S. Bureau of Mines have listed selected mineral resources and reserves for the United States projected through the year 2000 (Table 10.4).

THE ORIGIN AND DISTRIBUTION OF MINERAL RESOURCES

The origin and distribution of mineral resources is intimately related to the history of the biosphere and to the entire geologic cycle. Nearly all aspects and processes of the geological cycle are involved to some extent in producing local concentrations of useful materials.

Why are there local concentrations of minerals? It is now believed by planetary scientists that the Earth, like the other planets in the solar system, formed by condensation of matter surrounding the sun. Gravitational attraction brought together matter that was dispersed around the forming sun. As the mass of the proto-Earth increased, the material condensed and was heated by the process. The heat was sufficient to produce a molten liquid core. This core consists primarily of iron and other heavy metals, which sank toward the center of the Earth. The crust formed of generally lighter elements and is a mixture of many different elements. The crust does not have a uniform distribution of elements because geological processes and some biological processes selectively dissolve, transport, and deposit elements and minerals differently.

For example, it is now believed that plate tectonics is responsible for the formation of some mineral deposits.

TABLE 10.2
Comparison of the use of selected metals in 1967 and the predicted use by the year 2000.

| | Year 1967 | | | | Year 2000 | |
Metal	United States per Capita	World per Capita	World Production (millions)	Bring World to U.S. Standard	Double Population as Today	Double Population World to U.S. Standard
Iron	0.9 ton	0.15 ton	495.0 tons	×6	×2	×12
Copper	8.1 kg	1.4 kg	4.7 tons	×5.5	×2	×11
Lead	5.4 kg	0.7 kg	2.2 tons	×8	×2	×16

SOURCE: Data from Park, 1968.

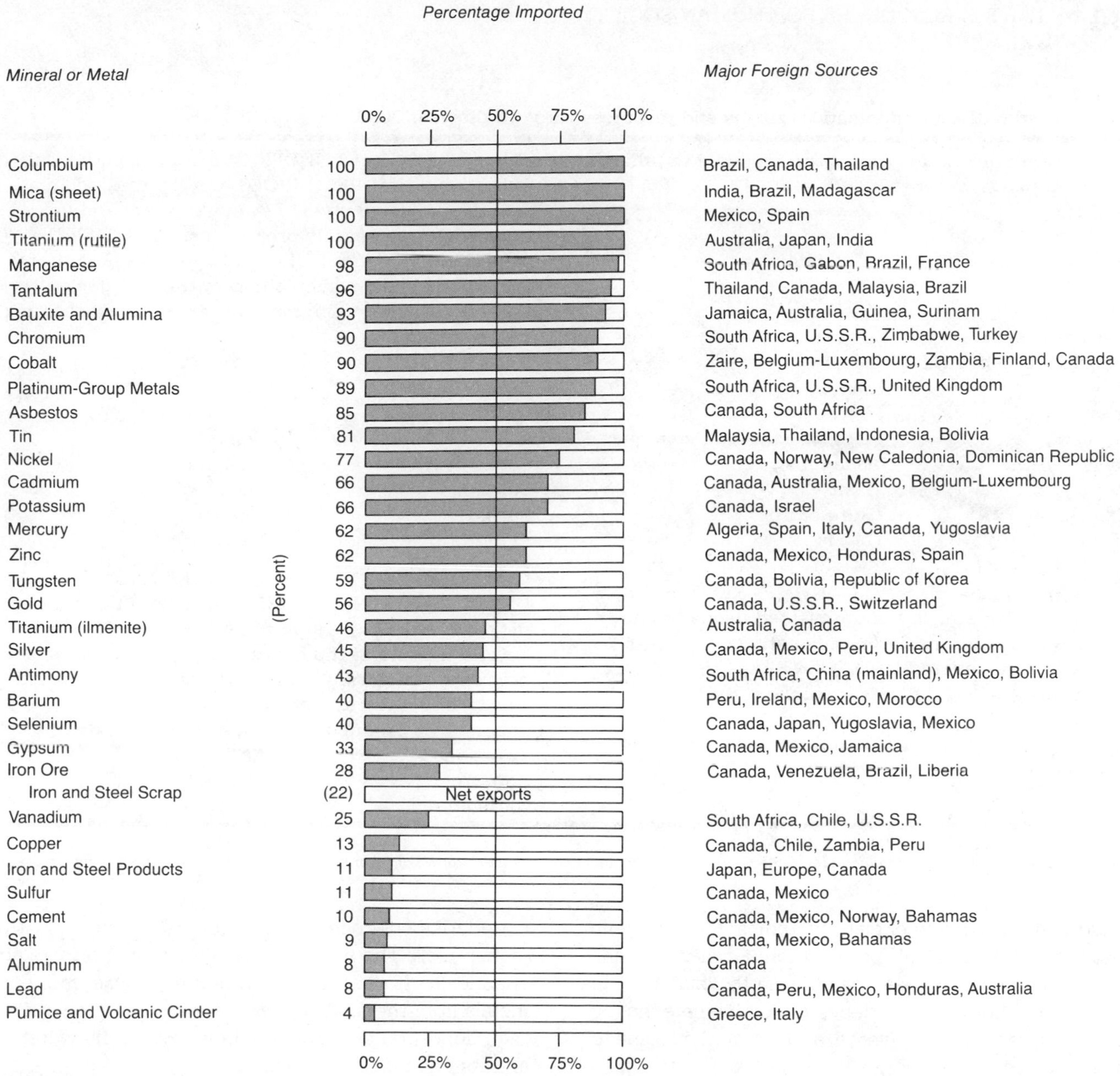

Mineral or Metal   *Major Foreign Sources*

Mineral or Metal	Percentage Imported	Major Foreign Sources
Columbium	100	Brazil, Canada, Thailand
Mica (sheet)	100	India, Brazil, Madagascar
Strontium	100	Mexico, Spain
Titanium (rutile)	100	Australia, Japan, India
Manganese	98	South Africa, Gabon, Brazil, France
Tantalum	96	Thailand, Canada, Malaysia, Brazil
Bauxite and Alumina	93	Jamaica, Australia, Guinea, Surinam
Chromium	90	South Africa, U.S.S.R., Zimbabwe, Turkey
Cobalt	90	Zaire, Belgium-Luxembourg, Zambia, Finland, Canada
Platinum-Group Metals	89	South Africa, U.S.S.R., United Kingdom
Asbestos	85	Canada, South Africa
Tin	81	Malaysia, Thailand, Indonesia, Bolivia
Nickel	77	Canada, Norway, New Caledonia, Dominican Republic
Cadmium	66	Canada, Australia, Mexico, Belgium-Luxembourg
Potassium	66	Canada, Israel
Mercury	62	Algeria, Spain, Italy, Canada, Yugoslavia
Zinc	62	Canada, Mexico, Honduras, Spain
Tungsten	59	Canada, Bolivia, Republic of Korea
Gold	56	Canada, U.S.S.R., Switzerland
Titanium (ilmenite)	46	Australia, Canada
Silver	45	Canada, Mexico, Peru, United Kingdom
Antimony	43	South Africa, China (mainland), Mexico, Bolivia
Barium	40	Peru, Ireland, Mexico, Morocco
Selenium	40	Canada, Japan, Yugoslavia, Mexico
Gypsum	33	Canada, Mexico, Jamaica
Iron Ore	28	Canada, Venezuela, Brazil, Liberia
Iron and Steel Scrap	(22) Net exports	
Vanadium	25	South Africa, Chile, U.S.S.R.
Copper	13	Canada, Chile, Zambia, Peru
Iron and Steel Products	11	Japan, Europe, Canada
Sulfur	11	Canada, Mexico
Cement	10	Canada, Mexico, Norway, Bahamas
Salt	9	Canada, Mexico, Bahamas
Aluminum	8	Canada
Lead	8	Canada, Peru, Mexico, Honduras, Australia
Pumice and Volcanic Cinder	4	Greece, Italy

(Percent)

FIGURE 10.6

Imports supplied a significant percentage of total U.S. demand for minerals and metals in 1979. (From Secretary of the Interior, 1980b.)

TABLE 10.3

Ratio of U.S. production to consumption for selected materials in 1978. A low ratio (import-dependent) means that domestic production is much less than consumption, whereas a high ratio (export potential) means domestic production exceeds consumption.

Import-Dependent		U.S. Production Matches Consumption		Export Potential	
Manganese	0.02	Steel	0.94	Ammonia	1.14
Cobalt	0.03	Aluminum	0.95	Soda ash	1.23
Bauxite	0.11	Lead	0.96	Phosphates	1.57
Tin	0.20	Copper	1.00	Magnesium	1.63
Nickel	0.25	Iron ore	1.00	Boron	1.69
Potassium	0.43	Cement	1.03	Molybdenum	2.21
Zinc	0.51	Lime	1.06		

SOURCE: Secretary of the Interior, 1979.

TABLE 10.4
A generalized outlook of domestic reserves and resources through the year 2000.

Group 1: RESERVES in quantitites adequate to fulfill projected needs well beyond 25 years.		*Group 3:* Estimated UNDISCOVERED (hypothetical and speculative) RESOURCES in quantities adequate to fulfill projected needs beyond 25 years and in quantities significantly greater than IDENTIFIED SUBECONOMIC RESOURCES. Research efforts for these commodities should concentrate on geologic theory and exploration methods aimed at discovering new resources.	
Coal	Phosphorus		
Construction stone	Silicon		
Sand and gravel	Molybdenum		
Nitrogen	Gypsum		
Chlorine	Bromine		
Hydrogen	Boron		
Titanium (except rutile)	Argon	Iron	Platinum
	Diatomite	Copper[a]	Tungsten
Soda	Barite[a]	Zinc[a]	Beryllium[a]
Calcium	Lightweight aggregates	Gold	Cobalt[a]
Clays	Helium	Lead[a]	Cadmium[a]
Potash	Peat	Sulphur	Bismuth[a]
Magnesium	Rare earths[a]	Silver[a]	Selenium
Oxygen	Lithium[a]	Fluorine[a]	Niobium[a]
Group 2: IDENTIFIED SUBECONOMIC RESOURCES in quantities adequate to fulfill projected needs beyond 25 years and in quantities significantly or slightly greater than estimated UNDISCOVERED RESOURCES.		*Group 4:* IDENTIFIED SUBECONOMIC and UNDISCOVERED RESOURCES together in quantities probably not adequate to fulfill projected needs beyond the end of the century; research on possible new exploration targets, new types of deposits, and substitutes is necessary to relieve ultimate dependence on imports.	
Aluminum	Vanadium		
Nickel[a]	Zircon[a]		
Uranium	Thorium		
Manganese		Tin	Antimony[a]
		Asbestos	Mercury[a]
		Chromium	Tantalum[a]

SOURCE: U.S. Geological Survey, 1975.
[a]Those commodities that may be in much greater demand than is now predicted because of known or potentially new applications in the production of energy.

According to the theory of plate tectonics, the continents "float" on the material below them—called tectonic plates—and move slowly across the Earth's surface. Metallic ores are thought to be deposited both where the tectonic plates separate, or *diverge*, and where they come together, or *converge* (Fig. 10.7). At divergent plate boundaries cold ocean water comes in contact with hot molten rock. The heated water is lighter and more active chemically. It rises through fractured rocks and leaches metals from them. The metals are carried in solution and then deposited as metal sulphides when the water cools and its chemistry changes.

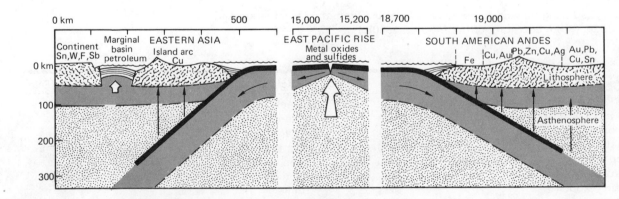

FIGURE 10.7
Idealized diagram showing the relationship between the East Pacific Rise (divergent plate boundary), Pacific margins (convergent plate boundaries), and metallic ore deposits. (After NOAA, 1977.)

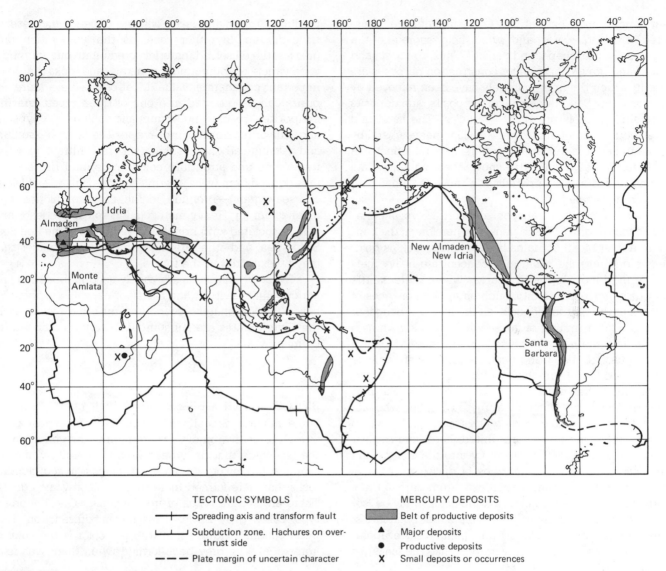

TECTONIC SYMBOLS

⊢—⊢ Spreading axis and transform fault

⌐—⌐ Subduction zone. Hachures on over-thrust side

––– Plate margin of uncertain character

MERCURY DEPOSITS

▒ Belt of productive deposits

▲ Major deposits

● Productive deposits

X Small deposits or occurrences

FIGURE 10.8
Relation between mercury deposits and recently active subduction zones. (From
Brobst and Pratt, 1973.)

At convergent plate boundaries, rocks saturated
with sea water are forced together, heated, and subjected
to intense pressure, which causes partial melting. The
combination of heat, pressure, and partial melting mobi-
lizes metals in the molten rocks. For example, most major
mercury deposits are associated with volcanic regions,
which occur close to convergent plate boundaries (Fig.
10.8). Geologists believe that the mercury is distilled out
of the tectonic plate as it moves downward, and the mercury moves upward and is deposited as the plate cools.

IGNEOUS PROCESSES
Ore deposits may form when molten rocks cool. As molten rock cools, heavier minerals that crystallize early may
slowly sink or settle toward the bottom, while lighter min-

erals that crystallize later are left at the top. Deposits of an
ore of chromium, called *chromite*, are thought to be
formed in this way. When molten rocks containing carbon
under very high pressure cool slowly, diamonds may be
produced.

Hot waters moving within the crust are perhaps the
source of most ore deposits. It is speculated that circulating groundwater is heated and enriched with minerals
upon contact with deeply buried molten rocks. This water
then moves up or laterally to other, cooler rocks, where
the cooled water deposits the dissolved minerals.

SEDIMENTARY PROCESSES
Sedimentary processes are often significant in concentrating economically valuable earth materials in suffi-

cient amounts for extraction. As sediments are transported, running water and wind help segregate the sediment by size, shape, and density. Thus, the best sand or sand and gravel deposits for construction purposes are those in which the finer materials have been removed by water or wind. Sand dunes, beach deposits, and deposits in stream channels are good examples. The sand and gravel industry amounts to over $1 billion per year, and by volume mined it is the largest nonfuel mineral industry in the United States. Figure 10.9 shows the production data for sand and gravel compared with those for other construction materials [4].

Stream processes transport and sort all types of materials according to size and density. Therefore, if the bedrock in a river basin contains heavy metals such as gold, streams draining the basin may concentrate heavy metals to form **placer deposits** in areas where there is little water turbulence or velocity, such as in open crevices of fractures at the bottom of pools, on the inside of bends, or on riffles. Placer mining of gold—which was known as a "poor man's method" because a miner needed only a shovel, a pan, and a strong back to work the streamside claim—helped to stimulate settlement of California, Alaska, and other areas of the United States. Furthermore, the gold in California attracted miners who acquired the expertise necessary to locate and develop other resources in the western United States and Alaska.

Rivers and streams that empty into the oceans and lakes carry tremendous quantities of dissolved material derived from the weathering of rocks. From time to time, geologically speaking, shallow marine basins may be isolated by tectonic activity that uplifts its boundaries, thereby restricting circulation and facilitating evaporation. In other cases, climatic variations like the Ice Ages

produce large inland lakes with no outlets which essentially dry up. In either case, as the evaporation progresses, the dissolved materials precipitate out, forming a wide variety of compounds, minerals, and rocks that have important commercial value. Most of these *evaporite deposits* can be grouped into one of three types: **marine evaporites** (solids)—potassium and sodium salt, gypsum, and anhydrite; **nonmarine evaporites** (solids)—sodium and calcium carbonate, sulphate, borate, nitrate, and limited iodine and strontium compounds; and **brines** (liquids derived from wells, thermal springs, inland salt lakes, and sea waters)—bromine, iodine, calcium chloride, and magnesium [5]. Heavy metals (such as copper, lead, and zinc) associated with brines and sediment in the Red Sea, Salton Sea, and other areas are important resources that may be exploited in the future.

Evaporite materials are widely used in industrial and agricultural activities, and their annual value is about $1 billion [5]. Fortunately, evaporite and brine resources in the United States are substantial, assuring no shortages for many years.

BIOLOGICAL PROCESSES

Some mineral deposits are formed by biological processes, and many minerals are formed under conditions of the biosphere that have been greatly altered by life. For example, the major iron ore deposits exist in sedimentary rocks that were formed more than 2 billion years ago [6]. There are several types of iron deposits. One important type, called **grey beds**, contains unoxidized iron. **Red beds** contain oxidized iron (the red color is the color of iron oxide). The grey beds formed when there was rela-

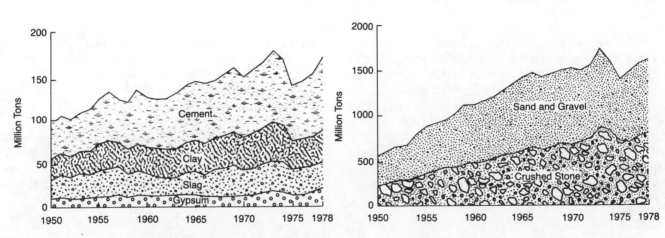

FIGURE 10.9
Supplies of major construction materials in the United States from 1950 to 1978.
(From Secretary of the Interior, 1979.)

TABLE 10.5
Minerals of biological origin. (An X indicates that at least one species of the
biological kingdom produces the mineral.)

Mineral	Prokaryotes[a] (bacteria and blue-green algae)	Protists[a]	Fungi	Multicellular Animals	Plants
Carbonates					
Calcite	X	X		X	X
Aragonite	X	X		X	X
Vaterite		X		X	X
Monohydrocalcite	X			X	
Amorph. hydr. carb.		X		X	
Phosphates					
Dahllite	X	X	X	X	
Francolite				X	
$Ca_3Mg_3(PO_4)_4$				X	
Brushite				X	
Amorph. dahllite precursor				X	
Amorph. brushite precursor				X	
Amorph. whitlockite precursor				X	
Amorph. hydr. ferric phosphate				X	
Halides					
Fluorite				X	
Amorph. fluorite precursor				X	
Oxalates					
Whewellite		X	X	X	X
Weddelite			X	X	
Sulphates					
Gypsum				X	X
Celestite		X			
Barite		X			
Silica					
Opal		X		X	X
Iron oxides					
Magnetite	X			X	
Maghemite	?				
Goethite				X	
Lepidocrocite				X	
Ferrihydrite	X		X	X	X
Amorph. ferrihydrates		X		X	
Manganese oxides					
Todorokite	X				
Iron sulphides					
Pyrite	X				
Hydrotroilite	X				

[a]The prokaryotes are organisms whose cells have no nucleus; the protists are organisms with true cell nuclei, but are primarily single-celled. See Appendix A for further discussion of the classification of species.

tively little oxygen in the atmosphere, and the red beds formed when there was relatively more oxygen. Although the processes are not completely understood, it appears that major deposits of iron stopped forming when the atmospheric concentration of oxygen reached its present level. This suggests that early life was important in beginning and ending the ore-forming processes for iron [7].

Organisms are able to form many kinds of minerals, such as the calcium minerals in shells and bones. Some of these minerals cannot be formed inorganically in the biosphere. Thirty-one different biologically produced minerals have been identified. Minerals of biological origin contribute significantly to sedimentary deposits [8] (Table 10.5).

WEATHERING PROCESSES

Weathering is responsible for concentrating some materials to the point that they can be extracted with reasonable effort. Insoluble ore deposits such as native gold are generally residual and unless removed by erosion will accumulate in the soil. Accumulation is favored where the parent rock is relatively soluble, such as limestone (Fig. 10.10). Intensive weathering of certain soils derived from aluminum-rich igneous rocks may concentrate oxides of aluminum and iron, while the more soluble elements, such as silica, calcium, and sodium, are selectively removed by soil and biological processes. If sufficiently concentrated, residual aluminum oxide forms an ore of aluminum known as *bauxite*. Important nickel and cobalt deposits are also found in such soils developed from ferromagnesium-rich igneous rocks.

Weathering is also involved in secondary enrichment processes to produce sulphide ore deposits from a low-grade primary ore. Near the surface, primary ore containing minerals such as iron, copper, and silver sulphides is in contact with slightly acid soil water in an oxygen-rich environment. As the sulphides are oxidized, they are dissolved, forming solutions that are rich in sulphuric acid and silver and copper sulphate that migrate downward, producing a leached zone devoid of ore minerals (Fig. 10.11). Below the leached zone and above the groundwater table, oxidation continues, and sulphate solutions continue their downward migration. Below the water table, if oxygen is no longer available, the solutions are deposited

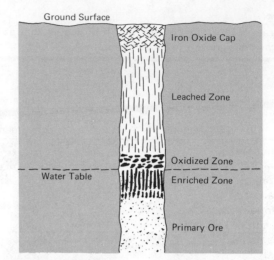

FIGURE 10.11

Idealized diagram of the typical zones that form during secondary enrichment processes. Sulphide ore minerals in the primary ore vein are oxidized and altered, and then are leached from the oxidized zone and redeposited in the enriched zone. The iron oxide cap is generally a reddish color and may be helpful in locating ore deposits that have been enriched. (After Foster, 1979.)

as sulphides, enriching the metal content of the primary ore by as much as ten times. In this way, low-grade primary ore is rendered more valuable, and high-grade primary ore is made even more attractive [9,10].

Several disseminated copper deposits have become economically successful because of secondary enrichment, which concentrates dispersed metals. For example, secondary enrichment of a disseminated copper deposit at Miami, Arizona, increased the grade of the ore from less than 1 percent copper in the primary ore to as much as 5 percent in some localized zones of enrichment [10].

MINERALS FROM THE SEA

Mineral resources in sea water or on the bottom of the ocean are vast, and, in some cases, such as magnesium, are nearly unlimited. In the United States, magnesium was first extracted from sea water in 1940. By 1972, one company in Texas produced 80 percent of domestic magnesium, utilizing sea water as its raw material source. Companies in Alabama, California, Florida, Mississippi, and New Jersey are also extracting magnesium from sea water.

The deep-ocean floor may be the site of the next big mineral rush. It is now known that manganese oxide nodules, which contain manganese (24%) and iron (14%) with

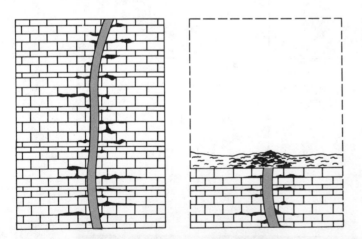

FIGURE 10.10

Idealized diagram showing how an ore deposit of insoluble minerals might form due to weathering and formation of a residual soil. As the limestone that contained the deposit weathered, the ore minerals became concentrated in the residual soil. (From Foster, 1979.)

secondary copper (1%), nickel (1%), and cobalt (0.25%), cover vast areas of the deep-ocean floor. The nodules are found in the Atlantic Ocean off Florida, but the richest and most extensive accumulations occur in large areas of the northeastern, central, and southern Pacific, where the nodules cover 20 to 50 percent of the ocean floor [11].

The average size of the manganese nodules varies from a few millimeters to a few tens of centimeters in diameter. Composed primarily of concentric layers of manganese in iron oxides mixed with a variety of other materials, each nodule formed around a nucleus of broken nodules, fragments of volcanic rock, and sometimes fossils. The estimated rate of growth is 1 to 5 millimeters per million years. The nodules are most abundant in those parts of the ocean where sediment accumulation is at a minimum, generally at depths of 2500 to 6000 meters [12].

The origin of the nodules is not well understood. Most likely, material weathered from the continents and transported by rivers to the oceans is carried by ocean currents to deposition sites in the deep-ocean basins. Both inorganic precipitation and bacteria-induced precipitation are probably important in the formation of nodules.

Expenditures for mining and metallurgical research to recover the nodules have surpassed $190 million, and proposed expenditures through the early 1980s are expected to approach $800 million. At least 20 corporations in several countries are examining metallurgical systems to process the nodules. Some would produce cobalt, copper, nickel, and manganese, while others would produce combinations of only copper and nickel [13].

Actual mining involves lifting the nodules off the bottom and up to the mining ship. French and Japanese researchers are experimenting with a system with a continuous line bucket dredge, in which a continuous rope with buckets attached at prescribed intervals is strung out between two ships. The buckets drag along the bottom as the two ships move, and the nodules are dumped into one of the ships as the rope loop is reeled from one ship to the other. Other methods of recovery being examined are hydraulic lifting and use of airlift in conjunction with hydraulic dredging [13]. It is a fair statement that the whole manganese oxide nodule industry is in its infancy, and considerable change in methods and technology is likely. Nevertheless, from current data it has been determined that mining of the nodules is technologically feasible and potentially profitable. However, prospective interest groups must cooperate in the management of the resource and carefully evaluate the environmental impact of the mining on the ecology of the ocean bottom so that the sea bed is not degraded.

ENVIRONMENTAL IMPACT OF MINERAL DEVELOPMENT

The impact of mineral exploitation on the environment depends upon such factors as mining procedures, local hydrologic conditions, climate, rock types, size of operation, topography, and many more interrelated factors. Furthermore, the impact varies with the stage of development of the resource. For example, the exploration and testing stage involves considerably less impact than the mining and processing stages.

Exploration activities for mineral deposits vary from collecting and analyzing remote-sensing data gathered from airplanes or satellites to field work involving surface mapping, drilling, and gathering of geophysical data. Generally, exploration has a minimal impact on the environment provided care is taken in sensitive areas, such as some arid lands, marsh lands, and areas underlain by permafrost. Some arid lands are covered by a thin layer of pebbles over fine silt several centimeters thick. The layer of pebbles, called **desert pavement** (Fig. 10.12), protects the finer material from wind erosion. When the pavement is disturbed by road building or other activity, the fine silts may be eroded, impairing physical, chemical, and biological properties of the soil in the immediate environment and scarring the land for many years if damage extends over a wide area. Marsh lands and other highly organic areas, such as the northern tundra, are also very sensitive to even light traffic.

Mining and processing of mineral resources generally have a considerable impact on the land, water, air, and biologic resources as well as initiating social impacts because of increased demand for housing and services in mining areas. As it becomes necessary to use lower and lower grade ores, we are faced with the problem of how to minimize mining's negative effects on the environment.

One of the major practical issues is whether surface or subsurface mines should be developed in any area. Surface mining is cheaper but has more direct environmental effects. The trend in recent years has been away from subsurface mining and toward large, open-pit (surface) mines such as the Bingham Canyon copper mine in Utah (Fig. 10.13) and the Liberty Pit near Ruth, Nevada (Fig. 10.14). The Bingham Canyon mine is one of the world's largest artificial excavations, covering nearly 8 square kilometers to a maximum depth of nearly 800 meters.

Surface mines and quarries today cover less than one half of 1 percent of the total area of the United States. Even though the impact of these operations is a local phenomenon, numerous local occurrences will eventually

FIGURE 10.12
Desert pavement. The upper layer, composed of rather coarse gravel often only one grain thick, overlies much finer material. If this upper layer is disturbed, considerable erosion by wind or running water may occur. (Photos by E. A. Keller.)

(a)

(b)

FIGURE 10.13
The Bingham Canyon copper mine, one of the largest artificial excavations in the world. (Photo courtesy of Kennecott Copper Corporation.)

FIGURE 10.14
The Liberty Pit surface mine near Ruth, Nevada. (Photo courtesy of Kennecott Copper Corporation.)

constitute a larger problem. Environmental degradation tends to extend beyond the excavation and surface plant areas of both surface and subsurface mines. Large mining operations disturb the land by directly removing material in some areas, which changes topography, and by dumping waste in others. At best these actions produce severe aesthetic degradation. Dust at mines may affect air resources, even though care is often taken to reduce dust production by sprinkling roads and other sites that generate dust. Water resources are particularly vulnerable to degradation even if drainage is controlled and sediment pollution reduced. Surface drainage is often altered at mine sites, and runoff from precipitation (rain or snow) may infiltrate into waste material, leaching out trace elements and minerals. Trace elements (cadmium, cobalt, copper, lead, molybdenum, and others), when leached from mining wastes and concentrated in water, soil, or plants, may be toxic or may cause diseases in people and other animals who drink the water, eat the plants, or use the soil. Specially constructed ponds to collect such runoff help, but cannot be expected to eliminate all problems.

The white streaks in Figure 10.15 are mineral deposits apparently leached from tailings from a zinc mine in Colorado. Similar looking deposits may cover rocks in rivers for many kilometers downstream from some mining areas. Thus, a potential problem associated with mineral resource development is the possible release of harmful trace elements to the environment.

Groundwater may also be polluted by mining operations when waste comes into contact with slow-moving subsurface waters. Surface water infiltration or groundwater movement causes leaching of sulphide minerals such as pyrite (FeS_2) in mine waste, often producing acid

FIGURE 10.15
Tailings from a zinc mine in Colorado. (Photo by E. A. Keller.)

water that may pollute groundwater and eventually seep into streams to pollute surface water. Groundwater problems are particularly troublesome because reclamation of polluted groundwater is very difficult and expensive.

Physical changes in the land, soil, water, and air associated with mining directly and indirectly affect the biological environment. Direct kills caused by mining activity or contact with toxic soil or water are examples of direct impacts. Indirect impacts include changes in nutrient cycling, total biomass, species diversity, and ecosystem stability due to alterations in ground and/or surface water availability or quality. Periodic or accidental discharge of low-grade pollutants through failure of barriers, ponds, or water diversions or through breach of barriers during floods, earthquakes, or volcanic eruptions also damages local ecological systems to some extent.

Because the demand for mineral resources is going to increase, we must minimize both the on-site and off-site problems by controlling sediment, water, and air pollution through good engineering and conservation practices. Although these actions will raise the cost of mineral commodities and hence the price of all items produced from these materials, they will yield other returns of equal or higher value to future generations. We must realize, however, that even the most careful measures to control environmental disruption associated with mining will occasionally fail.

Social impacts associated with large-scale mining result from a rapid influx of workers into areas unprepared for growth. Stress is placed on local services: water supplies, sewage and solid waste disposal systems, schools, and rental housing. Land use shifts from open range, forest, and agriculture to urban patterns. More people also increase the stress on nearby recreation and wilderness areas, some of which may be in a fragile ecological balance. Construction activity and urbanization affect local streams through sediment pollution, reduced water quality, and increased runoff. Air quality is reduced due to more vehicles, dust from construction, and generation of power.

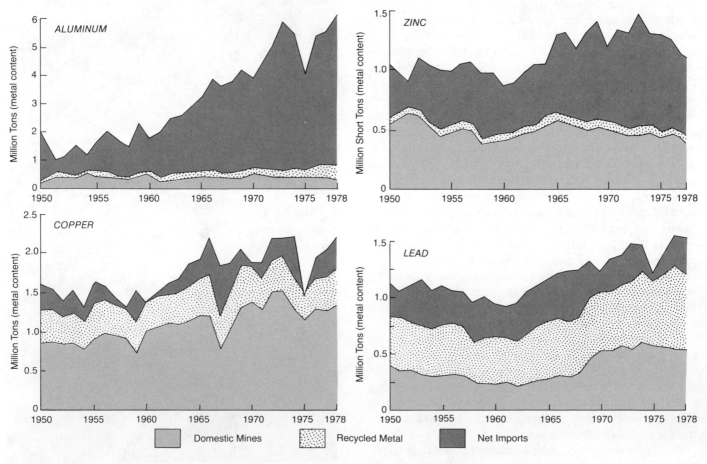

FIGURE 10.16
U.S. consumption and recycling of selected metals. (From Secretary of the Interior, 1979.)

RECYCLING OF RESOURCES

Over 180 million metric tons of household and industrial waste are collected each year in the United States alone. Of this, about 27 million metric tons are incinerated, generating over 6 million metric tons of residue. The tremendous tonnage of waste not incinerated and the residues from burning are usually disposed of at sanitary landfill sites or open dumps. These materials are sometimes referred to as **urban ore** because they contain many materials that could be recycled and used again to provide energy or useful products [14,15].

The notion of re-using waste materials is not new, and metals such as iron, aluminum, copper, zinc, and lead have been recycled for many years (Figs. 10.16 and 10.17). About 40 percent of steel is recycled in the United States [3]. Of the millions of automobiles discarded annually, nearly 90 percent are now dismantled by auto wreckers and scrap processors for metals to be recycled [15]. Recycling metals from discarded automobiles is a sound conservation practice, considering that nearly 90 percent by weight of the average automobile is metal (Fig. 10.18).

Recycling may be one way to delay or partly alleviate a possible crisis caused by the convergence of a rapidly rising population and a limited resource base. However, recycling the wide variety of materials found in urban waste is not an easy task. Before recycling can become a widespread practice, improved technology and more economic incentives are needed. Figure 10.19 shows a flow chart of necessary equipment to recycle urban refuse. The results of experiments conducted at a pilot plant by the U.S. Bureau of Mines using these techniques have been encouraging, and while refinements are necessary, urban refuse has been successfully separated into concentrates of light gauge iron, massive metals, glass, paper, plastics, and organic waste and other combustible wastes [14].

SUMMARY AND CONCLUSIONS

Mineral resources are generally extracted from naturally occurring, anomalously high concentrations of earth materials. These natural deposits allowed early peoples to exploit minerals while slowly developing technological skills.

Availability of mineral resources is one measure of the wealth of a society. In fact, modern technological civilization as we know it would not be possible without exploitation of mineral resources. However, it is important to recognize that mineral deposits are not infinite and that we cannot maintain exponential population growth on a finite resource base. The United States and many other affluent nations have insufficient domestic supplies of many mineral resources for current use and must supplement them by imports from other nations. In the future, as other nations industrialize and develop, such imports may be more difficult to obtain, and affluent countries may have to find substitutes for some minerals or use a smaller portion of the world's annual production.

An important concept in analyzing resources and reserves are that resources are not reserves! Unless discov-

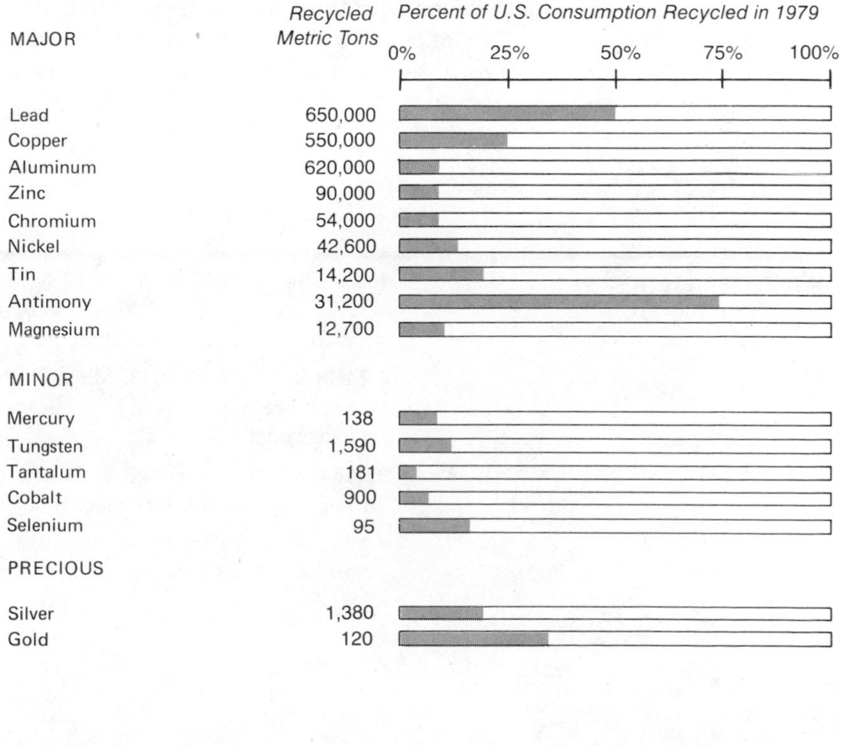

FIGURE 10.17

Amount of metals recycled in the United States in 1978. (Data from Secretary of the Interior, 1979.)

	Recycled Metric Tons	Percent of U.S. Consumption Recycled in 1979
MAJOR		
Lead	650,000	
Copper	550,000	
Aluminum	620,000	
Zinc	90,000	
Chromium	54,000	
Nickel	42,600	
Tin	14,200	
Antimony	31,200	
Magnesium	12,700	
MINOR		
Mercury	138	
Tungsten	1,590	
Tantalum	181	
Cobalt	900	
Selenium	95	
PRECIOUS		
Silver	1,380	
Gold	120	

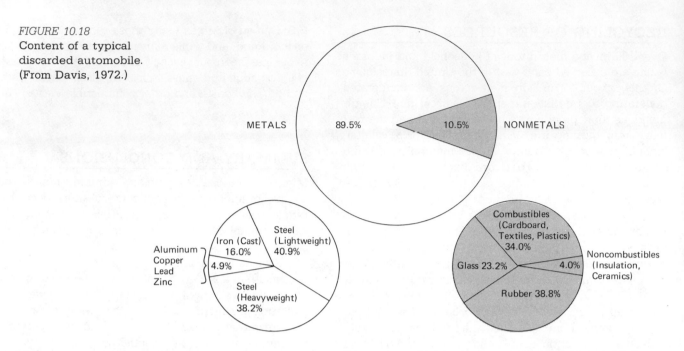

FIGURE 10.18
Content of a typical discarded automobile. (From Davis, 1972.)

eīed ānd cäptured, resources cannot be used to solve present shortages.

The origin and distribution of mineral resources are intimately related to the history of the biosphere as well as to the entire geologic cycle. Nearly all aspects and processes of the geologic cycle are involved to some extent in producing local concentrations of useful materials.

The enviromental impact of mineral exploitation depends upon many factors, including mining procedures, local hydrologic conditions, climate, rock types, size of operation, topography, and many more interrelated factors. In addition, the impact varies with the stage of development of the resource. In general, the mining and processing of mineral resources greatly affect the land, water, air, and biological resources as well as initiate certain social impacts due to increasing demand for housing and services in mining areas. Because the demand for mineral resources is going to increase, we must strive to minimize both on-site and off-site problems by controlling sediment, water, and air pollution through good engineering and conservation practices.

Recycling of mineral resources appears to be one way to delay or partly alleviate a possible crisis caused by the convergence of a rapidly rising population and a limited resource base. However, recycling a wide variety of materials found in urban waste is not an easy task, and innovative refinements in recycling methods will be necessary to insure that the recycling trend continues.

REFERENCES

1 GULBRANDSEN, R. A.; RAIT, N.; DRIES, D. J.; BAEDECKER, P. A.; and CHILDRESS, A. 1978. *Gold, silver, and other resources in the ash of incinerated sewage sludge at Palo Alto, California—A preliminary report.* U.S. Geological Survey Circular 784.

2 MCKELVEY, V. E. 1973. Mineral resource estimates and public policy. In *United States mineral resources,* eds. D. A. Brobst and W. P. Pratt, pp. 9–19. U.S. Geological Survey Professional Paper 820.

3 BROBST, D. A.; PRATT, W. P; and MCKELVEY, V. E. 1973. *Summary of United States mineral resources.* U.S. Geological Survey Circular 682.

4 YEEND, W. 1973. Sand and gravel. In *United States mineral resources,* eds. D. A. Brobst and W. P. Pratt, pp. 561–65. U.S. Geological Survey Professional Paper 820.

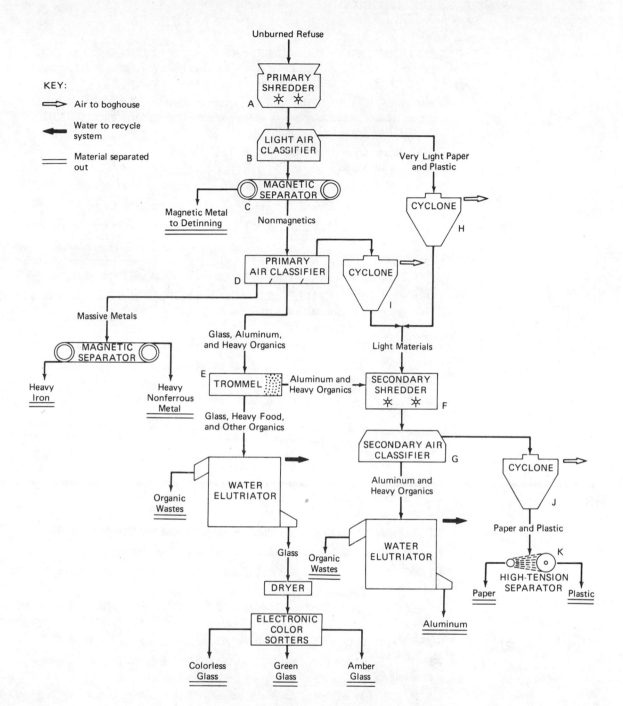

FIGURE 10.19

Flow chart showing how raw refuse may be separated for recycling of resources. (From Sullivan et al., 1973.)

5 SMITH, G. I.; JONES, C. L.; CULBERTSON, W. C.; ERICKSON, G. E.; and DYNI, J. R. 1973. Evaporites and brines. In *United States mineral resources,* eds. D. A. Brobst and W. P. Pratt, pp. 197–216. U.S. Geological Survey Professional Paper 820.

6 AWRAMIK, S. A. 1981. The pre-Phanerozoic biosphere—Three billion years of crises and opportunities. In *Biotic crises in ecological and evolutionary time,* ed. M. H. Nitecki, pp. 83–102. New York: Academic Press.

7 MARGULIS, L., and LOVELOCK, J. E. 1974. Biological modulation of the Earth's atmosphere. *Icarus* 21: 471–89.

8 LOWENSTAM, H. A. 1981. Minerals formed by organisms. *Science* 211: 1126–30.

9 BATEMAN, A. M. 1950. *Economic ore deposits.* 2nd ed. New York: Wiley.

10 PARK, C. F., Jr., and MACDIARMID, R. A. 1970. *Ore deposits.* 2nd ed. San Francisco: W. H. Freeman.

11 CORNWALL, H. R. 1973. Nickel. In *United States mineral resources,* eds. D. A. Brobst and W. P. Pratt, pp. 437–42. U.S. Geological Survey Professional Paper 820.

12 VAN, N.; DORR, J.; CRITTENDEN, M. D.; and WORL, R. G. 1973. Manganese. In *United States mineral resources,* eds. D. A. Brobst and W. P. Pratt, pp. 385–99. U.S. Geological Survey Professional Paper 820.

13 SECRETARY OF THE INTERIOR. 1975. *Mining and mineral policy, 1975.*

14 SULLIVAN, P. M.; STANCZYK, M. H.; and SPENDBUE, M. J. 1973. *Resource recovery from raw urban refuse.* U.S. Bureau of Mines R.I. 7760.

15 DAVIS, F. F. 1972. Urban ore. *California Geology,* May 1972: 99–112.

FURTHER READING

1 AMERICAN CHEMICAL SOCIETY. 1971. *Solid wastes.* Washington, D.C.: American Chemical Society.

2 BATEMAN, A. M. 1950. *Economic ore deposits.* 2nd ed. New York: Wiley.

3 ———. 1951. *The formation of mineral deposits.* New York: Wiley.

4 BROOKINS, D. G. 1981. *Earth resources, energy, and the environment.* Columbus, Ohio: Charles E. Merrill.

5 COMMITTEE ON MINERAL RESOURCES AND THE ENVIRONMENT. 1975. *Mineral resources and the environment.* Washington, D.C.: National Academy of Sciences.

6 KESLER, S. E. 1976. *Our finite mineral resources.* New York: McGraw-Hill.

7 PARK, C. F., Jr., and MACDIARMID, R. A. 1970. *Ore deposits.* 2nd ed. San Francisco: W. H. Freeman.

8 SKINNER, B.J., ed. 1981. *Use and misuse of Earth's surface.* Los Altos, Calif.: William Kaufmann.

STUDY QUESTIONS

1 What is the difference between a *resource* and a *reserve?*

2 Under what circumstances might sewage sludge be considered a mineral resource?

3 If surface mines and quarries cover less than one-half of 1 percent of the land surface of the United States, why is there environmental concern about them?

4 When is recycling of a mineral a viable option?

5 What is the difference between a renewable and a nonrenewable resource?

6 Which biological processes influence mineral deposits?

7 You meet a deep-sea diver who claims that the oceans can provide all of our mineral resources with no negative environmental effects. Do you agree or disagree?

8 What factors determine the availability of a mineral resource?

9 How does plate tectonics affect the nature and occurrence of mineral resources?

10 Making available a mineral resource involves three phases: (a) exploration; (b) recovery; and (c) consumption. Which phase has the greatest environmental effects?

11 Energy Resources

INTRODUCTION

It is fashionable today to think that an energy shortage is something new. In fact, people for thousands of years have had to deal with different types of energy shortages, going back to at least the early Greek and Roman cultures. The climate in coastal areas of Greece today is characterized by warm summers and cool winters, much as it was 2500 years ago. However, at that time the Greeks did not have any artificial method of cooling their houses during the summer, and their small charcoal-burning heaters were undoubtedly not very efficient in warming their homes during the winter. Wood was the primary source of energy, as it is today, for approximately half the people living on Earth.

By the fifth century B.C. fuel shortages became common, and much of the forest in many parts of Greece was destroyed for firewood. As local supplies of wood were depleted, it became necessary to import wood from further and further away. Valuable olive groves in the area became sources for wood to be made into charcoal for burning, reducing a valuable resource. Probably because they were aware of this dwindling resource, by the fourth century B.C. the city of Athens banned the use of olive wood for fuel.

About this time the Greeks recognized that they did have an alternate source of energy, namely, the sun. They began to build their houses to face toward the south so that the low-angled winter sun penetrated areas to be heated and the high summer sun shaded areas to be cooled. Recent excavations of ancient Greek cities suggest that large areas were planned so that individual homes could take maximum advantage of *passive solar energy*. Undoubtedly this helped to conserve local fuel supplies and considerably reduced the individual home-owner's fuel costs. (Fuel was still needed for cooking and other purposes.) Thus the Greeks' use of solar energy to assist in heating homes was a logical answer to their energy problem [1].

The use of wood in ancient Rome is somewhat analogous to the use of oil and gas in the United States today. Wealthy Roman citizens about 2000 years ago had central heating in their large homes, burning as much as 125 kilograms of wood every hour. Not surprisingly, local wood supplies were exhausted quickly, and the Romans had to import wood from outlying areas, as had the Greeks. Eventually wood had to be imported from as far away as 1600 kilometers.

Thus, for the same reasons that the Greeks eventually sought out solar energy, so did the Romans. However, in the Roman era solar technology became advanced. The Romans used windows to increase the effectiveness of solar heating, developed greenhouses to raise food during the winter, and oriented large public bathhouses (some of which accommodated up to 2000 people) to use passive solar energy. The Romans believed that sunlight in bath-houses was healthy, and it also greatly saved on fuel costs, allowing more wood to be available for heating the bath waters and steam rooms. The use of solar energy in ancient Rome was evidently quite widespread and resulted in the establishment of laws to protect a person's right to solar energy. As evidence of this, in some areas it was illegal for one person to construct a building that shaded another's [1].

The energy situation facing the United States and the world today is much like that which faced the early Greeks and Romans. The use of wood in the United States peaked in the 1880s when coal became abundant. The use of coal in turn began to decline after 1920, when oil and gas started to become available. Today we are facing the peak of oil and gas utilization. Fossil fuel resources, which took millions of years to form, may be essentially exhausted in just several hundred years. The decisions we make today will affect energy use for generations. Should we choose complex, centralized energy production methods, or use simpler, widely dispersed energy production methods? Which renewable sources of energy should be emphasized? How can we rely on nonrenewable energy sources? There are no easy answers. In this chapter we will cautiously explore various aspects of energy, including patterns of use, energy sources, hard path versus soft path, and environmental consequences of the various alternatives in energy utilization and development.

ENERGY, WORK, AND POWER

When we buy electricity by the kilowatt-hour, what are[a] we buying? We say we are buying energy, but what does that mean? Energy is an abstract concept. You can't see it or feel it, even if you have to pay for it.

To understand the concept of energy, it is easiest to begin with the idea of a **force.** We all have had the experience of exerting force—of pushing or pulling. The strength of a force can be measured by how much it accelerates an object. Suppose your car stalls while you are going up a hill, and you get out and push it to the side of the road (Fig. 11.1). You apply a force against gravity; or if the brake is left on, the brakes, tires, and bearings might heat up from friction. The longer the distance over which you exert the same force, the greater is the change in the car's speed, position uphill, or the heat in the brakes, tires, and bearings—that, in a physicist's terms, is the

---

[a]Based on H. J. Morowitz, *Energy Flow in Biology*, Oxbow Press, New Haven. 1979.

FIGURE 11.1

Idealized diagram showing some basic energy concepts, including force, potential energy, kinetic energy, and heat energy. See text for further explanation.

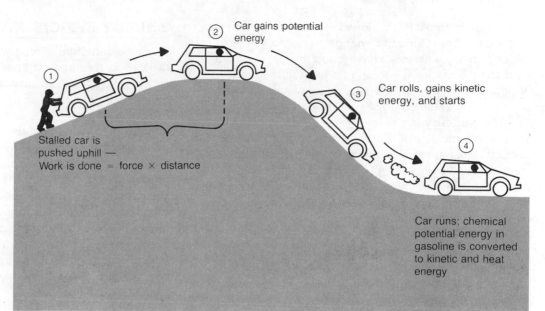

Car gains potential energy

Car rolls, gains kinetic energy, and starts

Stalled car is pushed uphill — Work is done = force × distance

Car runs; chemical potential energy in gasoline is converted to kinetic and heat energy

work done. **Work** is exerting a force over a distance; that is, work is the product of force times distance. If you push hard, but the car doesn't move at all, you have not done any work (according to its definition), even if you feel tired and sweaty.

In pushing your stalled car, you have done three things: changed its speed, moved it against gravity, and heated parts of it. These three things have something in common: they are all forms of *energy*. You have converted chemical energy in your body to the energy of motion of the car (called *kinetic energy*), the gravitational (or *potential*) energy of the car, and heat energy. Energy is never created or destroyed; it is always conserved. It can be transformed from one kind to another. The principle of the conservation of energy is known as the **first law of thermodynamics.**

The conservation and conversion of energy can be illustrated by the example of a clock pendulum (Fig. 11.2). When the pendulum is held in its highest position, it is neither moving nor getting hotter. It does, however, contain energy (due to its position). We refer to the stored energy as **potential energy,** which is converted to other forms when it is released. Other examples of potential energy are the gravitational energy in water behind a dam; chemical energy in coal, fuel oil, and gasoline as well as in the fat in your body; and nuclear energy, which is related to the forces binding the nuclei of atoms.

When the pendulum is released, it moves downward. At the bottom of the swing the speed is the greatest, and there is no potential energy. At this point all of its energy is in the energy of motion, which is called **kinetic energy.** As the pendulum swings back and forth, the energy

continually changes between the two forms, potential and kinetic. But at each cycle the pendulum slows down because of the friction of the pendulum moving through air and the friction at its pivot. The friction generates heat. Eventually all the energy is converted to heat and the pendulum stops.

This example illustrates another property of energy— its tendency to dissipate and end up as heat. It is rela-

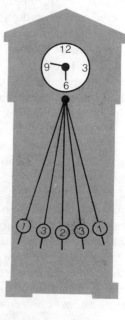

① Energy is all potential

② Energy is all kinetic

③ Energy is potential and kinetic

FIGURE 11.2

Idealized diagram of a grandfather (pendulum) clock illustrating relations between potential and kinetic energy. See text for further explanation.

tively easy to transform various forms of energy into low-grade heat, but difficult to change heat into energy with high efficiency. Physicists have found that we can change all of the gravitational energy in a pendulum to heat, but we cannot change all the heat energy thus generated back into potential energy.

Heat energy is the energy of the random motion of atoms and molecules, and there is something special about it. The tendency of energy to become randomly distributed as the kinetic energy of molecules forms the basis of the second law of thermodynamics. The **second law of thermodynamics** states that in any real process energy always tends to go from a more usable form to a less usable form.

Let us return to the example of your stalled car, which you have now pushed to the side of the road. Having pushed the car a little way uphill, you have increased its potential energy. You can convert this to kinetic energy by letting it roll back downhill. You engage the gears to restart the car. As the car idles, the potential chemical energy to move the car is converted to waste heat energy and various amounts of other energy forms, including electricity to charge the battery and play the radio. According to the first law of thermodynamics, the total amount of energy is always conserved. If this is true, why should there ever be any energy problem? Why could we not collect that wasted heat and use it to run the engine? Here we discover the importance of the second law of thermodynamics. According to that law, energy is always degraded to heat, which can never with 100 percent efficiency be reconverted to useful work. When we refer to low-temperature heat energy as *low grade*, we mean that relatively little of it is available to do useful work. High-grade energy, such as gasoline or sunlight, is largely available to do useful work. The biosphere continuously receives high-grade energy from the sun and radiates heat to the depths of space [2].

The unit of work used in the metric system is the *joule*. In talking about the energy used by the world or the United States, we use the unit *exajoule*, which is equivalent to 10^{18} (a billion billion) joules. The exajoule is roughly equivalent to one quadrillion, or 10^{15}, British thermal units (Btu). To put this in perspective, the United States now consumes about 80 exajoules of energy per year, and world consumption is about 250 exajoules.

In many instances, we are particularly interested in the rate of doing work, which is known as **power**. In the metric system, power is defined as joules per second, which is called a *watt*. When larger units of power are needed, we may use the *kilowatt* (1000 watts) or a *megawatt* (1 million watts). The *kilowatt-hour*, the unit by which we pay an electric bill, is 1000 watts applied for 1 hour (3600 seconds), which is equivalent to 3,600,000 joules.

ENERGY EFFICIENCY

A simple definition of efficiency is the ratio of the output of work to the input of energy. This definition is sometimes called the **first-law efficiency** because it refers only to the amount of energy and does not consider the quality or availability of energy. Another measure is the **second-law efficiency,** defined as the ratio of the minimum quantity of available work needed to perform a particular task to the actual quantity of available work used to perform that task. Table 11.1 lists some of the first- and second-law efficiencies for selected energy forms and processes. Of particular importance are the low second-law efficiencies because they indicate where potential improvements in energy planning (such as matching supplies to end uses) can be expected.

The first law-efficiency can also be expressed as a thermal efficiency:

$$\text{Efficiency} = \left[1 - \frac{\text{Temperature out (}^{\circ}\text{K)}}{\text{Temperature in (}^{\circ}\text{K)}} \right] 100$$

where degrees Celsius ($^{\circ}$C) = degrees Kelvin ($^{\circ}$K) $-$ 273. This relationship, when examined closely, reveals that we

TABLE 11.1
Selected examples of first- and second-law efficiencies.

First-Law Efficiency:

Energy Form	First-Law Efficiency	Waste Heat
Incandescent light bulb	5%	95%
Fluorescent light	20%	80%
Automobile	20–25%	75–80%
Power plants (electric): fossil fuel and nuclear	30–40%	60–70%
Burning fossil fuels (used directly for heat)	65%	35%
All energy (USA)	50%	50%

Second-Law Efficiency:

Energy Form	Second-Law Efficiency	Potential for Savings
Automobile	10%	Moderate
Water heating	2%	Very high
Space heating and cooling	6%	Very high
Power plants (electric)	30%	Low to moderate
All energy (USA)	10–15%	High

will always have an intermediate range of relatively moderate efficiencies. This results because the temperature coming out of the system can never be 0K* and the temperature going into the system has finite limits. For example, if steam enters a power plant at 800K (527°C) and exhausts at approximately 400K (100°C), then the theoretical upward limit of the efficiency is about 54 percent, with the actual efficiency about 40 percent. Nuclear reactors operating at an ingoing temperature of approximately 620K (350°C) and an exhaust of about 373K (100°C) have a theoretical upward limit efficiency of about 40 percent, with an actual operating efficiency of about 30 percent. For one final example, an automobile with the combustion temperature of about 3255K (2982°C) and an output of about 1433K (1160°C) provides an upper limit efficiency of about 56 percent, with an actual efficiency of about 25 percent [2].

ENERGY AND PEOPLE

Before the present energy shortage, energy seemed unlimited. People in industrialized countries, while a relatively small percentage of the world's population, consume a disproportionate share of the total energy produced in the world. In fact, there is a direct relationship between a country's standard of living (as measured by the gross national product) and the energy consumption per capita. In the next several decades, as petroleum and natural gas become more scarce and expensive, developing and developed countries are going to have to find innovative ways to obtain energy.

Figure 11.3 compares the projected population increase for the United States with projected energy needs. If energy demand continues to increase faster than population, one has to wonder where the energy will come from. For the last 30 years most of the energy consumed in the United States has come from natural gas and petroleum (Fig. 11.4a), with moderate amounts from coal and small amounts from hydropower and nuclear power. The United States has huge reserves of coal, but major new sources of natural gas and petroleum are becoming scarcer, few new large hydropower plants can be expected, and planning and construction of a large number of new nuclear power plants in the next few years have become uncertain for a variety of economic, social, political, and environmental reasons.

Supply and demand for energy are difficult to predict because technical, economic, political, and social assumptions underlying projections are constantly changing. Furthermore, large annual variations in energy consumption must also be considered. Figure 11.4b shows

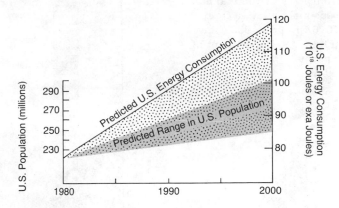

FIGURE 11.3

Comparison of projected (high) U.S. energy consumption with population to the year 2000. (Data from several U.S. agency sources.)

that energy consumption peaks during the winter months, with a secondary peak in the summer. Future changes in population or intensive conservation measures may change this pattern as might better design of buildings and more reliance on solar energy. One recent prediction states that U.S. energy consumption in the year 2010 may be as high as 137 exajoules or as low as 63 exajoules. (Energy consumption in 1979 was about 78 exajoules.) The high value assumes no change in energy policies, whereas the low value assumes very aggressive energy conservation policies. Figure 11.5 shows the recent energy flow in the United States.

The average first-law efficiency of 50 percent emphasizes the "thermal bottleneck"; as chemical energy is converted to thermal and mechanical uses, large amounts of energy are lost in producing electricity and in transporting people and goods. New innovations in energy production are likely to unplug part of the "thermal bottleneck." Of particular importance will be energy uses with applications below 100°C because a large portion of the total U.S. energy consumption (for uses below 300°C) is for space heating and water heating (Fig. 11.6).

We will now discuss selected geologic and environmental aspects of fossil fuels, geothermal energy, nuclear energy, and alternative energy sources, including hydropower (river and tidal), solar power, wind power, and biomass.

NONRENEWABLE ENERGY SOURCES

FOSSIL FUELS: GEOLOGY, DISTRIBUTION, AND SUPPLY

The fossil fuels (crude oil, natural gas, coal, oil shales, and tar sands) are all forms of stored solar energy. They were created from incomplete biologic decomposition of dead

---

*Formerly known as degree Kelvin (°K), the kelvin is now symbolized K.

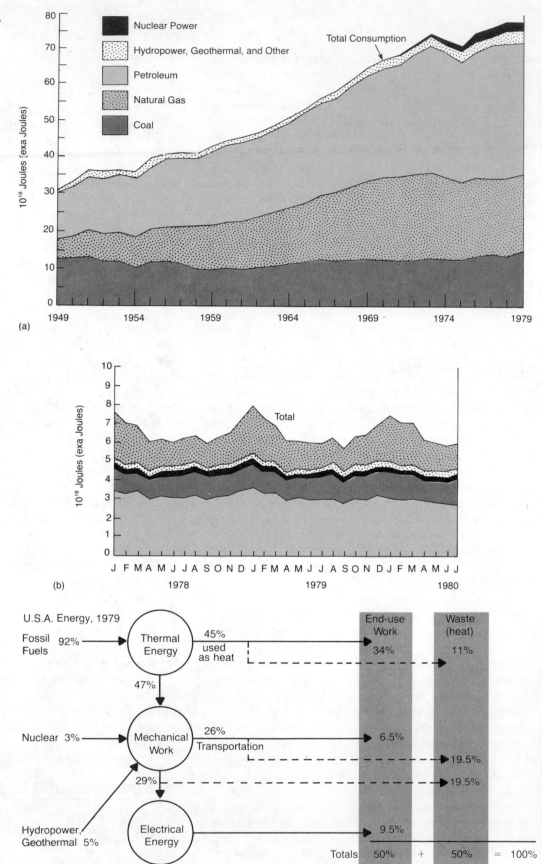

FIGURE 11.4
U.S. consumption of energy from 1949 to 1980 (a) and seasonal variation from 1976 through 1980 (b). (From U.S. Department of Energy.)

FIGURE 11.5
Highly generalized flow of energy in the United States. (Modified after Fowler, 1978.)

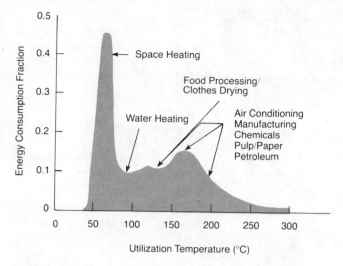

FIGURE 11.6

Spectra of energy use below 300°C in the United States.
(From Los Alamos Scientific Laboratory, 1978.)

organic matter; some organic material was buried, escaped oxidation, and was converted by complex chemical reactions in the geologic cycle to a fossil fuel. The biologic and geologic processes in the part of the geologic cycle that produces sedimentary rocks are responsible for the formation of fossil fuels. The fossil fuel resource is finite, taking millions of years of Earth history to form and a few hundred years of human history to burn. Using even the most optimistic predictions, we can expect the total fossil fuel epoch in human history to be only about 500 years (Fig. 11.7). Thus, while fossil fuels were extremely significant to the development of modern civilization, they must be an ephemeral event in the span of human history. Table 11.2 summarizes important information concerning the fossil fuels compared with other energy sources discussed in other parts of this chapter.

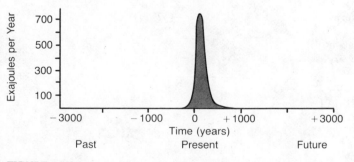

FIGURE 11.7

A perspective of the fossil fuel utilization epoch in human history. Note that the expected use of the fossil fuels will last only about 500 years. (Modified after Hubbert, 1962.)

Crude Oil and Natural Gas   Crude oil and natural gas are found primarily along young tectonic belts at plate boundaries (Fig. 11.8). Most of the known reserves (about 600 billion barrels) are located in a few fields (65 percent of the reserves are in 1 percent of the fields), the largest of which are in the Middle East. Although new oil and gas fields have recently been and are still being discovered in the United States, Mexico, Alaska, South America, and other areas, it is apparent that the present world reserves will be depleted in the next few decades. Table 11.3 summarizes the global crude oil situation, and Figure 11.9 shows the projected world maximum production rate of crude oil with demand (consumption) curves assuming 2, 3, and 5 percent growth rates. If these projections are correct, then demand for crude oil will exceed production sometime between 1985 and 2000.

World resources and reserves of natural gas are not well known, but total production will probably be less than 300 trillion cubic meters. The actual world reserves are about 72 trillion cubic meters, or about a 50-year supply at the present rate of consumption. In the United States domestic supplies (reserves) are only projected to last a few years, but vigorous exploration will probably add considerably to this.

Coal   Partially decomposed vegetation deeply buried in sedimentary environments may be slowly transformed into solid, brittle carbonaceous rock coal as idealized in Figure 11.10. Coal is by far the world's most abundant fossil fuel, with a total recoverable resource of about 6000 billion tons. Because the annual world consumption is about 4 billion tons, the resource should last many hundreds of years.

Coal is classified according to its carbon and sulphur content (Fig. 11.11 and Table 11.4). Energy content is greatest in bituminous coal, which has relatively few volatiles (oxygen, hydrogen, and nitrogen) and a low moisture content, and least in lignite, which has a high moisture content.

The distribution of the common coals (bituminous, subbituminous, and lignite) in the conterminous United States is shown in Figure 11.12. Most of the low-sulphur coal in the United States is relatively low-grade, low-energy lignite and subbituminous coal found west of the Mississippi River. The location of low-sulphur coal has environmental significance because low-sulphur coal causes less air pollution and therefore is more desirable as a fuel for power plants. Therefore, to avoid air pollution, thermal power plants on the East Coast will have to continue to treat some of the local coal to lower its sulphur content. Although it is expensive, treating the coal would

TABLE 11.2
Environmental effects of electrical power generation.

Energy Source	Physical Environment	Supply	Extraction Methods	Effects on Land
Crude oil and natural gas	Sedimentary rocks in young, tectonically active areas (plate boundaries). Most of resource is concentrated in a few areas (85% in 5% of the oil fields).	Few decades at most (USA)	Wells	Oil field waste water; pipeline construction; leaching by acid rain
Coal	Sedimentary rocks of various geologic ages. Original environment where organic material accumulated was swamps associated with estuaries, lagoons, and coastal plains or deltas. Rapid sedimentation and burial were necessary.	Hundreds of years (USA)	Surface and subsurface mining	Disruption of land; disposal of solid waste (ash); leaching by acid rain
Oil shale	Variable, but U.S. resource is concentrated in ancient lake beds of Colorado, Utah, and Wyoming. Oil shale is a fine-grained sedimentary rock containing significant amounts of relatively insoluble hydrocarbons.	Depends on future use pattern, but U.S. reserves are very large, equivalent to several hundred billion barrels of crude oil	Surface and subsurface mining; in place retorting	Disruption of land; disposal of spent oil shale; leaching by acid rain
Tar sands	Shales, sandstones, and limestones impregnated with tar, oil, asphalt, or other petroleum material that is difficult to remove because it is very viscous and will not readily flow	Depends on future use pattern, but because recovery is difficult, reserves will probably last many years	Surface mining	Disruption of land; disposal of wastes; leaching by acid rain
Geothermal	Young, tectonically active areas associated with plate boundaries and volcanic activity	Enormous potential resource	Wells	Waste water

256

Effects on Water	Effects on Air	Biological Effects	Overall Advantages	Overall Disadvantages
Oil spills; thermal pollution; acid rain	Oxides of nitrogen; carbon monoxide; hydrocarbons; acid rains; thermal pollution	Respiratory problems from air pollution; damage to plants from air pollution and acid rain	High-quality energy source; easy to recover, transport, and use	Resource is finite; causes pollution problems; is becoming expensive; domestic reserves are limited; contributes to CO_2 problem and global climate change as well as acid rain problem
Acid water draining from mines; thermal pollution; acid rain	Oxides of sulphur and nitrogen; particulates; some radioactive gases; thermal pollution	Respiratory problems and plant damage from air pollution; damage from acid rain	High-quality energy source; easy to use for a variety of purposes; very large domestic reserves	Resource is finite; causes pollution problems; disposal of ash; difficult to transport and remove sulphur; must disturb land; contributes to CO_2 problem and global climate change as well as acid rain problem
Same as for oil and gas	Same as for oil and gas	Same as for oil and gas	High-quality energy source; large domestic reserves; easy to transport and use	Resource is finite; is difficult and expensive to recover; causes air pollution; will contribute to CO_2 problem and global climate change as well as acid rain problem
Same as for oil and gas	Same as for oil and gas	Same as for oil and gas	High-quality energy source; easy to transport and use	Resource is finite; is difficult and expensive to recover; causes air pollution; contributes to air pollution, CO_2 problem, and acid rain
Possible pollution of near-surface waters	Release of some gases	Very little	Large potential domestic resource base	Resource is finite; waste disposal problem; may be expensive at some sites

TABLE 11.2 (continued)
Environmental effects of electrical power generation.

Energy Source	Physical Environment	Supply	Extraction Methods	Effects on Land
Nuclear fission	U^{235} in sedimentary and igneous rocks	U^{235} is very limited. Development of breeder reactor will extend supplies for hundreds of years.	U^{235} surface and subsurface mines	Disturbed land; waste disposal of mine tailings; waste disposal of high-level radioactive waste
Nuclear fusion	Fuels readily obtained from ocean and land environments	Virtually unlimited	Not developed	Very little
Solar	Best in southern exposures in mid- to low latitudes, in areas with many sunny days	Renewable	Passive and active	Very little
Wind	Areas that vertically or horizontally convert the flow of air	Renewable	Windmills	Aesthetic
Hydropower	River valleys	Renewable	Construction of dams	Loss of land to reservoirs
Biomass	Agricultural areas, forests	Renewable	Farming, silviculture, collection of urban trash	Use of land for production

TABLE 11.3
Crude oil: resources, reserves, and consumption.

	Billions of Barrels
World resources (ultimate production)	2000
World reserves	646
U.S. reserves	30
Already consumed	340
Expected future discoveries	1014
Annual world consumption	22
Depletion of present reserves (assuming constant consumption)	Year 2010
Peak in world production	1985–2000

SOURCE: Council on Environmental Quality and the Department of State, 1980.

be more economical than transporting low-sulphur coal from the western states.

Oil Shale   Oil shale is a fine-grained sedimentary rock containing organic matter. Upon heating (destructive distillation), oil shale yields significant amounts of hydrocarbons that are otherwise relatively insoluble. The best known oil shales in the United States are those in the Green River Formation, which underlies approximately 44,000 square kilometers of Colorado, Utah, and Wyoming (Fig. 11.13).

Total identified world shale-oil resources are estimated to contain about 3 trillion barrels of oil. However, evaluation of the oil grade and the feasibility of economic

Effects on Water	Effects on Air	Biological Effects	Overall Advantages	Overall Disadvantages
Thermal pollution; some radioactive liquids	Some radioactive gases	None detectable in normal operation; accidents can be extremely dangerous	High-quality energy	Radioactivity; waste disposal; nuclear proliferation; vulnerable to terrorist attack; becoming very expensive
Thermal pollution	Some radioactivity	Probably none	Potential source of nearly unlimited energy	Needs more work to define—research now on-going
Thermal pollution for solar power tower	Very little	None	Is renewable; uses conventional technology on a variety of scales; few pollution problems	Is expensive now; source varies in time and space
None	None	None	Is renewable, produces a high-quality energy	Is expensive now; source varies in time and space; good sites are limited
None	None	Fish kills due to dissolved nitrogen in water falling over spillways	Renewable, high-quality energy with few environmental problems	Sites are limited
Thermal pollution	Some air pollution	Respiratory problems from air pollution; thermal pollution	Some varieties use inexpensive fuel such as wood; renewable	Causes air pollution; some forms quite expensive; variable-quality energy

recovery with today's technology is not completed. Shale-oil resources in the United States amount to about 2 trillion barrels of oil, or two thirds of the total identified in the world; of this, 90 percent, or 1.8 trillion barrels, is located in the Green River oil shales.

With today's technology, approximately 80 billion barrels of oil can be extracted at a profit [3]. This is equivalent to about four years of total world crude oil consumption. As technology changes, more of the world's identified shale-oil resources may be recovered.

Tar Sands   Tar sands are sedimentary rocks or sands impregnated with tar oil, asphalt, or bitumen. Petroleum cannot be recovered from them by usual commercial methods such as wells because the oil is too viscous to

flow easily. Seventy-five percent of the world's known tar sand deposits are in Canada. The total resource is about 1000 billion barrels, but it is not known how much of this will eventually be recovered.

NUCLEAR ENERGY: FISSION

The first controlled nuclear fission, demonstrated in 1942, led to the development of the primary uses of uranium—that is, in explosives and as a heat source to provide steam for generation of electricity. One kilogram of uranium oxide produces a heat equivalent of approximately 16 metric tons of coal, making uranium an important source of energy in the United States and the world for at least a few more years [4].

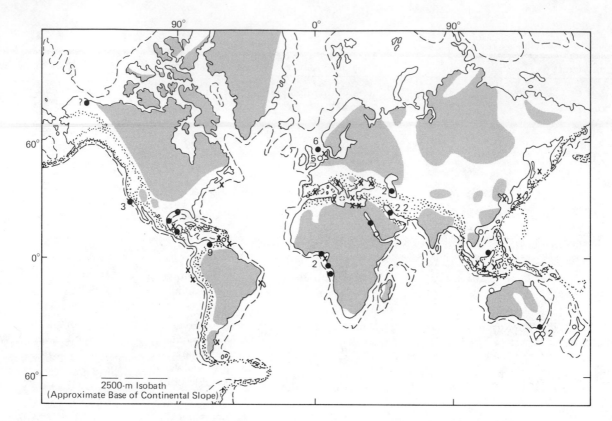

FIGURE 11.8

Giant oil and gas fields of the world relative to generalized tectonic belts. Active tectonic areas are shown by the stippled pattern, and regions not subject to tectonic activity over the last 500 million years are shaded. Giant oil fields are denoted by solid circles; gas fields by the open circles; and general areas of discovery either of less than giant size or areas under field development are indicated by an X. Numbers next to the symbols indicate the number of fields at a particular site. (After U.S. Geological Survey Circular 694, as adapted from Drake, 1972.)

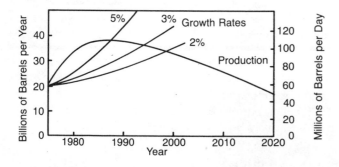

FIGURE 11.9

Maximum oil production for the world at a moderate depletion rate and three growth rates for oil production. If these projections are correct, then demand will exceed production sometime between 1985 and the year 2000. (From Council on Environmental Quality and the Department of State, 1980.)

Three types of uranium occur in nature: uranium-238, which accounts for approximately 99.3 percent of all natural uranium; uranium-235, which makes up about 0.7 percent; and uranium-234, which makes up about 0.005 percent. Uranium-235 is the only naturally occurring fissionable material and therefore is essential to production of nuclear energy. Processing uranium to increase the amount of uranium-235 from 0.7 percent to about 3 percent produces *enriched uranium,* which is used as the reactor fuel.

Fission reactors split uranium-235 by neutron bombardment (Fig. 11.14). The reaction produces three more neutrons released from uranium, fission fragments, and heat. The released neutrons each strike other uranium-235 atoms, releasing more neutrons, fission products, and heat. As the process continues, a chain reaction develops

(a) *Coal swamp forms.*

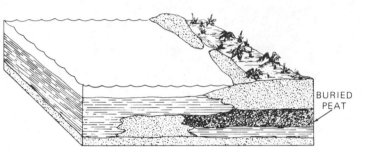

(b) *Rise in sea level buries swamp in sediment.*

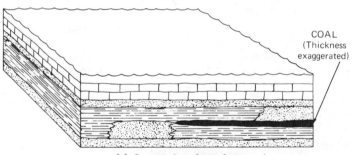

(c) *Compression of peat forms coal.*

FIGURE 11.10
Idealized diagram showing the processes by which buried plant debris (peat) is transformed into coal.

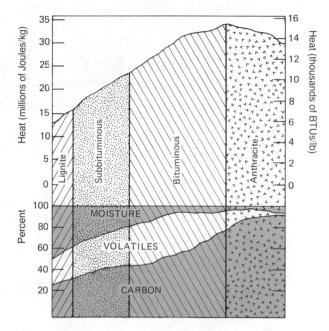

FIGURE 11.11
Generalized classification of different types of coal based upon their relative content (in percentage) of moisture, volatiles, and carbon. The heat values of the different types of coal are also shown. (After Brobst and Pratt, 1973.)

TABLE 11.4
Distribution of U.S. coal resources according to their rank and sulphur content.

	Sulphur Content (percent)		
Rank	Low 0–1	Medium 1.1–3.0	High 3+
Anthracite	97.1	2.9	—
Bituminous coal	29.8	26.8	43.4
Subbituminous coal	99.6	0.4	—
Lignite	90.7	9.3	—
All ranks	65.0	15.0	20.0

SOURCE: From Murphy, 1966.

as more and more uranium is split, releasing more neutrons.

Most reactors now in use consume more fissionable material than they produce and are known as *burner reactors*. The reactor itself is part of the nuclear steam supply system, which produces the steam to run the turbine generators that produce the electricity [5]. Therefore, the reactor has the same function as the boiler which produces the heat in coal- or oil-burning power plants (Fig. 11.15).

The main components of the reactor shown in Figure 11.16 are the core, control rods, coolant, and reactor ves-

sel. The *core* of the reactor is enclosed in a heavy stainless steel reactor vessel. Then, for extra safety and security, the entire reactor is contained in a reinforced concrete building [5].

Fuel pins, consisting of enriched uranium pellets placed into hollow tubes with a diameter less than 1 centimeter, are packed together (40,000 or more in a reactor) into fuel subassemblies in the core. A stable fission chain reaction in the core is maintained by controlling the num-

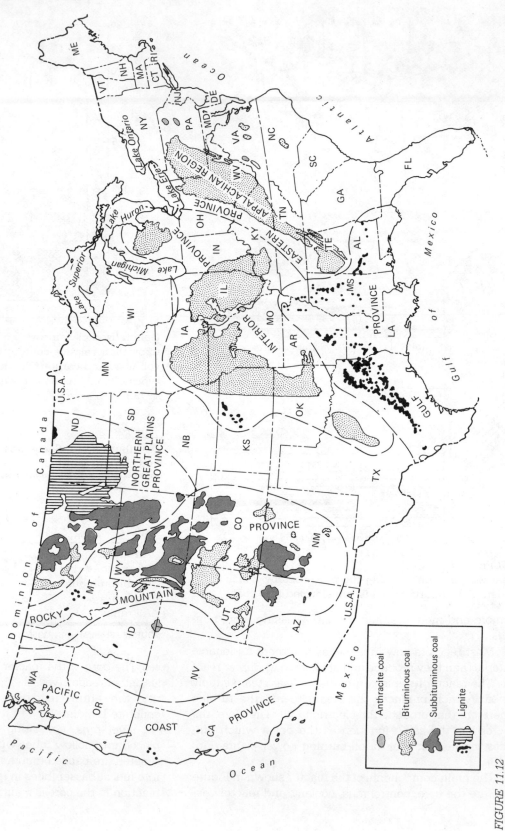

FIGURE 11.12
Coal fields of the United States. (After U.S. Bureau of Mines, 1971.)

FIGURE 11.13
Distribution of oil shale in the Green River Formation of Colorado, Utah, and Wyoming. (After Duncan and Swanson, 1965.)

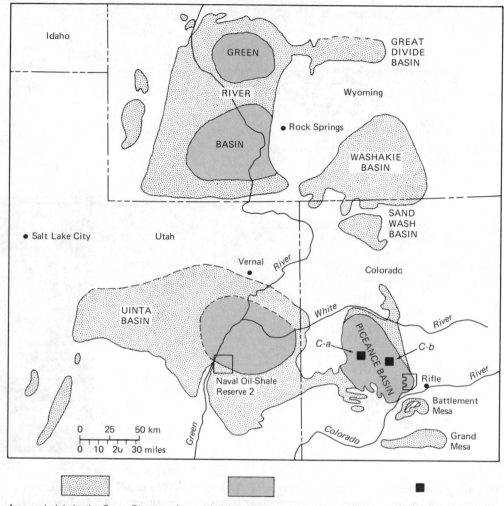

Area underlain by the Green River Formation in which the oil shale is unappraised or of low grade.

Area underlain by oil shale more than 3 m thick which yields 0.1 m³ or more oil per ton of shale.

Location of oil shale development discussed in this text.

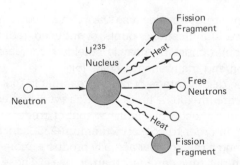

FIGURE 11.14
Idealized diagram showing fission of uranium-235. A neutron strikes the uranium-235 nucleus, producing fission fragments and free neutrons and releasing heat. The released neutrons may then each strike another uranium-235 atom, releasing more neutrons, fission fragments, and energy. As the process continues, a chain reaction develops.

ber of neutrons that cause fission as well as the fuel concentration. A minimum fuel concentration is necessary to keep the reactor *critical*—that is, to achieve a self-sustaining chain reaction. The *control rods,* which contain materials that capture neutrons, are used to regulate the chain reaction. If the rods are pulled out, the chain reaction speeds up; if they are inserted into the core, the reaction slows down [5].

The function of the *coolant* is to remove the heat produced by the fission reactions. When water is used as the coolant, it also acts as a moderator, slowing the neutrons down and facilitating efficient fission of uranium-235 [5].

Other parts of the nuclear steam supply system are the *primary coolant loops* and pumps which circulate a coolant (usually water) through the reactor, extracting heat produced by fission, and *heat exchangers* or *steam generators* which use the fission-heated coolant to make

FOSSIL FUEL POWER PLANT

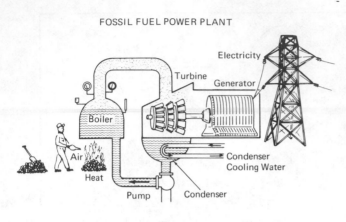

NUCLEAR POWER PLANT
Boiling Water Reactor (BWR)

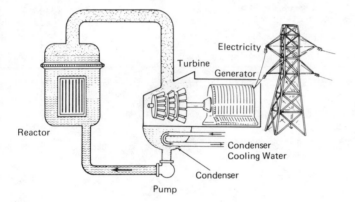

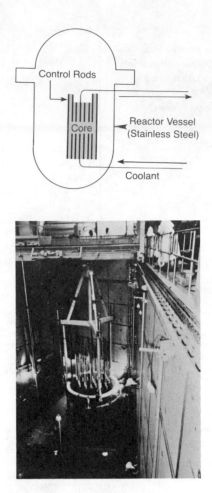

FIGURE 11.15
Idealized diagram comparing a fossil fuel power plant and nuclear power plant. Notice that the nuclear reactor has exactly the same function as the boiler in the fossil fuel power plant. (From American Nuclear Society, 1976.)

FIGURE 11.16
Idealized diagram showing the main components of a nuclear reactor. Below, loading fuel pins into the core. (From U.S. Energy and Resource Development Administration, 1976.)

steam. Figure 11.17a shows how these combine in a *pressurized water reactor* (PWR), which is called a **light-water reactor** (a type of burner reactor) because its coolant is water. Water heated by the reactor core is circulated in the primary coolant loop (a closed system) through a steam generator (heat exchanger), turning the water in the secondary loop to steam, which produces electricity [6].

A second type of light water reactor in use is the **boiling water reactor** (BWR), which is a direct-cycle system because there is no heat exchanger. The primary coolant loop goes through the reactor core directly to a turbine producing electricity. Examination of Figure 11.17b reveals a major disadvantage of the BWR. The steam that turns the turbine comes directly from the reactor core rather than first going through a heat exchanger, as it would in the PWR system. Therefore, particular care must be taken so that hazardous radiation does not leak into the turbine system, which is outside the main con-

tainment structure. Because the primary coolant (water) will have significant amounts of induced radioactivity, the turbines must be heavily shielded, increasing construction and maintenance costs [6].

One disadvantage of burner reactors is that they require uranium-235, a finite resource with limited potential for future recovery (similar to that of crude oil or natural gas). In order to conserve uranium-235, scientists are developing a reactor that actually produces more fissionable material (nuclear fuel) than it uses. These *breeder reactors,* utilizing a fuel core of fissionable material (plutonium-239) surrounded by a blanket of fertile material (uranium-238), will produce or breed additional plutonium-239 from the uranium-238. The transformation from uranium-238 to plutonium-239 occurs in the breeder reactor at the same time that the plutonium nuclei in the core are undergoing fission, providing heat that produces steam to drive turbines and generators which provide electricity. The breeder reactor will greatly extend the

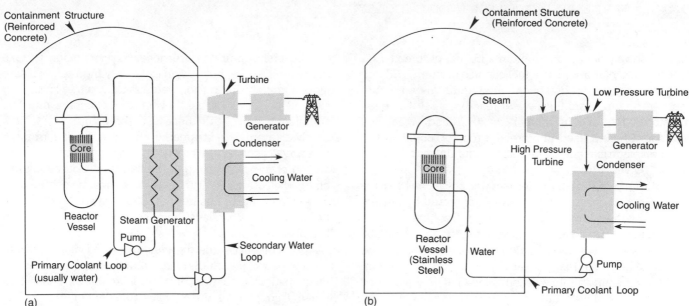

FIGURE 11.17

(a) Pressurized water reactor (PWR) showing the three main parts of the nuclear steam supply system: reactor, primary coolant loop, and steam generator (heat exchanger). (b) Boiling water reactor (BWR) showing the two main parts of the nuclear steam supply system: reactor and primary coolant loop. (Modified after U.S. Energy Research and Development Administration, 1976.)

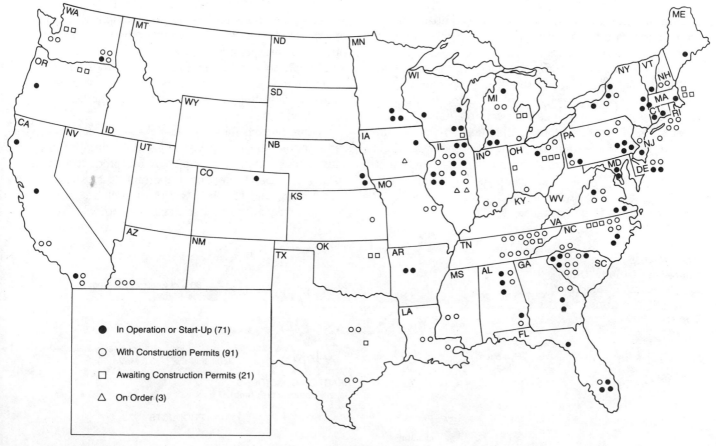

FIGURE 11.18

Nuclear power plants in the United States as of December 31, 1979. (From U.S. Department of Energy, 1979a.)

limited supply of natural uranium-235. In fact, breeder reactors using plutonium as a fuel can use uranium-238 tailings from the production of enriched uranium or uranium recovered from spent light-water reactor fuel. It has been estimated that the uranium now stockpiled as tailings and spent fuel, if used in a breeder reactor, could supply the total electrical energy demand in the United States for up to 100 years [6].

The growing demand for electric power will require an increase in the amount of power generated by nuclear energy. By the year 2000 it is expected that heat from nuclear reactors will have the capacity to produce about 200 gigawatts (10^9 watts) of electrical power in the United States. About 70 reactors are now in operation (Fig. 11.18), but many more will be needed if the projected generating capacity from uranium is to be realized. Because of increased costs, environmental considerations, and other factors, these projections are probably too high and will have to be revised. Therefore, the full impact of what began in 1942 is still to be determined.

NUCLEAR ENERGY: FUSION

In contrast to fission, which involves splitting heavy atoms such as uranium, fusion involves combining light elements such as hydrogen to form a larger element such as helium. As fusion occurs, heat energy is released (Fig. 11.19). Similar reactions are the source of energy in our sun and other stars. In a hypothetical fusion reactor, two isotopes of hydrogen (atoms with variable mass due to a different number of neutrons in the nucleus)—deuterium (D) and tritium (T)—are injected into the reactor chamber where the necessary conditions for fusion are maintained. Products of the D-T fusion include helium, carrying 20 percent of the energy released, and neutrons, carrying 80 percent of the energy released (Fig. 11.19) [7].

Several conditions are necessary for fusion to take place. First, there must be an extremely high temperature (approximately 100 million degrees Centigrade for D-T fusion). Second, the density of the fuel elements must be sufficiently high. At the necessary temperature for fusion, nearly all atoms are stripped of their electrons, forming a *plasma*. Plasma is an electrically neutral material consisting of positively charged nuclei, ions, and negatively charged electrons. Third, the plasma must be confined for a sufficient time to ensure that the energy released by the fusion reactions exceeds the energy supplied to maintain the plasma [7,8].

The potential energy available when and if fusion-reactor power plants (Fig. 11.20) are developed is nearly inexhaustible. One gram of D-T fuel (from a water and lithium fuel supply) has the energy equivalent of 45 barrels of oil. Deuterium can be extracted economically from ocean water, and tritium can be produced in a reaction with lithium in a fusion reactor. The reactor can therefore breed tritium, and lithium can be extracted economically from abundant supplies.

Many problems remain to be solved before nuclear fusion can be used on a large scale. Figure 11.21 lists major stages in fusion energy development with an optimistic timetable. Research is still in the first stage, which involves basic physics, testing possible fuels (mostly D-T), and magnetic confinement of plasma. Progress in fusion research has, in recent years, been steady, so there is optimism that useful power will eventually be produced from controlled fusion.

Energy from fusion, once released, has a variety of applications, including heating and cooling of buildings and production of synthetic fuels, but production of electricity is probably the most important. It is expected (but not proven) that fusion power plants will be competitive economically with other sources of electrical energy.

FIGURE 11.19
Deuterium-tritium (D-T) fusion reaction. (Modified from U.S. Department of Energy, 1980.)

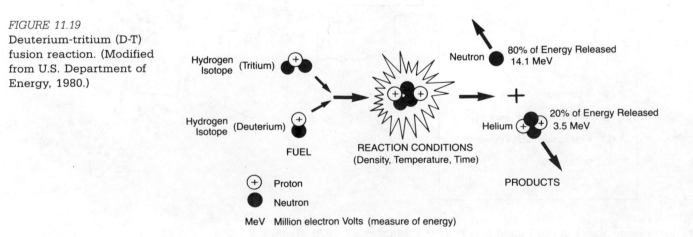

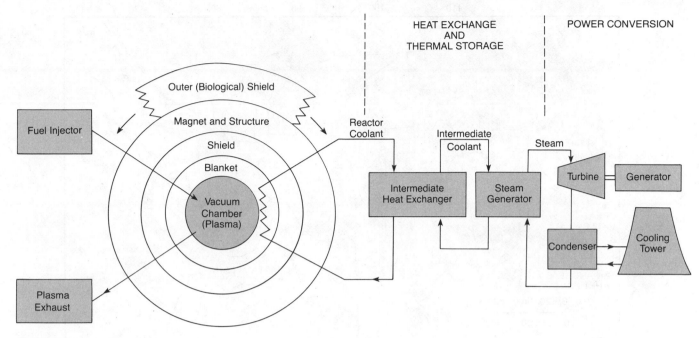

FIGURE 11.20
Schematic diagram of a fusion reactor and power plant. The vacuum chamber is
where plasma is confined by strong magnetic fields and fusion reactions take
place. (Modified after U.S. Department of Energy, 1979b.)

GEOTHERMAL ENERGY

The useful conversion of natural heat from the interior of
the Earth to heat buildings and generate electricity is an
exciting application of geologic knowledge and engineering technology. The idea of harnessing the Earth's internal heat is not new. As early as 1904, geothermal power

utilizing dry steam was developed in Italy, and natural internal heat is now being used to generate electricity in
the USSR, Japan, New Zealand, Iceland, Mexico, and California. However, existing geothermal facilities constitute
only a small portion of the total energy that might eventually be tapped from the Earth's reservoir of internal heat.

FIGURE 11.21
Stages in fusion energy
development. (Modified
after U.S. Department of
Energy, 1978.)

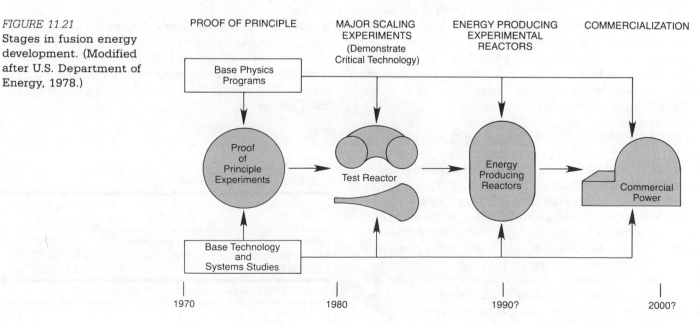

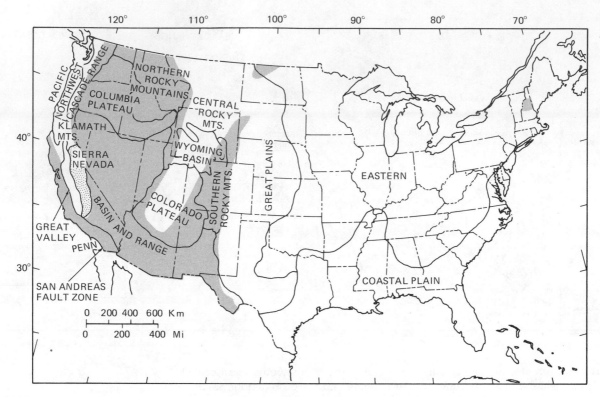

FIGURE 11.22
Map showing the generalized extent of hot (shaded), normal (white), and cold (stippled) crustal regions of the United States. Also shown are the major physiographic provinces, such as the Colorado Plateau. (After White and Williams, 1975.)

Natural heat production within the Earth is only partly understood. We do know that some areas have a higher flow of heat from below than others, and that for the most part these locations are associated with the tectonic cycle. Oceanic ridge systems (divergent plate boundaries) and convergent plate boundaries, where mountains are being uplifted and volcanic island arcs are forming, are areas where this natural heat flow from the Earth is anomalously high.

The relative distribution of heat flow from within the Earth for the United States is shown in Figure 11.22. The data for this generalized classification of hot, normal, and cold crustal areas are not sufficient to accurately define the limits of the region [9], and therefore the map has limited value for locating specific sites where geothermal energy resources could be developed. It is interesting to note that the region of high heat flow is concentrated in the western United States, where tectonic activity and volcanic activity have been recent.

Based on geologic criteria, several geothermal systems may be defined: hydrothermal convection systems, hot igneous systems, and geopressured systems [10,11].

Each of these systems has a different origin and different potential as an energy source, but the total resource base is very large (Table 11.5).

Hydrothermal convection systems are characterized by the circulation of steam and/or hot water that transfers heat at depths to the surface. An example is The Geysers, 145 kilometers north of San Francisco, California, which produces several hundred megawatts of electrical energy (Fig. 11.23).

TABLE 11.5
U.S. geothermal resources.

Type	Quantity of Energy (exajoules)
Hydrothermal convection systems	12,000[a]
Hot igneous systems (dry rock)	13,000,000[a]
Geopressurized	190,000[a]

SOURCE: Data from Los Alamos Scientific Laboratory.
[a]The present total energy used in the United States is about 80 exajoules per year.

FIGURE 11.23
The Geysers power plant north of San Francisco, California, with a 663-megawatt installed capacity, is the world's largest geothermal electricity development. (Photo courtesy of Pacific Gas and Electric Company.)

Hot igneous systems involve hot, dry rocks with or without the presence of near-surface molten rock (*magma*). These systems contain a tremendous amount of stored heat (Table 11.5), and recent experiments in New Mexico are investigating the scientific feasibility of extracting energy from them.

Geopressured systems exist where the normal heat flow from the Earth is trapped by impermeable clay layers that act as an effective insulator. Perhaps the best known regions where geopressurized systems develop are along the Gulf Coast of the United States. These systems have the potential to produce large quantities of electricity for three reasons. First, they contain hot-water thermal energy that could be extracted. Second, mechanical energy from the high-pressure water could be used to turn hydraulic turbines to produce electricity. Third, the waters contain considerable amounts of dissolved methane gas (up to 1 cubic meter per barrel of water) that could be extracted and used to generate electricity [11,12].

At present, geothermal energy supplies only a small fraction of 1 percent of electrical energy produced in the United States. With the exception of the unusual vapor-dominated systems such as The Geysers in California, the production of electricity from geothermal reservoirs is still rather expensive, generally more so than from other sources of energy. For this reason, commercial development of geothermal energy will not develop rapidly until the economic picture improves. Even if commercially available, geothermal sources would only supply 10 percent of the total electrical energy output in the near fu-

ture, even in states like California, where it has been produced and where expanding facilities are likely [12].

RENEWABLE ENERGY SOURCES

DIRECT SOLAR ENERGY

Broadly defined, solar energy includes a number of different energy sources, as shown in Figure 11.24.

The total amount of solar energy reaching the Earth's surface is tremendous. For example, on a global scale, two weeks of solar energy is roughly equivalent to the energy stored in all known reserves of coal, oil, and natural gas on Earth. In the United States, on the average, 13 percent of the sun's original energy entering the atmosphere arrives at the ground. However, the actual amount at a particular site is quite variable, depending upon the time of the year and the cloud cover. The average value of 13 percent is equivalent to approximately 177 watts per square meter on a continuous basis [13].

Direct use of solar energy may either be through passive solar systems or active solar systems. **Passive solar systems** often involve architectural design to enhance and take advantage of natural changes in solar energy that occur throughout the year without requiring mechanical power. Many homes and other buildings in the southwestern United States as well as other parts of the country now use passive solar systems for at least part of their energy needs. **Active solar systems** require mechanical power, usually pumps and other apparatus, to

FIGURE 11.24
Idealized diagram showing
the routes of the various
types of renewable solar
energy.

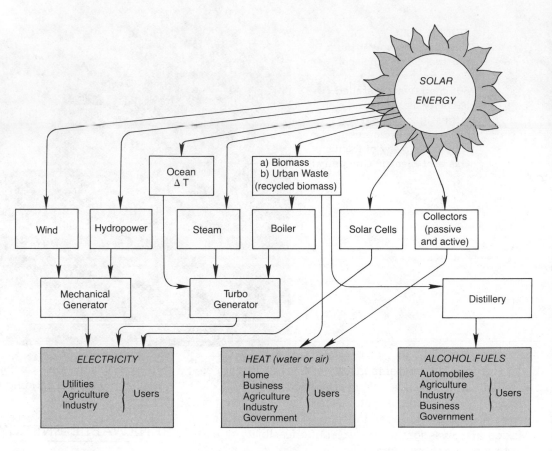

circulate air, water, or other fluids from solar collectors to a heat sink, where the heat is stored until used. **Solar collectors** are usually flat panels consisting of a glass cover plate over a black background upon which water is circulated through tubes. Short-wave solar radiation enters the glass and is absorbed by the black background. As longer-wave radiation is emitted from the black material, it cannot escape through the glass, so it heats the water in the circulating tubes to 38° to 93°C (100–200°F) [13]. Thus, solar collectors act as greenhouses. The number of systems using these collectors in the United States exceeds 100,000 and is rapidly growing.

Another potentially important aspect of direct solar energy involves **solar cells,** or **photovoltaics,** that convert sunlight directly into electricity. Although such cells might be used in unique situations, electricity produced from them generally costs about 20 times as much as that produced from traditional fossil fuel sources [14]. On the other hand, because the field of solar technology is changing so quickly, low-cost solar cells may become widely available.

Questions often asked about direct solar energy are, Where can it be used, and how much energy will it save? These questions are not easily answered. Figure 11.25 shows in a very general way the estimated year-round usability of solar energy in the United States. However, this view of solar energy is analogous to painting a picture

with a paint roller. Solar energy is site-specific, and detailed observation in the field is necessary to evaluate the solar energy potential in a given area. The energy that might be saved by converting to direct solar power is also quite variable. Optimistic people hope that by the year 2000 direct solar energy may supply up to 20 percent of the total energy demand. Considering how fast the solar energy industry is growing and the short lead time necessary for conversion to some solar energy technology, these hopes may be realistic. On the other hand, the conversions may be much slower if other sources of energy at cheaper prices become available in the next few years. This, however, is unlikely, and the future of solar energy seems quite secure.

Two other interesting types of solar energy are the *solar power tower* and *solar ponds.* The concept of a **power tower** is shown in Figure 11.26. The system works by collecting solar energy as heat and delivering this energy in the form of steam to turbines which produce electric power. An experimental 10-megawatt power tower near Barstow, California, is approximately 100 meters high and is surrounded by approximately 2000 mirror modules, each with a reflective area of about 40 square meters. The mirrors will adjust continually to reflect as much sunlight into the tower as possible. When excess steam is available, it will be stored so that it can be extracted during periods when no sun is available. The

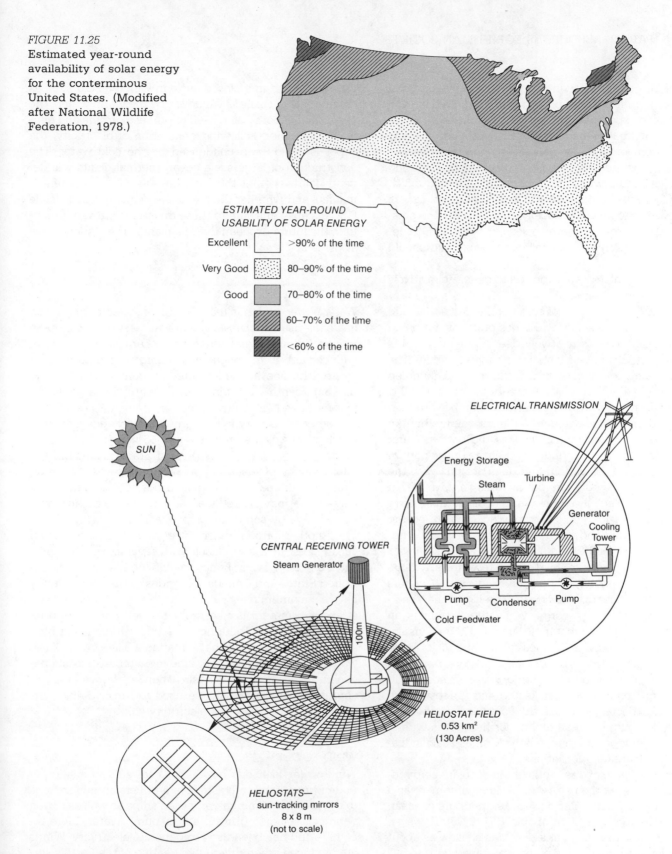

FIGURE 11.25
Estimated year-round availability of solar energy for the conterminous United States. (Modified after National Wildlife Federation, 1978.)

ESTIMATED YEAR-ROUND
USABILITY OF SOLAR ENERGY

Excellent   >90% of the time

Very Good   80–90% of the time

Good   70–80% of the time

60–70% of the time

<60% of the time

ELECTRICAL TRANSMISSION

SUN

Energy Storage

Steam

Turbine

Generator

Cooling Tower

CENTRAL RECEIVING TOWER
Steam Generator

Pump

Condensor

Pump

Cold Feedwater

100m

HELIOSTAT FIELD
0.53 km²
(130 Acres)

HELIOSTATS—
sun-tracking mirrors
8 x 8 m
(not to scale)

FIGURE 11.26
Idealized diagram showing the solar power tower being tested near Barstow, California, in the Mojave Desert. This power plant is expected to produce as much as 10 megawatts of electricity and is a cooperative effort between the U.S. Department of Energy and the Southern California Edison Company. (Modified after a drawing prepared by Southern California Edison Company.)

Japanese are currently trying to produce 1 megawatt of electricity from a tower approximately 75 meters high. If the Japanese experiment proves successful, they hope to eventually build a 60-megawatt plant. Although there is certainly room for optimism concerning power towers, the electricity generated is going to be very expensive during the first few years when the technology is being developed. In fact, it is probably safe to state that power towers may prove not to be economically viable in all sites presently being considered. Nevertheless, the research to develop the technology is certainly worthwhile and should continue.

The use of shallow **solar ponds** to generate relatively low-temperature water of about 68°C (180°F) is an interesting prospect for sources of commercial, industrial, and agricultural heat. Presently two types of ponds are being developed. One type is relatively deep (3 meters) and is approximately 20 meters long by 10 meters wide. The pond is designed to collect heat during the summer, building up to a bottom water temperature of about 68°C in the fall. During the winter, heat may then be extracted from the bottom. The heated water is kept on the bottom by the addition of salt, which makes the bottom water heavier. Circulation is restricted so that the dense bottom water remains on the bottom and does not mix with the water above. Reportedly, the pond works very well, even in areas with a relatively severe winter. For example, up to 25 centimeters of ice and snow may accumulate on the surface of the pond and the bottom water will still be approximately 38°C (100°F) throughout the winter [14].

The second type of solar pond resembles a large waterbed approximately 3.5 meters wide by 60 meters long and 5 to 10 centimeters deep. The top of the "bed" is transparent to solar energy, while the bottom is an energy-absorbing black material. The pond is then insulated from below to prevent heat loss to soil and rock. These ponds are essentially very large solar collectors, much like the smaller panels used for heating water in homes. To date, several ponds have been tested and the results are optimistic, especially because the cost is presently competitive with burning fossil fuel oil for heat. This type of solar pond would work well in relatively small-scale industrial and agricultural applications [14].

A last example of direct utilization of solar energy involves using part of the natural oceanic environment as a gigantic solar collector. The surface temperature of ocean water in the tropics is often about 28°C (82°F). However, at the bottom of the ocean at a depth as shallow as about 600 meters, the temperature of the water may be 1° to 3°C (35–38°F). Low-efficiency heat engines can be designed to exploit this temperature differential by either directly using the sea water or employing an appropriate heat exchange system in a closed cycle in which a fluid such as ammonia or propane is vaporized by the warm water. The expanding vapor is then used to propel a turbine and generate electricity. Following generation of electricity, the vapor is cooled and condensed by the cold water. The construction of large-scale ocean thermal plants will depend upon whether they can be built close to potential markets and whether the plants are economically feasible [13]. Because answers to these questions are very uncertain, the future of ocean thermal plants is very speculative at present.

WATER POWER

Water power, which is really a form of stored solar energy, has been used successfully since at least the time of the Roman Empire. Water wheels that harness water power and convert it to mechanical energy were turning in western Europe in the seventeenth century, and during the eighteenth and nineteenth centuries large water wheels provided energy to power grain mills, sawmills, and other machinery in the United States. Today, hydroelectric power plants provide about 15 percent of the total electricity produced in the United States. Although the total amount of electrical power produced by running water will increase somewhat in the coming years, this percentage may be reduced as nuclear, solar, and geothermal energy sources increase faster.

Another form of water power might be derived from ocean tides in a few places with favorable topography, such as the Bay of Fundy region of the northeastern United States and Canada. The tides in the Bay of Fundy have a maximum rise of about 15 meters. A minimum rise of about 8 meters is necessary to even consider developing tidal power. To harness tidal power, dams are built across the entrance to a bay, creating a basin on the landward side such that a difference in water level exists between the ocean and the basin. Then, as the water fills or empties the basin, it can be used to run hydraulic turbines which will produce electricity [15].

WIND POWER

Wind power, like solar power, has evolved over a long period of time, from early Chinese and Persian civilizations to the present. Wind has propelled ships as well as driven windmills used to grind grain or pump water. More recently wind has been used to generate electricity. Winds are produced when differential heating of the Earth's surface creates air masses with differing heat contents and densities. The potential energy that might be eventually derived from the wind is tremendous, and yet there will be problems because winds tend to be highly variable in time, place, and intensity [16].

Wind prospecting will become an important endeavor in the future. On a national scale, regions with the greatest potential for development of wind energy are the Pacific Northwest coastal area, the coastal region of the northeastern United States, and a belt extending from northern Texas northward through the Rocky Mountain states and the Dakotas. However, there are many other good sites, such as the mountain areas in North Carolina and the northern Coachella Valley in southern California.

At a particular site the direction, velocity, and duration of the wind may be quite variable, depending on local topography and regional to local magnitude of temperature differences in the atmosphere [16]. For example, the wind velocity often increases over hilltops or mountains, or wind may be funneled through a broad mountain pass (Fig. 11.27). The increase in wind velocity over a mountain is due to a vertical convergence of the wind, whereas in a pass the increase is partly due to a horizontal convergence as well. Because the shape of a mountain or a pass is often related to the local or regional geology, prospecting for wind energy is in part a geologic as well as a geographic and meteorological problem.

Significant improvements in the size of windmills and the amount of power they produce occurred from the late 1800s through approximately 1950, when many European countries and the United States became interested in large-scale generators driven by the wind. In the United States, thousands of small wind-driven generators have been used on farms. Most of the small windmills generated approximately 1 kilowatt of power, which is much too small to be considered for central power generation needs. Interest in wind power declined for several decades prior to the 1970s because of the abundance of cheap fossil fuels, but now there is revived interest in building larger windmills that may supply a few megawatts of electrical power each. Pilot projects in the Blue Ridge Mountains of North Carolina and the deserts of southern California are now in progress.. Figure 11.28 shows the large windmill nothwest of Palm Springs, California. Its three-bladed rotor is approximately 55 meters in diameter and is designed to generate as much as 3 megawatts of electricity, or enough to supply nearly 1000 homes. Mounted on a tower approximately 33 meters high, the windmill is located in an area where the winds average 24 to 27 kilometers per hour. It is expected that the windmill will operate approximately two thirds of the time, but will not run at full power for more than about 6 percent of the time.

ENERGY FROM BIOMASS

Biomass fuel is a new name for the oldest human fuel. Biomass is organic matter that can be burned directly as

FIGURE 11.27
Idealized diagram showing some of the ways that the geologic environment may converge wind in the vertical or horizontal direction, thus increasing the speed of the wind.

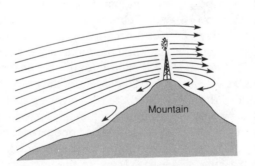

(a) Vertical Convergence of Wind (cross section)

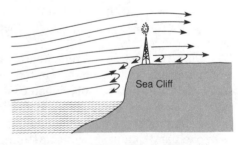

(b) Small Vertical Convergence of Wind with Severe Turbulence (cross section)

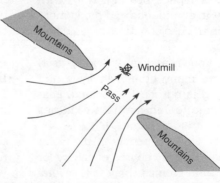

(c) Horizontal Convergence of Wind (plan view)

FIGURE 11.28
Giant windmill located near Palm Springs, California. The windmill is designed to generate as much as 3 megawatts of electricity, which is sufficient to supply nearly 1000 homes. The height of the tower is approximately 100 feet, or 33 meters. (Photo courtesy of Mel Manalis.)

a fuel or converted to a more convenient form and then burned. For example, we can burn wood in a stove or convert it to charcoal and then burn it. Biomass has provided a major source of energy for human beings throughout most of the history of civilization. When North America was first settled, there was more wood fuel than could be used. The forests were cleared for agriculture often by girdling trees (cutting through the bark all the way around the base of a tree), which killed them, and then burning the forests.

Until the end of the nineteenth century, wood was the major fuel source in the United States. During the early mid-twentieth century, when coal, oil, and gas were plentiful and high-grade mines in the United States provided abundant cheap energy, burning wood became old-fashioned and quaint; it was something done for pleasure in an open fireplace that conducted more heat up the chimney than it provided for space heating. Now, with other fuels reaching a limit in abundance and production, there is renewed interest in the use of natural organic materials for fuel.

Firewood is the best known and most widely used biomass fuel, but there are many others. In India and other countries, cattle dung is burned for cooking. Peat provides heating and cooking fuel in northern countries like Scotland, where the peat is abundant.

On a global scale over 1 billion people in the world today still use wood as their primary source of energy for heat and cooking. Energy from biomass may take several routes: direct burning of biomass to either produce elec-

tricity or heat water and air, and distillation of biomass to produce alcohol for a fuel. Today in the United States, various biomass sources supply nearly 2 percent of the entire energy consumption (approximately 1.5 exajoules per year), primarily from the use of wood in the forestry products industry for home heating [17].

There are two major sources of biomass fuels in North America: forest products and otherwise unused agricultural products. It has been estimated that by the year 2000 energy from biological processes could supply up to 15 to 20 percent of the current energy consumption in the United States. However, this estimate is dependent upon a number of factors, including availability of cropland, increased yields, good resource management, and, most importantly, the development of efficient conversion processes. Most of the increase is expected to result from burning a greater amount of wood. Figure 11.29a summarizes U.S. energy use in 1979, and Figure 11.29b shows two possible scenarios—one high and one low—for potential bioenergy supplies.

Biomass may also be recycled. Today there are 20 facilities in the United States that process urban waste to be used to generate electricity or to be used as a fuel. Presently, only about 1 percent of the nation's municipal solid wastes is being recovered for energy. However, if all the plants were operating at full capacity and if the additional 20 plants or so under construction were completed and operating, then about 10 percent of the country's waste, or 18 million tons per year, could be used to extract energy. At processing plants such as those used in Bal-

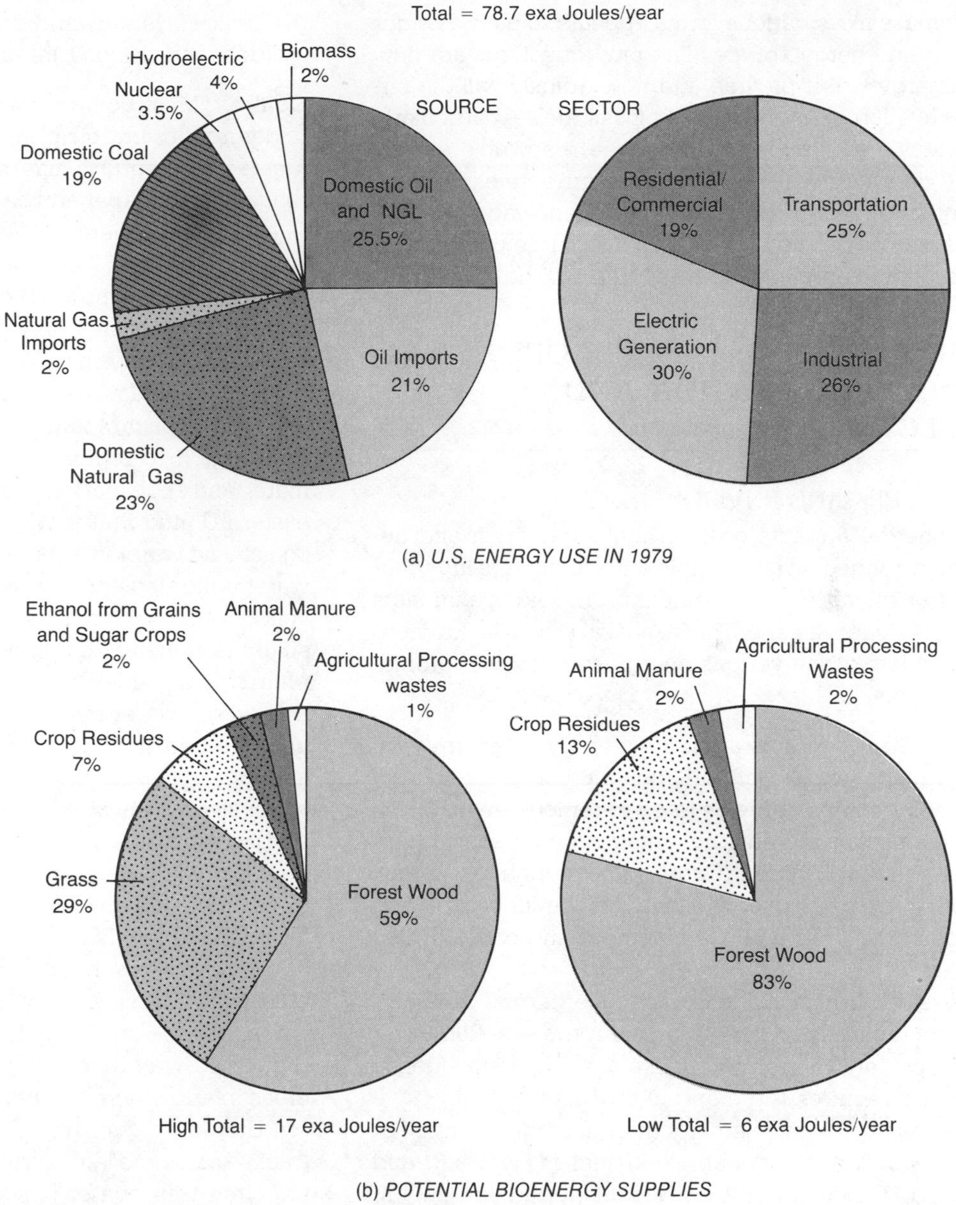

FIGURE 11.29

Energy use in the United States for 1979 (a), and two scenarios of potential bioenergy supplies (b), excluding speculative sources and municipal waste. (From Office of Technology Assessment, 1980.)

Total = 78.7 exa Joules/year

SOURCE

Biomass 2%
Hydroelectric 4%
Nuclear 3.5%
Domestic Coal 19%
Natural Gas Imports 2%
Domestic Natural Gas 23%
Domestic Oil and NGL 25.5%
Oil Imports 21%

SECTOR

Residential/ Commercial 19%
Transportation 25%
Electric Generation 30%
Industrial 26%

(a) *U.S. ENERGY USE IN 1979*

Ethanol from Grains and Sugar Crops 2%
Animal Manure 2%
Agricultural Processing wastes 1%
Crop Residues 7%
Grass 29%
Forest Wood 59%

High Total = 17 exa Joules/year

Animal Manure 2%
Agricultural Processing Wastes 2%
Crop Residues 13%
Forest Wood 83%

Low Total = 6 exa Joules/year

(b) *POTENTIAL BIOENERGY SUPPLIES*

timore County, Maryland; Chicago, Illinois; Milwaukee, Wisconsin; Tacoma, Washington; and Akron, Ohio, municipal waste is burned and the heat energy used to make steam for a variety of purposes, from space heating to industrial generation of electricity. The United States has been slower to utilize urban waste as an energy source than other countries. For example, in western Europe a number of countries now utilize between one third to one half of their municipal waste for energy production. With the end of cheap, available fossil fuels, certainly more and more energy recovery systems utilizing urban waste will be forthcoming [18].

A problem with biomass fuels is their net yield of energy. Because the density of vegetation production is comparatively low, and the density of unused forest and agricultural residues even lower, considerable energy is required to collect the biomass for fuel. Processing the biomass into more convenient fuels also requires energy. Biomass fuels provide the greatest net gain in energy when they are used locally. For example, in the production of sugar from sugar cane, unused parts of the cane can be burned to provide energy for local processing. When fuels are consumed far from their source, and if much conversion is required, there may be no net gain in energy. For example, converting corn stalks to alcohol and then transporting alcohol long distances to be used as a fuel for cars and trucks may produce little or no net gain in energy.

Biomass in its various forms appears to have a bright future as an energy source. The only questions are how much energy will it provide and how quickly will the energy be available. As the price of fossil fuels—particularly oil and gas—continues to rise, these renewable energy sources certainly will become more attractive. Because the transition from fossil fuels to renewable energy sources is now just underway, it is difficult to predict how long and how complete the transition will be.

ENVIRONMENTAL EFFECTS OF ENERGY EXPLORATION AND DEVELOPMENT

FOSSIL FUELS AND ENVIRONMENT

Environmental impacts of fossil fuel exploration and development range from negligible for remote-sensing techniques in exploration, to significant, unavoidable impacts for projects such as the Trans-Alaska Pipeline. Environmental effects of the various energy sources are summarized in Table 11.2.

Oil and Gas   Exploration for energy often involves building roads, exploratory drilling, and constructing supply lines (for camps, airfields, etc.) to remote areas. These activities, except in sensitive areas such as some semiarid to arid environments and some permafrost areas, generally have few adverse effects on the landscape and resources compared with development and consumption activities.

Development of oil, gas, and geothermal fields involves drilling wells on land or beneath the sea. Removing oil, gas, or hot waters may cause a surface subsidence hazard and requires the disposal of waste waters.

The most extensive environmental problems associated with oil and gas occur when fuel is delivered and consumed. Crude oil is mostly transported on land in pipelines or across the ocean by tankers, and both methods have the potential to produce oil spills. Marine oil spills are best known, and although the effects are relatively short-lived, they have killed thousands of sea birds, temporarily spoiled beaches, and caused loss of tourist revenue.

The Trans-Alaska Pipeline, completed in 1977, provides a good example of the possible spectrum of effects of an oil transport system on a sensitive environment. Main unavoidable impacts include disturbance of land and fish and wildlife habitats; human use of the land during construction, operation, and maintenance of the pipeline itself, access roads, and other support facilities; discharge of oil and other effluents into Port Valdez from the tanker system; and increased pressure on services, utilities, and the culture of the native population [19].

Coal   Most coal mining in the United States is done underground, but strip mining, which started in the late nineteenth century, has steadily increased because it tends to be cheaper and easier than underground mining. The increased demand for coal will lead to more and larger strip mines. There are 40 billion metric tons of coal reserves that are now accessible to surface mining techniques. In addition, approximately another 90 billion metric tons of coal within 50 meters of the surface is potentially available for strip mining.

The impact of large strip mines for fossil fuels varies from region to region, depending upon topography, climate, and reclamation practices. In humid areas of the eastern United States with abundant rainfall, mine drainage of acid water is a serious problem (Fig. 11.30). Surface water infiltrates into the spoil banks (material left after the coal or other minerals are removed) where it reacts with sulphide minerals such as pyrite (FeS_2) to produce sulphuric acid, which then runs into and pollutes streams and groundwater resources. Acid water also drains from underground mines and roadcuts and areas where coal and pyrite are abundant, but the problem is magnified when large areas of disturbed material remain exposed to surface waters. Acid drainage can be minimized by channeling surface runoff and groundwater before they enter a mined area and diverting them around the potentially polluting materials [20].

In arid and semiarid regions, water problems associated with mining are not as pronounced as in wetter regions, but the land may be more sensitive to mining activities such as exploration and road building. In some areas in arid environments of the western and southwestern United States, the land is so sensitive that even tire tracks across the land survive for years. Furthermore, soils are often thin, water is scarce, and reclamation work is difficult.

All strip mining has the potential to pollute or destroy scenic, water, biologic, or other land resources. However, good reclamation practices can minimize the damage (Fig. 11.31). Although reclamation is often site-specific, a case history of a modern coal mine in Colorado will emphasize its important principles.

The Trapper Mine on the western slope of the Rocky Mountains in northern Colorado is a good example of a new generation of large coal strip mines. The main operation is designed to minimize environmental degradation during the mining and reclaim the land for dryland farming and grazing of livestock and big game without artificial application of water.

FIGURE 11.30
The stream in the foreground is flowing through waste piles of a Missouri coal mine. The water reacts with sulphide minerals and forms sulphuric acid, a serious problem in coal mining areas. (Photo by J. D. Vineyard, courtesy of Missouri Geological Survey.)

Over a 35-year period the mine will produce 68 million metric tons of coal from 20 to 24 square kilometers to be delivered to a 800-megawatt power plant located adjacent to the mine. Four coal seams, varying from about 1 to 4 meters thick, each separated by various depths of overburden, will be mined. Depth of overburden to the coal varies from zero to about 50 meters.

The following steps are involved in the actual mining. First, the vegetation and the topsoil are removed with dozers and scrapers, and the soil is stockpiled for reuse. Second, the overburden along a cut up to 1.6 kilometers long and 53 meters wide is removed with a 23-cubic-meter dragline bucket (Fig. 11.32). The exposed coal beds are then drilled and blasted to fracture the coal, which is removed with a backhoe and loaded into trucks. Finally, the cut is filled, the topsoil replaced, and the land is either planted in a crop or returned to rangeland.

At the Trapper Mine the land is reclaimed without artificially applying water. The precipitation (mostly snow) is about 35 centimeters per year, which is sufficient to reestablish vegetation provided there is adequate topsoil. This emphasizes an important point—namely, that reclamation is site-specific. Therefore, what works at one location may not be applicable to other areas.

(a) (b)

FIGURE 11.31
A small open-pit mine before (a) and after (b) reclamation. (Photos by J. D. Vineyard, courtesy of Missouri Geological Survey.)

FIGURE 11.32
The Trapper Mine. Shown here is the 23-cubic-meter dragline bucket that removes the overburden prior to extraction of the coal. (Photo by E. A. Keller.)

Water and air quality are closely monitored at the Trapper Mine. Surface water is diverted around mine pits and groundwater intercepted while pits are open. Settling basins constructed downslope from the pit trap suspended solids prior to discharging water into local streams. Air quality at the mine is degraded by dust produced from blasting, hauling, and grading of the coal. Dust is minimized by regularly watering or otherwise treating roads and other surfaces.

Reclamation at the Trapper Mine has been successful during the first four years of operation. Although reclamation increases the cost of the coal by as much as 50 percent, it will pay off in the long-range productivity of the land as the land is returned to farming and grazing uses. It might be argued that the Trapper Mine is unique in that the fortuitous combination of geology, hydrology, and topography allows for successful reclamation. On the other hand, the success of the mine operation demonstrates that with careful site selection and planning strip mining is not incompatible with other land uses.

Federal guidelines governing strip mining of coal in the United States require that mined land be restored to support its premining use. The new regulations also prohibit mining on prime agricultural land and give farmers and ranchers the opportunity to restrict or stop mining on their land, even if they do not own the mineral rights. Reclamation includes disposing wastes, contouring the land, and replanting vegetation. Unfortunately, reclamation is difficult, and it is unlikely that it will be completely successful.

The real shortages of oil and gas are still a few years away, but when they do come, they will put tremendous pressure on the coal industry to open more and larger mines in both the eastern and the western coal beds of the United States. An increased number of mines may have tremendous environmental impacts for several reasons. First, more and more land will be strip mined and thus will require careful restoration. Second, unlike oil and gas, burned coal leaves ash (5 to 20 percent of the original amount of the coal) that must be collected and disposed of. Some ash can be used for landfill or other purposes, but most (85 percent) is useless at present. Third, handling of tremendous quantities of coal through all stages—mining, processing, shipping, combustion, and final disposal of ash—will have potentially adverse environmental effects such as aesthetic degradation, noise, dust pollution, and, most significant, release of trace elements into the water, soil, and air likely to cause serious health problems [21].

The transport of large amounts of coal, or energy derived from coal, from production areas with low energy demand to large population centers is a significant environmental issue. Coal may be converted on site to electricity, synthetic oil, or synthetic gas, which are relatively easy to transport, but with few exceptions these alternatives all have problems. Transmission of electricity over long distances is expensive, and there may not be sufficient water for cooling in power plants in semiarid, coal-rich regions. Conversion of coal to synthetic oil or gas, while possible, is expensive, and the technology is primitive. Furthermore, the conversion requires a tremendous amount of water, placing a significant demand on local water supplies in the coal regions of the western United States [22].

Methods of transporting large volumes of coal for long distances include freight trains and coal slurry pipelines. Trains will continue to be used because of relatively low costs for new capital expenses. Coal slurry pipelines, designed to use water to transport pulverized coal, have an economic advantage over trains provided [22]:

1 Transport distance is long and the volume of coal shipped is large.

2 Rates of inflation are high and interest rates are low.

3 Mines are large and have customers that will purchase large volumes of coal over a long period of time.

4 Sufficient, low cost water is available.

The economic advantages of the slurry pipeline are therefore rather tenuous, especially in the western United States, where large volumes of water will be difficult to obtain. For example, a pipeline transporting 30 million tons of coal requires about 20 million cubic meters of water per year—an amount which is sufficient to meet

FIGURE 11.33
Idealized diagram showing the coal slurry pipeline system and a map showing the locations of existing or planned pipelines in the United States. (From Council on Environmental Quality, 1979.)

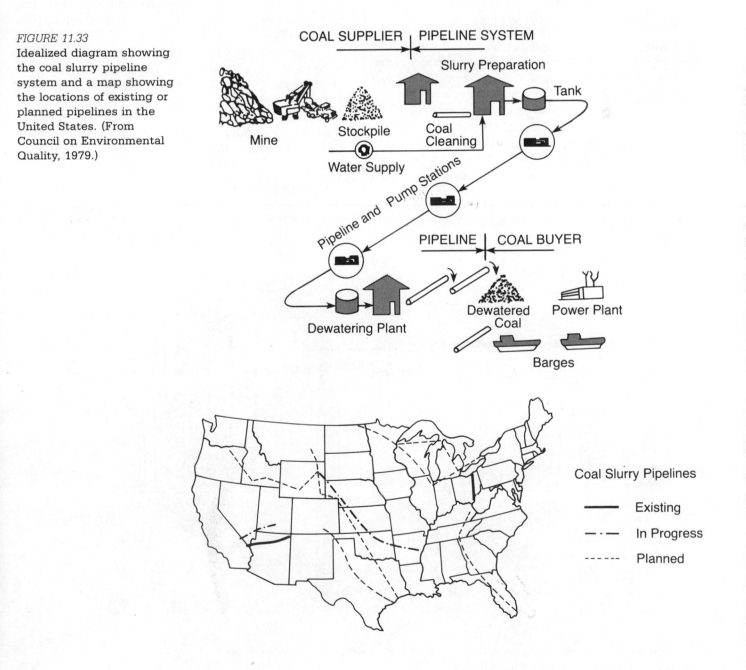

the water needs for a city of about 85,000 people or irrigate up to 40 square kilometers of farmland. In spite of these problems, some slurry pipelines will probably be constructed. Figure 11.33 shows an idealized slurry pipeline system and locations of existing, in progress, and planned pipelines in the United States.

Environmental problems associated with coal, while significant enough to cause concern, are not necessarily insurmountable, and careful planning could minimize them. At any rate, there may be few alternatives to mining tremendous quantities of coal to feed thermoelectric power plants, and to provide oil and gas by gasification and liquefaction processes in the future.

Oil Shale   Environmental impact of developing oil-shale resources will vary with the recovery technique used. At present surface and subsurface mining as well as in place (*in situ*) techniques are being considered.

Surface mining, whether open pit or strip mine, is attractive because nearly 90 percent of the shale-oil can be recovered compared with less than 60 percent for underground mining. However, waste disposal will be a major problem with any mining, surface or subsurface, which requires that oil shale be processed at the surface for **retorting** (crushing and heating raw oil shale to about 540°C to obtain crude shale oil). The volume of waste will exceed the original volume of shale mined by 20 to 30 percent.

Therefore, the mine from which the shale was removed will not be able to accommodate the waste, and it will have to be piled up or otherwise disposed of. The impact of the waste disposal can be considerable. For example, if surface mining is used to produce 100,000 barrels of shale oil per day for 20 years, the operation will produce 570 million cubic meters of waste. If 50 percent of this is disposed of on the surface, it could fill an area 8 to 16 kilometers long and 600 meters wide to a depth of 60 meters. Therefore, if large-scale mining is used, we will have to determine ways to contour and vegetate oil-shale waste to minimize the visual and pollutional impacts [23,24]. Experiments to learn more about how to accomplish this are now being conducted.

A process of oil-shale recovery being very seriously tested is known as *modified in situ,* or MIS, in which part of the oil shale (about 20%) is mined and the remainder is highly fractured or rubbled to increase the permeability. A block of rubblized shale known as the *retort block* is then ignited, and released oil and gas are recovered through wells (Fig. 11.34).

NUCLEAR ENERGY AND ENVIRONMENT

Throughout the entire nuclear cycle, from mining and processing of uranium to controlled fission, reprocessing of spent nuclear fuel, and final disposal of radioactive

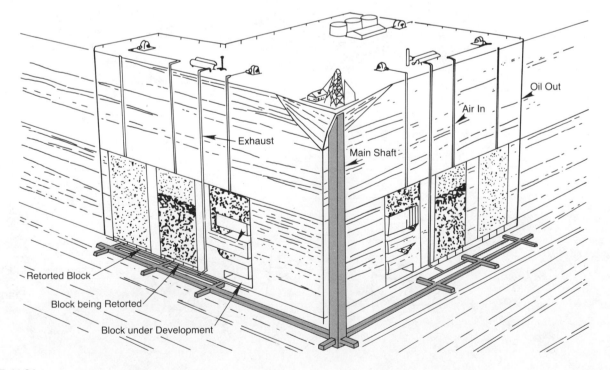

FIGURE 11.34
Idealized diagram showing the modified *in situ* method of development of oil-shale resources. (From U.S. Department of Energy.)

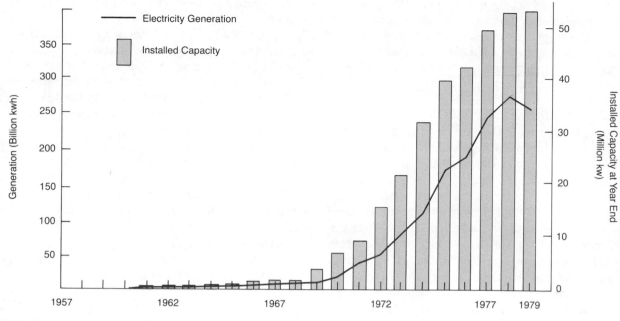

FIGURE 11.35

Nuclear power plant capacity and annual generation from 1957 to 1979. The
reduction in power generation in 1979 was due to the Three Mile Island accident
in Pennsylvania. (From U.S. Department of Energy, 1979a.)

waste, various amounts of radiation enter and affect the
environment.

The chance of a disastrous nuclear accident is estimated to be very low. Nevertheless, the probability of an
accident occurring increases with every reactor put into
operation. The 1979 accident at Three Mile Island nuclear
power plant near Middletown, Pennsylvania, involved a
chain of what were believed to be highly improbable
events resulting from both mechanical failure and human
error. Although a major disaster was avoided at Three
Mile Island, the incident raised important questions
about reactor safety; five other reactors were temporarily
shut down as a result of potential design-safety problems,
reducing the total amount of nuclear power produced in
the United States for the first time since 1960 (Fig. 11.35).

There are potential hazards associated with transporting and disposing nuclear material as well as supplying other nations with reactors. Terrorist activity and
the possibility of irresponsible persons in governments
add a risk that is present in no other forms of energy production. Nuclear energy may indeed be an answer to our
energy problems, and perhaps someday it will provide unlimited cheap energy. However, with nuclear power must
come an increased responsibility to insure that nuclear
power is used for, not against, people and that future generations inherit a quality environment and are free from
worrying about hazardous nuclear waste.

Fusion appears attractive from an environmental
point of view. First, land-use and transportation impacts
are small compared with fossil fuel or fission energy
sources. Second, compared with fission breeders, fusion
reactors produce no fission products and little radioactive
waste and are less likely to be involved in an accident
[25]. On the other hand, fusion power plants probably will
use materials that are toxic to people; lithium, for example, is toxic when inhaled or ingested. Other potential
hazards include strong magnetic fields and microwaves
used in confining and heating plasma, and short-lived radiation emitted from the reactor vessel [7].

GEOTHERMAL ENERGY AND ENVIRONMENT

Although the potentially adverse environmental impact of
intensive geothermal energy development is perhaps not
as extensive as that of other sources of energy, it is nevertheless considerable. Geothermal energy is developed at
a particular site, and environmental problems include on
site noise, emissions of gas, and industrial scars. Fortunately, development of geothermal energy does not require the extensive transportation of raw materials or
refining typical of the fossil fuels. Furthermore, geothermal energy does not produce the atmospheric particulate
pollutants associated with burning fossil fuels, nor does it
produce any radioactive waste. However, geothermal development, except for the vapor-dominated systems, does
produce considerable thermal pollution from hot waste
waters, which may be saline or highly corrosive. These
waters can be disposed of by reinjecting them back into

the geothermal reservoir, but injection of fluids may activate fracture systems in the rocks and cause earthquakes. In addition, the original withdrawal of fluids may compact the reservoir, causing surface subsidence. It is also feared that subsidence might occur when the heat in the system is extracted and the cooling rocks contract [11].

RENEWABLE ENERGY AND ENVIRONMENT

There is little doubt that the beginning of the energy transition from fossil fuels to renewable energy sources has begun. In the United States today, oil and gas still supply approximately three fourths of our energy needs; but as oil and gas become scarce and/or more costly, we will slowly change to new energy sources. The three major sources of energy likely to be used are fossil fuels, such as coal, oil shale, and tar sands; nuclear energy (fission today and perhaps fusion sometime in the future); and renewable energy, broadly defined to include solar energy, hydropower, biomass, and wind. Geothermal energy might also be added to this list, because in specific instances the natural heat flow may be sufficiently high that it may be replenished relatively quickly, but in many instances it is better thought of as a nonrenewable resource. In many parts of the world renewable energy sources may be the only energy sources indigenous to a given region.

Solar energy, when broadly defined, includes a number of different potential energy sources, as shown in Figure 11.24. These sources have the advantage of being inexhaustible and generally associated with minimal environmental degradation. With the exception of burning biomass or urban waste, solar energy sources do not pose a threat of increasing atmospheric carbon dioxide and thus modifying the climate. One major disadvantage is that most forms of solar energy—with the possible exceptions of hydropower, biomass, and ocean thermal conversion—are intermittent and spatially variable. Furthermore, some of these sources (solar cells, in particular) are presently much more expensive than fossil fuel or nuclear energy sources. Various aspects of renewable and nonrenewable energy sources, including advantages and disadvantages, are summarized in Table 11.2.

Another important aspect of renewable resources is the lead times necessary to implement the technology. Table 11.6 summarizes the technology available and lead time necessary for several of the renewable resources. Many of these are commercially available, and the construction lead time is often quite short relative to develop-

TABLE 11.6
Commercial status and lead times of renewable energy technologies.

Technology	Commercially Available	Regulatory Plus Construction Lead Time (years)
Biomass		
Direct combustion, electric	Yes	2–4.5
Direct combustion, nonelectric	Yes	2
Gasification	Early 1980s	2
Anaerobic digestion	Yes	1–2
Municipal solid waste combustion	Mid 1980s	5–8
Wood stoves	Yes	Less than 1
Wind turbines[a]		
Large	Mid 1980s	0.5–2
Small	Mid 1980s	0.5
Solar		
Passive	Yes	Less than 1
Water heating	Yes	Less than 1
Active space heating	Yes	Less than 1
Photovoltaics	Dispersed—mid 1980s	0.5–2
	Central station—after 2000	5
Thermal electric	Late 1980s or 1990s	9
Industrial heat	Some applications	2

SOURCE: California Energy Commission, 1980.
[a]Lead times given do not include resource verification, which requires a minimum of one year.

ment of new sources or construction of power plants to utilize fossil or nuclear fuels.

Water Power   Water power is clean power; it requires no burning of fuel, does not pollute the atmosphere, produces no radioactive or other waste, and is efficient. However, there is an environmental price to pay. Water falling over high dams may pick up nitrogen gas, which enters the blood of fish, expands, and kills them. Nitrogen has killed many migrating game fish in the Pacific Northwest. Furthermore, dams trap sediment that would otherwise reach the sea and replenish the sand on beaches. In addition, for a variety of reasons, many people do not want to turn all of the wild rivers into a series of lakes. For these reasons, and because many good sites for dams are already utilized, the growth of water power in the future appears limited.

Wind Power   The use of wind power will not solve all of our energy problems, but as one more alternative energy source it can be used in particular sites to reduce our dependency on fossil fuel. Wind energy does have a few disadvantages. First, demonstration projects have suggested that vibrations from windmills may produce objectionable noise. Second, windmills may interfere with radio and television broadcasts. Finally, many windmills may degrade an area's scenic resources. Still, everything considered, wind energy has a very low environmental impact, and its continued use should be carefully researched and evaluated. In fact, a number of large demonstration units are now being tested in Ohio, New Mexico, Hawaii, North Carolina, Washington, Rhode Island, and California.

Solar Energy   Solar energy has a relatively low impact on the environment. Its major disadvantage is that solar energy is relatively dispersed and a large land area is required to generate a large amount of energy. This problem is negligible when solar collectors can be combined with existing structures, as with the addition of solar hot-water heaters on the roofs of existing houses. Highly centralized and high-technology solar energy units, such as solar power towers, have a greater impact on the land. By locating these centralized systems in areas not used for other purposes and making use of dispersed solar energy collectors on existing structures wherever possible, this impact can be minimized.

The increased development of active solar energy technology may result in the widespread use of exotic materials which may themselves present environmental problems. Passive solar collectors, which use common materials like water and rock, pose negligible pollution problems.

Energy from Biomass   Use of biomass fuels can pollute the air and degrade the land. For most of us, the odor of smoke fumes from a single campfire is part of a pleasant outdoor experience, but wood smoke from many chimneys in narrow valleys under certain weather conditions can lead to unpleasant and dangerous air pollution. In recent years, the renewed use of wood stoves in homes has led to reports of such air pollution in Vermont.

The use of biomass as fuels places another pressure on already heavily used resources. The world shortage of firewood is adversely affecting natural areas and endangered species. For example, the need for firewood caused problems in the Gir forest in India, the last home of the Indian lion. The forest contains other rare species of large mammals who feed on the woody vegetation and were, in the past, food for the lion. Although the forest is set aside as a preserve, the nearby residents are taking badly needed firewood (as well as vegetation for cattle) and slowly destroying the forest trees. If our need for forest products and forest biomass fuel exceeds the productivity of our forests, they will also decrease. Any use of biomass fuel must be part of the general planning for all uses of the land's products.

ENERGY FOR TOMORROW

Prediction of future energy supplies is extremely difficult. It is not yet clear how the transition from fossil fuels to other energy sources will be accomplished. What is generally agreed upon is that modern societies waste a great deal of energy and could get by on less if the energy leaks were plugged. Technological advances that allow better conservation of existing energy supplies will improve energy efficiency. A simple example is the evolution of the automobile toward more efficient utilization of fuel. Buildings and other structures may also be designed to use less energy and thus save a substantial portion of the current energy use in the United States. Another example is **cogeneration** of electricity, which is the production of electricity as a by-product from industrial processes that normally produce and use steam as a part of their regular operations. It has been estimated that by as early as 1985, U.S. industry might be able to meet approximately one half of its electricity needs by cogeneration [26].

Energy policy in developing countries today seems to be at a crossroads. One road leads to further development of so-called hard technologies, which involves finding ever greater amounts of fossil fuels and building ever larger power plants. Following such a "hard path" means continuing as we have been for a number of years. In this respect the hard path is more comfortable; that is, it requires no new thinking or realignment of political, economic, and social conditions. The other road has been designated as the "soft path" [27]. The champion of this choice has been Amory Lovins, who states that the soft path involves energy alternatives that are renewable, flexible, and environmentally benign. As defined by Lovins, these alternatives have several characteristics:

1 They rely heavily on the renewable energy resources such as solar, wind, and biomass.

2 They are diverse, individually tailored for maximum effectiveness under specific circumstances.

3 They are flexible; that is, they are relatively low technologies that are accessible and understandable to many people.

4 They are matched in both geographical distribution and scale to prominent end-use needs (the actual use of energy).

5 There is good agreement or matching between energy quality and end-use.

This last point is of particular importance when, as pointed out by Lovins, people are not particularly interested in having a certain amount of oil, gas, or electricity delivered to their homes; rather, they are interested in having comfortable homes, adequate lighting, food on the table, and energy for transportation [27]. These latter uses are called *end-uses,* and only about 5 percent of the end-uses really require high-grade electricity. Nevertheless, a lot of electricity is used to heat homes and water. For some purposes, continual use of electricity is very important; but for others, Lovins shows that there is an imbalance when one uses nuclear reactions at extremely high temperatures or burns fossil fuels at high temperatures to simply meet needs where the temperature increase necessary may be only a few tens of degrees. Such large discrepancies are thought wasteful.

Assuming that in the future we will be able to get by on a lower rate of energy consumption, it is interesting to speculate on low-energy scenarios. One, published by Lovins (Fig. 11.36), involves a moderate decrease in energy consumption following a maximum consumption of about 100 exajoules in the year 2000. The eventual reduction accompanies the shift from dependence on fossil fuels to the soft technologies (renewable energy resources). Another recently published low-energy scenario for the United States involves a 40-percent reduction of the total energy used by the year 2020. The authors of this scenario (Fig. 11.37) emphasize, as does Lovins, that such a reduction does not have to be associated with a lower quality of life, but rather with increased conservation of energy and a more energy-efficient distribution of urban

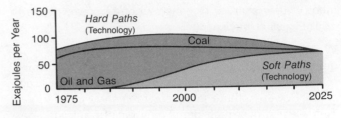

FIGURE 11.36
Low-energy scenario for the future as envisioned by A. B. Lovins. See text for further explanation. (Modified after Lovins, 1979.)

populations, agriculture, and industry. These authors offer the following alternatives: new, more energy-efficient settlement patterns that maximize accessibility of services and minimize the need for transportation; new agricultural practices that emphasize locally grown and consumed food and that require less total energy than for diets more dependent on beef and pork; and new indus-

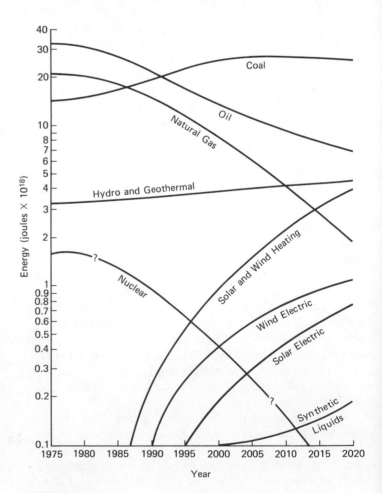

FIGURE 11.37
Energy supply for low-energy scenario of the United States. (Data from Steinhart et al., 1978.)

trial guidelines that promote energy conservation and minimize production of consumer waste [26]. To some extent these alternatives are already being practiced in some areas; that is, in some instances, highway construction has a lower priority than mass transit systems; agricultural lands near urban centers are preserved; and industry is more receptive to recycling and decreased production of consumer waste (such as unnecessary packaging).

The low-energy scenario shown in Figure 11.37 is different from that of the soft technology because it still relies heavily on fossil fuels, primarily coal. Both scenarios, however, do speculate that nuclear energy will not play an important part in the total energy picture by the year 2020. Both assumptions—the eventual demise of nuclear energy and the increase in coal production—are speculative, as is the notion that we really can obtain most of our energy needs from renewable energy sources in the next 50 years.

From an energy planning viewpoint, the next 30 years will be crucial to the United States and the rest of the industrialized world. The energy decisions we make in the very near future will greatly affect both our standard of living and quality of life. Optimistically, we have the necessary information and technology to insure a bright, warm, lighted, and moving future—but time is short, and we need action now. We can either continue to take things as they come and live in the year 2000 with the results of our present inaction, or we can build for the future an energy picture that we can be proud of [26].

SUMMARY AND CONCLUSIONS

The world's ever-increasing population and appetite for energy are staggering. However, industrialized nations must seriously question the need and desirability of an increased demand for electrical and other sources of energy. Quality of life is not necessarily directly related to greater consumption of energy.

The origin of fossil fuels (coal, oil, and gas) is intimately related to the geologic cycle. These fuels are essentially stored solar energy in the form of organic material that has escaped total destruction by oxidation. The environmental disruption associated with exploration and development of these resources must be weighed against the benefits gained from the energy. However, this development is not an either-or proposition, and good conservation practices combined with pollution control and reclamation can help minimize the environmental disruption associated with fossil fuels.

Nuclear fission will remain an important source of energy. However, the growth of fission reactors as an energy source in the United States will not be as rapid as projected because of concern for environmental hazards and increasing costs to construct large nuclear power plants. In order to avoid an eventual nuclear fuel shortage, work should continue on the breeder reactor.

Nuclear fusion, a potential energy source for the future, may eventually supply a tremendous amount of energy from a readily available and nearly inexhaustible fuel supply.

Use of geothermal energy will become much more widespread in the western United States, where natural heat flow from the Earth is relatively high. Although the electrical energy produced from the internal heat of the Earth will probably never exceed 10 percent of the total electrical power generated, it nevertheless will be significant. However, geothermal energy also has an environmental price; withdrawal of fluids and heat may cause surface subsidence, and injection of hot waste water back into the ground may produce earthquakes.

Renewable sources of energy are dependent upon solar energy and may take a variety of forms: wind power, hydropower, passive and active solar energy, and energy from biological processes (including recycled biomass from urban waste). Each of these energy sources generally causes little environmental disruption. Because these sources will not be depleted, they are dependable for the long term.

Because our fossil fuel resources are finite and expensive and may have unacceptable environmental consequences, we must explore a mixture of energy options, including coal, nuclear energy, and renewable energy resources. Of particular importance in the next few years will be the continued growth in direct solar energy as well as in recycling of urban waste to produce energy.

Hydropower will undoubtedly continue to be an important source of electricity, but it is not expected to grow much due to lack of potential sites and environmental considerations. Use of wind power, solar cells, and various biomass alternatives is more speculative, and therefore their growth cannot be predicted. We live in an interesting time, from an energy standpoint, and many changes and innovations will undoubtedly be forthcoming.

Today we are apparently at a crossroads and must choose between "hard path," centralized, high-technology sources and the "soft-path," more decentralized, lower technologies. Perhaps the *best path* will be a mixture of the old and the new, insuring a rational, smooth shift from the fossil fuels, which, with the exception of coal and perhaps oil shale, will be in short supply and much more expensive by the early years of the twenty-first century.

REFERENCES

1 BUTTLI, K., and PERLIN, J. 1980. *A golden thread.* Palo Alto, Calif.: Cheshire Books.

2 FOWLER, J. M. 1978. Energy and the environment. In *Perspectives on energy,* 2nd ed., eds. L. C. Ruedisili and M. W. Firebaugh, pp. 11–28. New York: Oxford University Press.

3 DUNCAN, D. C., and SWANSON, V. E. 1965. *Organic-rich shale of the United States and world land areas.* U.S. Geological Survey Circular 523.

4 FINCH, W. I., et al. 1973. Nuclear fuels. In *United States mineral resources,* eds. D. A. Brobst and W. P. Pratt, pp. 455–76. U.S. Geological Survey Professional Paper 820.

5 DUDERSTADT, J. J. 1978. Nuclear power generation. In *Perspectives on energy,* 2nd ed., eds. L. C. Ruedisili and M. W. Firebaugh, pp. 249–73. New York: Oxford University Press.

6 ENERGY RESEARCH AND DEVELOPMENT ADMINISTRATION. 1976. *Advanced nuclear reactors: An introduction.* ERDA–76–107.

7 U.S. DEPARTMENT OF ENERGY. 1980. *Magnetic fusion energy.* DOE/ER–0059.

8 ———. 1979. *Environmental development plan for magnetic fusion.* DOE/EDP–0052.

9 SMITH, R. L., and SHAW, H. R. 1975. Igneous-related geothermal systems. In *Assessment of geothermal resources of the United States—1975,* eds. D. F. White and D. L. Williams, pp. 58–83. U.S. Geological Survey Circular 726.

10 WHITE, D. F., and WILLIAMS, D. L. 1975. Introduction. In *Assessment of geothermal resources of the United States—1975,* eds. D. F. White and D. L. Williams, pp. 1–4. U.S. Geological Survey Circular 726.

11 MUFFLER, L. J. P. 1973. Geothermal resources. In *United States mineral resources,* eds. D. A. Brobst and W. P. Pratt, pp. 251–61. U.S. Geological Survey Professional Paper 820.

12 WORTHINGTON, J. D. 1975. *Geothermal development.* Status Report—Energy Resources and Technology, a report of the Ad-Hoc Committee on Energy Resources and Technology, Atomic Industrial Form, Incorporated.

13 EATON, W. W. 1978. Solar energy. In *Perspectives on energy,* 2nd ed., eds. L. C. Ruedisili and M. W. Firebaugh, pp. 418–36. New York: Oxford University Press.

14 RALOFF, J. 1978. Catch the sun. *Science News* 113: 16.

15 COMMITTEE ON RESOURCES AND MAN, NATIONAL ACADEMY OF SCIENCES. 1969. *Resources and man.* San Francisco: W. H. Freeman.

16 NOVA SCOTIA DEPARTMENT OF MINES AND ENERGY. 1981. *Wind power.*

17 OFFICE OF TECHNOLOGY ASSESSMENT. 1980. *Energy from biological processes.* OTA–E–124.

18 COUNCIL ON ENVIRONMENTAL QUALITY. 1979. *Environmental quality.*

19 BREW, D. A. 1974. *Environmental impact analysis: The example of the proposed trans-Alaska Pipeline.* U.S. Geological Survey Circular 695.

20 U.S. ENVIRONMENTAL PROTECTION AGENCY. 1973. *Processes, procedures and methods to control pollution from mining activities.* EPA–430/9–73–001.

21 COMMITTEE ON ENVIRONMENT AND PUBLIC PLANNING. 1974. Environmental impact of conversion from gas or oil to coal for fuel. *The Geologist,* Newsletter of the Geological Society of America, supplement to vol. 9, no. 4.

22 COUNCIL ON ENVIRONMENTAL QUALITY. 1978. *Progress in environmental quality.*

23 COMMITTEE ON ENVIRONMENTAL AND PUBLIC PLANNING. 1974. Development of oil shale in the Green River Formation. *The Geologist,* Newsletter of the Geological Society of America, supplement to vol. 9, no. 4.

24 OFFICE OF TECHNOLOGY ASSESSMENT. 1980. *An assessment of oil shale technologies.* OTA–M–118.

25 U.S. DEPARTMENT OF ENERGY. 1978. *The United States magnetic fusion energy program.* DOE/ET–0072.

26 STEINHART, J. S.; HANSON, M. E.; GATES, R. W.; DEWINKEL, C. C.; BRIODY, K.; THORNSJO, M.; and KAMBALA, S. 1978. A low energy scenario for the United States: 1975–2000. In *Perspectives on energy,* 2nd ed., eds. L. C. Ruedisili and M. W. Firebaugh, pp. 553-80. New York: Oxford University Press.

27 LOVINS, A. B. 1979. *Soft energy paths: Towards a durable peace.* New York: Harper & Row.

FURTHER READING

1 BERG, P., and TUKEL, G. 1980. *Renewable energy and bioregions.* San Francisco: Planet Drum Foundation.

2 COMMITTEE ON ENERGY AND COMMERCE. 1972. *Building a sustainable future.* Washington, D.C.: Solar Energy Research Institute.

3 COMMITTEE ON NUCLEAR AND ALTERNATIVE ENERGY SYSTEMS. 1980. *Energy and the fate of ecosystems.* Washington, D.C.; National Academy of Sciences.

4 DAVIS, W. K. 1974. *U.S. energy prospects: An engineering viewpoint.* Washington, D.C.: National Academy of Engineering.

5 HOYLE, F., and HOYLE, G. 1980. *Common sense in nuclear energy.* San Francisco: W. H. Freeman.

6 HUBBERT, M. K. 1971. The energy resources of the Earth. *Scientific American* 224: 60–70.

7 INGLIS, D. R. 1978. *Windpower and other energy options.* Ann Arbor: University of Michigan Press.

8 KREIDER, J. F.; KREITH, F.; and ENVIRONMENTAL CONSULTING SERVICES, INC. (Boulder, Colo.) 1975. *Solar heating and cooling: Engineering, practical design, and economics.* New York: McGraw-Hill.

9 LINDSAY, B. R. 1975. *Energy: Historical development of the concept.* Stroudsburg, Pa.: Dowden, Hutchinson, & Ross.

10 ——————. 1977. *The control of energy.* Benchmark Papers on Energy 6. Stroudsburg, Pa.: Dowden, Hutchinson, & Ross.

11 LOVINS, A. B. 1979. *Soft energy paths: Towards a durable peace.* New York: Harper & Row.

12 MOROWITZ, H. J. 1978. *Foundation of bioenergetics.* New York: Academic Press.

13 ROMER, R. H. 1976. *Energy, an introduction to physics.* San Francisco.: W. H. Freeman.

14 RUEDISILI, L. C., and FIREBAUGH, M. W., eds. 1978. *Perspectives on energy.* 2nd ed. New York: Oxford University Press.

15 STOBOUGH, R., and YERGIN, D. 1979. *Energy future.* New York: Random House.

16 TELLER, E. 1979. *Energy from heaven and Earth.* San Francisco: W. H. Freeman.

STUDY QUESTIONS

1 What evidence supports the notion that our present energy shortage is not the first in human history but may be unique in other ways?

2 Distinguish between *energy, work,* and *power.*

3 Explain the difference between nuclear fission and nuclear fusion. What are the similarities and differences in the potential environmental problems associated with each?

4 Compare and contrast the environmental problems associated with extracting oil and gas and those associated with geothermal energy.

5 Discuss the various solar energy options. Which is preferred—when and why?

6 Which has greater future potential for energy production, wind or water power? Which has more environmental problems?

7 What are some of the problems associated with energy from biomass?

8 What are some of the principal issues associated with reclamation of large coal strip mines?

9 Compare and contrast potential advantages and possible disadvantages of a major shift from "hard path" to "soft path" energy development.

10 Compare and contrast environmental problems associated with nuclear power with the more conventional sources such as oil, gas, and coal.

11 You have just purchased a wooded island in Puget Sound. Your house is uninsulated and built of raw timber. Although the island receives some wind, trees over 40 meters tall block most of this wind. You have a diesel generator for electric power, and hot water is produced by an electric heater which is run by the diesel generator. Oil and gas can be brought in by ship. Discuss the steps you would take in the next five years to reduce your costs for energy with the least damage to the island.

12 It is the year 2500. Natural oil and gas are rare curiosities which people see in museums. Considering the technologies available to us today, what would seem to be the most sensible fuel for airplanes? How would this fuel be produced to minimize adverse environmental effects?

12 Biological Resources: What Is Renewable and How To Keep It That Way

INTRODUCTION

The American whooping crane (*Grus americana*) and the California condor (*Gymnogyps californianus*) are two of North America's largest birds (Fig. 12.1). The whooping crane is the tallest, measuring more than 1.5 meters or approximately 5 feet. The condor has the greatest wing span—almost 3 meters, or 9 feet. Both are rare and endangered, but both are protected and have large preserves set aside for them. The two species seem to be suffering different fates. The whooping crane population was reduced to 14 individuals in 1937, but has since recovered and now numbers over 100. It is a current success story in the attempt to preserve endangered species [1]. The California condor population, on the other hand, is declining. Historical records suggest that its numbers rapidly diminished after 1840, but there were still about 60 until the 1940s; the population now numbers about 25 [2].

The extinction of species is not new; in a world of chance, the eventual fate of every species is extinction. Nor is the extinction of "the biggest" new. As early as 1876, Alfred Wallace, an English biological geographer, noted that "we live in a zoologically impoverished world, from which all of the hugest, and fiercest, and strangest forms have recently disappeared." Given that history, why is the whooping crane recovering while the California condor is not? The explanations can be found in the principles of environmental studies described earlier in this book. Briefly, the differences are ecological, as affected by human history.

At no time in their existence on the Earth was either species very abundant. Whooping crane fossils have been dated at 3.5 million years. During the last Ice Ages, the whooping crane lived throughout much of North America, but these very shy birds retreated rapidly when North America was colonized. By the mid-nineteenth century their population was reduced to about 1300 or 1400, most of which were shot or lost through habitat disruption.

Migratory birds were first placed under the protection of U.S. law in 1913. In 1937 wintering grounds for the whooping crane were set aside on the Blackjack Peninsula of Texas as the Aransas National Wildlife Refuge. The summer breeding grounds were discovered in 1955 in a remote area of muskeg bogs in Wood Buffalo Park in the Northwest Territories of Canada.

Although the whooping crane and California condor are both large and long-lived, they differ greatly in habitat and food. Unlike the crane, which migrates thousands of miles, the condor lives year-round in the Sierra foothills and coastal ranges of southern and central California. The crane feeds in wetlands on small aquatic animals; the condor, a member of the vulture family, feeds on carrion, primarily large mammals.

The decline of both species is in part due to human activities. The condor feathers were used by California Indians; Russian settlers in California found entire capes made of condor feathers. In the nineteenth century the condor, like the crane, was shot for sport, and their eggs were taken by collectors. The inability of these populations to respond by greatly increasing their reproduction is a consequence of their life history, shared by many of the largest and hugest; that is, large individuals tend to live long, but they reproduce at a low rate. Adult pairs of both species raise only one young at a time, and on the average do not raise a chick every year.

The recovery of the crane is possible because its habitat is in a good condition; food and nesting sites are still sufficient and plentiful. The condors, on the other hand, exist in a greatly altered habitat. The number of big game mammals such as deer and elk is greatly reduced in the mountains, and some areas that once burned often and were relatively open grasslands have been protected from fires and seem more heavily grown with dense chaparral and woodlands. Not only is food scarce, it is hard to find. Few active nest sites are found even far from human disturbance. Thus, although the condor was more abundant than the crane in the 1940s, the future is brighter for the crane than for the condor.

There is a lesson in this comparison: For an endangered species, it is better to have an ecosystem in "good" condition and a small population than to have an abundant population in an ecosystem in "poor" condition. The histories of the condor and crane also illustrate that the successful management of endangered species requires the application of the principles of environmental studies. This is true of all biological resources. In this chapter we will explore these major resources and the principles by which they can be managed.

THE KINDS OF BIOLOGICAL RESOURCES

The management of biological resources includes forestry, range and agricultural lands, fisheries, wildlife, and rare, threatened, and endangered species. Each has unique problems and a unique management history, but all share important characteristics. Biological resources are renewable because populations, communities, and ecosystems can be regrown, but they are vulnerable. Overuse or mismanagement can lead to extinction or to a reduction in the potential for sustained growth.

Biological resources provide food, fuel, and many kinds of economically useful materials—clothing, timber, paper, lubricants, and complex chemicals. For example,

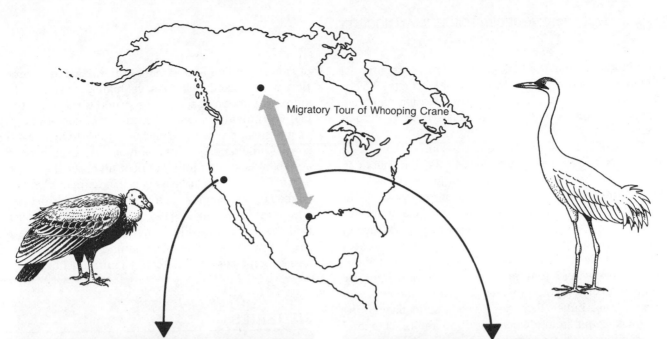

Migratory Tour of Whooping Crane

(b)

(a)

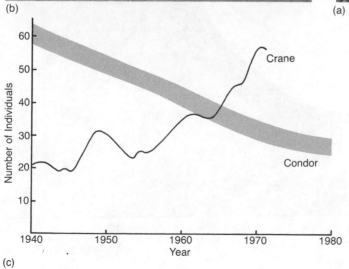

(c)

FIGURE 12.1
The whooping crane in its habitat (a) and the California condor (b). While the condor population has been decreasing, the whooping crane population has been increasing generally along an exponential curve (c). See text for further explanation. Photo (a) courtesy of Aransas National Wildlife Refuge; photo (b) courtesy of John Borneman, National Audubon Society.

291

many medicines have a biological origin, or are artificial derivatives based on natural medicinal products. Biological resources also are important in recreation and have aesthetic, religious, and cultural value for every human society.

Generally, for every biological resource there is at least one factor that we would like to be as abundant or as productive as possible. For example, a goal for fisheries is to make the annual harvest of fish as large as possible. In forestry, a goal is to maximize the annual yield of timber.

In many cases, a biological resource fulfills several functions simultaneously. In addition to timber, forests provide a watershed as a source of drinking water, recreation for hikers and hunters, and food and habitat for wildlife. In a city, a park provides recreation and may also act as a noise buffer. Fisheries provide sport fishing, income for commercial fishers and operators of sportfishing shops, and food for marine mammals which are themselves of aesthetic or cultural importance to people.

The management of biological resources is a complex subject and a growing professional field. It has an ancient history in human civilization. Managers of biological resources often talk about maximization and optimization of production and abundance. **Maximization of production** means exactly what it says: managing so that the average production in some time period is the largest possible. **Optimization** has a broader and sometimes more vague meaning and is usually employed when there are several simultaneous considerations. For example, a manager of a fishery might talk about an *optimal* yield of fish as a yield that is as large as possible without significantly increasing the chances of a major decline in the fish or in other organisms (like seals) that depend on the fish for food. An optimal yield of tuna would be one that is as large as possible without threatening with extinction species of porpoises that are caught in the fishing nets.

FISHERIES

Fisheries are a major source of the world's food and are particularly important as a source of protein. Many species of fish and shellfish are caught, but fewer than 20 species provide two thirds of the catch and 90 percent of the value. In the 1970s the worldwide annual harvest of fish was 70 million metric tons per year (60 from marine

FIGURE 12.2
Annual catch of oceanic fish (lower line) and all marine animals. The graph shows a downward trend for the catch of oceanic fish after 1970. Catch is estimated for the period shown by a dashed line. (b) The catch by major fishing countries. [Part (a) from Council on Environmental Quality and the Department of State, 1980; part (b) modified from Bell, 1978.]

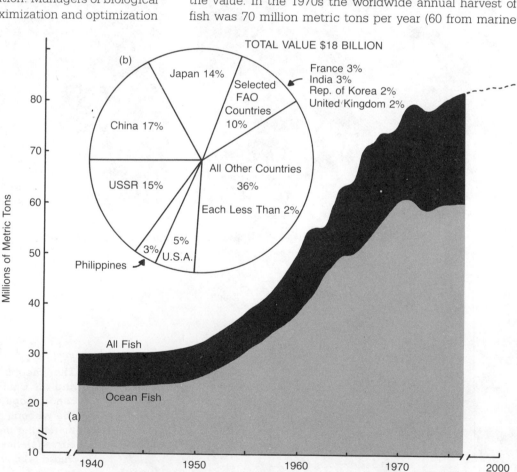

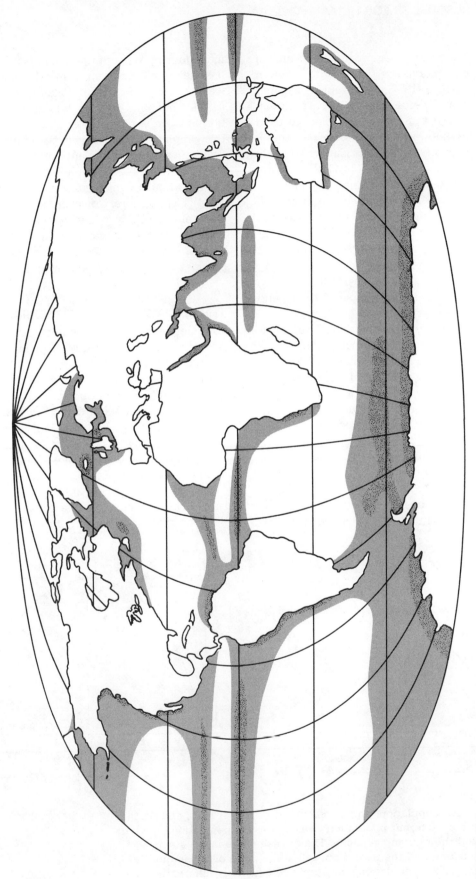

FIGURE 12.3
The Earth's major fishery regions. Heavier shading indicates regions of upwelling or high productivity. The major fisheries occur wherever the productivity of marine algae is high: near continents, in areas of upwelling, near estuaries, and along the continental shelves. The central ocean, which occupies the largest area, is relatively a desert in terms of economically valuable fish. (From Mackintosh, 1965. Reproduced by permission of the Buckland Foundation from *The stocks of whales*, published by Fishing News Books Ltd.)

fisheries, 10 from freshwater) [3]. Approximately 6 million metric tons were from *aquaculture,* the farming of fish in artificial ponds and tanks. In 1970, directly edible fish and shellfish provided about 5 percent of the world's protein and 8 percent when adjusted for quality. Fish are used also as feed for poultry and pigs, which supply another 5 or 6 percent of the world's protein. Thus fish may provide 10 to 13 percent of the world's protein [4], but they provide only 0.6 percent of the world's human food calories.

In Japan, Iceland, and Scandinavian countries, fish provide 25 percent of the protein. In the United States the percentage is much less; the annual U.S. per capita intake of fish is 5 kilograms. Moreover, in America the use of fish as food decreases with the distance from the shore; people in coastal areas consume 83 percent more fish than in other areas. For Peru, Iceland, Japan, Canada, and Denmark, fish are a major export [4].

Between 1950 and 1970, the world fishery catch expanded from 21 million to 70 million metric tons, or 12 percent per year, but since 1970 the catch has declined (Fig. 12.2). The value of the world's fisheries increased between 1963 and 1973 from $6.3 billion to $18 billion. Fish vary greatly in value, from tens of dollars to a thousand dollars per ton depending on the species and port.

Fishing is international, but a few countries dominate. Four countries—Japan, China, Russia, and the United States—caught 51 percent of all the world's catch in 1973 [4].

We tend to think of the oceans as a vast, inexhaustible resource. However, from the viewpoint of fisheries

management, most of the ocean is a desert. Commercial fisheries are concentrated in relatively few areas of the world's oceans (Fig. 12.3). Fish are abundant where their food is abundant—that is, where there is a high production of algae. Algae is most abundant in areas with relatively high concentrations of the chemical elements necessary for life, particularly nitrogen and phosphorus. These areas occur most commonly along the continental shelf, particularly in regions of upwelling and sometimes quite close to shore. Upwelling areas are rich in chemical elements necessary for life because these elements are returned from the deep ocean to the surface in these areas. Ten percent of the oceans—the continental shelves—provides more than 90 percent of the fishery harvest.

THE MANAGEMENT OF FISHERIES

To understand the management of fisheries, we need to review the growth of single populations as described in Chapter 4. A fish population is referred to as a **stock.** A stock of fish has traditionally been managed as if it grew independently of other species and as if it were supplied with a constant amount of food. A fish stock is often assumed to grow following the logistic growth curve (Fig. 12.4a). Its **carrying capacity,** or maximum population size, is the average maximum number of individuals the population can sustain over a long time.

Because fishermen are interested in obtaining as large a catch as possible, the fisheries manager wants to keep the annual net growth rate of a fish stock as large as

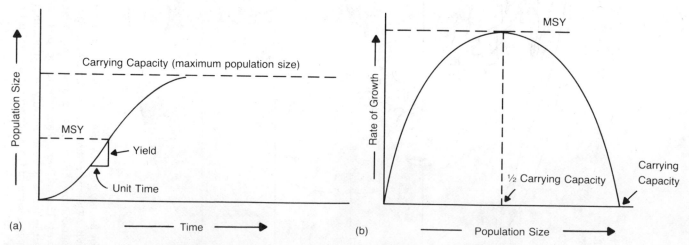

FIGURE 12.4
Idealized growth of a fish population. Part (a) shows growth over time, and part (b) shows the growth rate at different population sizes. A logistic population grows to a maximum size, called the carrying capacity. The growth curve is the steepest when the population is one-half the maximum, called the point of maximum sustainable yield (MSY). Real populations may grow considerably differently from this idealized graph.

possible without adversely affecting the population or its environment. Overfishing can lead, in subsequent years, to a decrease in the rate of growth.

How rapidly can a fish stock grow? If we assume that a fish stock grows according to the logistic curve, we can determine the population size that gives the maximum rate of growth. In Figure 12.4a we see the logistic growth curve for a fish population which shows the change in the number of fish over time [5]. We can regraph this relation to show the rate of growth plotted against the population size (Fig. 12.4b). In this graph we see that when the number of fish is small, the total number added is also small. We also see that at the carrying capacity the net growth is zero. This must be true by definition; if the net growth were not zero, the population would be changing and thus would not be at its carrying capacity. The rate of growth increases between the smallest population and the carrying capacity. For a population growing according to the logistic curve, the maximum rate of growth occurs when the population is exactly one-half its carrying capacity.

If a fish population really did grow according to this curve, then the management strategy would be straightforward. The strategy would be to harvest enough fish every year to bring the population to one-half its carrying capacity. Then the amount added the next year would be the maximum that could be added by that species. This point on the logistic growth curve where the maximum growth rate is obtained has become known as the point of **maximum sustainable yield** or **MSY** [6].

For the MSY concept to be true, and for management to be based on it, three things must hold:

1 The population must have an exact and single carrying capacity, and its growth must be determined exactly by the classical logistic curve.

2 It must be possible to know both the carrying capacity and the present population size exactly.

3 It must be possible to obtain complete cooperation from all fishermen of all countries so that they manage their individual catches to make the sum add up exactly to the amount required.

In reality, none of these conditions is met. Rarely do we know what the true carrying capacity might be, and therefore estimates of one-half the carrying capacity are dubious and subject to great error. It is extremely difficult to count all the fish of a species in the ocean. Elaborate international sampling schemes are used to estimate the populations and determine annual allowed catches on some fishing grounds, but these estimates are subject to large errors.

Furthermore, as we have learned in Part One, many factors interact to cause population size changes over time. This is just as true for fish as it is for other organisms, and in reality we can expect the carrying capacity and the actual size of a population to vary over time whether or not there are people catching fish. We cannot expect to harvest the same amount every year and maintain a population at exactly its point of maximum growth.

Suppose we assume a fish stock is at MSY and proceed to harvest it at the MSY level every year. If we follow the logistic exactly, we can get ourselves into a disastrous situation. Even the slightest mistake leading to an overestimate of the MSY will push the population to a smaller size the second year. If we continue to maintain the harvest levels, expecting the population to be at MSY, then we will drive the population to lower and lower levels and finally to extinction, unless we recognize our error.

In reality, we rarely fish a species to complete extinction because it is not worthwhile or practical economically to maintain fishing until the last fish is caught. Extinction of fish species occurs when one population is taken fortuitously—that is, whenever it happens to be found with another species being fished—or when a population is driven to such a low point that other events, including random variations, allow the remaining small population to decline even after fishing has ceased.

It is not possible to know the exact size of a fish population, and therefore fishing limits cannot be set at exactly the right number to obtain maximum yield. Nor is it possible to set up a cooperative agreement in which fishermen hunt exactly to the right level, although in certain cases, as in the tuna fisheries of the United States, there has been considerable cooperation among the fishermen. There is always some variation in the population and in the fishing effort.

For all of these reasons, the logistic equation and the point of maximum sustainable yield cannot be taken exactly and literally. In an attempt to avoid the pitfalls of this approach, fisheries managers have defined an **optimal sustainable yield (OSY)**. In current practice, OSY is defined as the population size that is 10 percent larger than the population size that provides maximum yield. This OSY is thought to provide some cushion against errors in estimating the carrying capacity, the point of MSY, and the current population size. Although not really optimal in any sense, this OSY provides a certain amount of protection against our lack of information. This definition of OSY deals with only one of the three classes of problems in fisheries management—that of the error in the estimate of population levels and other data.

During the last 200 years, with the development of our modern technological civilization, one fishery after another in the world has been developed and has crashed. The failure of the world's fisheries continues today, even in those areas with the most conscientious and active management [4,7].

A classic case of a managed failure is the history of the Peruvian anchovy fisheries. Once the Peruvian anchovy was the world's largest commercial fishery. In 1970 12.3 million metric tons were caught, but by 1972 only 1.8 million metric tons were caught—15 percent of the 1970 peak. The anchovy are found in the upwelling regions, where nutrient-rich waters move upward. As we discussed in Chapter 3, they are one of the principal foods of the birds that produce guano and thus are an important link in the global phosphorus cycle.

Why did the Peruvian anchovy fishery crash? Some claim that overfishing was responsible. The fishery was actively managed by the Peruvian government with assistance from the U.N. Food and Agriculture Organization (FAO). They tried to manage the anchovies to keep the population at the level that would provide the maximum sustainable yield (MSY). Unintentionally, the population may have been brought to a level below MSY, and as long as the catch was kept at the MSY level, the anchovy could only decrease.

Others blamed the decrease on long-term environmental fluctuations, which have been known for centuries to occur, particularly the condition called *El Niño*. When the winds change, the upwelling ceases and the fish become uncommon. Historical records about the guano deposits from birds who eat anchovies show that these failures in upwelling had occurred in the past. The management policies based on MSY therefore were bound to fail since they assumed a constant environment.

The real cause is not known, but the failure of this major fishery is a blow to the world's protein supply and to the management of fisheries.

TABLE 12.1
Major commercial fishery species groups in world catch (millions of metric tons).

Species Group	1970	1975
Herrings, sardines, anchovies	21.6	13.7
Cods, hakes, haddocks	10.5	11.8
Redfishes, basses, congers	3.9	5.0
Mackerels, cutlassfishes	3.1	3.6
Jacks, mullets, sauries	2.6	3.5
Salmons, trouts, smelts	2.1	2.8
Tunas, bonitos, billfishes	2.0	1.9
Shrimps, prawns	1.0	1.2
Squids, octopuses	0.9	1.1
Flounders, halibuts, soles	1.3	1.1
Total	49.0	45.7

SOURCE: Council on Environmental Quality and the Department of State, 1980.

TABLE 12.2
Problems of some major fisheries.

Anchovy:
Reached peak in 1970 (10,000,000 tons), then declined.

Atlantic herring:
Exploitation so high that recruitment was decreased.

Arctonorwegian cod:
High fishing level followed by 4 years of poor recruitment.

Downs' stock of herring in the North Sea:
Managers failed to grasp stock and recruitment problems.

North Atlantic haddock:
Catch averaged 50,000 tons for many years; increased to 155,000 in 1965, 127,000 in 1966, then fell off to 12,000 in 1971–74. In 1973, the International Commission for the Northwest Atlantic Fisheries (ICNAF) established a quota of 6000 tons. Apparently, haddock could sustain a 50,000-ton catch, but when this was tripled, the population was so decreased that only a smaller catch could be sustained.

Atlantic menhaden:
Peak catch was 712,000 metric tons in 1956; in 1969, it was 161,400. Fisheries experts believe the drop was due to overfishing.

Pacific sardines:
Declined catastrophically in the 1950s through the 1970s.

SOURCE: Cushing, 1975.

In the last 20 years, a number of other major fisheries have experienced similar crashes under intentional management regimes, including the collapse of the salmon fisheries off the California coast (Tables 12.1 and 12.2).

The U.S. National Marine Fisheries defines the status of a fish stock in terms of MSY. A stock is **depleted** if it has been so reduced through overfishing or any other human-induced or natural cause that fishing effort must be substantially curtailed to allow the stock to replenish itself. A stock is in **imminent danger** when it has been reduced to MSY, and the available fleet can catch enough fish to reduce the population below MSY. A stock is in **intensive use** when the population is being reduced to the estimated MSY.

Some marine resources are abundant and relatively unexploited. One of the most important occurs in the Southern Oceans, which lie off the coast of Antarctica. This area is well known for the abundance of a small crustacean, a shrimplike animal called krill (*Euphausia superba*), and for the abundance of other animal life—birds,

FIGURE 12.5

(a) The major Antarctic whaling grounds (shaded areas) lie in the Southern Oceans which surround the Antarctic continent. The whale catches occur in areas of dense patches of krill. Much of the open ocean whale catch since the end of World War II has been from this region. (b) The krill (*Euphausia superba*), a shrimplike crustacean, is one of the most productive animals in the oceans. They feed on small algae, and thus grow rapidly in the nutrient-rich water of the Southern Oceans. Krill are an important food for many animals who compete for it, including baleen whales, penguins, and various true fish. In recent years, several nations have begun harvesting the krill with large trawlers. [Part (a) modified from Mackintosh, 1965.]

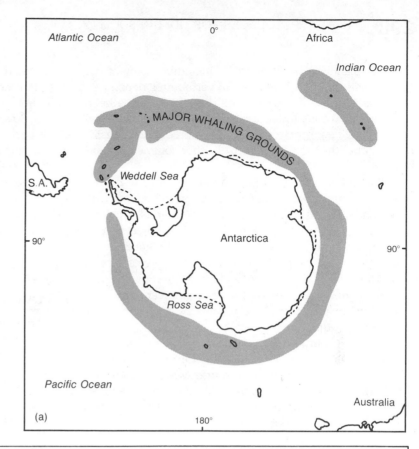

(a)

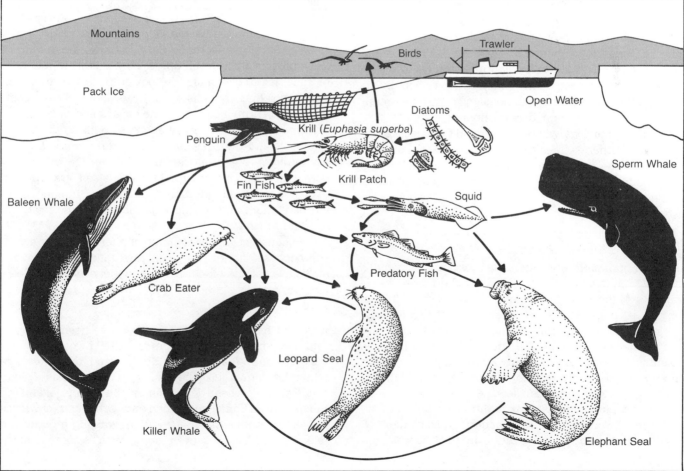

(b)

297

penguins, and marine mammals, including many of the great whales and seals. All of the vertebrates depend either directly or indirectly on krill (Fig. 12.5). It has been estimated that the production of krill is on the order of 100 or more million metric tons per year, a production considerably larger than the yield from all other commercial fisheries.

The krill provide food for species of the great baleen whales, which were reduced in number by the nineteenth- and twentieth-century commercial whalers. Krill occur in large, dense patches. When the great baleen whales found one of these large patches (which may be one or more kilometers in diameter), they would swim back and forth, leaving a row of cleared water much like a farmer plowing a hayfield. In fact, the nineteenth-century whalers were able to pursue whales by following these trails through the "whale feed."

Now the Southern Oceans have become a focus for a new major source of protein. Some countries, like Russia, have in recent years begun to send trawlers down to the Southern Oceans to harvest the krill. In the next decades we will see other attempts to exploit relatively unused oceanic areas as the better known fisheries become less and less productive.

The history of commercial fisheries and their management is mainly one of failure, but there is a bright side. Of all the areas of natural resource management, none has exceeded fisheries in the attempt to use formal mathematics, statistics, and scientific methods to formulate appropriate policies. This conscientious scientific attempt has provided a new dimension in resource management. Managers trained in economics, mathematics, statistics, and biology will be in demand for careers associated with fisheries. Like the other areas of environmental studies, the management of fisheries must also take into account culture and human history as they influence peoples' appreciation of and desire to use fisheries as a resource.

THE FUTURE OF FISHERIES

Although optimistic projections suggest that the world's fisheries may provide greater yields in the future, this is doubtful. Many fisheries have declined in yield in this century, and the harvest of most kinds of fish appears near or beyond a maximum sustainable level. Not only are many fisheries overharvested, but the world's fisheries are subject to pollutants of many kinds which may, in the long run, decrease the productivity of the fish and the organisms which provide their food.

Aquaculture could provide additional sources of fish. At the World Conference on Aquaculture held by the U.N. Food and Agriculture Organization in 1976, it was con-

cluded that the harvest from this method could increase 5 to 10 times in the next several decades if sufficient money and resources were available [3].

Several challenges face the managers of fisheries. The primary ones are to prevent further crashes in major fisheries and to protect the world's oceans as living systems. Because fisheries are generally international, the management of fish populations requires a degree of cooperation among countries that is unusual in recent history. We will discuss the legal aspects of fisheries in Chapter 18.

WHALES AND OTHER MARINE MAMMALS

Marine mammals have long fascinated people. Drawings dated as old as 2200 B.C. show whales [5], and there is evidence that Eskimos used whales for food and clothing as long ago as 1500 B.C. In the ninth century A.D., whaling by Norwegians was reported by travelers whose accounts were written down in the court of the English King Alfred. In the early maritime cultures of Western civilization, dolphins were among the first animals to be used symbolically.

During the last 80 million years, several separate groups of mammals have undergone adaptations to marine life. All of these groups were originally inhabitants of the land, and today each group shows a different degree of transition to ocean life (see box on p. 299). Some, such as the dolphins, porpoises, and the great whales, complete their entire life cycle in the oceans and have organs and limbs which are highly adapted to life in the water. The seals and their relatives, called the *pinnipeds* (literally, fin-foot), breed on the land and can move about in both water and land environments. The sea otter appears to be a more recent addition to marine environments, living near shore and having legs that seem still relatively adapted to land.

Because of the great interest these animals evoke, and because of their past economic utility, marine mammals have been the focus of a number of classic conflicts.

THE GREAT WHALES

In recent history, the great whales have been an important food and economic resource and a major concern in conservation. Whales are said to be used for *subsistence* when they form part of the basic food or materials of a culture, as with present-day Eskimos, the Japanese, and the inhabitants of Tonga and Greenland. Other uses of whales are called *commercial*. By tradition, whaling grounds are referred to as fisheries, even though whales are mammals.

THE BIOLOGY OF WHALES AND OTHER MARINE MAMMALS

CETACEANS

Whales are mammals; they have hair, nurse their young with milk, are warm-blooded, and breathe air. The whales are divided into two groups: the *Mysticeti* (whalebone or baleen whales) and the *Odontoceti* (toothed whales). These evolved separately from land animals; fossils suggest the first whales evolved about 50 million years ago. Of the 10 species of great whales, only one, the sperm whale, is toothed. It feeds on squid and other fish below the surface. The baleen whales strain small planktonic animals through their baleen plates and feed at the surface. Comparatively long-lived, whales give birth usually to a single calf once in 2 to 4 or more years. Some whales have highly developed social behavior and communicate by means of complex sounds.

PINNIPEDS

Pinnipeds are the flippered marine mammals, seals, and their relatives which have evolved from land carnivores. Three orders are included: the eared seals (sea lions and fur seals); walruses; and true seals. Pinnipeds have been especially important to native peoples of North America, providing food, oil for fuel, and skins for clothing for Eskimos and Indians. There are about 30 million pinnipeds in the world; one half of them are one species—the crabeater seal of Antarctica. The biomass of all pinnipeds is approximately 5 million metric tons.

SIRENIANS

The sirenians include the dugong and sea cow, docile herbivores of warm coastal waters. They are the only completely aquatic herbivorous mammals, feeding on large aquatic plants. There are only four living species in two families; only one, the Caribbean manatee, is native to the United States, living in the coastal rivers of Florida. Overexploited for meat, oil, and hides, these docile slow-moving creatures are killed inadvertently by power boats.

MARINE OTTERS

Marine otters include only two species, the sea otter of the Northern Hemisphere and the South American marine otter. Sea otters are related to other otters (order *Mustelidae*) but are larger (adults weigh about 30 kilograms). They live near shore in depths of 40 meters or less and have articulating digits which they use to hold food. Unlike other marine mammals, sea otters lack a layer of fat or "blubber" under their skin to help keep them warm. They have a thick coat of fur as insulation and must eat a great deal to replace the energy they expend in maintaining their metabolic rate. Sea otters need to eat 20 to 30 percent of their body weight per day.

The earliest hunters killed whales from the shore or from small boats near shore, but gradually whale hunters ventured further out from shore. In the eleventh and twelfth centuries, the Basques hunted the Atlantic right whale in open boats in the Bay of Biscay. Medieval whaling was still *shore-based;* the whales were hunted from open boats and brought ashore for processing (Table 12.3).

Whaling became *pelagic*—that is, in the open ocean —with the invention of furnaces and boilers (tryworks) for extracting whale oil at sea. With this invention, whaling grew as an industry. The English and Dutch sought right and bowhead whales in the Atlantic in the seventeenth and eighteenth centuries, sailing from Spain to Hudson Bay. American fleets developed in the eighteenth century in New England, and in the next century the United States dominated the industry, providing most of the ships and even more of the crews for whaling. For exam-

ple, the pelagic Northern Pacific bowhead fishery began in 1848 with the accidental discovery of that species off the Pacific coast of North America. Five years later, more than 200 ships sought this species in the northern waters. The bowhead industry continued until 1914, and was composed of almost entirely U.S. crews and primarily U.S. ships [8,9].

The pelagic whaling ships at first sought whales that were large, slow moving, and easy to process. The right whale got its name because it was the "right" whale to catch; that is, it floated rather than sank after dying, yielded great quantities of oil, and was relatively docile and easy to catch. Bowhead, gray, and sperm whales were the other species sought in the nineteenth century (Fig. 12.6).

The early pelagic whalers used sharp, hand-thrown, but nonexplosive harpoons. In the 1860s a Norwegian, Svend Foyn, invented a harpoon fired by a cannon, allow-

TABLE 12.3
The history of whaling and the management of marine mammals.

11th to 12th century:	Basques hunt right whales in Bay of Biscay.
17th to 18th century:	English and Dutch conduct pelagic whaling in the Atlantic.
1788:	Beginning of Pacific whaling.
18th century:	Rise of American whaling.
1848:	Beginning of bowhead whaling in the North Pacific.
1864:	Svend Foyn invents the cannon-fired harpoon.
1870s:	Explosive harpoon is used.
1880s:	Steam whalers introduced.
1900s:	Beginning of Antarctic whaling.
1911:	Convention on fur seals leads to the first international agreement emphasizing conservation; signed by USA, Britain, Japan, and Russia.
1914:	Steel springs replace baleen for corset stays; bowhead whaling ends.
1920s:	Rapid expansion of whaling.
1931:	League of Nations' "Convention for the Regulation of Whaling" signed in Geneva. Limited protection for immature whales.
1937:	International Whaling Conference in London sets minimum size limits for blue, fin, humpback, and sperm whales; prohibits killing grey, right, and bowhead.
1946:	International Whaling Conference in Washington establishes International Whaling Commission (IWC).
1961–62:	Whaling catch peaks at 66,090. IWC sets up scientific subcommittees.
1972:	IWC introduces observers on whaling ships; U.N. Conference on the Human Environment calls for a 10-year moratorium on whaling. U.S. Congress passes the Marine Mammal Protection Act.
1975:	New management procedures place major emphasis on findings of the IWC's scientific committee.
1976:	IWC establishes permanent offices in Cambridge, England.

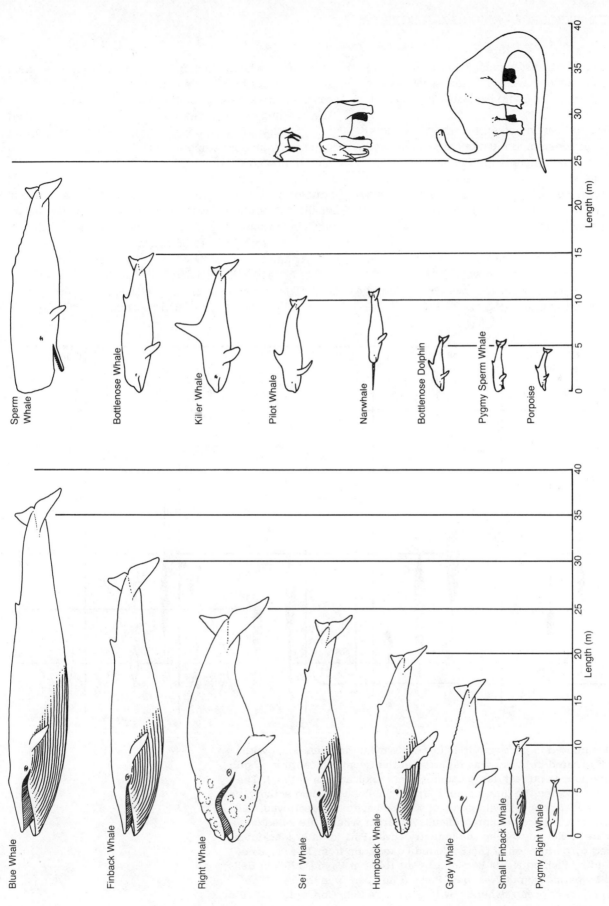

FIGURE 12.6

There are two major groups of whales: the baleen whales, which have large, comblike teeth used to filter small food items out of sea water; and the toothed whales. Except for the sperm whale, all of the great whales are baleen whales. (From *Mammals in the seas*, vol. 1, Food and Agriculture Organization of the United Nations, 1978.)

301

ing whalers to catch the faster great whales, including the blue, fin, sei, and Bryde's whales. Steam whalers were introduced in the 1880s, and the explosive harpoon followed [8,9].

Whales provided many nineteenth-century products. The oil was used for cooking, lubrication, and lamps, and whales provided the main ingredients for the base of perfumes. The elongated teeth of the baleen whales (whalebone, or baleen), which enabled them to strain the ocean waters for food, are flexible and springy and were used for corset stays and other products before the invention of inexpensive steel springs.

Although the nineteenth-century whaling ships are the most famous, made popular by novels like *Moby Dick,* more whales have been killed in the twentieth century than in the nineteenth century (Fig. 12.7). For example,

the entire bowhead whaling industry was supported in its 67-year history by a landed catch of only 16,000 whales, less than an average of 300 per year. In the twentieth century, whaling catch increased. Factory ships, introduced in 1925, allowed even more efficient catching; the catch increased from approximately 23,000 from 1924 to 1925 to more than 43,000 from 1930 to 1931 [5].

Whaling was never a major world industry in terms of gross or net economic return. In 1959, the gross value of whaling production was approximately $500 million, while in the same year the gross value of all other fisheries was approximately $4 to $5 billion.

Attempts to control whaling by international agreements began with the League of Nations in 1924 (Table 12.3). The first agreement, the "Convention for the Regulation of Whaling," was signed by 21 countries in 1931. In

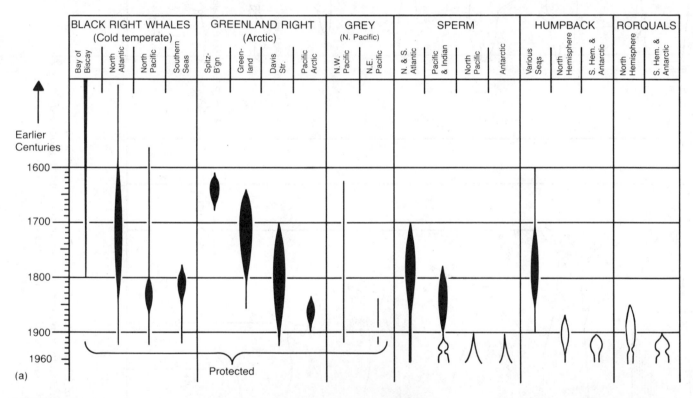

FIGURE 12.7
Species of whales hunted in past centuries (a). Old whaling in open boats with hand harpoons (indicated in black) was replaced in the twentieth century by gun-fired explosive harpoons and fast, large catcher boats (indicated in outline). The width of the lines indicates the development and decline of a whaling activity, not the catch. In the modern era (b), wild herds of whales have been destroyed, and finding the survivors has become more difficult. As larger whales are killed off, smaller species are exploited to keep the industry alive. [Part (a) after Mackintosh, 1965. Reproduced by permission of the Buckland Foundation from *The stocks of whales,* published by Fishing News Books Ltd. Part (b) after Payne, 1968, in Bell, 1978. Reprinted by permission of Westview Press from *Food from the sea: The economics and politics of ocean fisheries* by Frederick W. Bell. Copyright © 1978 by Westview Press, Boulder, Colorado.]

1946 a conference in Washington, D.C., initiated the International Whaling Commission (IWC), which has since regulated the take of whales by voluntary agreement among countries [4,10].

Whatever its success or failures, the IWC was a major landmark in wildlife conservation and the management of biological resources. It represented a major attempt by nations to agree on a reasonable annual catch of whales. The annual meeting of the IWC has become a forum to discuss international conservation, to work out basic concepts of maximum and optimal sustainable yields, and to formulate a scientific basis of commercial harvesting.

The original members of the IWC were Australia, Brazil, Canada, Denmark, France, Iceland, Japan, Mexico, the Netherlands, New Zealand, Norway, Panama, South Africa, USSR, the United Kingdom, and the United States. In 1979 only eight countries were involved in whaling: Australia, Brazil, Denmark, Iceland, Japan, Norway, USSR, and the United States. A few countries that are not members of the IWC (Peru, Chile, Portugal, Spain, and South Korea) carry out shore whaling operations [5].

THE MANAGEMENT OF WHALES AND OTHER MARINE MAMMALS

The basis for the management of whales and other marine mammals is the same as the basis of fisheries management. As with fish populations, each marine mammal population is treated as if it were isolated with a constant supply of food, subject only to the effects of human harvesting. It is assumed that the growth of a marine mammal population follows the logistic curve, and that the point of maximum sustainable yield occurs when the population size is equal to one-half the carrying capacity.

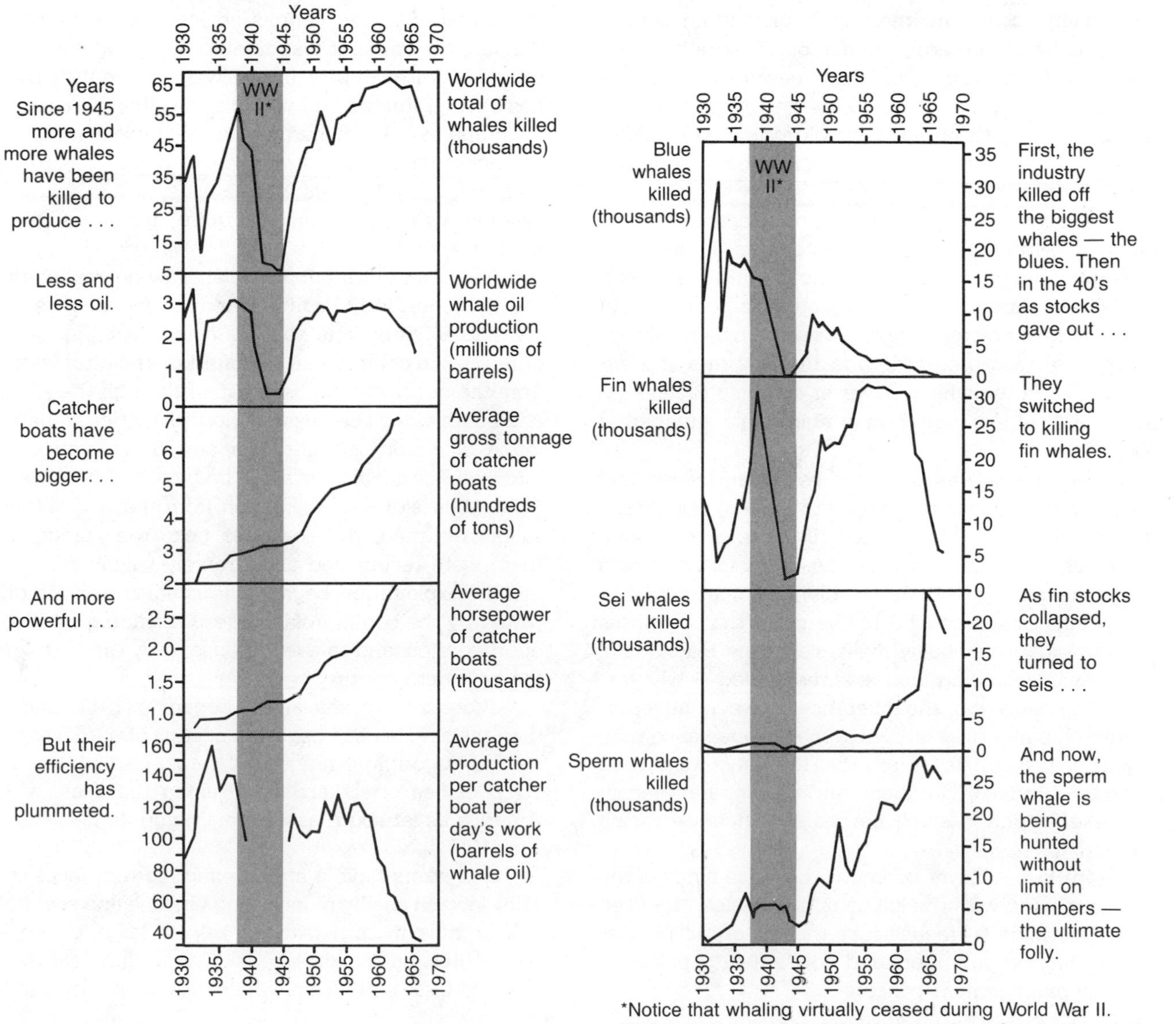
Years
1930 1935 1940 1945 1950 1955 1960 1965 1970
WW II*
Years Since 1945 more and more whales have been killed to produce . . .
Worldwide total of whales killed (thousands)
Less and less oil.
Worldwide whale oil production (millions of barrels)
Catcher boats have become bigger. . .
Average gross tonnage of catcher boats (hundreds of tons)
And more powerful . . .
Average horsepower of catcher boats (thousands)
But their efficiency has plummeted.
Average production per catcher boat per day's work (barrels of whale oil)
Blue whales killed (thousands)
First, the industry killed off the biggest whales — the blues. Then in the 40's as stocks gave out . . .
Fin whales killed (thousands)
They switched to killing fin whales.
Sei whales killed (thousands)
As fin stocks collapsed, they turned to seis . . .
Sperm whales killed (thousands)
And now, the sperm whale is being hunted without limit on numbers — the ultimate folly.

(b)

*Notice that whaling virtually ceased during World War II. That time of turmoil for people was a time of peace for whales.

TABLE 12.4
Whale abundances: past and present.

Species	Before Commercial Whaling (number of individuals)	Recent Estimates (number of individuals)
Blue	200,000[a]	10,000
Bowhead	30,000[b]	2,400 + [b]
Humpback	100,000 +	5,000
Fin	450,000	102,000
Right	50,000[c]	4,000[c]
Gray	15,000[c]	11,000[c]
Sperm	950,000	600,000 +
Sei	200,000	74,000
Minke	250,000	200,000 +

[a]All values from Australia, 1979, unless otherwise noted.
[b]Bockstoce and Botkin, 1980.
[c]Sheffer, 1976.

The goal of whaling management, however, is different from that of fisheries. The international agreements about whaling are concerned with preventing extinction and maintaining large population sizes rather than maximizing production. For this reason, the Marine Mammal Protection Act enacted by the United States in 1972 defines an **optimal sustainable population (OSP)** as its goal rather than a maximum or optimal sustainable yield (MSY or OSY).

Since the formation of the IWC, no species has gone extinct, the total take of whales has decreased, and harvesting of species considered endangered has been halted. The endangered species protected from hunting have had a mixed history (Table 12.4). The blue whale appears to have recovered little and remains rare and endangered. The grey whale, however, appears to have increased and is now relatively abundant, numbering 11,000.

No one knows why one species has recovered and another has not. One theory about the blue whale is that it has suffered from competition. It feeds in the same range of ocean depths as the sei whale, whose numbers appear to have increased. Some scientists speculate that the sei whale has increased to the point that it utilizes most of the food previously available to the blue whale. The blue whale, lacking food, has responded slowly.

Others believe that the difference between the recovery of the blue and grey whales is more appearance than reality. These scientists believe that the greys were never reduced to very small numbers, and the apparent recovery represents only a small percentage increase in the population.

The truth may never be known because many of the facts disappeared with the killing of the whales. However, research methods to reconstruct the history of populations may help us to understand how whale populations respond to management policies.

SEA OTTERS

The sea otter (*Enhydra lutris*) is an important example of the conflicts among competing interests. It became endangered because it was hunted almost to extinction in the nineteenth century. In the twentieth century the otter has been protected and its numbers have increased; now it competes with fishermen for valuable shellfish. This conflict is particularly significant because the otter plays an important role in its marine ecosystems; that is, ecosystems with sea otters are quite different from those without them.

The sea otter inhabits the shallow nearshore areas of the cold northern Pacific where it feeds on shellfish, sometimes using one shell as a tool to break open another. Before commercial exploitation, the otter was found from the northern islands of Japan through the Aleutians and south along the coast of North America to Baja California. The story is told that about 1870 a group of shipwrecked Russian sailors survived by eating sea otters and using their skins as protection from the cold. When the sailors returned, the sea otter pelts they brought with them were recognized as extremely valuable, and commercial exploitation began. The species was rapidly depleted by the commercial harvest, declining from several hundred thousand to a few thousand by the beginning of the twentieth century.

Protection of sea otters began in 1911, and since then the species has begun to occupy about 30 to 40 percent of its original range and now numbers more than 100,000. Sea otters are found along the coast of North America as far south as central California, near San Simeon]10].

Sea otters have a strong effect on their ecosystems. They feed on shellfish, including abalone and sea urchins. Sea urchins in turn feed on kelp, a large brown algae which grows along the shores. The urchins eat the base or "hold-fast" of the kelp, which kills the plant. Kelp forms dense beds, which are anchored to the ocean bottom by their hold-fast and grow in long strands to the surface. These beds are important habitats and breeding areas for many species. Thus a nearshore area in the north Pacific with many sea otters also has abundant kelp and associated species, but relatively few sea urchins and abalone. An area from which sea otters have been removed typically has more urchins and abalone, little kelp, and a greatly different group of associated species.

Although sea otters represent a recreational and aesthetic resource for people, the sport catch of abalone and clams has been reduced in areas where sea otters are present. Commercial fishermen claim that the otters are a major cause in the reduction of abalone; others argue that the fishermen are the major cause of this reduction and that the otters play a lesser role.

This conflict illustrates the problem of *multiple use* of a natural renewable resource, one in which any management policy will have several effects. Some suggest moving otters to areas where they do not compete directly with commercial fishermen. If we learn how to balance all the conflicting interests in the sea otter situation, we may find a key to managing other natural renewable resources with multiple uses.

DOLPHINS AND OTHER SMALL CETACEANS

There are many species of small ''whales'' or *cetaceans,* including dolphins and porpoises. More than 40 of these species have been hunted commercially or been killed inadvertently as a part of other fishing efforts [10]. A classic case in this regard is that of the spinner, spotted, and common dolphins of the eastern Pacific. Because these carnivorous, fish-eating mammals often feed with yellowfin tuna, a major commercial fish, commercial tuna fishing has inadvertently netted and killed large numbers of these marine mammals. More than 100,000 of these dolphins have been killed in recent years [10].

The attempt to reduce dolphin mortality is an important example of the cooperation among fishermen, conservationists, and government agencies and of the role of scientific research in the management of renewable resources. The U.S. Marine Mammal Commission and commercial fishermen have cooperated in seeking methods to reduce the dolphin mortality. Research conducted on dolphin behavior helped to change netting procedures and make them less likely to trap the mammals. These new procedures have been adopted by the fishermen and have greatly reduced the trapping of dolphins in the tuna nets. Thus we can see that cooperation is possible among apparently competing interests and that scientific research can be applied to help balance the multiple uses of our renewable biological resources.

THE MANATEE

The manatee and its relatives, known as the *sirenians, are* the only herbivorous marine mammals, and they are all relatively rare. In North America, the manatee is one of the rarest and most endangered large mammals; only a few hundred remain in Florida. While these docile creatures were once caught and eaten by people, they are now protected. Even so, they still suffer from destruction of their habitat and inadvertent killing. The manatee feed on aquatic plants which grow in shallow streams and brackish inlets. These areas have been greatly altered by channelization and other development, and in many areas the aquatic plants have died because of water pollution or because they have been treated intentionally with herbicides in order to clear channels for boats. Thus there are few remaining places where the manatee can feed and reproduce. In addition, the animals are occasionally cut by propellers from motorboats. Although the incidence of these accidents is small, the reproductive rate of the manatee is so low that a few additional deaths a year could cause an overall decline. Although much is being done to try to protect these animals, including the posting and patroling of waterways, the future of this species is in doubt. The case of the manatee—like the case of the whooping crane and the California condor—illustrates that it is important to have a habitat in a good condition in order to help save an endangered species.

These examples are only a few of the many cases involving the management and conservation of marine mammals. It is clear that many people attach great significance to whales and other marine mammals compared with other wildlife. For example, whale watching, particularly of the grey whale off the southern California coast and of sperm, fin, and other whales off Cape Cod, has become quite popular. Few other issues in environmental studies have raised as much public attention and debate as the harvesting of marine mammals. Comparatively large amounts of money have been raised by organizations like the Whale Protection Fund and Greenpeace for the protection of whales. In the 1970s, the conservation of whales became a subject of considerable and varied political activity, ranging from the support of education about whales and the support of research by organizations like the Whale Protection Fund to direct interference in the process of whaling by Greenpeace, whose members have placed their ship and themselves between whales and whaling ships.

For people interested in the cultural, aesthetic, religious, and emotional aspects of conservation, the history of marine mammals is an important case. It reveals that many people attach a significance other than an economic one to renewable resources and the environment. (The economics of whaling are discussed in Chapter 15.)

There have been several major conflicts between the protection of marine mammals and other interests. One of these conflicts, which we have already discussed, concerned the commercial harvesting of the blue whale versus its protection from extinction. Another concerned the question of the Eskimos' rights to harvest the bowhead whale following cultural traditions versus the protection of the whale from extinction. In the sea otter–abalone fishers controversy, there is competition between an endangered species (the otter) and fishermen for the same resource (abalone). When porpoises were being killed inadvertently during the netting of tuna, commercial interests and conservationists clashed, but cooperation between these groups has been shown to yield a solution. Thus marine mammals have been the focus of many different conflicts in environmental studies.

FORESTRY

Forests are one of the world's most abundant resources, but they are rapidly being depleted. It is estimated that forests covered one quarter of the Earth's entire land area in 1950 and one fifth in 1980 [3]. There are approximately 2.5 billion hectares of closed forest (with tree canopies overlapping or touching so that the ground is continuously shaded) and 1.2 billion hectares of open woodlands and savannahs (Table 12.5) [3]. About one third (300 million hectares, or 740 million acres) of the United States is forested to the extent that it is 10 percent stocked with trees. Another third (340 million hectares, or 840 million acres) is rangeland and includes natural grasslands, savannahs, shrubland, most deserts, tundra, coastal marshes, and meadows. Another 340 million hectares are classified as commercial timberland and are capable of producing at least 1.4 cubic meters of wood per hectare per year (Fig. 12.8) [11].

Nearly 30 percent of U.S. rangeland (110 million hectares, or 270 million acres) is in Alaska. Commercial timberland occurs in many parts of the United States: nearly 75 percent in the eastern half (about equally divided north and south); the rest occurs in the west (Oregon, Washington, California, Montana, Idaho, and Colorado) and in Alaska and the other Rocky Mountain states.

Of these forest lands, 62 percent are privately owned, 19 percent are on U.S. Forest Service lands, and 19 percent are on other federal lands. The publicly owned forests (120 million hectares, or 290 million acres) are primarily in the Rocky Mountain and Pacific Coast states on sites of poor quality and high elevation. Because much has never been harvested from these areas, they hold a large part of the nation's quantity of timber.

Forestry is big business. The 1975 worldwide harvest of forest products, including that used in firewood, construction, paper, and industrial processes, was at least 2.4 billion cubic meters [3]. In 1974 Canada, the largest net exporter, had a net export of almost $5 billion worth of forest products. Sweden, the second largest, had a net export of $3.6 billion. In that year the United States, although a large producer of timber, imported $746 million more than it exported. Japan, the world's largest importer, imported $4.4 billion more than it exported (Table 12.6).

Russia has the largest forest resource (785 million hectares) in the world—one third more than the United States and Canada combined. Europe has 135 million hectares.

TABLE 12.5
World forested area (millions of hectares) by region.

	Forest Land	Closed Forest	Open Woodland	Total Land Area	Closed Forest (% of land area)
North America	630	470	(176)	1,841	25
Central America	65	60	(2)	272	22
South America	730	530	(150)	1,760	30
Africa	800	190	(570)	2,970	6
Europe	170	140	29	474	30
USSR	915	785	115	2,144	35
Asia	530	400	(60)	2,700	15
Pacific area	190	80	105	842	10
World	4030	2655	(1200)	13,003	20

SOURCE: Council on Environmental Quality and the Department of State, 1980.
NOTES: Data on North American forests represent a mid-1970s estimate. Data on USSR forests are a 1973 survey by the Soviet government. Other data are from Persson (1974); they represent an early-1970s estimate. Forest land is not always the sum of closed forest plus open woodland, as it includes scrub and brushland areas which are neither forest nor open woodland, and because it includes deforested areas where forest regeneration is not taking place. In computation of total land area, Antarctica, Greenland, and Svalbard are not included; 19 percent of arctic regions are included.

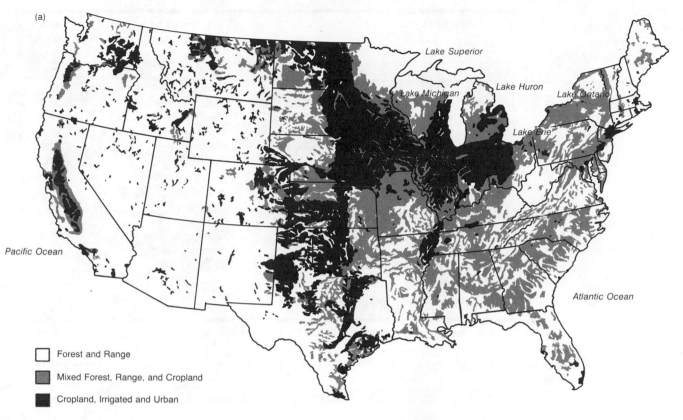

(a)

- ☐ Forest and Range
- ▨ Mixed Forest, Range, and Cropland
- ■ Cropland, Irrigated and Urban

FIGURE 12.8
Forest and range land in the conterminous United States. (a) Land use by area. (b) Sawtimber volume. Three fourths of this country's commercial forest land is in the east. However, in terms of sawtimber volume, almost the reverse is true: 70 percent in the west and 30 percent in the east. [Part (a) from U.S. Forest Service, 1980; part (b) from U.S. Forest Service and Stoddard, 1978.]

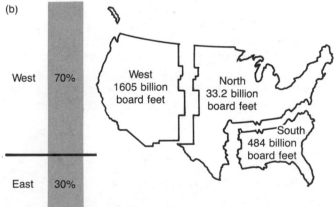

(b)

West 70%

East 30%

West 1605 billion board feet

North 33.2 billion board feet

South 484 billion board feet

THE MANAGEMENT OF FORESTS

The management of forests is called **silviculture.** Silviculture has a long history, and forestry is a profession with its own schools and undergraduate and graduate degrees. A goal of forest management is **sustained yield,** which is defined as a sustained harvest without damage to the land or the trees. This goal is analogous to the fishery manager's concept of optimal sustainable yield. Some forests are managed like mechanized farms; a single species is planted in straight rows, and the land is fertilized, sometimes by helicopter. Modern machines make harvesting rapid, and some remove the entire tree, root and all. Intensive management is characteristic of Europe and parts of the northwestern United States, particularly in Washington and Oregon. In forests that are managed less actively, such as those of New England, trees are allowed to reseed themselves and little is done to the forest except to cut trees. Which approach works best depends on the forests, the environment, and the characteristics of the commercially valuable species.

Foresters call an area of forest a **stand,** and they classify stands on the basis of tree composition. The two major kinds of commercial stands are **even-aged,** where all trees began growth from seeds and roots planted in the soil in the same year, and **uneven-aged stands,** which have at least three distinct age classes. Even in even-aged stands, trees differ in height, girth, and vigor. Foresters give trees four classifications: the tallest, most vig-

TABLE 12.6
Major traders of forest products in 1974.

Major Net Exporters	Exports Less Imports (millions $)	Major Net Importers	Imports Less Exports (millions $)
Canada	4921	Japan	4365
Sweden	3601	United Kingdom	3795
Finland	2273	Italy	1442
USSR	1552	German Fed.	1439
Ivory Coast	706	France	1186
Indonesia	666	Netherlands	1085
Austria	540	U.S.A.	746
Malaysia, Sabah	373	Belgium-Lux.	504
Philippines	237	Spain	479
Romania	218	Denmark	472
Malaysia, Peninsula	200	Norway	416
Gabon	133	Australia	295
Chile	115	German DR	284
Portugal	112	Switzerland	259
New Zealand	94	Argentina	245
		Hungary	244
		Hong Kong	193

SOURCE: Council on Environmental Quality and the Department of State, 1980.

orous trees are called **dominants**; the next most vigorous and tallest are **codominants,** whose tops receive some sunlight from above; **intermediate** trees receive only a small amount of light from above; and **suppressed** trees are shaded completely by the other trees (Fig. 12.9).

How productive a forest is depends in part on the fertility of the soil, the supply of water, and the local climatic regime. Foresters classify sites by **site quality,** which is the maximum timber crop the land can produce in a given time. Site quality can decrease with poor management. For example, too frequent burning of forests de-

creases the potential for tree growth by lowering soil fertility. Foresters develop site indices for forested areas and derive yield tables to estimate future production. The management of forests also involves thinning to remove poorly formed and unproductive trees and to permit the larger trees to grow more rapidly; planting genetically-controlled seedlings; and fertilizing.

Managers attempt to protect trees from disease and insect infestations (Table 12.7). Insect outbreaks tend to occur infrequently, but when they do occur, they have devastating results. Some insect problems are due to *in-*

FIGURE 12.9
Tree crown classes. Foresters divide a forest stand into the *dominants,* the most vigorous and the best timber trees; the *codominants,* slightly less vigorous, but with some branches in the top of the forest; *intermediate;* and *suppressed.*

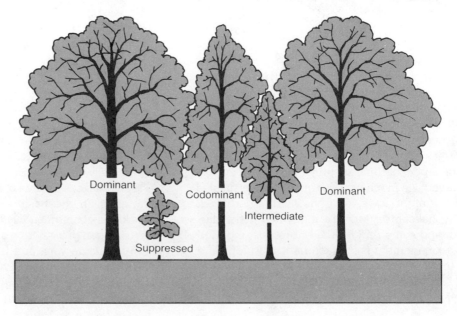

TABLE 12.7
Common forest tree insect problems in North America.

Name	Principal Species Affected	Type of Damage	Prevention	Control Season
Gypsy moth	Oaks, birch, aspen	Defoliation	Remove egg masses	May–June
Bark beetles	Pines	Girdles tree by killing cambium layer under bark	Salvage green, blown-down timber	Spring, summer, fall
Red turpentine beetles	Hard pines—south	Bores holes, reduces strength, making subject to blowdown	Remove infested trees	All seasons
Sawflies	Eastern and southern pines and tamarack	Reduces growth by defoliation; epidemics kill stands	Cut mature and overmature stands	Early summer at period of greatest activity
Spruce budworm	True firs, Douglas fir	Reduces growth by defoliation; kills older trees extensively	Cut mature and overmature stands	Early summer as insects emerge
White pine weevil	Eastern white pine, Norway spruce	Kills leaders, causes forked and crooked boles	Maintain shade where possible to 20 ft	Early summer as insects emerge
Tent caterpillar	Broadleaved trees (esp. northern hardwoods and aspen)	Defoliates; kills older and weaker trees	Remove tents by hand or burn	Spring
Scales and aphids (also spittle bugs)	All trees	Sucking insects reduce growth	—	All seasons
Hemlock looper	Western hemlock	Defoliation; kills oldest trees	—	June–July
Tussock moth	True firs, Douglas fir	Defoliation, all sizes	—	June–July
White grubs	Conifer seedlings	Destroys roots and kills trees	—	Summer
Pales weevil	Pine seedlings and saplings	Girdles bark of seedlings and defoliates saplings	—	Spring
European pine shoot moth	Pines and spruces (sapling stage)	Deforms trees	Plant only where temperature falls below 10°F	Fall–late spring

SOURCE: From Stoddard, 1978.

troductions. The gypsy moth, for example, was introduced intentionally into New England around the turn of the century as a source of silk, but it escaped and has spread through many eastern states. Other insect outbreaks appear to be naturally recurrent and to have existed for a long time. For example, a nineteenth-century gazetteer for Cheshire County in New Hampshire refers to a ''plague of loathesome worms'' that removed all the leaves from large areas of forest. Herbicides are sometimes used to combat these insects. Insects affect trees by *defoliating* them (removing the leaves), by eating the buds at the tops of the trees and destroying a straight form, by eating fruits, and by serving as a carrier of diseases.

Tree diseases are primarily fungal ones. Often, as with the Dutch elm disease, an insect spreads the fungus from tree to tree. There has been relatively little success in controlling diseases in forests.

Forests are much easier to manage than fisheries for obvious reasons. Trees stay in one place and can be easily counted and measured, and the factors that make them grow are relatively easy to study. In addition, the age and

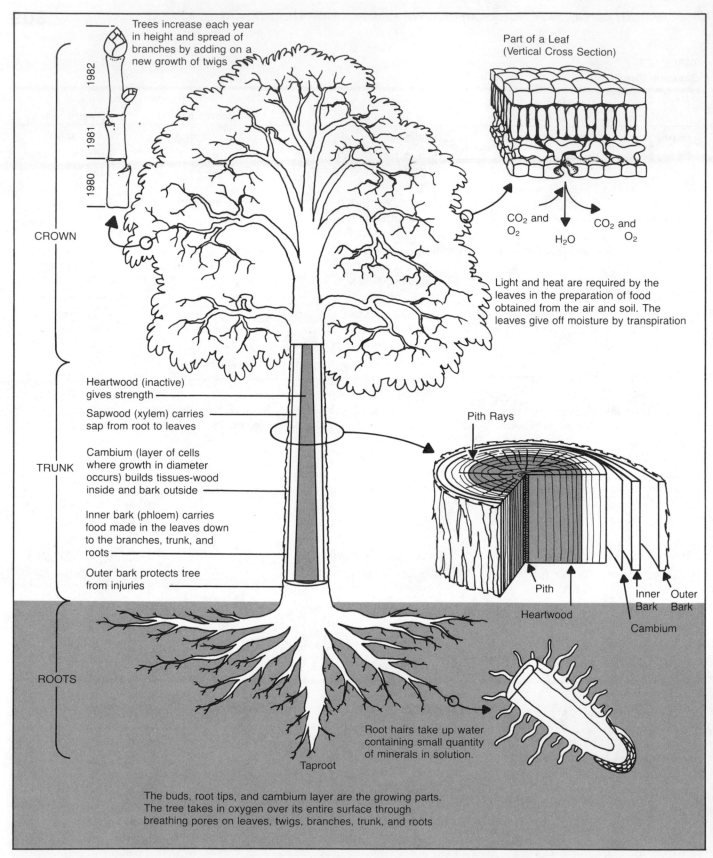

Trees increase each year in height and spread of branches by adding on a new growth of twigs

1982
1981
1980

CROWN

Part of a Leaf
(Vertical Cross Section)

CO_2 and O_2 CO_2 and O_2

H_2O

Light and heat are required by the leaves in the preparation of food obtained from the air and soil. The leaves give off moisture by transpiration

Heartwood (inactive) gives strength

Sapwood (xylem) carries sap from root to leaves

Cambium (layer of cells where growth in diameter occurs) builds tissues-wood inside and bark outside

Inner bark (phloem) carries food made in the leaves down to the branches, trunk, and roots

Outer bark protects tree from injuries

TRUNK

Pith Rays

Pith

Heartwood

Inner Bark

Outer Bark

Cambium

ROOTS

Taproot

Root hairs take up water containing small quantity of minerals in solution.

The buds, root tips, and cambium layer are the growing parts. The tree takes in oxygen over its entire surface through breathing pores on leaves, twigs, branches, trunk, and roots

FIGURE 12.10
How a tree grows. (From Stoddard, 1978.)

growth rate of trees can be measured from tree rings (Fig. 12.10). In temperate and boreal forests, trees produce one growth ring per year, and a tree can be aged by counting the number of rings. As a result, forestry science has proceeded in general beyond that of the study of fisheries and marine mammals and has become highly scientific. Forest geneticists breed new strains of trees like agricultural geneticists have bred new strains of corn, wheat, tomatoes, and most other crop plants. New "super trees" are supposed to be able to maintain a high rate of growth and increase the total production of forests.

Foresters tend to take an ecosystem approach to management. They must, because the success of trees depends on soils, climates, competition, and the abundance of parasites and herbivores—in short, on the ecosystem.

Forests are an ancient kind of ecosystem. Not long after green plants began to occupy the land, treelike forms evolved. The earliest forests occurred more than 300 million years ago. They were made up of ancestors of horsetails, which today are small, obscure plants of waste places. Fossils of these ancient forests formed coal found in Pennsylvania and other areas.

The oldest relatives of modern tree species are ancestors of the conifers, which are found in Paleozoic fossils. Coniferous trees dominated forests in the Mesozoic era. Angiosperms (flowering plants) evolved later, but were common by the Cretaceous period. The first angiosperms evolved on temperate tropic uplands in the ancient Gondwanaland [12]. With the breaking of that hypothetical landmass into separate continents, many new species evolved. In the Tertiary period, forests became widely distributed, and vast areas of the Earth's land were covered by forests for millions of years. The history of forests during the last 2 million years is one of great change. During the great glacial ages of the last 2 million years, the continental ice sheets have covered large areas of the Earth that were previously, and subsequently, forested [12]. Modern studies suggest that the forests are still in a process of recovery following the last glaciation: tree species are still migrating northward and have not reached a static distribution [13].

Although the forests are subject to change, the pace of that change is long compared with our lives and forests seem to us relatively permanent aspects of the landscape. Early travellers to North America noted impressive stands of large trees. For example, the coastal forests of New England in 1634 were described by William Wood in *New England's Prospect,* where he wrote "the timber of the country grows straight and tall, some trees being twenty, some thirty foot high before they spread for their branches" [14]. However, there is a tendency to believe

that forests really are constant in time when undisturbed by people. In reality, the North American forests that the European colonists found in the seventeenth century were "in a state of constant change wrought by forest succession, climatic change, fire, wind, insects, fungi, browsing animals and Indian activity" [12].

THE FUTURE OF FORESTS

Homo sapiens appears to have evolved in savannahs near the borders between forests and grasslands, and forests have always been important as a resource and as a source of religious and aesthetic inspiration. Forests pose some of the most important and difficult conflicts for the future of our societies.

The Clearcutting Controversy   There are a number of issues concerning forests that have been important and will continue to be so in the future. One of these issues is the method of cutting (Figs. 12.11 and 12.12). A controversy arose in the 1960s and 1970s over **clearcutting,** the practice of cutting all trees in a stand at the same time. Because clearcuts often looked like disaster areas, those wanting to use forests for recreation have opposed this practice. Others criticized clearcutting as harmful to the forest ecosystem and merely economically expedient. These critics claimed that clearcutting led to erosion, loss of soil and soil fertility, degradation of the drainage patterns, and permanent decreases in the site quality. Clearcutting has been defended as necessary for the regeneration of some desirable species, especially when those species are early successional ones.

In the mid-1960s major experiments initiated by the U.S. Forest Service and university scientists tested the effects of clearcutting. The most famous of these experiments, and the first to introduce measurements of chemical cycling and the loss of chemical elements necessary for life in the stream waters, was done at the Hubbard Brook Experimental Forest in New Hampshire (Fig. 12.12) [15]. Several experiments were carried out, each involving an entire forest watershed with an area of 25 to 50 hectares.

In one experiment, an entire watershed was clearcut and herbicides applied to prevent regrowth for two years. In another treatment, the forest was cut gradually in strips horizontal to the slope. Three cuts were made the first year at three elevations, removing one third of the timber. Two years later another third was cut, and the last third cut two years after that. The results were dramatic. In the complete clearcut (Fig. 12.12), erosion increased and the pattern of water runoff changed substantially. The bare soil decayed more rapidly, and the concentra-

(a)

tions of nitrates in the stream water exceeded public health standards so that the waters in this forest stream were no longer drinkable. Defenders of clearcutting pointed out that water flow increased during August, the low point of flow; since these forests served as sources of water, they claimed the clearcutting had at least one beneficial effect.

Experiments like Hubbard Brook demonstrated that clearcutting could be a poor practice on steep slopes in areas of moderate to heavy rainfall. However, where there is little slope and less rainfall, and where the desirable species require open areas for growth, clearcutting may be desirable. **Strip cutting,** as in the second experiment, provides a compromise by opening the forest enough to let early successional species regenerate but preventing erosion.

Harvesting All of the Trees   The technology used in clearcutting, or any cutting, has great effects on the ecosystem, so harvesting techniques are also of concern. Traditional timber harvesting removes only the main stems and largest branches of the trees—those parts with the straightest, most valuable timber. The remaining parts—leaves, bark, small stems, and roots—contain the bulk of the chemical elements necessary for life that are most likely to limit forest growth. The stem wood is composed mainly of carbon, hydrogen, and oxygen, all readily available.

Newer practices, with modern machinery, remove all of the above-ground parts of the tree, much of which is chipped into small fragments for making paper. This is called **whole-tree harvesting.** Other machines remove roots and thus even more valuable nutrients from the for-

(b)

(c)

FIGURE 12.11
Development of forestry in the nineteenth century (a). Devices like the large, steel-wheeled log carriers greatly increased the speed and efficiency of timber cutting. Technological advances have continued throughout the twentieth century (b). Modern forestry involves use of heavy machines and research, as illustrated by a weir installed in a forestry stream to monitor water flow under experimental cutting regimes (c). (Photos by D. B. Botkin.)

FIGURE 12.12

Forest cutting practices. Forests are cut in a variety of ways, including clearcutting and strip cutting (a). A clearcut looks unpleasant (b), but if done properly, it can lead in some forests to growth of desirable species. However, in the nineteenth and twentieth centuries, logging of white pine in Michigan was done destructively, with clearcutting and repeated accidental burning. Some of these areas, last cut in the 1920s, have yet to recover; the cut stumps remain as evidence of the poor management (c). Photos (a) and (b) were taken at the U.S. Forest Service's Hubbard Brook Experimental Forest in New Hampshire. (Photos by D. B. Botkin.)

(a)

(b)

(c)

est. The profit is a short-term benefit, but in the long run these practices are detrimental to the forest [16]. Whole or complete tree harvesting over long periods of time will require the addition of considerable fertilizers to replace the chemicals lost. Forests subject to these kinds of harvests will be run much like the large "agribusiness" farms, with highly mechanized and energy consumptive operations. As the shortage of wood and paper becomes more intense worldwide, the pressure to use these techniques will grow and major conflicts will develop among the various users of forest lands.

Managing with and for Fire   For much of the twentieth century it was the practice to try to suppress all fires. As we discussed in Part One, some tree species and some forest animals depend on fire and only grow in areas that have burned. Areas with very high forest fire danger, like the chaparral of California, may best be managed through the intentional introduction of frequent, light fires, which clears the ground of fuel and prevents conditions that lead to fires that destroy homes, property, and life. Because burned-over areas are usually not pleasant to look at, and because occasionally controlled fires become uncontrolled, the use of fire as a management tool will continue to be controversial.

International Aspects and Multiple Use   Much of the world's temperate zone forests have been cut at least once and are now second-growth forests. The major uncut valuable timber lies in the tropical rain forests, which are being rapidly cut. To the countries in which these rain forests lie, the timber represents a valuable resource and source of foreign exchange. To people in countries like the United States, which has experienced the effects of the cutting of virgin forests, the removal of timber from the rain forests is the destruction of a global resource. This is another case where national and international interests conflict over environmental issues.

Before 1900, wood provided more energy than any other source in the United States; by 1970, wood provided less than 1 percent of the nation's energy. A cord of wood

(3.6 cubic meters) has the energy equivalent of 1 ton of coal or 150 gallons of fuel oil [17].

In addition to timber, the U.S. Forest Service manages its lands for the purposes of recreation, wildlife, and water supply. It also maintains research stations from which some of the major research in ecology has been conducted during the last several decades.

The global shortage of timber and firewood means that more and more pressure will be exerted on all of the world's forest lands. During the next decades, pressure to open more federal lands to commercial cutting will increase. The question our society faces is, Can we manage our forests to provide us with a sustained yield of timber products, without permanent disruption of the forest ecosystems, and at the same time meet our needs for water supply, recreation, wilderness, wildlife management, and the protection of rare and endangered species?

The World Firewood Shortage   The shortage of firewood all over the world poses a particularly poignant problem. For many of the people in less developed countries, firewood is the only fuel, and much wood is used. For example, in Tanzania fuel consumption is 1.8 tons per person per year; in Thailand, it is 1.1 tons per person per year [18]. The cost of firewood is increasing rapidly, and in countries like Niger, families spend one fourth of their income on firewood. As the human population increases, the removal of wood can only increase; and as the abundance of forest land decreases, the proportion removed each year can only grow larger and larger.

The Spread of Deserts and the Loss of Soil Fertility   The pressing need for firewood is an example of factors that are leading to overuse of land and a degradation of the landscape: Good farmland can be converted to poor farmland by practices which lead to soil erosion; poor farmland can be made unusable; marginal farmland, rangeland, and forests can be converted to deserts. These are global problems.

Certain regions of the Earth are highly susceptible to desertification, and poor climate and poor land use are

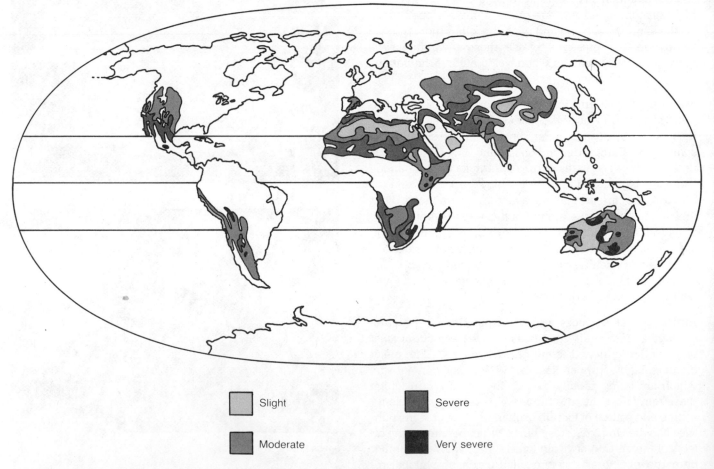

Slight   Severe

Moderate   Very severe

FIGURE 12.13

The creation of deserts. This map shows regions of the Earth that are highly sensitive to intensive use—areas where new deserts are likely to occur or are already being produced. (From Council on Environmental Quality and the Department of State, 1980.)

TABLE 12.8
Annual soil loss from various crops in different regions.

Crop	Location	Slope (%)	Soil Loss (ton/acre)	Year
Corn (continuous)	Missouri (Columbia)	3.68	19.7	1935
Corn (continuous)	Wisconsin (LaCrosse)	16	89	1937
Corn	Mississippi (northern)		21.8	1965
Corn	Iowa (Clarinda)	9	28.3	1967
Corn (plow-disk-harrow)	Indiana (Russell, Wea)		20.9	1967
Corn (plow-disk-harrow)	Ohio (Canfield)		12.2	1967
Corn (conventional)	Ohio (Coshocton)		2.8	1967
Corn (conventional)	South Dakota (eastern)	5.8	2.7	1972
Corn (continuous chem.)	Missouri (Kingdom City)	3	21	1973
Corn (contour)	Iowa (southwestern)	2 to 13	21.4	1974
Corn (contour)	Iowa (western)		24	1974
Corn (contour)	Missouri (northwestern)		24	1974
Cotton		2 to 10	19.1	1939
Cotton	Georgia (Watkinsville)		20.4	1965
Wheat	Missouri (Columbia)	3.68	10.1	1935
Wheat (black fallow)	Nebraska (Alliance)	4	6.3	1960
Wheat	Pacific Northwest (Pullman)		5 to 10	1960
Wheat-pea rotation	Pacific Northwest (Pullman)		5.6	1961
Wheat (following fallow)	Washington (Pullman)		6.9 to 9.9	1968
Bermuda grass	Texas (Temple)	4	0.03	1939
Native grass	Kansas (Hays)	5	0.03	1939
Forest	North Carolina (Statesville)	10	0.002	1939
Forest	Central New Hampshire	20	0.01	1974

SOURCE: From Pimentel et al., 1976.

producing new deserts (Fig. 12.13). Today, 8 million square kilometers of the Earth are desert. According to *The Global 2000 Report,* if all areas identified now as having a high risk of becoming deserts are converted to deserts, by the year 2000 deserts would occupy three times as much land as they now occupy [3].

Soil is lost from all forms of agriculture, but the rate of loss varies with the crop and methods of agriculture (Table 12.8). In the United States, the best land for crops is already in production. Although about 50 million hectares (125 million acres) of cropland have been added since 1900 through irrigation and drainage of wetlands, 80 million hectares (200 million acres) have been either totally ruined by soil erosion or made only marginally productive. The arable land per person is decreasing. In the management of our biological resources, we must pay more attention to soils and their loss through forest and agricultural practices [19].

WILDLIFE

The profession of wildlife management has traditionally focused on terrestrial wild animals that were hunted for food, commercial products, or sport, that is, large mammals and birds. Aldo Leopold, the father of modern wildlife management, defined game (or wildlife) management as "the art of making land produce sustained annual crops of wild game for recreational use" [20].

Wildlife management can be traced back to ancient times, to the tamed wild cheetahs and birds shown in Egyptian murals. Restrictions on the taking of game can be found in the Bible in Deuteronomy 22:6. Many wildlife preserves began as private, protected preserves owned by nobility in medieval times. The story of Robin Hood is the story of a twelfth-century noble's private game reserve and the irritation that it caused to others. In the thirteenth century, Marco Polo described the first system of game management for conservation in the Mongol Empire: "There is an order which prohibits every person throughout all the countries subject to the Great Khan from daring to kill hares, roebucks, fallow deer, stags or other animals of that kind, or any large bird, between the months of March and October. This is that they may increase and multiply."

In Europe, laws establishing closed seasons can be traced back to Henry VIII. Early laws protected game for the royalty. In the United States, the management of wildlife began as the control of hunting. In 1776, 12 of the original 13 colonies had closed seasons on certain species.

In the United States, much of the management is carried out by state fish and game departments and by the Fish and Wildlife Service of the Department of the Interior. Wildlife management traditionally had as its goal the maintenance of large numbers of individuals of an age, size, and vigor that hunters most desired. Recently, wildlife management has broadened to include the management of endangered animals or those, like the moose at Isle Royale National Park, that are naturally abundant but are managed to maintain large populations for viewing in national parks and preserves.

Wildlife management techniques are regulating the kill, restricting the numbers taken by sex and size or age, and manipulating the habitat. Wildlife managers sometimes take a single-species approach similar to that of much fisheries management, but more often wildlife managers maintain an ecosystem perspective.

Every state has enthusiasts for wildlife, both those who hunt and those who look. In the more densely populated states, the manipulation of habitat is intense. For example, in New Jersey fields are planted with plants liked by and nutritious for deer; hunters sometimes shoot the deer near or in these planted fields.

Although wildlife do not obey state boundaries, states differ markedly in their laws and regulations for wildlife. Most states issue licenses allowing hunters a fixed take which must be recorded. The record provides scientific data on the population which are then used to set subsequent levels of hunting. Connecticut, however, allows no hunting of deer except by landowners on their own land. Some migratory birds are also regulated by state laws, even though they cross many states during their annual migrations.

THE ECONOMIC VALUE OF WILDLIFE

In the United States, the use of wildlife (hunting, fishing, bird watching, hiking, etc.) is big business. In 1974 more than $10 billion were spent on these activities [21].

In one year (1965) more than 18 million hunters spent more than $1 billion on 2.25 million big game animals yielding about 225,000 metric tons (500 million pounds) of meat [22]. By 1979 there were more than 26 million hunting licences, permits, tags, and stamps issued to more than 32 million hunters at a cost of $200 million. That same year there were more than 35 million paid individual sport fishing licenses, permits, tags, and stamps ob-

TABLE 12.9
Outdoor recreation in the United States.

Activity	Numbers (millions)	Percentage of U.S. Population
Camping		
Developed areas	51.8	30%
Undeveloped areas	36.0	21%
Hunting	32.6	19%
Fishing	91.0	53%
Riding off-road vehicles	43.6	25%
Hiking and backpacking	48.1	28%
Sailing	19.1	11%
Canoeing, kayaking, etc.	26.9	15%
Other boating	57.3	33%
Scuba diving	0.2	—
Skiing	11.9	7%
Snowmobiling	13.8	8%
Rock climbing	0.2	—

SOURCE: Based on U.S. Bureau of the Census, 1979.
NOTE: Data for persons 12 years and older participating from June 1970 to June 1977.

tained, costing $174 million. There were almost 7 million bird watchers and 4.5 million wildlife photographers in 1970, and that year hikers, bird watchers, and photographers spent more than 786 million days in the field (Table 12.9). Bird watchers spend more than $500 million per year in the United States [23].

The pursuit of fur-bearing animals, a major activity in the eighteenth and nineteenth centuries, played an important role in the exploration of North America. In recent years, annual income from furs in the United States was $100 to $125 million [24]. Beaver, muskrat, oppossum, mink, weasels, and otters are among the major sources of fur.

Large mammals managed for hunting purposes include deer, caribou, moose, sheep, buffalo, and elk. The principal birds include geese and duck, quail, and other birds hunted for sport.

In addition to recreation, wildlife and plants have other uses. Some people depend on wildlife for some of their essential foods, clothing, and materials. Throughout human history, plants have been the source of medicines and are the source of many of our modern drugs. Digitalis, a drug used to treat heart disease, is obtained from foxglove, a small flowering herb. Quinine, obtained originally from cinchona bushes, is used to treat malaria. The use of wild animals and plants in medicine is just beginning, however. Recently, scientists have begun to study marine organisms as potential sources of powerful drugs.

THE MANAGEMENT OF ENDANGERED SPECIES

We began this chapter with the stories of the American whooping crane and the California condor, and we have discussed whales and other endangered species in the remaining sections. The problem of endangered species runs through all aspects of the management of biological resources; extinction is the rule of nature. Although extinction is a species' ultimate fate, the rate of extinctions has varied greatly over geologic time, and has increased rapidly since the Industrial Revolution. Most species in evolutionary history are now extinct, as we would expect from millions of years of environmental change and biological experiment. Natural extinctions often appear to follow understandable patterns, such as the replacement of one form by a more successful one [1]. This was not the case, however, at the end of the Pleistocene epoch, that is, at the end of the last great continental glaciation. At that time, massive extinctions of large birds and mammals occurred for no immediately obvious reason (Fig. 12.14). It was this loss that led A. R. Wallace to observe that "all of the hugest, and fiercest, and strangest forms have recently disappeared" [25].

At the end of the Pleistocene epoch in North America, there was a loss of 33 genera of large mammals (those weighing 50 kilograms or more), while only 13 genera had become extinct in the preceding 1 or 2 million years. Smaller mammals were not so affected, nor were marine mammals, which would have become extinct if the cause had been a biospheric environmental catastrophe. Wallace also observed that these sudden extinctions did not seem to be correlated with major environmental change [26]. Instead, they seemed to coincide with the arrival, on different continents, at different times of Stone Age human beings (see Chapter 7.)

People have been a major cause of extinction. For example, because New Zealand was isolated from the continents for a long period which included the rise of mammals, the land vertebrate fauna of New Zealand consisted almost entirely of birds. Many of the birds had evolved to fit ecological roles, or niches, which on the continents had been occupied by mammals. Fossils suggest that more than 150 species of large, flightless grazing birds lived there when the first Polynesians arrived about 950 A.D. [27]. By the time Captain Cook sailed to New Zealand in 1769, 20 species of moas and several other birds had

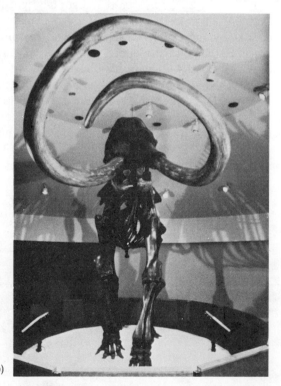

(a)

(b)

FIGURE 12.14
Extinction is the ultimate fate of all species, but the rate of extinctions has increased rapidly. At the end of the last Ice Age, many species of large animals became extinct in North America, including the errant eagle (a) and the imperial mammoth (b). (Photos from Los Angeles County Museum of Natural History, George C. Page Museum.)

been driven to extinction. European settlers increased the extinction rate even further, and the last of the world's moas were killed in the late eighteenth century, as well as 2 species of flightless geese, a great swan, a great eagle, and all of the flightless rails [27].

Modern civilization has greatly increased the rate of extinction. The International Union for the Conservation of Nature estimates that three fourths of the extinctions since 1600 of birds and mammals were caused by human beings through hunting, the introduction of foreign predators, and habitat disruption. Hunting is estimated to cause 42 percent of the extinction of birds and one third of that of the mammals.

In the United States, 47 species of wildlife became extinct between 1700 and 1970, but 25 of these were lost in the last 50 years. It is estimated that the current rate of extinction among most groups of mammals is 1000 times greater than the "high" extinction rate at the end of the Pleistocene [28].

Today there are 59 endangered species of deer, antelope, and their relatives that are hunted as game; 45 species of carnivores, endangered because they are predators and considered to be harmful; 35 endangered species of primates, mostly because of their use in medical research; and 33 endangered marsupials, partly because they are killed for hides and partly because they do not compete successfully with introduced mammals. These are in addition to the endangered marine mammal species described earlier in this chapter.

There are somewhere between 3 and 10 million species on the Earth, many yet undiscovered and unnamed. Thus any measure of the actual rate of extinction is likely to result in an underestimate. The best current estimates (Table 12.10) suggest that between 14 and 19 percent of the species now in existence will be extinct by the turn of the century.

As our discussions of condors, whooping cranes, and other large animals suggest, species that are likely to be-

come endangered through human activities tend to share certain traits. Knowledge of these common traits helps us to protect and manage such species. Easily endangered species are generally long-lived and large, particularly so for vertebrates. Such species tend to have low reproductive rates and recover slowly. There is a great difference, for example, in the potential growth rate of rats, who may have several litters of several offspring each year, and a pair of condors, who raise one offspring once in several years. The biggest and largest also require the largest territories and the most food per individual.

Carnivores are particularly subject to extinction because they are higher on food chains and require a larger base in the net primary production of vegetation than herbivores, and thus require large home ranges or habitats to maintain a viable population. Carnivores are also vulnerable because they are usually viewed as dangerous pests of domestic animals.

As we discussed in Chapter 6, the minimum population that can be sustained for a long period with a small risk of genetic problems resulting from a small gene pool is 500 individuals or more [29]. Wolves require about 26 square kilometers each, so a preserve that would maintain this minimum population would have to be approximately 13,000 square kilometers. In contrast, 500 field mice could persist in a preserve that is a fraction of one square kilometer.

Because few natural preserves can be large enough for the persistence of some of the major predators without active management, the future survival of such species requires active human manipulation. For example, problems of **genetic drift**—the fixation of deleterious genes in small populations—can be avoided by occasionally transporting individuals among preserves.

Since larger organisms are more subject to extinction, we must be particularly careful in our management of them. Undoubtedly, choices will have to be made. One leading conservationist, Norman Myers, argues that we

TABLE 12.10
Estimates of the present number of species and the number likely to become extinct by the turn of the century.

Location	Present Number of Species	Projected Extinctions
Tropical forests		
Latin America	300,000– 1,000,000	100,000– 500,000
Africa	150,000– 500,000	20,000– 250,000
S. and S.E. Asia	300,000– 1,000,000	130,000– 500,000
Subtotal	750,000– 2,500,000	250,000–1,250,000
All other areas		
Oceans, fresh waters, non-tropical forests, islands, etc.	2,250,000– 7,500,000	190,000– 625,000
Total	3,000,000–10,000,000	440,000–1,875,000

SOURCE: Modified from Council on Environmental Quality and the Department of State, 1980.

cannot help all species that are endangered because the number is too great for our resources. "Considering all species on Earth," he has written, "it is not unrealistic to suppose that we are losing at least one species per day. By the end of the 1980s we could lose as many as one million species" [30]. He urges that we adopt a "triage strategy," taken from the French medical practice in World War I. In that war, doctors found that there were more wounded than they could care for, so they assigned each wounded soldier to one of three categories: those who would be helped by medical attention and might not survive without it; those who would survive without medical attention; and those who would die no matter what aid they were given. The doctors concentrated on the first, and Myers suggests that we apply the same strategy to endangered species.

THE FUTURE OF THE MANAGEMENT OF ENDANGERED SPECIES

Our knowledge of environmental studies provides us with basic tools to promote the conservation and wise management of endangered species. The needs, however, exceed our capacities; the choices are ours. We can no longer rely on nature left "alone" to protect the endangered species, nor to provide all the answers as to whom we should save. We must choose to save those species most important to our survival according to our sense of right and wrong and according to our sense of what is desirable in nature.

Some species must be preserved from extinction because they perform essential biosphere functions; bacteria and blue-green algae, for example, fix nitrogen. Other microorganisms carry out chemical transformations necessary for chemical cycling in the biosphere. These "public service" functions of organisms can only be replaced at great expense and investment of energy and materials. Other species provide important genetic pools. Relatives of domestic crops hold genetic information that may provide hybrids resistant to new strains of diseases. Organisms used for medicinal purposes should be preserved, as well as organisms which provide other useful products.

Beyond these practical uses, the diversity of life is important to our culture and to each of us. The diversity of life has always been a source of wonder and beauty.

Earlier in this century, the approach to conservation was a hands off attitude; that is, since human activities have increased extinctions, we should remove all human interference. More knowledge, however, has given us a new perspective: Only by active management can we prevent the extinction of some species. Active management varies in degree, from studying the endangered species to establishing preserves, improving their habi-

tats, providing new sources of food and other resources, and even captive breeding. All of these methods have been suggested for the California condor.

On the one hand, we face a sad truth: The wilderness—in the sense of a place for wild creatures never touched by civilization—is passing away. Even those areas we believed to be pristine we recognize now have been under the hands of our ancestors for many centuries. On the other hand, our knowledge gives us great opportunities and great choices. Only with the perspective of environmental studies—the interrelated functions of the Earth's geology, climatology, oceanography, and biology—and an understanding of human history and values, can we learn to choose and learn to act wisely. In the end, we are the primary biological resource.

SUMMARY AND CONCLUSIONS

Biological resources include forests, range and agricultural lands, fisheries, wildlife, and rare, threatened, and endangered species. Each has unique problems and a unique management history, but all share important characteristics.

Biological resources are renewable because populations and communities can be regrown, but biological resources are vulnerable. They provide food, fuel, and many kinds of economically useful materials; they are important in recreation and have aesthetic, religious, and cultural value for every human society.

Every biological resource has at least one factor that we would like to be as abundant or as productive as possible; usually we can identify more than one such factor. Biological resource management, therefore, involves the concept of multiple use.

Managers have believed that an ecosystem or habitat has a certain carrying capacity for any species. While in the past managers have often sought to maximize the production of a useful resource (managing for maximum sustainable yield), the uncertainties in our knowledge, inevitable changes that occur over time, and the requirement for multiple use have led to a goal of optimal sustainable yields and optimal sustainable populations. Wise management of biological resources requires in most cases that we view the resource from an ecosystem perspective, and often from a global perspective. Thus biological resource management is a profession requiring training in many fields.

While many biological resources are essential, the growing human population and less than adequate management practices are leading to a loss of biological resources and a decrease in the productive capacity of the biosphere.

REFERENCES

1 MILLER, R. S., and BOTKIN, D. B. 1974. Endangered species: Models and predictions. *American Scientist* 62: 172–81.

2 WILBUR, S. R. 1978. *The California condor, 1966–1976: A look at its past and future.* U.S. Fish and Wildlife Service North American Fauna No. 72. Washington, D.C.: U.S. Department of the Interior.

3 COUNCIL ON ENVIRONMENTAL QUALITY AND THE DEPARTMENT OF STATE. 1980. *The global 2000 report to the President: Entering the twenty-first century.*

4 BELL, F. W. 1978. *Food from the sea: The economics and politics of ocean fisheries.* Boulder, Colo.: Westview Press.

5 AUSTRALIA, INQUIRY INTO WHALES AND WHALING. 1979. *The whaling question: The inquiry by Sir Sidney Frost of Australia.* San Francisco: Friends of the Earth.

6 MAY, R. M.; BEDDINGTON, J. R.; CLARK, C. W.; HOLT, S. J.; and LAWS, R. M. 1979. Management of multispecies fisheries. *Science* 205: 267–77.

7 CUSHING, D. 1975. *Fisheries resources of the sea and their management.* London: Oxford University Press.

8 BOCKSTOCE, J. R., and BOTKIN, D. B. 1980. *The historical status and reduction of the western Arctic bowhead whale* (Balaena mysticetus) *population by the pelagic whaling industry, 1848–1914.* New Bedford, Conn.: Old Dartmouth Historical Society.

9 BOCKSTOCE, J. R. 1978. *A preliminary estimate of the reduction of the western Arctic bowhead whale* (Balaena mysticetus) *population by the pelagic whaling industry: 1848–1914.* National Technical Information Service, Publ. PB–286 797. Washington, D.C.: National Technical Information Service.

10 U.N. FOOD AND AGRICULTURE ORGANIZATION. 1978. *Mammals in the seas.* Report of the FAO Advisory Committee on Marine Resources Research, Working Party on Marine Mammals. FAO Fisheries Series 5, vol. 1. Rome: U.N. Food and Agriculture Organization.

11 U.S. FOREST SERVICE. 1980. *An assessment of the forest and range land situation in the United States.*

12 SPURR, S. H., and BARNES, B. V. 1973. *Forest ecology.* New York: Ronald Press.

13 DAVIS, M. B. 1976. Pleistocene biogeography of temperate deciduous forests. *GeoSciences and Man* 13: 13–26.

14 WOOD, W. 1634. *New England's prospect.* Prince Society (eds.), Boston, 1865.

15 LIKENS, G. E.; BORMANN, F. H.; PIERCE, R. S.; EATON, J. S.; and JOHNSON, N. M. 1977. *The biogeochemistry of a forested ecosystem.* New York: Springer-Verlag.

16 ABER, J. D.; BOTKIN, D. B.; and MELILLO, J. M. 1979. Predicting the effects of different harvesting regimes on productivity and yield in northern hardwoods. *Canadian Journal of Forest Research* 9: 10–14.

17 STODDARD, C. H. 1978. *Essentials of forestry practice.* 3rd ed. New York: Wiley.

18 ECKHOLME, E. P. 1975. The firewood crisis. *Natural History* 84: 7–22.

19 PIMENTEL, D.; TERHUNE, E. C.; DYSON-HUDSON, R.; ROCAEREAU, S.; SAMIS, R.; SMITH, E. A.; DENMAN, D.; REIFSCHNEIDER, D.; and SHEPHERD, M. 1976. Land degradation: Effects on food and energy resources. *Science.* 194: 149–55.

20 LEOPOLD, A. 1947. *Game management.* New York: Scribners and Sons.

21 BROKAW, H. P. 1978. *Wildlife and America.* Washington, D.C.: Council on Environmental Quality.

22 U.S. FISH AND WILDLIFE SERVICE. 1979. *Federal aid in fish and wildlife restoration.* Washington D.C.: Wildlife Management Institute and Sport Fishing Institute.

23 MYERS, N. 1979. *The sinking ark.* Oxford, England: Pergamon Press.

24 BLACK, J. D. 1954. *Biological conservation.* New York: McGraw-Hill.

25 WALLACE, A. R. 1876. *The geographical distribution of animals*. New York: Hafner. (Reissued in 1962 in 2 vols.)

26 _____. 1911. *The world of life*. New York: Moffat-Yard.

27 FISHER, J.; SIMON, H.; and VINCENT, V. 1969. *Wildlife in danger*. New York: Viking Press.

28 EHRENFELD, D. W. 1972. *Conserving life on Earth*. New York: Oxford University Press.

29 SOULÉ, M. E. 1980. Thresholds for survival: Maintaining fitness and evolutionary potential. In *Conservation biology*, eds. M. E. Soulé and B. A. Wilcox, pp. 151–69. Sunderland, Mass.: Sinauer.

30 MYERS, N. 1981. How shall we choose which ones to save? *No Man Apart* (news magazine of Friends of the Earth) 2:9.

FURTHER READING

1 BELL, F. W. 1978. *Food from the sea: The economics and politics of ocean fisheries*. Boulder, Colo.: Westview Press.

2 CAUGHLEY, G. 1977. *Analysis of vertebrate populations*. New York: Wiley.

3 DASMANN, R. 1981. *Wildlife biology*. 2nd ed. New York: Wiley.

4 EHRENFELD, D. W. 1972. *Conserving life on Earth*. New York: Oxford University Press.

5 EHRLICH, P. R., and EHRLICH, A. 1981. *Extinction: The causes and consequences of the disappearance of species*. New York: Random House.

6 HENDEE, J. C.; STANKEY, G. H.; and LUCAS, R. C. 1978. *Wilderness management*. U.S. Forest Service Miscellaneous Publication 1365.

7 HOLT, S. J., and TALBOT, L. M. 1978. *New principles for the conservation of wild living resources*. Wildlife Monographs No. 59. Washington, D.C.: The Wildlife Society.

8 MURDOCH, W. 1980. *The poverty of nations*. Baltimore: John Hopkins University Press.

9 MYERS, N. 1979. *The sinking ark*. Oxford, England: Pergamon Press.

10 SEARS, P. 1935. *Deserts on the march*. Norman: University of Oklahoma Press.

11 STODDARD, C. H. 1978. *Essentials of forestry practice*. 3rd ed. New York: Wiley.

12 U.N. FOOD AND AGRICULTURE ORGANIZATION. 1978. *Mammals in the seas*. Report of the FAO Advisory Committee on Marine Resources Research, Working Party on Marine Mammals. FAO Fisheries Series 5, vol. 1. Rome: U.N. Food and Agriculture Organization.

STUDY QUESTIONS

1 Why are we so unsuccessful in making rats an endangered species?

2 Debate the following issue: The failure of the sardine fisheries along the California coast was not due to overfishing, but to environmental changes.

3 Distinguish between an "optimal" and a "maximum" yield of a biological resource. Why is it likely that a manager would never achieve a maximum yield for a long time?

4 What is meant by an "optimal" abundance of a biological resource such as trees or fish?

5 What is meant by *multiple use* of a biological resource? How would you apply this concept to (a) whales and (b) Douglas fir trees on U.S. Forest Service land?

6 What are the major differences in ownership which affect the management of (a) marine fisheries; (b) forests; and (c) deer?

7 In the next 50 years, are fish likely to provide a greater or lesser percentage of the human protein requirement?

8 We are all familiar with the story of Robin Hood and his "merry men" who illegally hunted game in the king's woods. As a game manager, would you view Robin Hood's hunting as a help or hindrance? Under what conditions does your answer apply?

9 You are hired by a sports club to manage wildlife on its lands. You are asked to increase the number of wild ducks that winter on the ponds. What actions would you take? Which of these actions affect (a) the ecosystem and (b) primarily the duck population?

10 Which of the following resources seem truly "renewable": (a) tropical rain forests; (b) mallard ducks; (c) white-tailed deer; (d) the wild turkey; (e) the California condor?

In the second part of this book, we learned about the biosphere's mineral, energy, and biological resources, as well as its air and water resources—the great fluid media that transport materials and energy, resources and wastes. We discovered the origin of many pollutants and the ways that they are transported through the biosphere.

In the third and last part, we consider people and the environment. First, we will discuss natural and human-produced physical, chemical, and biological processes/hazards. We will consider how pollutants affect human health, ecosystems, and the biosphere. We will learn how the effects of pollutants are assessed and the choices we have in dealing with hazards and pollutants.

Next, we will view human beings and their

PART 3

People and the Environment

environment within a social context, including the methods of planning, the legal basis for environmental decisions, and the economic factors that affect our decisions. We will briefly trace the history of human attitudes toward the environment so that at the end we will be able to view *Homo sapiens* and the environment within a holistic context—that is, one framed by physical, chemical, biological, and geological constraints, and governed by our economics, attitudes, ethics, and aesthetics. Finally, we will look briefly at our potential for a well-managed landscape and at explorations beyond our own biosphere.

13 Fires, Storms and Floods: Natural Hazards as a Part of the Environment

INTRODUCTION

People are basically optimistic in their outlook toward natural hazards. This attitude prevails in part because many processes that are hazardous to people occur relatively infrequently, and when a hazardous process does occur, it only affects a few people (relative to the world population).

The experience of about 5000 Icelandic people on the island of Heimaey is a good example of people responding to a serious, destructive hazard. In January of 1973, the dormant volcano Mt. Helgafell came alive, and subsequent eruptions nearly buried the town of Vestmannaeyjar in ash and lava flows. The harbor, a major fishing port, was nearly blocked (Fig. 13.1a), and with the exception of about 300 town officials, firefighters, and police, the inhabitants were evacuated. Those people who remained behind to fight the volcano began the world's most ambitious attempt to stop slow-moving lava flows. The method they used is known as *hydraulic chilling*—the application of large amounts of water to slowly advancing lava flows (Fig. 13.1b). Fortunately, the flow was slow enough and cool enough on top that heavy equipment could be driven up and pipes placed at strategic locations. Watering at each location lasted several weeks, or until the steam stopped coming out of the lava in that particular area. In the beginning, watering had little effect, but then that part of the flow began to slow down. Following the outpouring of lava, which stopped in June of 1973, the harbor was still usable. In fact, by fortuitous circumstances, the shape of the harbor was actually improved, since the new rock provided additional protection from the sea [1,2].

In July, 1973, the people returned to the island to survey the damage and to estimate the chances of rebuilding their lives and the town. The situation was grim, as ash had drifted up to 4 meters thick and covered much of the island and town (Fig. 13.1c). Furthermore, molten lava was still steaming near the volcano. The first task they undertook was to dig out their homes and shops, salvaging what they could. They then used the same volcanic debris that had buried their town to pave new roads and an area for an airport that would allow materials to be moved in and distributed. They decided to take advantage of the volcano, and by January of 1974 the first home was heated by heat from the cooling lava [1,2].

Unfortunately, all volcanoes are not as calm as Mt. Helgafell was in 1973. At 8:32 A.M. on May 18, 1980, an earthquake registering 5.0 on the Richter scale was recorded on Mt. St. Helens in the southwest corner of the state of Washington. That earthquake triggered a large landslide-avalanche which involved the entire north flank of the mountain, on which a bulge had been growing at the rate of about 1.5 meters per day for a period of several weeks. The avalanche shot down the north flank of the mountain, displacing water in nearby Spirit Lake, struck and overtopped a ridge 8 kilometers to the north, and then made an abrupt turn and moved for a distance of 18 kilometers down the Toutle River. The avalanche released internal pressure, and Mt. St. Helens erupted with a lateral blast directly from the area occupied by the bulge. At nearly the same time, a large vertical cloud quickly rose to an altitude of approximately 19 kilometers. Eruption of the volcano's vertical column continued for more than 9 hours, and large volumes of volcanic ash fell on a wide region of Washington, northern Idaho, and western and central Montana (Fig. 13.2). The total amount of volcanic ash ejected was several cubic kilometers, and during the 9 hours of eruption a number of pyroclastic flows (hot mixtures of gas, volcanic ash, and other debris) swept down the northern slope of the volcano.

On the northern slope of the volcano, the upper part of the north fork of the Toutle River basin was devastated as forested slopes were transformed into a grey, hummocky landscape consisting of volcanic ash, rocks, blocks of melting glacial ice, narrow gullies, and hot steaming pits (Fig. 13.3). Several mudflows, consisting of mixtures of water, volcanic ash, rock, and organic debris such as logs, occurred shortly after the start of the eruption. The flows and accompanying flood raced down the valleys of the north and south forks of the Toutle River at estimated speeds of 29 to 55 kilometers per hour (Fig. 13.4). Water levels in the river reached at least 4 meters above flood stage, and nearly all bridges along the river were destroyed. The hot mud quickly raised the temperature of the Toutle River to as high as 38°C. Mud, logs, and boulders were carried 70 kilometers downstream and deposited into the Cowlitz River, to be eventually deposited 28 kilometers further downstream into the Columbia River. Nearly 40 million cubic meters of material was dumped into the Columbia River, reducing the depth of the shipping channel from a normal 12 meters to 4.3 meters over a distance of about 6 kilometers [3].

When the volcano could again be viewed following the eruption, it was observed that the maximum altitude of the volcano was reduced by 350 to 400 meters and that the mountain, originally symmetrical, was now a huge, steep-walled amphitheatre facing northward (Fig. 13.5). The landslide-avalanche, horizontal blast, pyroclastic flows, and mudflows devastated a large area (nearly 400 square kilometers), killing 24 persons and leaving 44 others missing or presumed dead. Over 100 homes were destroyed by the flooding, and several billion board feet of timber were flattened by the blast (Fig. 13.6). The total damage is estimated to be several billion dollars, but long-term damage to fisheries and other resources is difficult to estimate [3].

FIGURE 13.1
Volcanic eruption on the island of Heimaey, Iceland, in 1973. (a) Overview of the eruption. (b) Close-up of the attempt to control the movement of a lava flow with water. (c) Volcanic ash covering buildings. (Photos courtesy of Icelandic Airlines.)

(a)

(b)

(c)

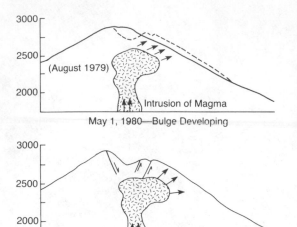

(August 1979)

Intrusion of Magma

May 1, 1980—Bulge Developing

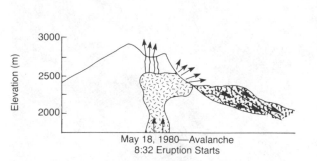

May 17, 1980—Bulge Area

May 18, 1980—Avalanche
8:32 Eruption Starts

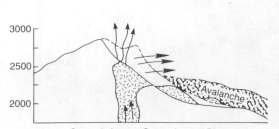

Avalanche

Seconds Later—Strong Lateral Blast

Elevation (m)

(a)

(b)

(c)

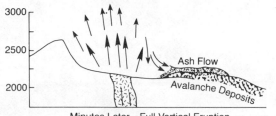

Ash Flow
Avalanche Deposits
3000
2500
2000
Minutes Later—Full Vertical Eruption

FIGURE 13.2

Idealized diagram showing the sequence of events for the May 18, 1980, eruption of Mt. St. Helens. [Photos (a), (b), and (c) © 1980 by Keith Ronnholm, the Geophysics Program, University of Washington, Seattle; photo (d) by Robert Krimmel, U.S. Geologic Survey. Drawings inspired by lecture by James Moore, U.S. Geological Survey.]

(d)

329

(a)

FIGURE 13.3

Aerial view (a) and ground view (b) of deposits from the May 18, 1980, landslide/debris-avalanche associated with the eruption of Mt. St. Helens. Notice the hummocky nature of the topography. The person in photograph (b) provides a scale by which the size of some of the large blocks of the debris can be estimated. (Photos courtesy of U.S. Geological Survey and Harry Glicken.)

(b)

330

FIGURE 13.4
Flooding of the Toutle River associated with the May 18, 1980, eruption of Mt. St. Helens. (Photo courtesy of U.S. Geological Survey.)

The two case histories of Mt. St. Helens and Mt. Helgafell are quite different in terms of the magnitude of the volcanic event involved. Certainly, planning for volcanic hazards in Iceland may be quite different than in the Cascades, where more explosive eruptions may be

likely. Mt. St. Helens is a valuable example of the kind of problems that can be expected during and after a high-magnitude physical event that disrupts a large area. As such, the experience should help in devising emergency plans for future volcanic eruptions and other hazardous events such as large earthquakes. On the other hand, the Icelandic event is encouraging because it points out the necessity to learn to live with natural processes/hazards, be they floods, earthquakes, or volcanic eruptions. How we live with these natural processes, however, will vary with the expected type of event. Thus evacuation of a very large area is necessary if an explosive eruption is likely. However, if an eruption is likely to be characterized by volcanic ash and slow-moving lava flow, then there may be time to partially control the process advantageously, as was done in Iceland. Throughout the rest of this chapter we will focus on how people adjust to natural hazards.

NATURAL HAZARD OR NATURAL PROCESS?

As we learned in Chapter 1, there have always been Earth processes that are hazardous to people. Notice in this statement that we use the term *Earth process*. This brings up an important point—namely, that most natural hazards are natural processes. These processes become

(a)

(b)

FIGURE 13.5
Mt. St. Helens before (a) and after (b) the May 18, 1980, eruption. As a result of the eruption, much of the north side of the volcano was blown away and the summit was reduced by approximately 450 meters. (Photos courtesy of U.S. Geological Survey and Harry Glicken.)

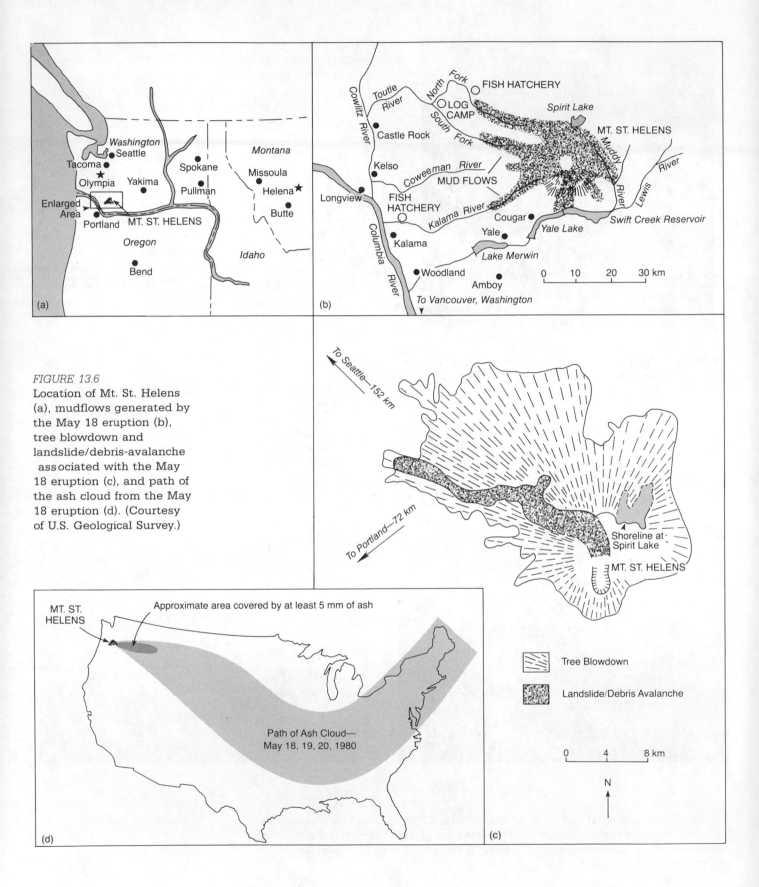

FIGURE 13.6
Location of Mt. St. Helens
(a), mudflows generated by
the May 18 eruption (b),
tree blowdown and
landslide/debris-avalanche
associated with the May
18 eruption (c), and path of
the ash cloud from the May
18 eruption (d). (Courtesy
of U.S. Geological Survey.)

332

hazardous to people when we live close to a potential danger or modify processes in such a way as to increase the hazard.

Many processes continue to cause loss of life and property damage, including flooding of coastal or flood-plain areas, landslides, earthquakes, expansive soils, volcanic activity, wind, drought, fire, and coastal erosion. The magnitude and frequency of these processes or events depend upon such factors as climate, geology, vegetation, and human use of the land. Therefore, Earth processes may be recognized or predicted by considering these factors. After the process has been identified and the potentially hazardous aspects studied, this information should be made available to planners and decision makers so that they may avoid these hazards or processes or minimize their threat to human life and property.

NATIONAL AND REGIONAL OVERVIEW

Evaluation of mean annual losses in the United States from several natural hazards provides insight into the magnitude of the problem. Table 13.1 summarizes losses from selected natural hazards or processes for the United States. The largest loss of life every year is associated with tornadoes and windstorms, but other processes such as lightning strikes, floods, and hurricanes also take a heavy toll in human life. In terms of cost per year, expansive soils (expansion and contraction of soils due to changing water content) are surprisingly the greatest hazard, about twice as costly as floods, landslides, and frost and freeze.

An important aspect of all natural hazards and processes is the potential to produce a **catastrophe**, defined as any situation in which the damages to people, property, or society in general are sufficient that recovery and/or rehabilitation is a long, involved process [4]. Those processes most likely to produce a catastrophe include floods, hurricanes, tornadoes, tsunamis (seismic sea waves), volcanoes, and large fires (Table 13.1). Other processes, such as landslides, generally cover a smaller area and may have only a moderate catastrophe potential. Drought, which may cover a wide area but generally involves plenty of warning time, also has a moderate catastrophe potential. Processes with a low catastrophe potential include coastal erosion, frost, lightning strikes, and expansive soils [4].

Loss of life and property damage in the United States from natural hazards shift with time due to changes in land-use patterns, which influence people to develop on marginal lands; urbanization, which changes the physical properties of earth materials; and increasing population. Table 13.2 summarizes in a qualitative way some of these changes. Damage from most hazards in the United States is increasing, but the number of deaths from many are decreasing due to better warning, forecasting, and prediction of some hazards.

HUMAN USE AND HAZARDS

Many natural processes can be influenced by people's activities, as shown in Table 13.1. In this section we will discuss a few specific cases where human activities affect these processes.

TABLE 13.1
Effects of selected natural hazards/processes in the United States.

Hazard	Deaths per Year	Cost per Year (million $)	Occurrence Influenced by Human Use	Catastrophe Potential[b]
Flood	86	1200	Yes	H
Earthquake[a]	50+?	130+?	Yes	H
Landslide	25	1000	Yes	M
Volcano[a]	<1	20	No	H
Coastal erosion	0	330	Yes	L
Expansive soils	0	2200	No	L
Hurricane	55	510	Perhaps	H
Tornado and windstorm	218	550	Perhaps	H
Lightning	120	110	Perhaps	L
Drought	0	792	Perhaps	M
Frost and freeze	0	1300	Yes	L

SOURCE: Modified after White and Haas, 1975.
[a]Estimate based on recent or predicted loss over 150-year period. Actual loss of life and/or property could be much greater.
[b]Catastrophe potential: high (H), medium (M), low (L).

TABLE 13.2
Recent trends in deaths and damages resulting from natural hazards.

Hazard	Damages	Deaths
Avalanche		
Coastal erosion		NA
Drought	?	NA
Earthquake		
Flood		
Frost		NA
Hail		NA
Hurricane		
Landslide		
Lightning	?	
Tornado		
Tsunami	NA	?
Urban snow		
Volcano	NA	
Windstorm		

SOURCE: From White and Haas, 1975.
NOTE: NA—Not applicable

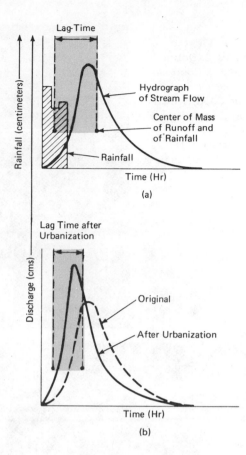

FIGURE 13.7
Generalized hydrographs before and following urbanization. Hydrograph (a) shows the typical lag time between the time when most of the rain falls and the time when the flood in the stream occurs. Hydrograph (b) shows the decrease in lag time due to urbanization. (From Leopold, 1968.)

URBAN FLOODING

Flooding in urban areas may be greatly increased by urbanization. Small drainage basins of a few square kilometers are most susceptible to hydrologic changes associated with urbanization. Hydrologically speaking, urbanization means an increase in the area rendered impermeable (covered by parking lots, streets, and roofs), increasing runoff because less water is able to infiltrate into the soil. Urban areas also may be drained by storm sewers that quickly deliver water to urban streams. Figure 13.7 shows generalized **hydrographs** (graphs of stream discharge versus time) before and after urbanization. **Lag time** (time between when most of the rain falls and the greatest runoff occurs) decreases due to quicker runoff after urbanization, and there is an increase in flood peak after urbanization. It is the increase in flood peak that causes urban flood problems; that is, compared with preurban conditions, urban areas for a given set of storms will flood more often and with higher discharge.

The effects of urbanization are also most pronounced with floods that can be expected to recur in 2 to 30 years. High-magnitude floods with large recurrence intervals (for example, 50 years) are little changed by urbanization. Because very large floods are produced by large, infrequent storms with rainfall intensities that greatly exceed the infiltration rate of water into a soil, it makes little difference if the land is urbanized or not.

People have also caused floods by constructing dams that have failed. The Buffalo Creek, West Virginia, flood in 1972 resulted from the failure of a coal-waste dam that was never designed to hold a great deal of water. The flood killed 118 people, destroyed 500 homes (Fig. 13.8), left 4000 people homeless, and caused over $50 million in property damage [5]. Failure of the Teton Dam in 1976 killed 14 people and inflicted about $1 billion in property damage. The cause of the failure was directly related to adverse geologic conditions, including highly fractured volcanic rock in the foundation which was not adequately treated during construction.

LAND USE AND LANDSLIDES

Urbanization often modifies slopes, leading to an increase in landslides, particularly on steep slopes. Four ways in which a stable slope or cut can be rendered unstable are steepening a slope; increasing the height of a slope by excavation; saturating a slope; and placing fill on top of a slope; all may decrease stability and produce a landslide. Because it steepens and lengthens a slope, construction of roads and highways may be particularly troublesome (Fig. 13.9).

FIGURE 13.8
Buffalo Creek, West Virginia, in 1972. Debris in the channel and floodplain was deposited by a catastrophic flood, which was caused by the failure of a coal-waste dam. (Photos courtesy of U.S. Army Corps of Engineers.)

Removing or changing vegetation may also decrease slope stability and produce landslides. Timber harvesting in conjunction with adverse geologic conditions is often associated with an increase in landslides usually several years following logging. The lag time between logging and landslides is thought to be due in part to the slow decay of tree roots. Live tree roots help hold the soil together (increases the soil strength), inhibiting landsliding. Following timber harvesting the roots slowly die and the soil strength decreases.

A mixture of human use and adverse geologic conditions resulted in a bizarre landslide in 1960 in Handlova, Czechoslovakia. A large coal-burning power plant in Handlova burned soft coal, emitting a large volume of ash that was deposited downwind. So much ash had accumulated that land used for grazing had to be plowed. Plowing allowed rain water to infiltrate at a greater rate, disturbing a delicate groundwater situation and initiating the slide. A well-organized program to stop the land motion by draining the slide material was successful, but

FIGURE 13.9
Landslide in Rio de Janeiro that demolished several houses and two apartment buildings. More than 132 people died as a result of the slide. The large slide was evidently facilitated by a smaller landslide associated with a highway cut that overloaded the slope. (Photo by F. O. Jones, courtesy of U.S. Geological Survey.)

not before 20 million cubic meters of earth had moved about 150 meters, destroying 150 homes [6]. The Handlova slide is a good example of our fundamental principle (Chapter 1) that everything affects everything else.

EARTHQUAKES CAUSED BY PEOPLE

Earthquakes have been caused by several types of human activity, including disposal of liquid waste deep in the Earth, building large reservoirs, and exploding nuclear devices underground. Several hundred earthquakes near Denver, Colorado, from 1962 to 1965 were apparently related to deep-well disposal of liquid chemical waste. The earthquakes, while not particularly large, were sufficient to knock bottles off shelves in stores. Correlation between the number of earthquakes and the rate of waste injected (Fig. 13.10) clearly shows that the waste disposal was responsible for the earthquakes.

Numerous small local earthquakes occurred during the first 10 years following the completion of Hoover Dam, which supplies Lake Mead in Arizona and Nevada. Most were very small, but one was at least moderate. Earthquakes induced by building large reservoirs can kill people, and one in India killed approximately 200 people. Evidently, fracture zones or faults are activated by the increased load of the water on the land and by the increased water pressure in the rocks below the reservoir. The problem of induced seismicity associated with large reservoirs is being intensively studied.

Numerous, generally small earthquakes have also been triggered by nuclear explosions at Nevada test sites. The information from these and other earthquakes produced by human use and interest in the land may eventually be applied to control or predict earthquake activity in the future.

COASTAL EROSION AND HUMAN ACTIVITY

The process of coastal erosion is definitely influenced by human use, particularly when structures are built in the coastal zone. In areas characterized by a **sea cliff** (Fig. 13.11), urban runoff may increase cliff erosion by delivering large amounts of water to the face of the cliff. Careful control of storm water runoff will help to minimize erosion (Fig. 13.12).

Along the eastern coast of the United States, particularly south of New York City and along the Gulf Coast, much of the coastal zone is characterized by barrier islands. **Barrier islands** are long, narrow strings or chains of sand which separate the mainland from the coast. Landward of the islands there is a salt marsh or lagoon (Fig. 13.13), and inlets (openings or breaks in the barrier island) with strong tidal currents allow water from the salt marsh or lagoon to exchange with the ocean. The islands have developed in the last few thousand years in response to a slow rise in sea level and tend to migrate shoreward through a number of processes, one of which is known as *overwash*. Overwash occurs when the frontal line of sand dunes is broken by waves during storms and sand and other sediment are washed toward the salt marsh behind the sand dunes. There has been a continued controversy as to the importance of overwash in the development of barrier islands. Although many kilometers of sand dunes in the past have been protected by artificial means, we now recognize that this program has disrupted the natural environment and may have actually increased the erosion.

Construction of breakwaters to form a small boat harbor or jetties to protect a river mouth has often caused

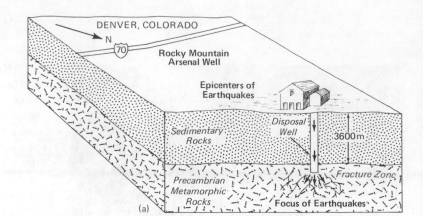

FIGURE 13.10
Idealized diagram showing
the Rocky Mountain
Arsenal well (a), and graph
showing the relationship
between earthquake
frequency and rate of liquid
waste disposal for five
characteristic periods (b).
[Part (b) from Evans, 1966.]

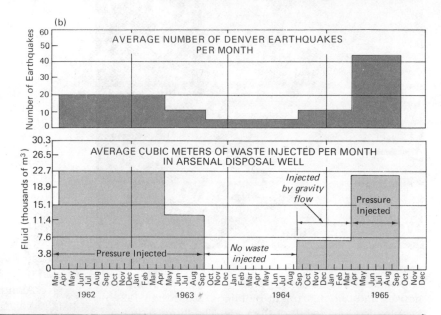

FIGURE 13.11
Sea cliff, beach, and wave-
cut platform in Santa
Barbara, California. (Photo
courtesy of Donald
Weaver.)

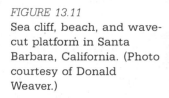

(a)

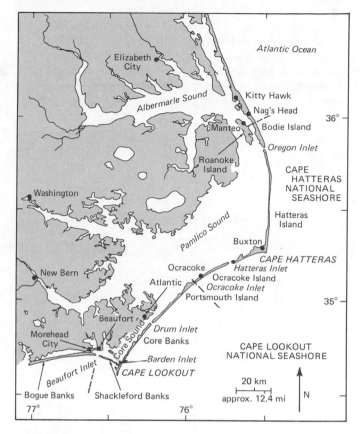

FIGURE 13.13

The prominent barrier island system of the Outer Banks of North Carolina. (From Godfrey and Godfrey, 1973.)

(b)

coastal erosion (Fig. 13.14). Erosion occurs because the breakwater or jetty interferes with the natural flow of sand along a beach. The flow of sand, known as *littoral drift*, is produced by waves as they strike the shore at an angle. As shown in Figure 13.14, deposition in one area of the coast is compensated for by erosion in the downdrift direction. Sand must be dredged where it accumulates and placed back in the littoral drift system downcoast beyond the breakwater or jetty, replenishing beach material and thereby minimizing erosion.

FIGURE 13.12

Erosion of sea cliff composed of soft compaction shale near Isla Vista, California. Uncontrolled runoff from storm drains on the sea cliff results in serious erosion (a). Controlled runoff released at the base of the sea cliff on the beach causes much less erosion (b). (Photos by E. A. Keller.)

FIGURE 13.14
Idealized diagrams
illustrating the effects of
breakwaters and jetties on
local patterns of deposition
and erosion along a
coastline.

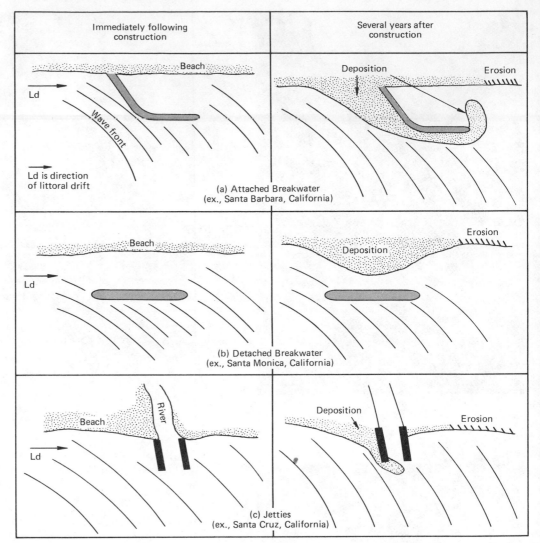

FIGURE 13.14
Idealized diagrams illustrating the effects of breakwaters and jetties on local patterns of deposition and erosion along a coastline.

SELECTED METEOROLOGIC PROCESSES AND PEOPLE

Processes that produce hurricanes, tornadoes, windstorms, lightning, and drought may perhaps be influenced by human activities. The connection, however, is rather tenuous; that is, local, regional, or even global climates would have to be changed by people's activities. Certainly there is reason to speculate that people may be changing the global climate by burning fossil fuels and introducing carbon dioxide into the atmosphere in ever greater quantities. Overgrazing and poor agricultural practices in semiarid regions may result in **desertification,** or the conversion of land from a productive state to that resembling a desert. Desertification is a serious worldwide problem (Fig. 13.15). In parts of India and Africa, desertification has been responsible for the starvation or malnutrition of millions of people. Less attention has been given to desertification in the United States, where loss of life is not a problem. Nevertheless, about 2.8 million square kilometers in North America, or 37 percent of the total arid lands (much of which is in the United States), have suffered from desertification characterized by significant invasion of brush, wind and water erosion, or a high salinity that has reduced crops. About 27,000 square kilometers in North America, nearly all of which is in the United States, has undergone desertification characterized by large-scale gully erosion, sand dunes, or salt crusts. Africa, by comparison, has about four times as much desertified area as North America. As in other parts of the world, desertification of North America has far-reaching consequences, including loss of food crops and livestock; reduced food exports, resulting in a less favorable balance of payments; and general deterioration or impoverishment of ecosystems [7].

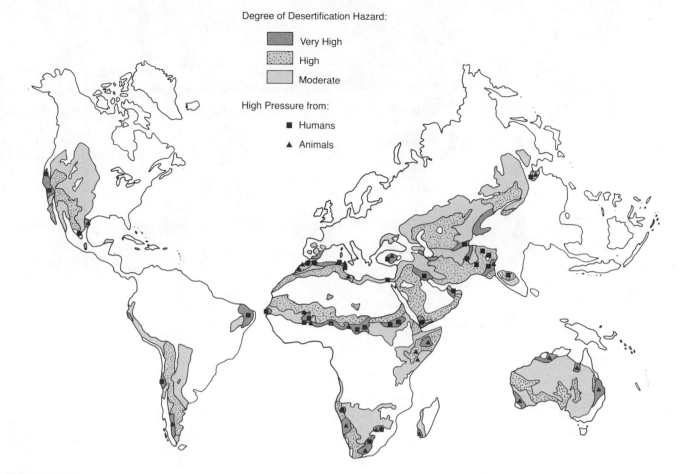

FIGURE 13.15
Degree of desertification (as a hazard) on a global scale. (From Council on Environmental Quality, 1978.)

HAZARDS NOT INFLUENCED BY HUMAN ACTIVITY

A few natural processes, such as volcanoes and expansive soils, are not influenced by human activity. Volcanoes, such as the Cascade Range in North America, are produced along major plate boundaries by deep-seated processes that are not affected by human use of the land (Fig. 13.16). Expansive soils are caused by a particular clay mineral resulting from weathering processes and thus are independent of human activity.

PREDICTION OF HAZARDS

Learning how to predict hazards so we can minimize human loss and property damage is an important endeavor. For each particular hazard or process, we have a certain amount of information; in some cases it is sufficient to predict or forecast events accurately. When there

is insufficient information to make accurate forecasts or predictions, the best we may be able to do is simply locate areas where hazardous events have occurred and infer where and when future events might take place. Thus, the prediction of hazards involves the following aspects: identified locations where a hazard occurs, probability of occurrence, precursor events, forecast, and warning.

For the most part, we know where a particular hazard is likely to occur. For example, the major zones for earthquakes and volcanic eruptions have been delineated satisfactorily on a global scale by mapping the major plate boundaries. On a regional scale, based on the past record of activity, areas likely to have a significant hazard from a large mudflow or ash eruptions associated with a volcanic eruption have also been delineated for several Cascade volcanoes, including Mt. Rainier (Fig. 13.17). On a local scale, detailed work with soils may easily identify slopes that are likely to fail (landslide) or where expansive soils

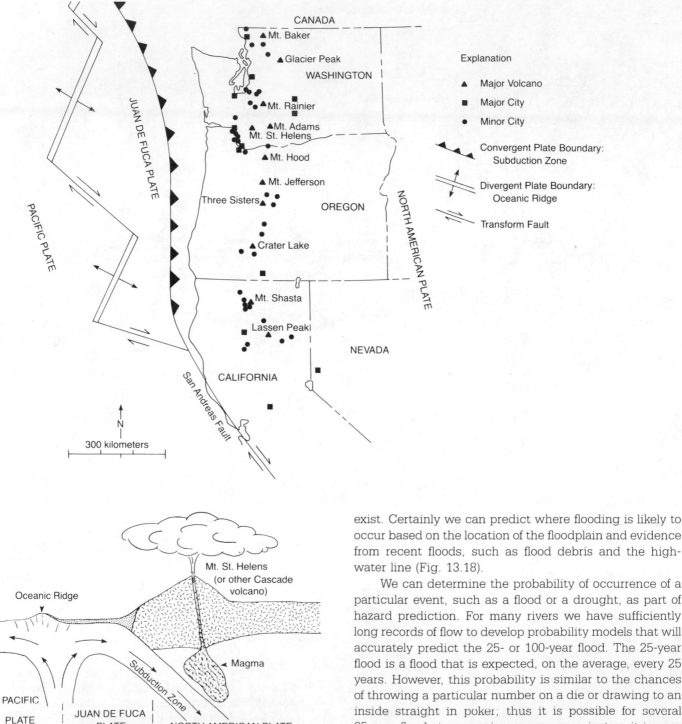

FIGURE 13.16

Idealized diagram showing the Cascade Range of volcanoes and their relationship to the plate tectonic model. (After Crandell and Waldron, 1969.)

exist. Certainly we can predict where flooding is likely to occur based on the location of the floodplain and evidence from recent floods, such as flood debris and the high-water line (Fig. 13.18).

We can determine the probability of occurrence of a particular event, such as a flood or a drought, as part of hazard prediction. For many rivers we have sufficiently long records of flow to develop probability models that will accurately predict the 25- or 100-year flood. The 25-year flood is a flood that is expected, on the average, every 25 years. However, this probability is similar to the chances of throwing a particular number on a die or drawing to an inside straight in poker, thus it is possible for several 25-year floods to occur in any one year, just as it is possible to throw two straight 6's with a die. Likewise, droughts may be assigned a probability based on past occurrence of rainfall in a particular region. Work with hazards such as earthquakes and volcanic events has not yet advanced to the point that probability of occurrence may

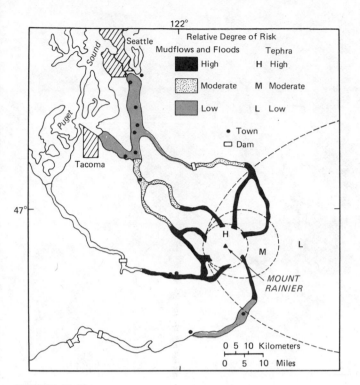

FIGURE 13.17
Rough map of Mt. Rainier and vicinity showing the relative degree of potential hazards from ashfalls, mudflows, and floods that might result from an eruption. (From Crandell and Mullineaux, 1975.)

be calculated accurately on a regular basis. Nevertheless, a national earthquake hazard map based on probability has been developed (Fig. 13.19). Although it is not useful

for small, site-specific studies, the map is valuable in regional planning.

Many hazardous Earth processes have precursor events. For example, the surface of the ground may creep or move slowly for a long period prior to an actual landslide. Often the rate of creep has increased up to the final failure and landslide. Volcanoes have been noticed to swell or bulge before an eruption (Fig. 13.20), and there often is a significant increase in the local seismic activity surrounding the volcano.

Precursor events associated with earthquakes are not particularly well known nor understood, but increases in emission of radon gas from wells, foreshock activity, unusual tilt or uplift of the land, and perhaps even strange animal activity may be precursor events. Anomalous tilt or uplift may begin months or even years prior to the earthquake (Fig. 13.21), whereas unusual animal activity evidently occurs very close to the time of the event. Seismic gaps (areas where earthquakes are expected but have not occurred) at both regional and local scales have also been valuable in predicting some earthquakes.

With some natural processes it is possible to accurately forecast when the event will arrive. For example, flooding of the Mississippi River, which occurs in the spring in response to snow melt or very large regional storm systems, is fairly predictable, and we can sometimes forecast when the river will reach a particular flood stage. When hurricanes are spotted far out to sea and tracked toward the shore, we can forecast when it will actually strike the land (Fig. 13.22). **Tsunamis,** or seismic sea waves generated by disturbance of ocean waters due

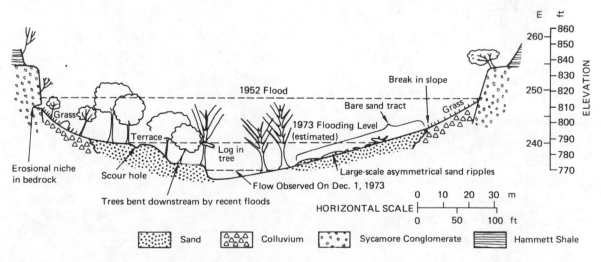

FIGURE 13.18
Schematic cross sections of the Pedernales River valley, Texas, illustrating the floodplain features used in estimating floods. (After Baker, 1976.)

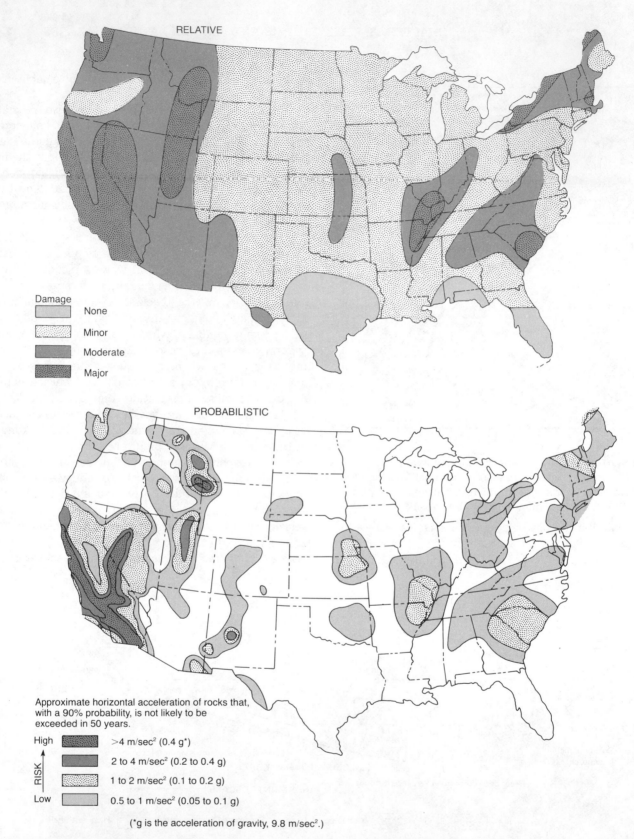

RELATIVE

Damage
None
Minor
Moderate
Major

PROBABILISTIC

Approximate horizontal acceleration of rocks that,
with a 90% probability, is not likely to be
exceeded in 50 years.

High
RISK
Low

>4 m/sec² (0.4 g*)

2 to 4 m/sec² (0.2 to 0.4 g)

1 to 2 m/sec² (0.1 to 0.2 g)

0.5 to 1 m/sec² (0.05 to 0.1 g)

(*g is the acceleration of gravity, 9.8 m/sec².)

FIGURE 13.19

A probabilistic approach to the seismic hazard in the United States. The darker the
area on the map, the greater is the hazard. (After Algermissen and Perkins, 1976.)

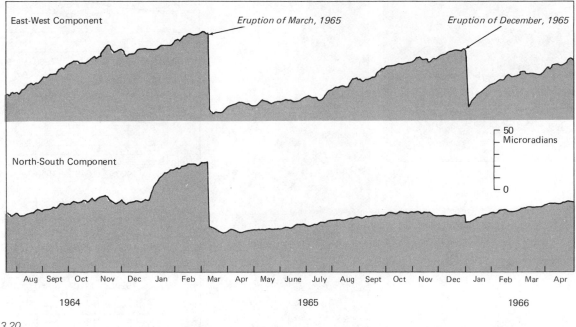

FIGURE 13.20

Graph showing the east-west component and the north-south component of ground tilt recorded frm 1964 to 1966 on Kilauea volcano in Hawaii. Notice the abrupt change in ground tilt before eruption. (From Fiske and Koyanagi, 1968.)

to earthquakes or submarine volcanoes, may also be forecast (Fig. 13.23). The tsunami warning system has been fairly successful in the Pacific Basin, and in some instances the time of arrival of the waves has been forecast precisely.

THE NATURAL SERVICE FUNCTION OF HAZARDOUS PROCESSES

It is ironic that natural processes or hazards, while taking human life and destroying property, also perform important service functions. For example, flooding supplies nutrients to floodplains, as in the case of the Mississippi River and the Nile Delta prior to building of the Aswan Dam. Flooding also causes erosion on mountain slopes, delivering sediment to beaches from rivers and flushing pollutants from estuaries in the coastal environment. Rapids in the Grand Canyon result in part from natural flood events which have delivered coarse material to the main channel of the Colorado River in the canyon. In the past, large floods have occasionally removed some of this large debris; but now that the Colorado River is tamed by dams, the rapids may eventually become more dangerous to river runners as more material piles up where tributary channels enter the Colorado River (Fig. 13.24).

Landslides also may perform some natural service functions, particularly in the formation of lakes. Landslide debris may form dams, making lakes in mountainous areas. These lakes provide valuable water storage and are an important aesthetic resource.

Volcanic eruptions, while having the potential to produce real catastrophes, perform numerous public service functions. New land can be created, as in the Hawaiian Islands, which are completely volcanic in origin. In addition, nutrient-rich volcanic ash may settle on soils and quickly become incorporated in them. Earthquakes also perform a number of natural service functions. For example, groundwater barriers may be created when rocks are pulverized against each other to form a clay zone known as **fault gouge** (Fig. 13.25). There are numerous cases where groundwater has been dammed upslope from a fault, producing a water resource. Earthquakes are also important in mountain building and thus are directly responsible for many of the scenic resources of the western United States.

ADJUSTMENTS TO HAZARDS

Major adjustments to natural hazards and processes include land-use planning, insurance, evacuation, disaster

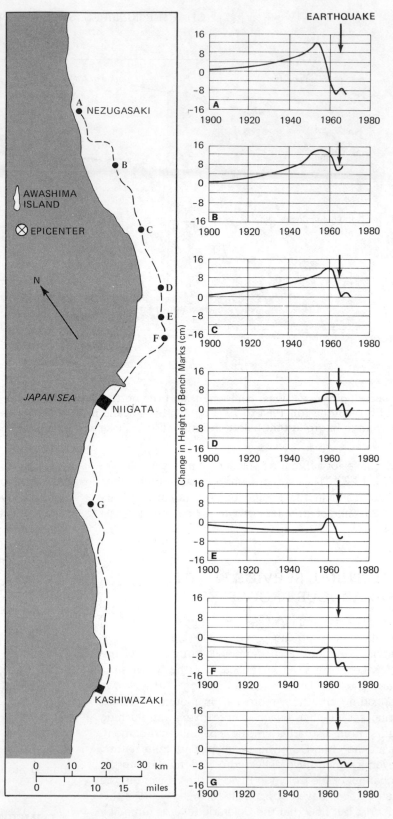

FIGURE 13.21

Anomalous uplift of the Earth's crust observed for approximately 10 years before the 7.5-magnitude earthquake that struck Niigata, Japan, in 1964. (From "Earthquake prediction" by F. Press. Copyright© 1975 by Scientific American, Inc. All rights reserved.)

FIGURE 13.22
Catastrophic hurricanes affecting the United States from 1964 to 1979. The track of Celia is not shown. (Modified after U.S. Department of Commerce, 1970. Updated by NOAA.)

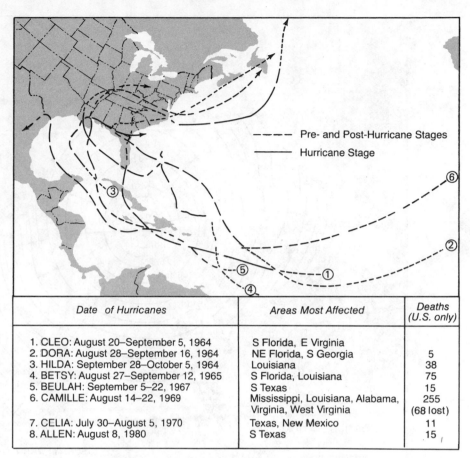

- - - - Pre- and Post-Hurricane Stages
———— Hurricane Stage

Date of Hurricanes	Areas Most Affected	Deaths (U.S. only)
1. CLEO: August 20–September 5, 1964	S Florida, E Virginia	
2. DORA: August 28–September 16, 1964	NE Florida, S Georgia	5
3. HILDA: September 28–October 5, 1964	Louisiana	38
4. BETSY: August 27–September 12, 1965	S Florida, Louisiana	75
5. BEULAH: September 5–22, 1967	S Texas	15
6. CAMILLE: August 14–22, 1969	Mississippi, Louisiana, Alabama, Virginia, West Virginia	255 (68 lost)
7. CELIA: July 30–August 5, 1970	Texas, New Mexico	11
8. ALLEN: August 8, 1980	S Texas	15

preparedness, and bearing the loss. Which option is chosen by an individual depends upon a number of factors, the most important of which is hazard perception.

In recent years a good deal of work has been done to try to understand how people perceive various natural hazards. This is obviously an important endeavor because the success of hazard reduction programs depends on the attitude of the people likely to be affected by the hazard. For example, it has been difficult to develop earthquake hazard reduction programs where strong earthquakes only occur once every few generations. Similarly, it is difficult to tell an individual who has lived many years in a particular home on the floodplain that he is living in a very dangerous area because the floodplain is inundated by water on the average of once every 100 years. Because flooding at a particular site may occur infrequently, the individual may not perceive flooding to be a serious hazard for him. While there may be an adequate perception of hazards at the institutional level, this may not filter down to the general population. This is particularly true for those hazards that occur infrequently.

Proximity to hazards such as volcanoes and coastal erosion seems to be very important. That is, people who live near volcanoes or in the coastal zone are more likely to

be aware of the hazard and possibly take steps needed to minimize potential damages.

People are more aware of hazards such as brush or forest fires, which may occur every few years. There may even be institutionalized as well as local ordinances to control damages resulting from these events. For example, homes built in some areas of southern California have roofs that are constructed with shingles that will not burn readily and may even have sprinkler systems, and the lots are often cleared of brush. Such measures are often noticeable during the rebuilding phase following a fire.

One of the most environmentally sound adjustments to hazards involves land-use planning. That is, people can avoid building on floodplains, in areas where there are active landslides or active fault traces, and in areas where coastal erosion is likely to occur. In many cities, floodplains have been delineated (Fig. 13.26) and zoned for a particular land use. Zoning associated with active and potentially active faults is also commonplace in California (Fig. 13.27). With respect to landslides, legal requirements for soils engineering and engineering geology studies at building sites may greatly reduce potential damages (Table 13.3). Although it may be possible to control physical processes in specific instances, certainly

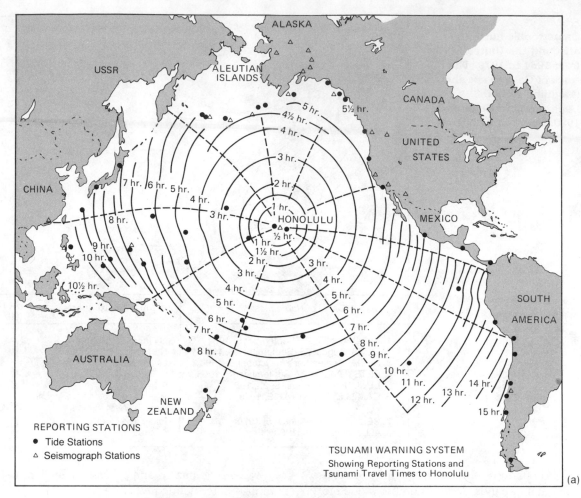

REPORTING STATIONS
• Tide Stations
△ Seismograph Stations

TSUNAMI WARNING SYSTEM
Showing Reporting Stations and
Tsunami Travel Times to Honolulu

(a)

FIGURE 13.23
The tsunami warning
system. Map shows
reporting stations and
tsunami travel times to
Honolulu, Hawaii (a). Sandy
Beach on the island of
Oahu moments before a
tsunami generated by an
earthquake in the vicinity
of the Aleutian Trench
struck the beach (b). (Map
from NOAA. Photo by Y.
Ishii, courtesy of *Honolulu
Advertiser.*)

(b)

(a)

(b)

FIGURE 13.24
Crystal Rapid of the Colorado River in the Grand Canyon before (a) and after (b) the construction of the Glen Canyon Dam. Photograph (a), taken in 1963, shows the rapid essentially as it existed prior to the building of the dam. Photograph (b), taken in 1967, shows the effect of a 1965 flash flood. The debris and rock delivered from Crystal Creek (left) form a fan-shaped deposit in the Colorado River. This deposit has made the rapid more difficult to negotiate, and the debris will remain in the river for a relatively long time as large floods that would normally remove some of the debris are controlled by the dam. [Bureau of Reclamation photos by Al Turner (a) and Mel Davis (b).]

TABLE 13.3
Landslide and flood damage to hillside homes in Los Angeles County, California, from before 1952 to 1969. Notice the lessening of damages following adoption of building codes.

Construction Dates and Legal Requirements	Number of Homes Built on Hillside Sites	Damaged Homes		Total Damage	Average Cost Prorated for Total Number of Homes
		Number	Percent of Total		
Pre–1952: No legal requirements for soils engineering or engineering geology studies	10,000	1040	10	$3,300,000	$300
1952–1963: Soils engineering studies required. Minimum engineering geology studies	27,000	350	1.3	$2,767,000	$100
1963–1969: Extensive engineering geology and soils engineering studies required	11,000	17	0.15	$ 80,000	$ 7

SOURCE: After Slosson and Krohn, 1977. Data from Los Angeles Department of Building and Safety.

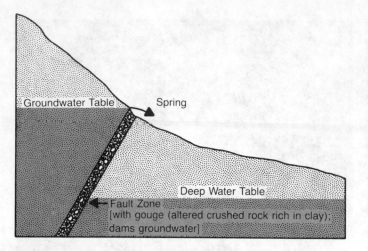

FIGURE 13.25
Idealized diagram showing how a fault zone with gouge (altered crushed rock rich in clay) may dam groundwater, forming a spring.

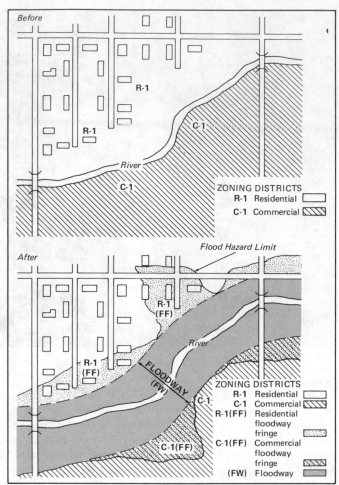

FIGURE 13.26
Typical zoning map before and after the addition of floodplain regulations. (From Water Resources Council, 1971.)

FIGURE 13.27
Example of the zoning associated with an active fault. Where the location of the fault trace is well known, no new buildings are allowed to be constructed within 15 meters of each side of the fault. Dwellings that house more than a single family are required to be 38 meters from the fault. Where the precise location of the fault trace is less well known, more conservative setbacks of 30 meters for single-family residences and 53 meters for higher occupancies are required. (From Mader et al., 1980, and Nichols and Buchanan-Banks, 1974.)

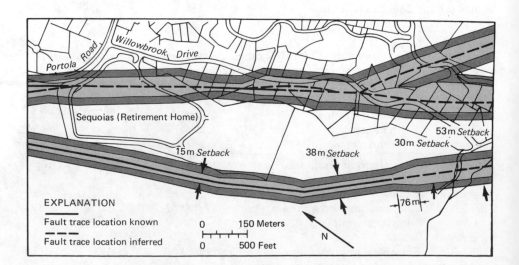

350

land-use planning to accommodate natural processes is preferable to a technological fix which may or may not work.

Insurance is another option that people may exercise in dealing with natural hazards. Flood insurance is relatively common in many areas, and earthquake insurance is also available. However, other than fire insurance or en-

forced insurance against flood, few people purchase extra policies. Only a small percentage of people in southern California, for example, have earthquake insurance.

Evacuation is an important option or adjustment to the hurricane hazard in the Gulf States and eastern coast of the United States. Often there will be sufficient time for people to evacuate provided they heed the predictions and warnings. However, if people do not react quickly and the affected area is a large urban region, then evacuation routes may be blocked by people panicking in the last minute and trying to evacuate. There is concern that a large hurricane in an urban area would cause catastrophic loss of life and property and that people either will not evacuate or will have trouble doing so. A great number of people were able to evacuate prior to hurricane Allen's approach to the Texas coastline in August of 1979, and by fortuitous circumstances the storm struck a relatively uninhabited stretch of coastline and did relatively little damage. If the storm had not stalled offshore for several hours and lost energy, a catastrophe would certainly have resulted if it had struck a large urban area.

Disaster preparedness is an option that individuals, families, cities, states, or even entire nations can implement. Of particular importance here is the training of individuals and institutions to handle large numbers of injured people or people attempting to evacuate an area after a warning is issued.

An option that all too often is chosen is bearing the loss caused by a natural hazard. Many people are optimistic about their chances of making it through any sort of natural hazard, and therefore will take little action in their

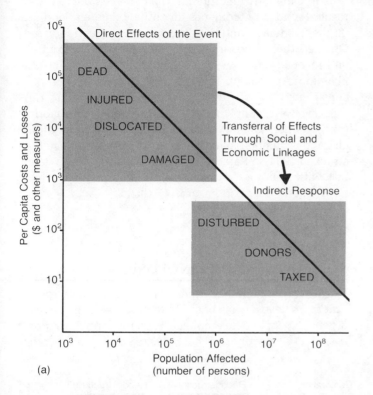

(a)

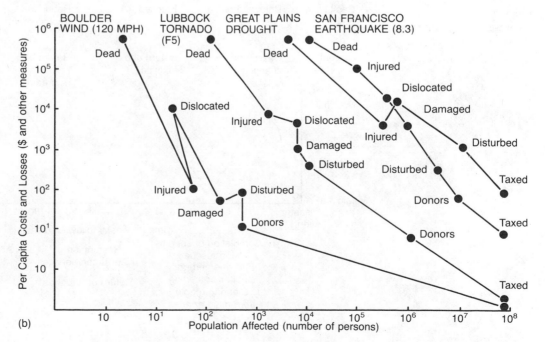

(b)

FIGURE 13.28
Impact of disaster in terms of the continuum of effects (a); and comparison with real data from four disasters (b). [From White and Haas, 1975. Part (a) adapted from Bowden and Kates, 1974.]

own defense. This is particularly true for those hazards that may recur only rarely in a particular area, such as volcanic eruptions and earthquakes.

IMPACT OF AND RECOVERY FROM DISASTERS

The impact of a disaster upon a population may be either direct or indirect. Direct effects include people killed, injured, dislocated, or otherwise damaged by a particular event; indirect impacts are generally responses to the disaster, including people who are generally bothered or disturbed by the event, people who donate money or goods, and the taxing of people to help pay for emergency services, restoration, and eventual reconstruction. These concepts are summarized in Figure 13.28a, which shows that as the general level of effects decreases, the size or percentage of the population affected increases. Figure 13.28b shows actual data for several types of natural disasters, thus demonstrating this continuum of effects [8,9].

The stages following a disaster are emergency work; restoration of services and communcation lines; and reconstruction. Figure 13.29 shows an idealized model of recovery, and Figure 13.30 shows actual recovery activities following the 1964 earthquake in Anchorage, Alaska, and the 1972 flash flood in Rapid City, South Dakota. Examination of Figure 13.30 shows that restoration following the earthquake in Anchorage began almost immediately

in response to a tremendous influx of dollars from federal programs, insurance companies, and other sources approximately one month after the earthquake. As a result, reconstruction was a hectic process, with everyone trying to obtain as much of the available funds as possible. In Rapid City, the restoration did not peak until approximately 10 weeks after the flood, and the community took time to carefully think through the best alternatives. As a result, Rapid City today has an entirely different land use on the floodplain, and the flood hazard is much reduced. On the other hand, in Anchorage the rapid restoration and reconstruction were accompanied by little land-use planning. Apartments and other buildings were hurriedly constructed across areas that had suffered ground rupture and were simply filled in and regraded. In ignoring the potential benefits of careful land-use planning, Anchorage is vulnerable to the same type of earthquake that struck in 1964. In Rapid City, the floodplain is now a green belt with golf courses and other such activities, which has reduced the flood hazard [4,8,9].

SUMMARY AND CONCLUSIONS

One of the fundamental principles outlined in Chapter 1 is that there have always been Earth processes that are hazardous to people. The emphasis is on the term *Earth process*. That is, most natural hazards are simply natural proc-

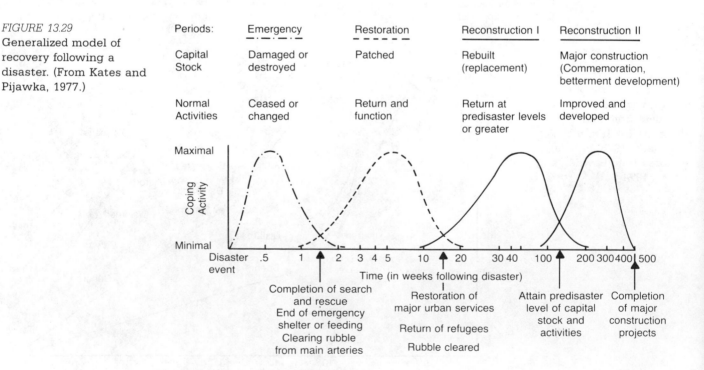

FIGURE 13.29
Generalized model of recovery following a disaster. (From Kates and Pijawka, 1977.)

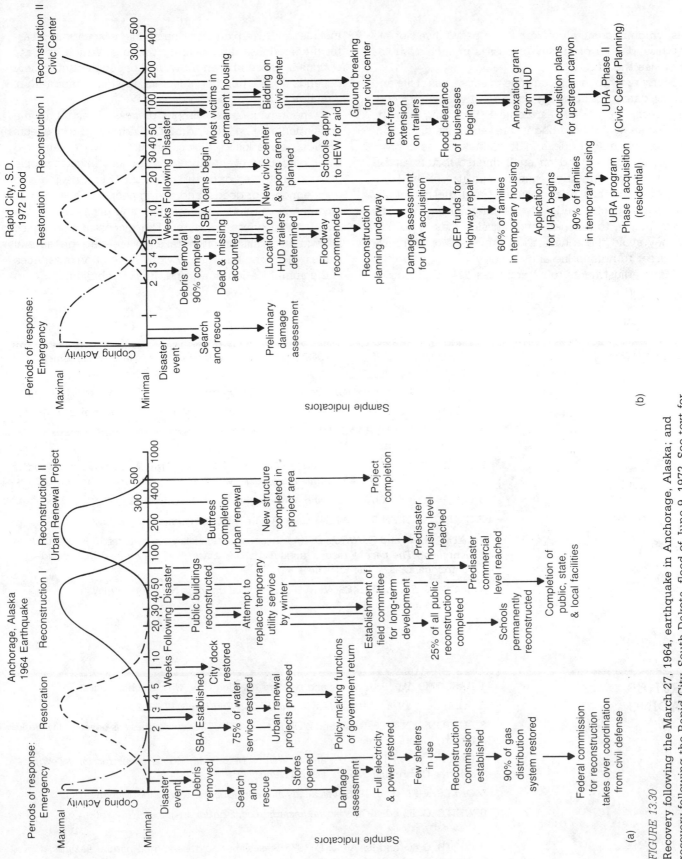

FIGURE 13.30

Recovery following the March 27, 1964, earthquake in Anchorage, Alaska; and recovery following the Rapid City, South Dakota, flood of June 9, 1972. See text for further explanation. (From Kates and Pijawka, 1977.)

353

esses which become a problem when people live close to a potential danger or modify processes in such a way as to increase the hazard.

Many processes will continue to cause loss of life and property damage, including flooding of coastal or flood-plain areas, landslides, earthquakes, volcanic activity, wind, expansive soils, drought, fire, and coastal erosion. However, the magnitude and frequency of these processes or events depend on such diverse factors as climate, geology, vegetation, and human use of the land. Once a process has been identified and the potentially hazardous aspects studied, this information must be made available to planners and decision makers so that they may avoid these hazards or processes or minimize their threat to human life and property.

Major adjustments to natural hazards and processes include land-use planning, insurance, evacuation, disaster preparedness, and bearing the loss. Which of these options is chosen by an individual or segment of society depends upon a number of factors, the most important of which is hazard perception.

The impact of a hazardous process (disaster) upon a population may be either direct or indirect. Direct effects include people killed, dislocated, or otherwise damaged by a particular event. Indirect impacts involve people generally bothered or disturbed by the event, people who donate money or goods, and taxing people to help pay for emergency services, restoration, and eventual reconstruction following a disaster. The reconstruction phase following a disaster often takes place through several stages, including emergency work, restoration of services and communication lines, and, finally, rebuilding.

REFERENCES

1 WILLIAMS, R. S., Jr., and MOORE, J. G. 1973. Iceland chills a lava flow. *Geotimes* 18: 14–18.

2 CORNELL, J., ed. 1974. *It happened last year—Earth events—1973.* New York: Macmillan.

3 HAMMOND, P. E. 1980. Mt. St. Helens blasts 400 meters off its peak. *Geotimes* 25: 14–15.

4 WHITE, G. F., and HAAS, J. E. 1975. *Assessment of research on natural hazards.* Cambridge, Mass.: The MIT Press.

5 DAVIES, W. E.; BAILEY, J. F.; and KELLY, D. B. 1972. *West Virginia's Buffalo Creek flood: A study of the hydrology and engineering geology.* U.S. Geological Survey Circular 667.

6 LEGGETT, R. F. 1973. *Cities and geology.* New York: McGraw-Hill.

7 COUNCIL ON ENVIRONMENTAL QUALITY. 1980. *Environmental quality.*

8 KATES, R. W., and PIJAWKA, D. 1977. Reconstruction following disaster. In *From rubble to monument: The pace of reconstruction,* eds. J.E. Haas; R.W. Kates; and M.J. Bowden. Cambridge, Mass.: The MIT Press.

9 COSTA, J. E., and BAKER, V. R. 1981. *Surficial geology.* New York: Wiley.

FURTHER READING

1 BASCOMB, W. 1980. *Waves and beaches.* Garden City, N.Y.: Anchor Books.

2 BOLT, B. A. 1978. *Earthquakes: A primer.* San Francisco: W. H. Freeman.

3 BURTON, I.; KATES, R. W.; and WHITE, G. F. 1978. *The environment as hazard.* New York: Oxford University Press.

4 HAAS, J. E.; KATES, R. W.; and BOWDEN, M. J., eds. 1977. *Reconstruction following disaster.* Cambridge, Mass.: The MIT Press.

5 OAKESHOTT, G. B. 1976. *Volcanoes and earthquakes.* New York: McGraw-Hill.

6 WHITE, G. F., ed. 1974. *Natural hazards: Local, national, global.* New York: Oxford University Press.

7 WHITE, G. F., and HAAS, J. E. 1975. *Assessment of research on natural hazards.* Cambridge, Mass.: The MIT Press.

**STUDY
QUESTIONS**

1 Why is it sometimes difficult to distinguish between a natural hazard and a natural process?

2 A two-year resident on the floodplain of a small urban stream says that upstream development is responsible for a recent flood that damaged his home. How would you respond to him?

3 If you pave over (with cement) the entire drainage area of the Mississippi River, the floods at New Orleans would not increase in magnitude or frequency. Do you agree or disagree? Why?

4 Discuss some of the ways human activity has caused earthquakes. Could any of these cause a really large earthquake, like the one that destroyed San Francisco in 1906?

5 Why is desertification considered to be a serious global threat? What can be done to minimize the hazard?

6 What are some of the ways that hazardous natural processes are predicted? Which hazards can we predict best? Why?

7 Discuss major adjustment strategies people use in dealing with hazardous processes. Which are best? Why?

8 What are the stages of recovery following a disaster? What factors affect how long recovery takes?

9 How can knowing something about the magnitude and frequency of a particular process help in planning to minimize possible damage?

10 What do you think was learned from the 1980 eruptions of Mt. St. Helens that may help planning for future high-magnitude (natural hazard) events in the United States?

14

The Hazards We Produce: Pollutants and Their Effects

INTRODUCTION

In 1976, in a residential area near Niagara Falls, New York, trees and gardens began to die. Children found the rubber on their tennis shoes and on their bicycle tires disintegrating. Dogs sniffing in a landfill area developed sores that would not heal. Puddles of toxic, noxious substances began to ooze to the soil surface; a swimming pool popped its foundations and was found to be floating on a bath of chemicals.

A study revealed that the residential area had been built on the site of a chemical dump, originally dug in 1892 as the beginnings of the Love Canal which was supposed to provide a transportation route between industrial centers. When that plan failed, the ditch was unused for decades and seemed a convenient dump for wastes. From the 1920s to the 1950s, more than 80 different chemicals were dumped there. Finally, in 1953, the company dumping the chemicals donated the land to the city of Niagara Falls for one dollar. Eventually 200 homes and an elementary school were built on and near the site (Fig. 14.1). Heavy rainfall and heavy snowfall during the winter of 1976–77 set off the events that made Love Canal a household word.

Residents of the area had higher than average rates of miscarriages, blood and liver abnormalities, birth defects, and chromosome damage. The old dumpsite contained a number of substances that were suspected of being carcinogens, including benzene, dioxin, dichloreth-

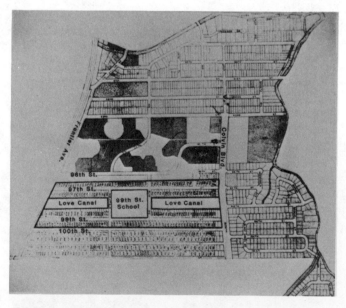

FIGURE 14.1
Map of the Love Canal census tract showing the canal dump site. (From Janerick et al., 1981.)

ylene, and chloroform. Although experts readily admitted that little was known about the impact of these chemicals and others at the site, there was grave concern for the people living in the area. It has been estimated that it may cost $100 million to clean up the Love Canal site and relocate residents (over 200 families have been evacuated). In addition, approximately $3 million per year may be needed simply to monitor the area in the future. What went wrong in the story of Love Canal? How can we avoid such disasters in the future? The real tragedy of Love Canal is that it is probably not an isolated incident. That is, there are probably many hidden "Love Canals" across the country, "time bombs" waiting to explode [1,2].

The Love Canal story illustrates many of the patterns we face with environmental pollution and environmental toxicology. Our society is producing chemicals at a rate faster than we can determine their environmental and health effects and faster than we can find proper methods of disposal; poor disposal methods and lack of knowledge of toxic effects produce environmental time bombs. In this chapter, we will first consider certain kinds of pollutants and then consider certain shared, general features.

THE MAJOR FORMS OF POLLUTION

A **pollutant** is most simply defined as any factor that has a harmful effect on living organisms or their environment. In addition to disease-carrying organisms, there are seven major categories of environmental pollutants: acids, toxic chemical elements (particularly heavy metals), radioisotopes, organic compounds, heat, particulates, and noise.

To understand the problems posed by these pollutants, we must understand how they are produced; how they affect individuals, populations, ecosystems, the biosphere, and human health; how they are transported (and what changes they undergo as they are transported through the environment); and how we may dispose of them.

Each pollutant has its own origins, pathways, and effects, but all are spread throughout the biosphere by air and water. We discussed the pathways of pollutants through the atmosphere and the water cycle in Chapters 8 and 9, respectively, so in this chapter we will concentrate on the other aspects of pollution: conversion of pollutants from one form to another, toxic effects, and disposal.

Almost every part of the human body is affected by one pollutant or another (Fig. 14.2a). For example, lead and mercury affect the brain, arsenic the skin, carbon monoxide the lungs, and chlorinated hydrocarbons concentrate in the fat. Similarly, the effects of pollutants on

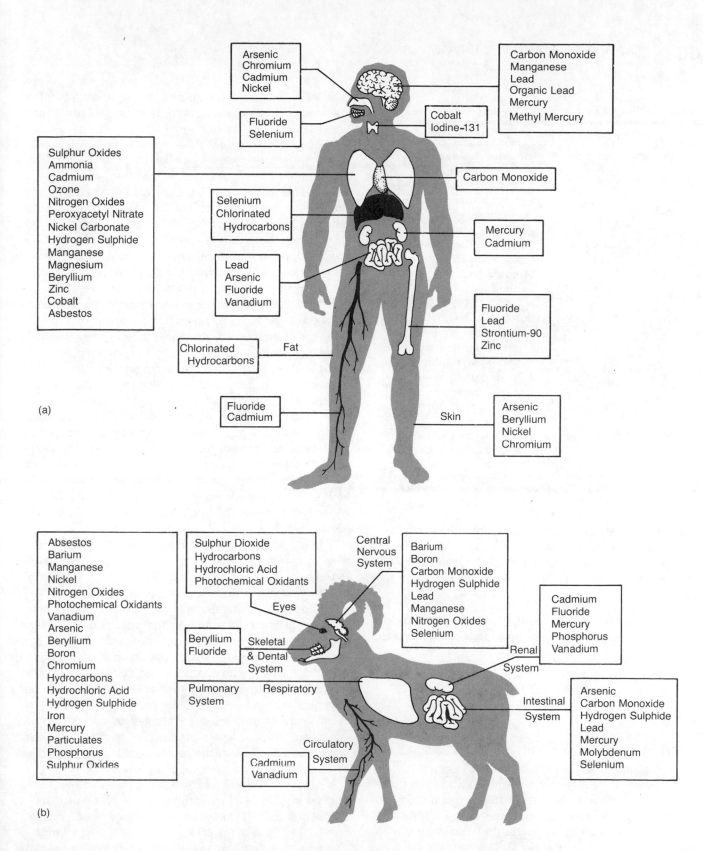

FIGURE 14.2
The site of effects of some major pollutants in human beings (a). Known sites of effects of some major pollutants in wildlife (b). [Part (a) from Waldbott, G. L., *Health effects of environmental pollutants,* 2nd ed. Copyright © 1978 by C. V. Mosby, St. Louis; part (b) from Newman, 1980.]

TABLE 14.1
Effects of pollutants on wildlife.

Effect on Population	Pollutants
Changes in abundance	Arsenic, asbestos, cadmium, fluoride, hydrogen sulphide, nitrogen oxides, particulates, sulphur oxides, vanadium
Changes in distribution	Fluoride, particulates, sulphur oxides
Changes in birth rates	Arsenic, lead, photochemicals, oxidants
Changes in death rates	Arsenic, asbestos, beryllium, boron, cadmium, fluoride, hydrogen sulphide, lead, particulates, photochemicals, oxidants, selenium, sulphur oxides
Changes in growth rate	Borium, fluoride, hydrochloric acid, lead, nitrogen oxides, sulphur oxides

SOURCE: From Newman, 1980.

wildlife have been documented for many organs and aspects of the life cycle (Table 14.1 and Fig. 14.2b).

If we decide that a certain substance is toxic and must be eliminated or reduced, we have the following choices:

1 Stop its production (either find a substitute or stop using the process and products that led to the toxin's production).

2 Learn to transform it or degrade it to a harmless material at a faster rate.

3 Find a safe repository for it.

ACID RAIN

In recent years, fish have disappeared from lakes in Sweden where they were once abundant and were used for food and recreation. Records of 15 years or more from Scandinavian lakes show an increase in the acidity accompanied by a decrease in fish (Fig. 14.3) and the loss of many other forms of life (Fig. 14.4). The death of the fish has been traced to acid rain, the result of industrial processes far away in other countries, particularly western Germany and Great Britain. These industries spew sulphur oxides into the atmosphere where they combine with water to form sulphate and sulphuric acid. Weather systems carry the pollutants to Sweden, where the acid water is deposited as rain.

Acid rain affects a lake ecosystem by dissolving chemical elements necessary for life, such as calcium, and keeping them in solution so that they leave the lakes with the water outflow. Elements that once cycled within the lake are thus lost from the lake. Without these nutrients, the lakes' algae do not grow, and the small animals that feed on the algae have little to eat. The fish, who are typically predators on the smaller invertebrate animals, also lack food. The acid water has other adverse effects on living organisms and their reproduction. For example, crayfish produce fewer eggs in the acid water, and those eggs produced often have malformed larvae.

Rain is naturally slightly acid (normal rain pH is 5.5) because dissolved carbon dioxide from the air produces a weak acid. When the bedrock or substrate under a lake is high in metallic elements like calcium and magnesium, the natural acidity can be neutralized. Lakes with high concentrations of such elements, called *hard-water lakes,* are said to have a high *buffering capacity.* Lakes on sand, sandstone, or igneous rocks usually lack enough of the metallic elements to neutralize the acid of the rain and thus are naturally acidic. However, their acidity is weak, with a pH of approximately 5.6 to 5.7. (A neutral body of water has a pH of 7.) Such lakes are called *soft-water lakes.*

In practice, a simple and fast index of a lake's hardness or buffering capacity is its electrical conductivity. Pure water is a poor conductor of electricity; water high in dissolved elements is a good conductor [3].

Bogs with abundant peat are another kind of naturally acidic body of water. As the peat and other dead vegetation decay, organic acids are added to the bog waters and color them brown. Tannic acid, the brownish acid that gives tea its characteristic color and is used in the tanning of hides, is one such acid. A transparent but brownish water is common in bogs and streams that drain such natural areas in Minnesota and Michigan. Acid bogs have a lower pH (higher acidity) than soft-water lakes because of the organic acids. The pH values in these bogs range from 5.4 to as low as 3.4.

Moderate acidity itself does not mean an end to life in a pond or lake. Some organisms, such as peat moss (*Sphagnum*), are favored by such acidic conditions [3]. Some lakes with a pH of 3.4 to 4.5 have many kinds of algae, some crustaceans, and insects. However, not only are fish less abundant, there are fewer species of them. Sunfish, bullhead, and pirate perch are among the fish that are found in some acid lakes. The best game fish—trout, for example—are not found in these waters.

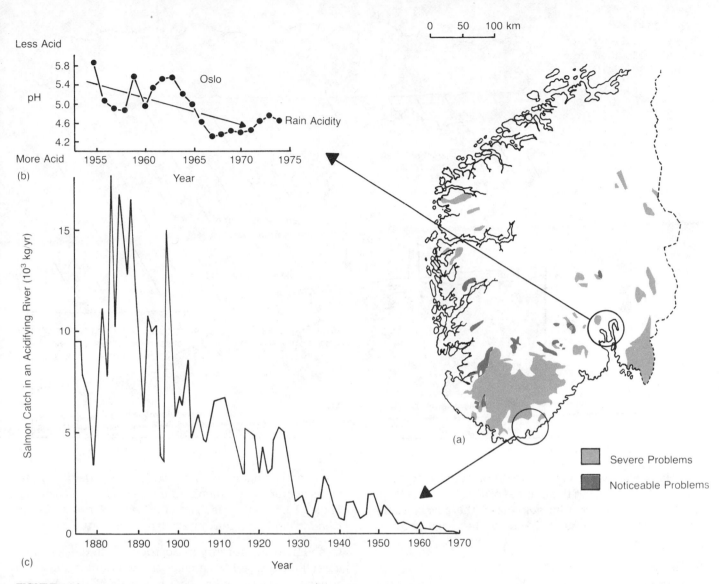

FIGURE 14.3

In Norway, many lakes in the south have severe problems with acid rain (a). The rain has become more acidic during the last 20 years, as measured at Oslo. Measurements at five other sites in southern Norway show the same trend (b). The catch of fish, as illustrated by the catch of salmon in the Tovdalselva River of southern Norway, has decreased dramatically (c). [Parts (a) and (c) from Wright et al., 1976; data for (a) from Muniz et al., 1976, and data for (c) from Snekvik, 1970. Part (b) from Odén, 1976.]

If the chemical elements necessary for life are available, life can continue in acid waters as a restricted, simplified community. Those green plants that survive in acid waters tend to be particularly efficient at obtaining chemical elements. Bogs and lakes that have long been acidic have species of plants that are adapted to these conditions. These plants produce compounds called *humates,* which help to retain and make available metals dissolved in the waters. Waters that are rapidly made acid may lose much of their chemical fertility before such plants can become established and thus will suffer a catastrophic decline in life [3]. Thus the rate of change of acidity, as well as the acidity itself, is important to the effects of acid rain on lakes. A lake that became acid over a period of centuries would maintain more forms of life than one that became acid in one year.

FIGURE 14.4
The ecological effects of acid rain. (Modified from Odén, 1976.)

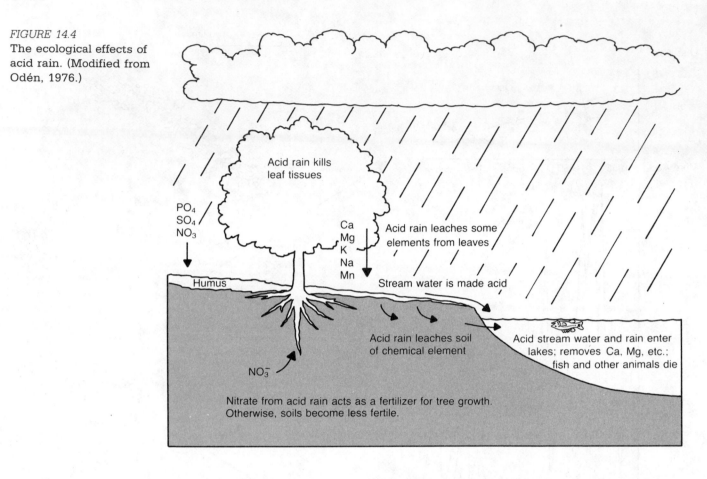

Acid rain kills leaf tissues

PO_4
SO_4
NO_3

Ca
Mg
K
Na
Mn

Acid rain leaches some elements from leaves

Humus

Stream water is made acid

Acid rain leaches soil of chemical element

Acid stream water and rain enter lakes; removes Ca, Mg, etc.; fish and other animals die

NO_3^-

Nitrate from acid rain acts as a fertilizer for tree growth. Otherwise, soils become less fertile.

Human activities continue to aggravate the acid rain problem by increasing the number of acid lakes, the kinds of substrates that contain acid lakes, the acidity of the acid lakes, and the rate of change of the lakes' acidity.

TOXIC HEAVY ELEMENTS

Some chemical elements are directly toxic to organisms. The major toxic elements tend to be heavy and to have a high atomic mass. Among the major heavy elements that pose hazards are mercury, lead, cadmium, nickel, gold, platinum, silver, bismuth, arsenic, selenium, vanadium, chromium, and thallium. Each of these has uses in our modern industrial society, and each is also a by-product of the mining, refining, and use of other elements. Heavy elements often have direct physiological toxic effects. Some are stored or incorporated into living tissue, sometimes permanently. The content of heavy metals in our bodies is referred to as the **body burden**. Figure 14.5 shows the average human body burden of some toxic heavy elements.

Mercury, thallium, and lead are the most toxic to humans. They have long been mined and used, and their toxic properties are well known. Mercury, for example, is the "Mad Hatter" element. At one time, mercury was used in making hats; because mercury damages the brain, hatters were known to act peculiarly in Victorian England. Thus, the Mad Hatter in Lewis Carroll's *Alice in Wonderland* had a real antecedent in history.

Case History: Mercury in Seafood   In the Japanese coastal town called Minamata, on the island of Kyushu, a strange illness began to occur in the 1950s. The ailment, which particularly afflicted families of fishermen, had subtle first symptoms: fatigue, irritability, headaches, numbness in arms and legs, and difficulty in swallowing. More severe symptoms involved the sensory organs; vision was blurred and the visual fields were restricted. Afflicted people became hard of hearing and lost muscular coordination. Some complained of a metallic taste in their mouths; their gums became inflamed and they suffered from diarrhea. Eventually, 43 people died and 111 were severely disabled; in addition, 19 babies were born with congenital defects. Those affected lived in a small area, and much of the protein in their diet came from fish from the Minamata Bay.

A plastics factory on the bay used mercury in an inorganic form in its production processes. The mercury was released in water effluent which flowed into the bay. What was not realized was that the industrial processes

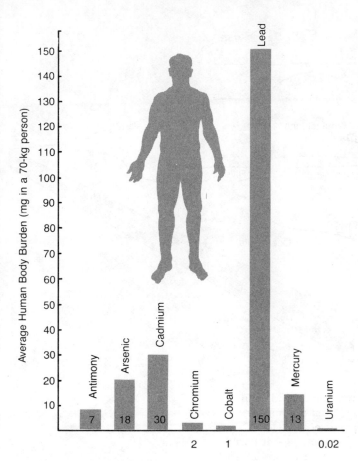

FIGURE 14.5

Average human body burden of major toxic heavy metals. (Data based on Lippmann and Schlesinger, 1979.)

symptoms appear. In one fatal case in Niigata, the individual had 1300 parts per billion. It is important to note that in all cases toxic effects occurred at very low concentrations of mercury.

The mercury episodes at Minamata and Niigata illustrate five major factors that must be considered in evaluating and treating environmental pollutants. First, *individuals vary in their response to exposure to the same dose or amount of a pollutant.* Not everyone in Minamata responded in the same way; there was variation even among those most heavily exposed. Because we cannot predict exactly how any single individual will respond, we need to find a way to state an average expected response of individuals in a population. Second, *some pollutants may have a threshold*—that is, a level below which the effects are not observable and above which effects become apparent. Symptoms appeared in individuals with 500 parts per billion of mercury; no measurable symptoms appeared in individuals with significantly lower concentrations in their bodies. Third, *some effects are reversible.* Some people recovered when the mercury-filled seafood was eliminated from their diet. Fourth, *the chemical form of the pollutant has a great effect on its toxicity.* Fifth, *the pollutant and its activity are changed markedly by ecological and biological processes.* In the case of mercury, its chemical form and concentration changed as the mercury moved along the food webs.

Lead · Lead is mined, refined, processed, and used in a variety of applications (Fig. 14.6). It affects the brain and other organs. Lead-based paints became well known as a cause of brain damage when small children in cities, particularly in older areas, ate lead-based paint that had peeled from the walls. Once an important constituent of household paints, it has been largely replaced since World War II by plastic- or rubber-based paints.

Lead is an important constituent of automobile batteries and other industrial products. When lead is added to gasolines, they burn more evenly in automobile engines. The lead in gasoline is emitted into the environment in the exhaust. In this way, lead has been spread widely around the world and has reached high levels in soils and waters along roadways.

Once released, lead can be transported through the air as particulates to be taken up by plants through the soil or deposited directly on plant leaves. Thus it enters terrestrial food chains.

When lead is carried by streams and rivers, deposited in quiet waters, or transported to the ocean or lakes, it is taken up by aquatic organisms and thus enters aquatic food chains.

The concentration of lead in Greenland glaciers (Fig. 14.7) shows that lead was essentially at zero concentra-

converted inorganic mercury into a much more toxic organic form. Inorganic mercury does not pass through cell membranes readily; it damages the intestines, liver, and kidneys. Organic methyl mercury, however, readily passes through cell membranes, is transported by the red blood cells throughout the body, and enters and damages brain cells [4]. Moreover, the organic form is rapidly incorporated into food chains. For example, fish absorb methyl mercury from water 100 times faster than they absorb in organic mercury. Once absorbed, methyl mercury is retained 2 to 5 times longer than inorganic mercury is retained [4]. The effects of the mercury are delayed from 3 weeks to 2 months. If mercury intake ceases, symptoms may gradually disappear [5].

Another similar incident occurred in 1965 in Niigata, Japan, when families ate fish containing 5 to 20 parts per billion of mercury up to 3 times a day. The normal level of mercury in humans is 5 parts per billion, and 50 parts per billion is the maximum safe level. At 500 parts per billion,

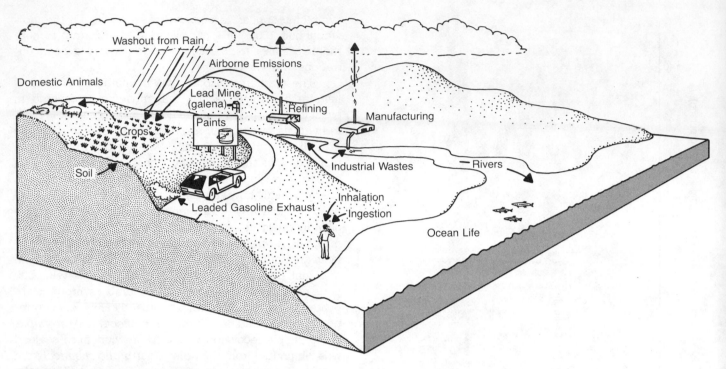

FIGURE 14.6

The pathways of lead in the environment. Lead is released from mines and from refining and industrial processes. There are airborne emissions from automobile exhaust, emissions to rivers from industrial wastes, and transport of lead-containing materials, such as paint, putty, toys, pottery, and gunshot. The airborne particles fall on the soil and edible plants. They are inhaled, eaten, or incorporated into the tissue of organisms, which are eaten by others. Lead affects the central nervous system, can cause anemia and malfunction of the kidneys, and can affect reproduction. (Modified from Lippmann and Schlesinger, 1979.)

tion in the ice around 800 A.D. and reached measurable levels with the beginning of the Industrial Revolution in Europe. The lead content of the glacial ice increased steadily from 1750 until the mid-twentieth century, when the rate of accumulation by the glaciers increased rapidly. This increase reflects the rapid growth in the use of the internal combustion engine in automobiles and trucks and the use of lead additives in gasoline. Lead reaches the Greenland glaciers as airborne particulates and by sea water. The accumulation of lead in these glaciers demonstrates that our use of heavy metals in this century has reached a point where the entire biosphere is affected.

Heavy metals may undergo **biological concentration**, also called **biomagnification**. Cadmium, for example, enters the environment in part in the ash from burned coal. Cadmium, a trace element in the coal, exists in a very low concentration of 0.05 part per million. The ash is spread widely from stacks and chimneys and falls on plants, where the cadmium is incorporated into the plant tissue and concentrated 3 to 7 times (Table 14.2). As the cadmium moves up the trophic levels through the food

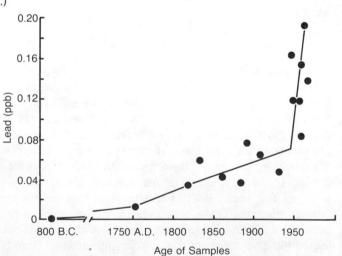

FIGURE 14.7

Concentrations of lead in Greenland glaciers. In the Middle Ages, essentially no lead fell on the Greenland glaciers. With the beginning of the Industrial Revolution, the amount of lead reaching Greenland began to increase. The steep rise after 1940 reflects the increasing use of automobiles and lead additives to gasoline. (From Murozumi et al., 1969.)

TABLE 14.2
Biological magnification of cadmium.

Source	Cadmium Content (ppm)	Concentration Factor[a]
Coal ash substrate	0.05 ± 0.01 SE (water extractable)	—
Coal ash substrate	0.12 ± 0.0002 SE (dry weight)	—
Producers (plants)	0.35 ± 0.03 SE	2.9
Herbivores	0.97 ± 0.11 SE	2.8
Carnivores	2.82 ± 0.34 SE	2.9

SOURCE: From Skinner et al., 1976.
[a]The concentration factor is the ratio of cadmium in one trophic level to that of the next trophic level. Each trophic level concentrates the cadmium approximately by a factor of 3. (Note that plants are compared with the dry weight coal ash substrate.)
SE = Standard Error

chains, each trophic level concentrates it approximately 3 times over what it was in the next lower trophic level. Herbivores have approximately 3 times the concentration of green plants, and carnivores approximately 3 times the concentration of herbivores.

RADIATION AND RADIOISOTOPES

There are three major kinds of radiation: alpha, beta, and gamma rays (Table 14.3). **Alpha rays** have the most mass and generally travel the shortest distance before being stopped by some other matter. They are stopped by very

TABLE 14.3
Radiation and radiation units.

There are three kinds of radiation:
Alpha rays, which are the nuclei of helium atoms.
Beta rays, which are electrons.
Gamma rays, which are electromagnetic radiation, like light, but with a much shorter wavelength.

A *radioisotope* is a form of a chemical element that spontaneously undergoes radioactive decay—that is, it changes from one isotope to another.

A measure of radiation at the source is the *rate of radioactive decay.* This is the number of atoms that decay in one second.

A *curie* is a standard unit for decay. One *curie* is 37 billion atoms decaying in one second.

The effect of radiation is the *dose.*

The *dose* in terms of energy received per unit of material is called a *rad.* A rad is 100 ergs (energy absorbed) per gram.

The *dose* in terms of health effects varies with the form and energy of the radiations and the nature of the absorbing material. The health effect is called the *dose equivalent* and is measured in units called *rems,* which is an empirically derived unit.

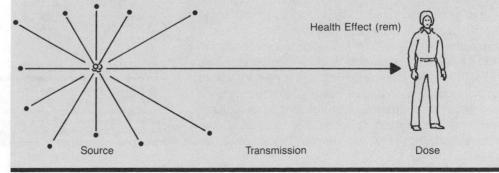

Disintegrations per Second (curies)   Dose (energy absorbed per gram, or rad)

Health Effect (rem)

Source   Transmission   Dose

thin material. **Gamma rays** are high-energy photons, travel the longest average distance, and penetrate thick shielding. **Beta radiation** is intermediate in mass and is stopped by moderate shielding.

A **radioisotope** is a form of a chemical element that spontaneously undergoes radioactive decay. It changes from one isotope to another and during the process emits one or more forms of radiation. Each isotope has its own characteristic emissions; some isotopes emit only one kind of radiation, and some emit a mixture.

Some radioisotopes, particularly those of very heavy elements, undergo a series of radioactive decay steps, finally reaching a stable, nonradioactive isotope. For example, uranium decays through a series of steps and ends up as a stable isotope of lead. There are three naturally occurring chains or series: the uranium, actinium, and thorium chains.

The three kinds of radiation have different toxicities. In terms of human health, or the health of any organism, alpha radiation is most dangerous when ingested. All of its radiation is stopped within a short distance by the body's tissues, and therefore all the damaging radiation is absorbed by the body. On the other hand, when an alpha-emitting isotope is stored in a container outside the body, it is relatively harmless. A gamma-ray emitter can be dangerous outside or inside the body, but when ingested, some of the radiation emitted passes outside the body. Beta radiation is intermediate in its effects, although most beta radiation will be absorbed by the body when a beta-emitter is ingested.

Tritium, the radioactive isotope of hydrogen that occurs in heavy water, is an alpha-emitter, and thus is dangerous to ingest but may be stored safely in containers. Carbon-14, the radioactive isotope of carbon that occurs naturally in the form of carbon dioxide, is a beta-emitter; it is dangerous to ingest but may be stored relatively safely.

Uranium emits gamma rays, which may be dangerous even from a long distance away. Uranium can be dangerous inside or outside the body and must be stored under heavy shielding.

Natural Radiation   Low levels of radiation occur naturally from several sources. Natural radiation can come from earth materials, originate in outer space, or result from the action of radiation on materials in the biosphere.

Some radioactive elements occur as natural minerals in the Earth's crust. Uranium, which is mined for use in atomic weapons and in nuclear power plants, is one such element. Minerals containing uranium and other radioactive elements emit radiation into the environment, but characteristically at very low levels. Potassium-40 is among the more common radioisotopes, and potassium is one of the elements required for all living things. Potassium-40 occurs in soils and rocks, particularly granite [6]. The distribution of radioisotope-bearing minerals, like the distribution of all mineral resources, is not uniform, and the background radiation from such sources varies considerably around the Earth. For example, in parts of Brazil there occur some of the highest levels of background radiation known anywhere in the world. The average human dose of radiation from natural sources is shown in Table 14.4.

The sun, other stars, and many astronomical phenomena produce very high-energy radiation, some of which enters the Earth's atmosphere and interacts with the materials in the atmosphere. One of the more important results is the production of radioactive carbon-14. Because naturally occurring radioactive carbon can be used to date objects up to 40,000 years old, it is an important tool in anthropology, geology, biology, and other sciences.

An important characteristic of a radioisotope is its **half-life**, which is the time required for one half of a given amount of the isotope to decay to another form. Every radioisotope has a unique, characteristic half-life. Radioactive iodine, for example, has a short half-life of 8 days. Radioactive carbon-14 has an intermediate half-life of 5570 years. Uranium-235 has a half-life of 700 million years.

TABLE 14.4

Average annual whole-body dose rates for U.S. population.

Source	Whole-body Exposure (mrems/yr)
Natural radiation	
Cosmic radiation	44
Radionuclides in the body	18
External gamma radiation	40
Total	102
Human-produced radiation	
Fallout	4
Occupational exposure	0.8
Nuclear power (1970)[a]	0.003
Total	4.803
Medical and dental	73

SOURCE: National Research Council, 1972a.
NOTE: The Radiation Protection Guide (Federal Radiation Council) recommends a maximum exposure for the general population of 170 mrems/yr (exclusive of medical radiation, for which no limit has been stated).
[a]By the year 2000, nuclear power is expected to account for 1 mrem/yr. It is hoped that the contribution from fallout will decline.

A radioisotope's degree of dangerousness, either to the environment or to human health, depends on several factors: the kind of energy of the radiation emitted, the half-life, and the ordinary chemical activity of the isotope. Other things being equal, low-energy, short half-lived isotopes are less dangerous than high-energy, long half-lived isotopes. However, isotopes that readily enter a gaseous phase and are released into the atmosphere where they can be inhaled may be extremely dangerous even with low energies and short half-lives. Tritium and carbon-14 are examples of such dangerous isotopes.

Radiation Doses   How much radiation is dangerous? Because there are three kinds of radiation, each with a range of energies, and because there is considerable chance involved in what is struck by emitted radiation, it is difficult to set down a universal and easily employed measure of a dose of radiation.

For these reasons, the damage per unit of radiation emitted must be determined by experiment. It is common to use as a measure the **LD-50,** which is the amount of radiation required to kill one half of an experimental population.

An important characteristic of any radioisotope is the **maximum permissible dose,** or the maximum amount of radiation an individual can be exposed to without any expectation of suffering appreciable injury within his lifetime. A 1972 National Academy of Sciences panel concluded that for the health of the population, exposure should be no more than 170 millirems per year from human-produced radiation (except medical). A millirem, or mrem, is one thousandth of a rem (see Table 14.3) [6].

From natural sources, a person receives approximately 100 millirems of radiation per year: 40 from cosmic radiation; 40 from the Earth's biosphere (soil, rocks, and atmosphere); and 20 internal (from the ingestion and incorporation of naturally occurring radioisotopes) [7].

Environmental Effects of Radioisotopes   Radioisotopes affect the environment in two ways: by emitting radiation that affects other materials, and by entering the normal pathways of mineral cycling and ecological food chains. The explosion of an atomic weapon does damage in both ways. At the time of the explosion, intense radiation of many kinds and energies is sent out which kills organisms directly. The explosion generates large amounts of radioactive isotopes, which are dispersed into the environment. Nuclear bombs exploded in the atmosphere produce a huge cloud that sends radioisotopes directly into the stratosphere, where the radioactive particles may be widely dispersed by winds. The atomic "fallout," the deposit of these radioactive materials around the world, was a major environmental problem in the 1950s and 1960s when the United States, Russia, China, France, and Great Britain were testing and exploding nuclear weapons in the atmosphere.

The pathways of some of these isotopes illustrate the second way that radioactive materials can be dangerous in the environment. One of the radioisotopes emitted and sent into the stratosphere by atomic explosions was cesium-137. This radioisotope was deposited in relatively small concentrations but widely dispersed in the Arctic region of North America. It fell on reindeer moss, a lichen that is a primary winter food of the caribou. A strong seasonal trend was found in the levels of cesium-137 in caribou; the level was highest in the winter when reindeer moss was the principal food, and was lowest in the summer. Eskimos who obtained a high percentage of their protein from caribou ingested the radioisotope from eating its meat, and their bodies concentrated the cesium. The more a group depended on caribou as its primary source of food, the higher was the level of the isotope in their bodies.

The cesium was moved long distances by biospheric phenomena. After entering specific ecosystems through the vegetation, it underwent biomagnification, or ecological food-chain concentration. That is, at each trophic level the concentration of the toxic material, relative to other materials in the bodies of organisms, increased. Cesium-137 concentrations approximately doubled with each trophic level. This biomagnification was unknown until radioisotopes and toxic organic compounds were found to be occurring at higher and higher concentrations at higher and higher trophic levels. Food-chain or trophic-level concentration is one of the major ecological factors in environmental toxicology.

The actual body burden of cesium varied within the Eskimo population. Adult males between the ages of 20 and 50 had the highest levels, apparently because their diet contained the most caribou meat. Concentration varied seasonally and increased over the years of intensive atmospheric bomb testing (Fig. 14.8) [8].

Another important factor in the toxicity of cesium-137 from fallout is the length of time the isotope remains in the body. This is measured by the **biological half-life,** the time for one half the concentration to be lost. In the caribou, the average biological half-life was 1.5 days. The biological half-life in the Eskimos, however, was approximately 65 days.

It is possible to predict the environmental pathways that radioisotopes will follow because we know the normal pathways of the nonradioactive isotopes with the same chemical characteristics. Our knowledge of biomagnification and of large-scale air and oceanic move-

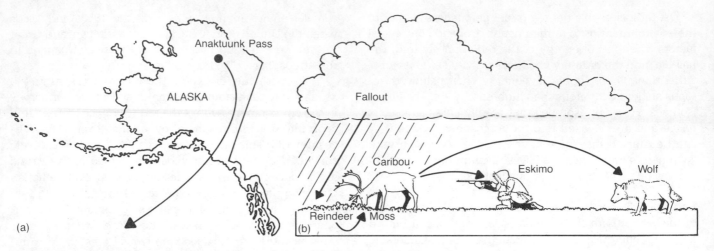

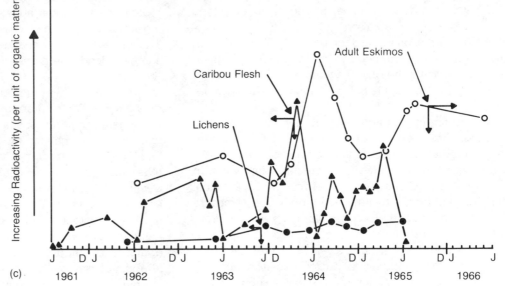

FIGURE 14.8
Cesium-137, released into the atmosphere by atomic bomb tests, was part of the fallout deposited onto the soil and plants. The cesium fell on lichens, which were eaten by caribou. The caribou were in turn eaten by the Eskimo (b). Measurements of cesium in the lichens, caribou, and Eskimo in the Anaktuvuk Pass of Alaska (a) show that the cesium was concentrated by the food chain. Peaks in concentrations occurred first in the lichens, then in the caribou, and last in the Eskimo (c). [Part (c) reprinted with permission from W. G. Hanson, Cesium-137 in Alaskan lichens, caribou, and Eskimos, *Health Physics* 13: 383–89. Copyright © 1967, Pergamon Press, Ltd.]

ments that transport radioisotopes throughout the biosphere will also help us to understand the effects of radioisotopes.

A massive thermonuclear war would have large-scale effects on the biosphere, on the Earth's biomes, and on individual ecosystems, communities, populations, and individuals which are beyond our imagination and knowledge to grasp. However, several general results can be expected.

First, many nuclear explosions in the atmosphere would disrupt the normal pattern of winds and the flux of air between the lower and upper atmosphere. While the effects of such disruptions are not well understood, it is possible that there would be a serious alteration, at least in the short run, of weather.

Second, the immediate destruction of much life by the emission of radiation and the later mortality due to fallout would have widespread effects. In general, based on studies such as those at Brookhaven and Oak Ridge National Laboratories and analyses of experimental nuclear explosions, we can say that there would be a great simplification of ecosystems and that the survivors would tend to be species already adapted to harsh, highly variable, early successional environments. Such organisms would have small body size, high reproductive potential, and short lifetime. We can also expect that such massive destruction would favor generalist species, such as rats, cockroaches, and crows, over specialist species, whose

NUCLEAR POWER PLANTS

One of the most dramatic events in the history of radiation pollution occurred on March 28, 1979, at Three Mile Island nuclear power plant near Harrisburg, Pennsylvania. Malfunctions in the nuclear plant resulted in a release of radioisotopes into the environment as well as intense radiation release within one of the nuclear facilities. The actual release into the environment was at a low level per person exposed. Exposure from the plume emitted into the atmosphere has been estimated at 100 millirems, which is low in terms of the amount of radiation required to cause acute toxic effects. However, radiation levels were much higher near the site. On the third day after the accident, 1200 millirems per hour were measured at ground level near the site.

The Three Mile Island incident made clear that there are many problems with the way that our society has dealt with nuclear power. Historically, nuclear power has been relatively safe, and the state of Pennsylvania was somewhat unprepared to deal with the accident. For example, there was no state bureau for radiation health, and the state department of health did not have a single book on radiation medicine (the medical library had been dismantled two years before for budgetary reasons). One of the major impacts of the incident was fear, yet there was no state office of mental health nor any authority to allow anyone from the department of health to sit in on briefing sessions.

Because the long-term chronic effects of exposure to low levels of radiation are not well understood, the effects of the Three Mile Island exposure—although apparently small—are difficult to estimate. This case illustrates that our society needs to improve its ability to handle the crises that could arise from the sudden releases of pollutants from our modern technology. It also shows our lack of preparedness and apparent readiness to treat a nuclear power plant as an acceptable risk [9].

ORGANIC COMPOUNDS

Organic compounds are compounds of carbon produced either by living organisms or artificially. It is difficult to generalize about the environmental and health effects of artificially produced organic compounds because there are so many of them, because they have so many uses, and because they have a potential for so many different kinds of effects. Artificial organic compounds are used primarily in industrial processes, pest control, pharmaceuticals, food additives, and other consumer products. A computer registry of chemicals maintained by the American Chemical Society has more than 4 million entries, and there are about 6000 additions per week. About

FIGURE 14.9
The production of artificial organic compounds has increased rapidly during the twentieth century. Current production exceeds 100 billion kilograms per year. (From White, 1980.)

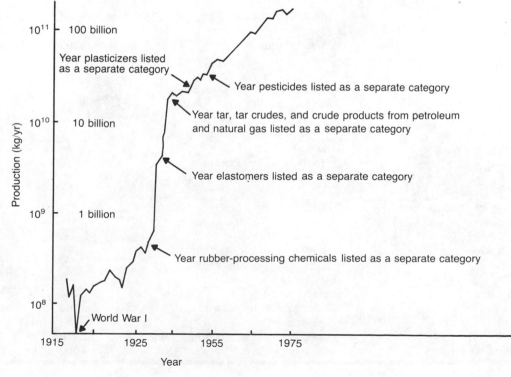

33,000 chemicals are in common use, and these do not include pesticides, pharmaceuticals, or food additives. According to the U.S. Food and Drug Administration, there are about 4000 active and 2000 inert ingredients in drugs. There are 2500 nutritional additives and 3000 additives to promote the life of products [10]. The production of artificial organic compounds has grown rapidly in the twentieth century (Fig. 14.9 and Table 14.5) [11].

Organic compounds can have physiological, genetic, or ecological effects. Because artificial organic compounds are new and organisms have had little time to adjust to them, they are more likely to be toxic to living organisms than natural organic compounds. Some organic compounds are potentially more hazardous than others. For example, rapidly degrading substances are less likely to cause problems as chronic hazards than slowly degrading ones. Fat-soluble compounds are likely to undergo biomagnification.

The likely problems posed by an organic compound depend on the pathways it will follow through the environment. We can determine the likely path for a chemical by testing its solubility in water, absorption by natural solids, leaching rates, volatility, and oil or fat solubility [10]. Artificial organic compounds are spread widely throughout the biosphere and can be found in many species. In fact, such compounds are found in nontarget marine animals far from the sources of production of the synthetic chemicals (Table 14.6).

Oil Spills Oil spills, one of the most spectacular forms of pollution, have attracted great attention. However, the primary effects of oil pollution appear to be short-term. For example, when an oil well blew out in the Santa Barbara Channel, California, in 1969, the beaches of the city were blackened with 4500 tons of crude oil, killing approximately 3600 birds. Public concern culminated in the establishment of an organization whose sole aim was to combat the oil pollution.

Major oil spills in the last several decades have resulted from ship accidents, such as the collision between the *Fort Mercer* and the *Pendleton* near Chaltham, Massachusetts, in 1952, as well as from blowouts like the

TABLE 14.5
Annual production of synthetic organic compounds.

General Type	Example	Production (millions of kg)	Use
Organohalogen	Ethylene dibromide	140	Gasoline additive
	Carbon tetrachloride	450	Specialty solvent
			Grain fumigant
	Hexachlorobutadiene	4	Solvent
			Heat transfer liquid
	Vinyl chloride	2300	Chemical intermediate
Organophosphorous	Tri(2-chloroethyl) phosphate	13	Flame retardant for plastics
			Synthetic lubricants
	Acyclic phosphorodithioate Lube oil additives [zinc di(butylhexyl)-phosphorocithioate]	31	Lube oil additive
Organosilicones	Silicone fluids (polydimethylsiloxane)	34	Release agents
			Antifoaming agents
			Polish and cosmetic ingredients
Oxygen-containing	Dioxane	6	Specialty solvent
			Inhibitor in chlorinated solvents
Nitrogen-containing pulp	(Ethylenedinitrilo) tetraacetic acid, tetrasodium salt	29	Agriculture, detergents
			Pulp and paper processing
	Ethyleneimine	2	Chemical intermediate
Plastics	Polyethylene	3400	Plastic
Detergents	Dodecylbenzenesulfonic acid, sodium salt	163	Detergent
	N,N-Dimethyldodecylamine oxide	16	Detergent
Hydrocarbons	Benzene	4000	Chemical intermediate

SOURCE: Modified from Freed et al., 1977.

TABLE 14.6
Some synthetic organic compounds in marine animals.

Species	Source	Chemical (mg/kg fresh tissue)		
		Dieldrin	DDT	PCB
Shellfish				
Mussel	Mediterranean	0.001–0.030	0.01–0.65	0.08–0.18
	Baltic	—	0.005–0.07	0.01–0.33
Fish (mg/kg muscle)				
Herring	North Sea	<0.001–0.034	0.035–0.17	<0.001–0.48
Sardine	Mediterranean	<0.001–0.14	0.001–0.63	0.03–6.9
Cod	North Sea	<0.001–0.023	<0.003–0.052	<0.001–0.099
Cod	Newfoundland	—	0.011	0.038
(liver)		—	2.7	22.
Haddock	Newfoundland	—	0.002	0.030
Shark (liver)	North Atlantic	—	4.8	5.8
Mammals (mg/kg blubber)				
Grey seal	Scotland	0.46–1.7	8.5–36.3	12–88
Common seal	Netherlands	<0.02–0.09	9.5	385–2530
Porpoise	Scotland	3.1–4.5	16.8–37.4	31–68
Sea lion	California	—	41–2678	
Various cetaceae	Atlantic Ocean	<0.1–3.0	1.1–268	0.7–114
Birds				
White-tailed eagle (muscle)	Baltic Sea	—	290–400	150–240
Cormorant (fat tissue)	Arctic Ocean	—	6.5–15	14–47
Various (liver)	North Atlantic	—	—	0.02–311

SOURCE: Compiled from Whittle et al., 1977, in *Chemical contamination in the human environment* by Morton Lippman and Richard Schlesinger. Copyright © 1979 by Oxford University Press, Inc. Reprinted by permission.
NOTE: Some areas are considered particularly heavily contaminated with DDT and/or PCB. These are the Baltic, W. Mediterranean, southern area of the North Sea, west coast of the United Kingdom, and coast of California.

Santa Barbara incident. Although ship spills are highly publicized, they are by no means the only source of pollution of waters by oil. Oil released by automobiles on roadways is transported by surface water runoff to rivers, which contribute the greatest amount of oil to the ocean of all land sources. Municipal wastes and coastal refineries also contribute a significant amount. Of the marine sources, tanker spillage contributes the most oil, but offshore seeps, the clearing of ships' bilges, and dry-docking procedures also contribute large amounts.

Oil's acute effects on the ocean are perhaps more well known than its chronic effects. Oil is a particular problem for commercially valuable ocean life because most major oil spills have occurred near shore where there is an abundance of shellfish and fin fish. Oil is a mixture of hydrocarbons, some of which are light and float and others which are heavy and sink. Generally, chronic effects are most pronounced from the heavier compounds. For example, gasoline evaporates quickly and is less of a chronic problem than crude oil.

Oil is directly toxic to some organisms when it is ingested, and it reduces the insulating properties of fur and feathers. When birds are covered with oil, their feathers no longer provide insulation against the cold water and air and the birds cannot fly. Because they cannot feed, they die of cold shock and starvation. Similarly, sea otters lose the insulating properties of their fur when they are covered with oil and can die relatively quickly when the insulating properties of the fur are reduced.

A layer of oil floating on the ocean surface can interfere with the exchange of oxygen and carbon dioxide and reduce the rate of photosynthesis of marine plankton and the respiration of marine animals. Fuel oil added to sea water in very low concentration (199 parts per billion) depresses photosynthesis.

The death of birds from oil spills has attracted much attention. In the Torrey Canyon incident in 1967, an estimated 40,000 to 100,000 birds died. The *Fort Mercer* and *Pendleton* collision in 1952 reduced the wintering population of eider ducks from 500,000 to 150,000 [12]. Some believe that the jackass penguin, which lives in South Africa, is endangered because floating oil from tankers rounding the Cape of Good Hope is killing hundreds of thousands of these birds each year.

Surprisingly, in spite of the great amount of publicity that surrounds oil spills, little is known about their long-term effects. Immediately after the spill, there is a great effort to save the lives of oil-soaked birds and to clean up the oil on beaches. In fact, special techniques have been developed solely for these purposes. However, once the initial problem has been dealt with, there has been little follow-up study. Only a few oil spills have been subject to long-term studies. The spill of No. 2 fuel oil off West Falmouth, Cape Cod, in 1969 occurred near the Woods Hole Oceanographic Institute, whose scientists began a study that continued for several years. In that spill, benthic organisms, including valuable shellfish, died rapidly and in great numbers. The Torrey Canyon spill, which took place near the Marine Biological Association Laboratories in Milford Haven, England, was also well studied. In that incident, larval fish concentrated near the ocean surface came in direct contact with the oil and were killed in massive numbers.

The chronic effects of oil spills appear less serious than the acute effects. Oil, a natural hydrocarbon, is food for some marine microorganisms and is degraded by them. When the oil spills are localized, then marine organisms can migrate back into the destroyed area once the oil has been degraded.

Perhaps the most serious long-term problem may be the slow, widespread leakage of oil at many points all over the oceans. The oil may contain chemicals that have specific toxic effects on certain species. If the chemicals are sufficiently widespread, they may be able to affect an entire species. Although the tanker oil spills and drilling rig blowouts are spectacular, their limited effects may cause less of a problem than nonpoint sources.

Pesticides   Few issues in environmental studies are simple; most involve a complex balancing of difficult and conflicting choices. This is nowhere better illustrated than by pesticides.

The conflicts were vividly brought to public attention by the medfly problem in California in 1981. The medfly, a pest which destroys fruits, was accidently introduced into the state's fruit-growing areas in 1981, and its population grew rapidly. The misuse of one kind of biological control—the release of supposedly sterile male flies, which turned out not to be sterile—made the problem worse. The fly population expanded so rapidly that local spraying from the ground of chemical pesticides did not work. The state government had to weigh the positive effects of aerial spraying (possible control of the medfly) against its negative effects, which ranged from damage to automobile paint to potential, unknown, long-term health hazards. Fortunately, recent technical advances allowed the pesticide to be sprayed in sugary droplets which attracted the fly but relatively few other species.

Thus, unable to control the fly in any other way and facing a billion-dollar effect on the state's agricultural industry, the state government opted for aerial spraying of the pesticide.

Ever since the beginnings of agriculture, pests have been the bane of farmers. In the United States, pests consume about one third of the total potential crop before harvest and 10 percent after harvest [13]. Pesticides have been used in one form or another for more than 2000 years, but pesticide use became widespread only in the last century.

The ideal pesticide would be a ''magic bullet,'' a chemical that affected only the pest and no other living thing or aspect of the environment. The history of pesticide development can be seen as an attempt to find chemicals that had more and more specific effects and that were effective in controlling pests.

The first modern chemical pesticides were arsenic compounds used on potatoes, cotton, and apples. Although these are effective on the insect pests of those crops, the compounds are also highly toxic to human beings and many other forms of life. Green plants produce some natural pesticides—nicotine, for example—which were used in some of the earlier compounds. Inorganic chemicals were most important until DDT and other chlorinated hydrocarbons were developed in the 1940s, leading to a revolution in pesticides. In the past 30 years, U.S. agriculture has come to depend more and more on chemical pesticides. During the last decade the use of pesticides has increased 12 times [13], and millions of kilograms are produced every year (Table 14.7).

Like their precursors, modern pesticides have been a mixed blessing and the center of major environmental controversies. Although it is commonly believed that pesticides are used primarily against insects, particularly those that eat leaves and fruit of crop plants, only about one third of pesticides sales in the United States in recent years is for this purpose. Almost 60 percent are *herbicides* (chemicals to control weed competitors of crops), and slightly less than 10 percent are *fungicides* (chemicals to control fungal plant diseases). The herbicides 2,4-D and 2,4,5-T and a contaminant of them, dioxin, are among the pesticides causing the most environmental concern.

Pesticides help to reduce some human diseases that are spread by insects. Prior to the use of modern insecticides, there were approximately 300 million cases and 3 million deaths annually worldwide from insect-borne diseases (primarily from malaria). These have been reduced to 120 million cases and 1 million deaths, in spite of the great increase in the world's human population [14].

The primary problem with pesticides is that they are intentionally toxic, and if they are toxic to undesirable or-

TABLE 14.7
Annual pesticide production.

Class	Common Name	Production (millions of kg A.I.)[a]
Insecticide	Methyl parathion	20
	Toxaphene	23
	Carbaryl	20
	Malathion	16
	Chlordane	11
	Parathion	7
	Methoxyclor	4.5
	Diazinon	4.5
	Carbofuran	3.6
	Disulfoton	3.6
	Phorate	3.6
Herbicide	Atrazine	41
	2,4-D	20
	MSMA-DSMA	16
	Sodium chlorate	14
	Trifluralin	11
	Propachlor	10
	Chloramben	9
	Alachlor	9
	CDAA	4.5
	2,4,5-T	2.7
Fungicide	PCP and salts	21[b]
	Dithiocarbamates	18[c]
	TCP and salts	9
	Captan	8
	PCNB	4
	Dodine	4

SOURCE: From Freed et al., 1977.
[a]Active ingredient.
[b]Includes use as herbicide, desiccant, molluscicide, and for termite control.
[c]Includes CDEC, Ditane M-45, Ditane S-31, Ferbam, Maneb, Metham, Nabam, Niacide, Polyram, Thiram, Zineb, and Ziram.

ganisms, they may also be toxic to desirable ones, including ourselves. This danger was brought to public attention in 1962 with the publication of Rachel Carson's *Silent Spring,* a book that popularized the possibility that DDT and other chemicals might be affecting the reproduction of birds [15].

DDT is a classic case of a pesticide thought to be safe and effective. The discoverer of its use as an insecticide, Paul Muller, was awarded a Nobel Prize. DDT came into widespread use after World War II and first appeared to be an amazing chemical. It contributed to the control of at least 27 diseases, and by 1953 it was credited with saving 5 million lives and preventing 100 million illnesses [16]. Early experiments, including those with human volunteers, indicated that DDT had little direct (i.e., acute) toxic effect on people, although it was recognized by 1948 that the compound was stored in body fat. DDT and other

chlorinated hydrocarbons are persistent. At first, this appeared to be an advantage; for example, they could be sprayed on the walls of a house and kill mosquitoes for a long time afterwards. However, their persistence in the environment became a problem.

The drainage of DDT into rivers, swamps, and coastal waters killed crabs; fish were killed in streams that drained forests where DDT had been sprayed to control the spruce budworm. Its wide distribution in the biosphere was evident when it was detected in Antarctic penguins at very low levels (1 part per billion) and in canned milk, fresh cows' milk, and human milk. In high concentrations, DDT was found to injure the liver of animals, and there is some controversial evidence that it might cause cancer.

In the 1960s, scientists recognized that widespread and rapid decrease in the weight of eggs and thickness of eggshells of birds had been occurring for the previous several decades in Britain and North America. Both effects appeared to correlate with the use of DDT and other persistent pesticides. The eggshell thinning was found in more than 20 species, and a decrease in shell weight of 20 percent or more was found in 9, including the peregrine falcon, the bald eagle, and the osprey [17]. Experiments on mallard ducks verified that DDT and its chemical derivatives, particularly DDE, caused eggshell thinning.

DDT is poorly soluble in water, but quite soluble in fats and organic oils. DDT was taken up by aquatic organisms and stored in their fatty tissues or, in the case of vegetation, in oils. When these aquatic organisms or plants were eaten, the metabolism of the predators tended to favor the retention of DDT in the body fat, and the predators concentrated it in their own bodies. Each step in the food chain resulted in an increased concentration of DDT. Finally, the carnivorous birds at the top of the food chain seem to have had concentrations sufficient to cause toxic effects.

As the history of DDT illustrates, fat-soluble organic compounds can have subtle, hidden, but important environmental effects. Because the compounds, poorly soluble in water, appear rare in the nonbiological environment, they are easily ignored. However, they do undergo trophic-level concentration or *biomagnification*, and thus negative effects can take place when the compound appears to be only a trace substance in the environment. The discovery of biomagnification has led to a better understanding of the pathways of pollutants through the environment.

In the 1960s, DDT fell from its earlier status as a miraculous saver of lives and crops to a symbol of all that was bad in our misuse of the environment. After decades of discussion, DDT still remains controversial, including its actual effects on human health and the extent to which it has caused environmental damage compared with other artificial organic compounds and other forms of pollutants. Defenders of the use of DDT have pointed out that chemicals used in the plastic industry, PCBs, have similar effects and are difficult to distinguish from DDT.

Although banned in the United States, DDT is still used in other countries. Birds migrating to the United States from areas where DDT is in use may become prey for North American hawks and other birds of prey. Thus DDT may still pose an environmental problem for North America.

In recent years, there has been an attempt to develop alternative methods of pest control. Because of the problems caused by the persistence of chlorinated hydrocarbons such as DDT, shorter-lived chemicals such as organic phosphates have been used more widely. These, however, tend to have a higher acute toxicity in people and must be handled carefully. As an alternative to these toxic compounds, chemical sex attractants have been used to attract one sex of an insect into a trap. This method has been used to control the bollworm in cotton.

Alternative pest control strategies, generally lumped under the name of **Integrated Pest Management (IPM)**, include chemical, cultural (mechanical cultivation of the soil, crop rotation, etc.), and biological (the use of predators and parasites of the pests) pest control and the development of pest-resistant plants. IPM has a number of advantages. In addition to reducing the use of dangerous chemicals, IPM can be used where pests have built up a resistance or tolerance to a chemical pesticide.

The use of parasites and predators of pests has proven effective. For example, the bacteria *Bacillus thuringiensis,* which causes diseases in many pests but is harmless to human beings or other mammals and birds, is now used extensively.

Industrial Chemicals   Among the artificial organic compounds similar to DDT is a group used in the production of plastic—polychlorinated biphenyls, or PCBs. PCBs are toxic to people and other organisms. An example of their effects involves an industrial plant near the Hudson River which used the chemical in producing electrical capacitors and released the wastes into the river. In 1971 striped bass in the Hudson River were found to have 11 parts per million in their eggs and 4 parts per million in their flesh. Fish with PCB concentrations above 5 parts per million exceed health safety levels; all fish in the lower Hudson River exceeded this value except shad, large sturgeon, and goldfish. The shellfish industry in the Hudson estuary, worth million of dollars, was closed. The manu-facturers of PCBs voluntarily restricted production as a result of these problems. Although PCBs are no longer released into the river, cleaning up the sediments remains an important issue. Dredging may do more harm than good by dispersing the chemical more widely.

PCBs are an example of a useful chemical thought to be safe but found to be toxic and harmful to the environment. Voluntary action has reduced its production, but the chemical remains in the environment [18].

THERMAL POLLUTION

When heat released into water or air produces undesirable effects, thermal or heat pollution occurs. Heat pollution can occur as sudden, acute events (Table 14.8) or as a long-term, chronic release. The major sources of chronic heat pollution are electric power plants that produce electricity in steam generators.

The release of large amounts of heated water into rivers changes the average water temperature and thus changes the rivers' species composition. Every species has a range of temperature within which it can survive and an optimal temperature. For some species of fish, this range is small, and a slight change in water temperature can disturb them. Lake fish move away when the water temperature rises more than 1.5C° above normal; river fish can withstand a rise of 3C° (Fig. 14.10) [19]. Thus heating the water can disturb the river ecosystem from its original conditions.

There are several solutions to thermal discharge into bodies of water. The heat can be released into the air by cooling towers, or the heated water can be temporarily stored in artificial lagoons until it is cooled to normal temperatures. Some attempts have been made to use the heated water to grow organisms of commercial value which require warmer water temperatures.

PARTICULATES

Particulates are small particles of dust that are released into the atmosphere by many activities. Modern farming adds considerable amounts of particulates to the atmosphere as do desertification and volcanic eruptions [6].

Many chemical toxins, like sulphur oxides and heavy metals, enter the biosphere as particulates. However, some particulates can cause environmental problems merely because they are small particles of solids suspended in the air, and not because of the specific toxic effects of the elements or compounds that compose them.

Particulates have effects on human health, ecosystems, and the biosphere. Dust that enters the lungs may lodge there and have chronic effects on respiration. Certain materials, such as asbestos, are particularly danger-

TABLE 14.8
Effects of large heat additions to the atmosphere.

Phenomenon	Energy Rate (Mw)	Area (km²)	Energy-Flux-Density (W/m²)
Large brush fire[a]	100,000	50	200
Consequences: (Relatively small energy-flux rate, very large area.) Cumulus cloud reaching to a height of 6 km formed over 0.1 km² area of fire. Convergence of winds into the fire area.			
Forest fire whirlwind[b]			
Consequences: Typical whirlwind: Central tube visible by whirling smoke and debris. Diameters few feet to several hundred feet. Heights few feet to 4000 ft. Debris picked up—logs up to 30 in. in diameter and 30 ft long.			
World War II fire storm[c]		12	
Consequences: Turbulent column of heated air 2.5 mi in diameter. Fed at base by inrush of surface air; 1.5 mi from fire, wind speeds increased from 11 to 33 mph. Trees 3 ft in diameter were uprooted.			
Fire at Hiroshima[d]			
Consequences: (10–12 hr after atomic bomb.) "The wind grew stronger, and suddenly—probably because of the tremendous convection set up by the blazing city—a whirlwind ripped through the park. Huge trees crashed down; small ones were uprooted and flew into the air. Higher, a wild array of flat things revolved in the twisting funnel." The vortex moved out onto the river, where it sucked up a waterspout and eventually spent itself.			
Surtsey volcano[e]	100,000	<1	100,000
Consequences: Permanent cloud extending to heights of 5–9 km. Continuous sharp thunder and lightning, visible 115 km away. (Phenomenon probably peculiar to volcano cloud with many small ash particles.) Waterspouts resulting from indraft at cloud base, caused by rising buoyant cloud.			
Surtsey volcano[f]	200,000	1	200,000
Consequences: Whirlwinds (waterspouts and tornadoes) are the rule rather than the exception. More often than not there is at least one vortex downwind. Short inverted cones or long, sinuous horizontal vortices that curve back up into the cloud, and intense vortices that extend to the ocean surface.			
French Meteotron[g]	700	0.0032	219,000
Consequences: ". . . artificial thunderstorms, even tornadoes, many cumulus clouds . . . substantial downpour." Dust devils.			
Meteotron[h]	350	0.016	22,400
Consequences: 15 min after starting the burners, observers saw a whirl 40 m in diameter . . . whirlwind so strong burner flames were inclined to 45°.			
Single large cooling tower	2,250	0.0046	484,000
Consequences: Plume of varying lengths and configurations.			
Array of large cooling towers [48,000 Mw(e) NEC; area 48,000 acres]	96,000	194	495
Consequences: Unknown.			

SOURCE: After Rotty, 1976, in Council on Environmental Quality and the Department of State, 1980.
[a]From Taylor et al., 1973. [e]From Borne, 1964.
[b]From Graham, 1955. [f]From Thorarinsson and Vonnegut, 1964.
[c]From Landsberg, 1947. [g]From Dessens, 1962.
[d]From Hersey, 1946. [h]From Dessens, 1964.

FIGURE 14.10
Acceptable temperatures for various species of fish. Trout require colder water than perch and bass, which require colder water than catfish. Heat pollution can change the species of fish which occur in a river, stream, or lake. (From Chanlett, 1979.)

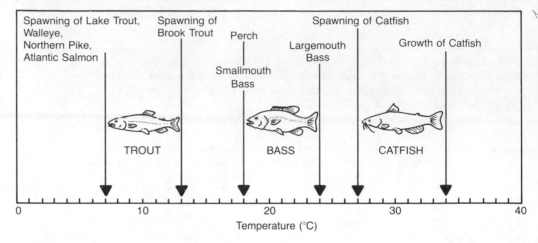

ous in this way. Dust raised by road building and plowing and deposited on the surface of green plants may interfere with their absorption of carbon dioxide and oxygen and their release of water, and heavy dust may affect the breathing of animals. Dust associated with large construction projects may therefore kill organisms and damage large areas, changing species composition, altering food chains, and generally affecting ecosystems.

During the twentieth century, our modern industrial society has added more and more dust to the atmosphere. Dust reflects the sunlight and, when present in sufficient amounts, may thereby cool the Earth's surface. In fact, the increase in atmospheric dust from human activities may offset the warming effects that may be occurring from the burning of fossil fuel and the addition of carbon dioxide to the atmosphere. Whether the cooling of the atmosphere from particulates will have a greater effect than the warming trends caused by carbon dioxide is unknown.

NOISE POLLUTION

Noise is unwanted sound. Sound, a form of energy, is measured in units of *decibels*. The environmental effect of noise depends not only on the total energy, but on the sound's *pitch* or *frequency* (or mixture of frequencies) and its time pattern. Very loud noises (more than 140 decibels) cause pain, and high levels can cause hearing loss. Lower levels interfere with human communication, cause annoyance, and can be unpleasant.

There is some evidence that noise can affect behavior and physiology. For example, one study indicates that monkeys exposed to high levels of noise had higher blood pressure than those not exposed. In North American deserts, there are few natural sources of noise louder than occasional thunder. The principal sounds are those of wind, rain, and the call of animals. For one species of desert

toad, which lives underground most of the time, thunder is a mating signal; thunder indicates rain, which means temporary pools of water above ground in which to rear young. An unnaturally loud noise with a decibel level similar to thunder can trigger the same mating behavior, causing the toad to migrate above ground, where it will perish from lack of water. It is possible that human activities in deserts which involve high noise levels could disrupt the behavior of such desert animals.

GENERAL EFFECTS OF POLLUTANTS

Although there are many kinds of pollutants, each with its own method of action and environmental pathways, there are certain features that are characteristic of most environmental toxins.

DOSE-RESPONSE CURVES AND THRESHOLD EFFECTS

Five centuries ago, Paracelsus wrote that "everything is poisonous, yet nothing is poisonous." That is, essentially anything in too great amounts can be dangerous, yet anything in extremely small amounts can be relatively harmless. Every chemical element has a spectrum of possible effects on a particular organism. For example, selenium is required in small amounts by living things, but may be toxic to or cause cancer in cattle when it is in high concentrations in soil. Copper, chromium, and manganese are other examples of chemical elements that are required in small amounts but are toxic in higher amounts.

It was recognized many years ago that the effect of a certain chemical or any toxic factor on an individual depends on the dose or concentration of the toxic factor. This **dose dependency** can be represented by a generalized **dose-response curve** (Fig. 14.11).

FIGURE 14.11
Generalized dose-response curve. See text for explanation.

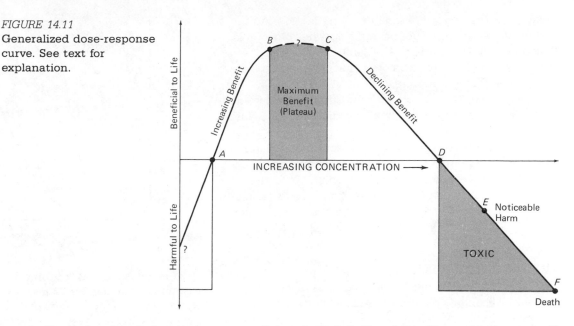

When various concentrations of a chemical or toxic factor present in a biological system are plotted against the effects on the organism, three things are apparent. First, while relatively large concentrations are toxic, injurious, and even lethal (*D-E-F* in Fig. 14.11), trace concentrations may be beneficial or necessary for life (*A-B*). Second, the dose-response curve has two maxima (*B-C*) forming a plateau of optimal concentration and maximum benefit. Third, the threshold concentration, where harmful effects to life start, is not at the origin (zero concentration) but varies with concentrations less than at point *A* (where there is too little of the factor) and greater than at point *D* (where there is too much of the factor) in Figure 14.11.

Points *A, B, C, D, E,* and *F* in Figure 14.11 are significant threshold concentrations. Unfortunately, points *E* and *F* are known only for a few substances for a few organisms, including people, and the very important point, *D*, is all but unknown. The width of the maximum-benefit plateau (points *B, C*) for a particular form of life depends on the organism's particular physiological state. The levels that are beneficial, harmful, or lethal may differ widely quantitatively and qualitatively and therefore are difficult to characterize.

Fluorine in rocks and soils illustrates this general dose-response curve. Fluorine is fairly abundant in rocks and soil, and industrial activity and application of fertilizers have also, on a limited basis, contributed locally to an

FIGURE 14.12
Dose-response curve for fluoride.

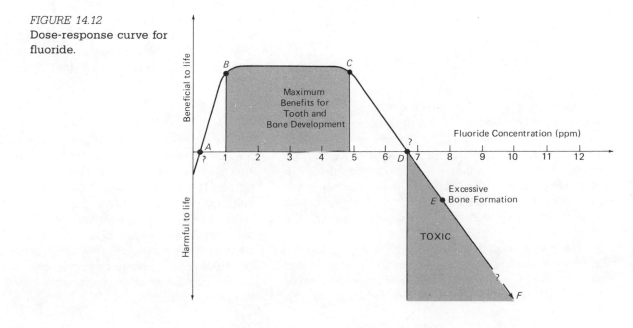

increase in the concentration of fluorine in soils. An important trace element, fluorine forms fluoride compounds, which prevent tooth decay. The same processes occur in bones, where fluoride promotes the development of a more perfect bone structure.

Relationships between the concentration of fluoride (in a compound of fluorine such as sodium fluoride, NaF) and health indicate a specific dose-response curve, as shown in Figure 14.12. The plateau for optimal fluoride concentration (point B to point C) for the reduction of dental caries is from about 1 part per million (100 parts per million = 0.01 percent) to just less than 5 parts per million. Toxic effects begin between 6 and 7 parts per million (point D in Fig. 14.12).

THE TOXIC DOSE-RESPONSE CURVE

The effect of an environmental pollutant is often described by a slightly different dose-response curve, which is actually a more detailed view of the generalized curve from point D to point F in Figure 14.11. Toxic dose-response curves show a negative response, either death or injury, plotted against the increasing intensity of exposure to a pollutant. The upper limit of such curves represents 100 percent of the population affected. The general dose-response curve suggests there is a threshold for negative effects.

A controversy exists over whether and when thresholds can be found for environmental toxins. A threshold is a level below which no effect occurs and above which effects begin to occur. If a threshold exists, then any level in the environment below that threshold would be safe. Alternatively, it is possible that even the smallest amount of toxin has some negative effect (see Fig. 14.15). Some data suggest the existence of a threshold and of a toxic dose-response curve.

For example, Figure 14.13 shows the dose-response curves for three species of trees exposed to chronic (long-term) radioactivity. These trees were in a forest that was part of an experiment at Brookhaven National Laboratory on Long Island, New York, to test the effects of radioactivity on a natural ecosystem. In this experiment, a source of gamma radiation, cesium-134, was placed in the center of an oak-pine forest. The cesium source was placed on a vertical shaft that could be lowered into a lead shield. The forest was irradiated for 20 hours every day for many years; during the 4 hours per day that the source was shielded, scientists could enter the forest to examine the site.

The damage to the forest by the gamma radiation varied with the species and among individuals within species. There were several clearly distinguishable zones of effects. Nearest the source, where radiation was the

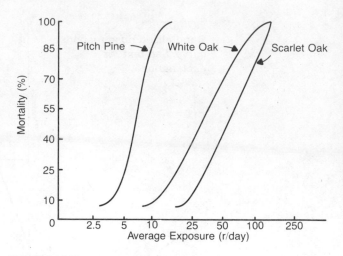

FIGURE 14.13

Dose-response curves for three species of trees exposed to gamma radiation for 32 months at Brookhaven National Laboratory. The experiment indicates that pitch pine is much more sensitive to radiation than oaks, and that scarlet oak is the least sensitive of the three species. The oaks appear to have a threshold level of about 10 roentgens per day before any mortality above that occurring in the normal forest can be measured. (From Woodwell and Rebuck, 1967.)

most intense, a completely devastated zone was produced in which all woody plants were killed. This occurred at dosages of 200 roentgens per day or greater (Fig. 14.14).

The dose-response curves (Fig. 14.13) show the percentage of tree mortality as a function of the amount of radiation they received. According to the graph, essentially 100 percent of pitch pine were killed at an exposure of 10 roentgens per day, the level at which mortality is first measurable for white oak. Both scarlet oak and white oak can withstand 10 times more radiation (100 roentgens per day) than the pine before suffering 100 percent mortality, suggesting that oaks have a higher threshold in regard to radiation than pitch pine.

LD-50

A concept closely linked to the dose-response curve is the amount of exposure required for 50 percent of the population to show a response. For death, this concept is known as the **LD-50,** or the amount of exposure to a toxin that results in the death of one half of the exposed population (Fig. 14.15). A similar concept can be used for the exposure required to produce any nonlethal symptom in 50 percent of the population. The LD-50 is useful because individuals are variable in their response, and it is difficult to estimate the exact dose that will cause a response in a particular individual.

FIGURE 14.14
In an experiment at Brookhaven National Laboratory, radioactive cesium was placed in a forest of pines and oaks. Vegetation was killed in a circle surrounding the source (a). Very near the source, all higher plants were killed. Further out, the only plants that survived were those of small size and short lifetimes (b). The large, longer-lived trees could only be found even further from the source (c). The pattern of vegetation from harshest environment to least harsh is similar to that observed as one travels down the side of a mountain. [Photo (a) courtesy of Brookhaven National Laboratory; photos (b) and (c) by D. B. Botkin.]

(a)

(b)

(c)

TOLERANCE

The determination of dose-response curves may be made more difficult because of the development of tolerance in individuals and in populations. **Tolerance** is an increase in resistance that results from exposure. It can develop for some pollutants in some populations, but not for all pollutants in all populations.

Tolerance may result from behavioral, physiological, or genetic adaptation. As an example of behavioral response, mice learn to avoid traps. Physiological tolerance

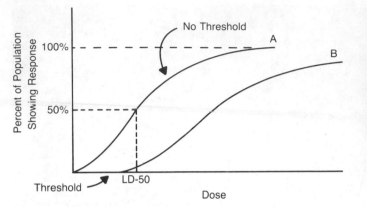

(a)

FIGURE 14.15
In this hypothetical toxic dose-response curve, toxin A has no threshold; even the smallest amount has some measurable effect on the population. The LD-50 for toxin A is the dose required to produce a response in 50 percent of the population.

means that the body of an individual adjusts to tolerate a higher level of a pollutant. For example, people become *tolerant* of low oxygen at high altitudes over a period of days because the body increases the number of red blood cells. Mice and rats exposed to small doses of lead show increased resistance to death from lead, but not to all injury [20]. There are many mechanisms for physiological tolerance, including detoxification, or the internal transport of the toxin to a part of the body where it is not harmful. Genetic adaptation results when those individuals who are more resistant survive an exposure to a toxin and have more offspring than others. These resistant individuals will prevail in later generations, as has been observed among some insect pests following exposure to some chemical pesticides. For example, certain strains of malaria-causing mosquitoes are now resistant to DDT.

(b)

ACUTE AND CHRONIC EFFECTS

Pollutants can have both acute and chronic effects. An acute effect is one that occurs soon after exposure, usually to large amounts of a pollutant. A chronic effect is one that takes place over a long period, often due to exposure to low levels of a pollutant. For example, a person exposed to a high dose of radiation at one time may be killed (an

FIGURE 14.16
Acid rain and heavy metals from smelters kill vegetation. At the Palmerton, New Jersey, zinc factory (a), nearly all the vegetation near the smelter is dead (b); even on a ridge nearby, the vegetation has been altered (c). (Photos by D. B. Botkin.)

(c)

acute effect), but that same total dose received slowly, in small amounts over an entire lifetime, may cause mutations, lead to cancer, or affect the person's offspring.

ECOLOGICAL GRADIENTS

The experiment about the effects of radiation on a forest at Brookhaven National Laboratory illustrates another general effect of pollutants. The species differences in dose-response effects produce a curious ecological result. The kinds of vegetation that persist nearest to the radiation source are small plants with relatively short lifetimes (grasses, sedges, and weedy species usually regarded as pests) and adapted to harsh and highly variable environments. Near the radiation source, a sedge (*Carex pennsylvanica*), persisted under radiation levels of 300 roentgens per day (400 to 450 roentgens is enough in a single dose to kill 50 percent of the human beings exposed to it). Trees occurred furthest from the site. The changes in vegetation with distance from the radioactive cesium are similar to the kinds of changes one finds in walking down a mountain from the summit, the harshest environment, to the valley floor, the most benign environment.

The radiation also simplified the forest, decreasing the number of species. The nearer to the radiation source, the fewer were the species. A similar effect occurred in old fields exposed to radiation where the diversity of vegetation was reduced 50 percent at radiation levels of 1000 roentgens per day [21].

The same patterns of disturbance are found around smelters and other industrial plants that discharge pollutants to the atmosphere from stacks. These patterns can be seen in the area around Sudbury, Ontario, which has one of the largest metal smelters in the world. Tall smokestacks emit toxic metals and sulphur oxides into the air. Near to the smelters, an area that was once forest is now a patchwork of bare rock and soil occupied by small plants. Vegetation characteristic of disturbed areas occurs nearer to the smelter than do forest trees. A similar pattern occurs at one of the world's largest zinc smelters in Palmerton, New Jersey (Fig. 14.16).

DISPOSAL OF TOXIC WASTES

Disposal of waste, particularly solid waste, is a major urban problem. In the United States alone, urban areas produce about 640 million kilograms of solid waste each day. That amount of waste is sufficient to cover more than 1.6 square kilometers every day to a depth of 3 meters [22]. Figure 14.17 summarizes major sources and types of solid waste, and Table 14.9 lists their compositions. Disposal or treatment of liquid solid wastes by state and municipal agencies costs billions of dollars every year (Fig. 14.18). It is one of the most expensive environmental expenditures that governments must bear, accounting for about 55 percent of the total environmental expenditures [23].

Waste disposal since the Industrial Revolution has changed from a practice of "dilution and dispersion" to a new concept of "concentrate, contain, and recycle." A fast-growing disposal practice for solid waste is the sanitary landfill. However, methods such as on-site disposal,

FIGURE 14.17

Types of materials or refuse commonly transported to a disposal site.

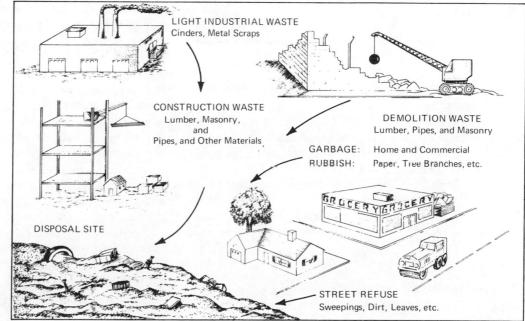

LIGHT INDUSTRIAL WASTE
Cinders, Metal Scraps

CONSTRUCTION WASTE
Lumber, Masonry,
and
Pipes, and Other Materials

DEMOLITION WASTE
Lumber, Pipes, and Masonry

GARBAGE: Home and Commercial
RUBBISH: Paper, Tree Branches, etc.

DISPOSAL SITE

GROCERY GROCERY

STREET REFUSE
Sweepings, Dirt, Leaves, etc.

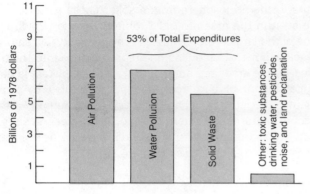

FIGURE 14.18
Pollution abatement and environmental quality expenditures for 1978. (Data from Council on Environmental Quality, 1979.)

TABLE 14.9
Generalized composition of solid waste likely to end up at a disposal site.

Type of Waste	Average (%)
Paper products	43.8
Food wastes	18.2
Metals	9.1
Glass and ceramics	9.0
Garden wastes	7.9
Rock, dirt, and ash	3.7
Plastics, rubber, and leather	3.1
Textiles	2.7
Wood	2.5
Total	100.0

SOURCE: Council on Environmental Quality, 1970.

feeding garbage to swine, composting, incineration, and open dumping are still used.

Hazardous chemical waste management may be the most serious environmental problem in the United States. Hundreds or even thousands of uncontrolled disposal sites may be time bombs that eventually will cause serious public health problems. A highly technological society will continue to produce hazardous chemical wastes. Therefore, it is important that carefully engineered secure landfills and other safe disposal methods be developed and used.

CHEMICAL WASTE

There are several ways that uncontrolled dumping of chemical wastes may pollute the surface, soil, and groundwater (Fig. 14.19). First, some compounds form gases which volatilize and enter the atmosphere. Methane, ammonia, hydrogen sulphides, and nitrogen oxides are examples of such gases. Second, heavy metals such as lead, chromium, and mercury are retained in the soil, making the local site hazardous. Third, soluble chemicals such as chlorides, nitrates, and sulphates are carried by water and enter the groundwater system. Above-ground runoff of water can also transport these soluble chemicals. Fourth, crops and other plants can selectively take up heavy metals and other toxic materials, which are then passed along food chains to other organisms. Decaying vegetation can return the toxins to the soil.

Furthermore, liquid chemical wastes are sometimes dumped in an unlined "lagoon," where the contaminated water may then percolate through the soil and rock, and eventually to the groundwater table. Liquid chemical wastes are sometimes illegally dumped in deserted fields or even along dirt roads.

There are numerous examples of problems associated with disposal of hazardous wastes. Waste from producing pesticides and organic solvents at Benderson, Nevada, may be contaminating Lake Mead, the water supply for Las Vegas. At a location near Louisville, Kentucky, known as the "Valley of the Drums," thousands of drums of waste chemicals, stored on the surface or buried, are oozing toxic materials.

Near Elizabeth, New Jersey, are the remains of about 50,000 charred drums, stacked four high in places, next to a brick and steel building once owned by a now bankrupt chemical corporation. The drums and other containers had been left to corrode for nearly 10 years, and many had been either improperly labeled or have been burned so that the nature of the chemicals cannot be determined simply from outside markings. Leaking barrels allowed unknown waste to seep into a stream which flows into the Hudson River. Cleanup is difficult because there are few precedents for disposing of unknown dangerous chemicals. Identification of some of the materials at the site showed that there were two containers of nitroglycerine; numerous barrels full of biological agents; cylinders of phosgene and pyrophoric gases, which are extremely volatile and ignite when exposed to the air; as well as a variety of heavy metals, pesticides, and solvents, many of which are extremely dangerous and potentially explosive. It took months of work with a large crew to remove most of the material from the New Jersey site. However, it is difficult to know if all the wastes have been removed. Additional material may be buried or stored at other sites not yet located [24].

Natural Disposal   Toxic organic compounds may be stored for certain periods, but this is a limited solution. It

FIGURE 14.19
Ways that uncontrolled dumping of chemical waste may pollute soil and/or groundwater.

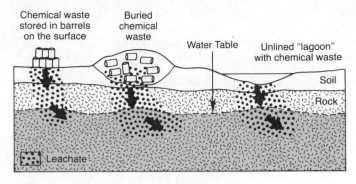

is more desirable to convert the compounds to their simple original inorganic forms, a process called **mineralization.** Mineralization, however, rarely occurs through inorganic processes at normal biospheric temperature and oxygen levels. Organic compounds may be burned at high temperatures, and the efficiency of this process is improved by increasing the oxygen levels (either by burning at high pressures or by adding oxygen). This is a comparatively expensive approach requiring carefully monitored and controlled technologies.

In nature, complete degradation is almost always carried out by enzymes of certain microbes [25]. **Microbial conversion,** a public service function of natural ecosystems, has been shown to occur for such compounds as DDT, aldrin, and hectachlor. In some cases, the microbes do not use the compounds, but merely convert them from toxic to nontoxic form. The opposite also occurs: Microbes may convert less toxic forms to more toxic ones. For example, microbes can convert inorganic mercury to methyl mercury, which, as we have already seen in the cases of Minamata and Niigata, is much more toxic. We can and do make use of these natural processes when we dump water into streams, marshes, and agricultural or forest soils, or when we use microbial activity in sewage treatment. These methods work as long as the concentrations do not exceed the capacities of the microbes.

RADIOACTIVE WASTES

Radioactive wastes can be divided into two types: low-level and high-level wastes. Low-level waste contains sufficiently low concentrations or quantities of radioactivity that they do not present a significant environmental hazard if handled properly. Included here are a wide variety of items, such as residuals or solutions from chemical processing; solid or liquid plant waste, sludges, and acids; and slightly contaminated equipment, tools, plastic, glass, wood, fabric, and other materials [26]. Some experiences suggest that low-level radioactive waste might be

buried safely in carefully controlled and monitored near-surface burial areas where the hydrologic and geologic conditions severely limit the migration of radioactivity [26]. However, careful planning and monitoring over long periods are required.

An example of high-level wastes is the used fuel assemblages from nuclear reactors. This spent fuel is contaminated with large quantities of fission products and must periodically be removed and reprocessed or disposed of. Present waste management problems involve removal, transport, storage, and eventual disposal of spent fuel assemblies [27]. The tailings from uranium mines and mills must also be considered hazardous (Fig. 14.20). The existence of over 20 million tons of abandoned tailings in the western United States that will produce radiation for at least 100,000 years emphasizes the problem.

Production of plutonium for weapons also generates high-level waste. At present, several hundred thousand cubic meters of liquid and solid high-level waste are being stored at U.S. Department of Energy repositories at Hanford, Washington; Savannah River, Georgia; and Idaho Falls, Idaho.

A sense of urgency surrounds the disposal of this extremely toxic high-level waste as the spent fuel assemblies slowly accumulate. It has been projected that, if there is no disposal program by the year 2000, several hundred thousand spent fuel elements from commercial reactors will be in storage awaiting disposal or eventual reprocessing to recover plutonium and unfissioned uranium. With reprocessing, solid high-level waste would only occupy several thousand cubic meters—a small volume that would not cover a football field to a depth of one meter [27].

Serious problems have resulted from liquid radioactive waste being buried in underground tanks. For example, at Hanford, Washington, 16 leaks involving 1330 cubic meters were found from 1958 to 1973. In 1973, a single leak released 437 cubic meters of low-temperature waste. Since then, improvements such as stronger, double-shelled storage tanks; reduction of the volume of

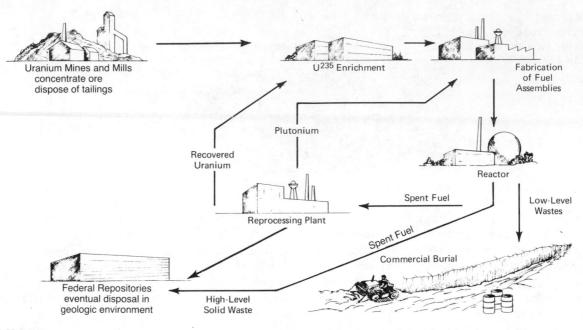

FIGURE 14.20
Idealized diagram showing the nuclear fuel cycle. The United States does not now reprocess spent fuel. Disposal of tailings, which because of their large volume may be more toxic than high-level waste, has been treated casually. (From Office of Industry Relations, 1974.)

liquid waste stored through a solidification program; and increased reserve capacity have been instituted to reduce the chance of future incidents [28].

Storage of high-level waste is at best a temporary solution that allows the federal government to meet its commitments for accepting waste. Regardless of how safe any ''storage'' program is, it requires continuous surveillance and periodic repair or replacement of tanks and vaults. Therefore, it is desirable to develop more permanent disposal methods in which retrievability may be possible but not absolutely necessary.

Certain geologic formations may provide the safest containment of high-level radioactive waste. Deep salt beds or domes received early attention for permanent disposal of such waste. There are several advantages to disposing of nuclear waste in salt: Salt is relatively dry and impervious to water; fractures that develop tend to be self-healing; salt permits the dissipation of larger quantities of heat than is possible in other rock types; salt is approximately equal to concrete in its ability to shield harmful radiation; salt has a high compressive strength (when dry) and generally is located in areas of low earthquake activity; and, because salt deposits are relatively abundant, using some for waste disposal will cause a negligible resource loss [28].

On the other hand, salt has recently been criticized because it has a relatively low ability to absorb radioactive nuclides from the waste in an insoluble form. Further-

more, pockets of brine (very salty water) found in some salt depositories would certainly create serious problems and jeopardize waste retrieval if a repository were accidentally flooded. The presence of brine or other groundwater will also decrease the strength of salt. But, on the positive side again, careful site evaluation might be able to identify areas where brine might cause problems, and waste could be wrapped in materials that readily absorb harmful nuclides from the waste [27].

The U.S. government has considered disposing of radioactive wastes in salts since 1955; and during 1966 and 1967 it initiated Project Salt Vault, in which high-level solid-waste disposal procedures were simulated in abandoned salt mines in Kansas. The project was considered a success, and in 1970 it was announced that the abandoned Carey Salt Mine near Lyons, Kansas, was to be used as a demonstration site for waste disposal. If successful, the site would eventually become the National Radioactive Waste Repository. However, the Lyons site was abandoned because of scientific, public, and state concern over two factors. First, there was geologic uncertainty; several thousand abandoned mine shafts and drill holes in the area might allow waste to migrate up the shafts or drill holes. Second, at that time the land was partly urbanized and used for farming and grazing, a land use that conflicted with the proposed disposal plan [29]. Investigation of sites in New Mexico is underway, and it is expected that a pilot plant to dispose of high-level waste in bedded salt may be licensed by the 1980s. Figure 14.21 is an idealized diagram of such a disposal operation.

Stable bedrock offers the most promise for radioactive waste disposal. Other possibilities that are being explored include disposal into polar ice caps and into sediment in deep ocean basins. Sites will be chosen on the basis of geologic and hydrologic conditions and presumed future changes in climate, groundwater flow, erosion, and tectonics. Whether a site is used depends on political, legal, and social decisions as to whether the risks are acceptable to society.

A major problem with the disposal of high-level radioactive waste remains, How credible are long-range (thousands to a few million years) geologic predictions [27]? There is no easy answer to this question, as geologic processes vary over both time and space. Climates change over long periods of time as do areas of erosion,

deposition, and groundwater activity. For example, large earthquakes hundreds or even thousands of kilometers away from a site may permanently change groundwater levels. The seismic record for most of the United States only extends back a few hundred years, and therefore estimates of future activity are tenuous at best. Environmental scientists can suggest sites that have been relatively stable in the geologic past, but they cannot absolutely guarantee future stability. Decision makers need to evaluate the uncertainty of prediction in light of political, economic, and social concerns [27].

SUMMARY AND CONCLUSIONS

There are seven major forms of pollutants: acids, toxic chemicals, radioisotopes, organic compounds, particu-

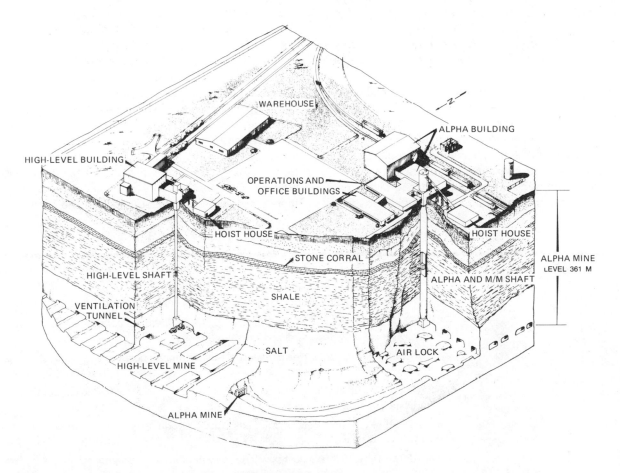

FEDERAL REPOSITORY

FIGURE 14.21
Schematic diagram showing how underground bedded salt may be used as a disposal site for solidified, high-level radioactive wastes. (Courtesy of U.S. Energy Research and Development Administration.)

lates, heat, and noise. All affect organisms and the environment.

The effect of a pollutant on a population is described by a dose-response curve. Individuals vary in their response to a pollutant because of behavioral, physiological, and genetic factors. The toxicity of a pollutant is generally measured by the amount that is required to cause injury and death to 50 percent of the population, an amount called the LD-50.

An important consideration for any pollutant is whether there is a threshold effect—a level below which effects are not observable and above which effects become apparent. Some effects of pollutants are reversible, and in some cases an individual population may become tolerant of a pollutant.

Pollutants tend to simplify ecosystems. They also tend to favor species that are adapted to variable environments and harsh conditions; organisms that are characteristic of early stages in ecological succession; and organisms that are hardy and have high reproductive rates and short lifetimes.

Acids have major effects on freshwater ecosystems. They reduce the availability of chemical elements necessary for life.

Toxic chemicals act directly on individuals and populations and are likely to be persistent in the environment. Their toxicity depends greatly on their particular chemical form, which may be altered biologically, especially by microbes in waters and soils. Heavy metals are a particularly important kind of environmental toxin.

Organic compounds are produced artificially for use in industry, in consumer products, as pesticides, as food additives, and as pharmaceuticals. Among organic chemicals, pesticides have caused the most environmental concern. Organic compounds can affect individuals, populations, ecosystems, and the biosphere. They can affect the physiology of individuals, cause mutations, promote the growth of cancer, or affect genetics. Ecosystem effects occur through biomagnification and the tendency of toxic organic compounds to simplify ecosystems.

Heat is a waste product of industries, especially the generation of electrical power. Its effects are most pronounced on bodies of water, particularly rivers and estuaries. The heat changes their average temperatures and thus affects the survival and persistence of fish.

Radioisotopes can affect organisms when incorporated into their bodies or when present outside of them. Radioisotopes emit one or more of three kinds of radiation: alpha, beta, or gamma. The environmental effects of a radioisotope depend on both its chemical activity and the forms of radiation emitted. Radioisotopes can kill individuals, cause mutations, and affect a population's genetics. Some radioisotopes undergo biomagnification and increase in concentration as they move along the food chains. Radioisotopes tend to simplify ecosystems and favor species adapted to disturbed areas, variable environments, and the early stages of succession.

Disposal of toxic wastes poses difficult technical and social problems. Our society has moved from "dilute and disperse" methods to a policy of "concentrate, contain, and recycle." Improper disposal can lead to a reintroduction of a toxin into the environment. The disposal of radioactive wastes will be one of the most difficult environmental problems our society faces in the future.

REFERENCES

1 ELLIOT, J. 1980. Lessons from Love Canal. *Journal of the Americal Medical Association* 240: 2033–34, 2040.

2 KUFS, C., and TWEDWELL, C. 1980. Cleaning up hazardous landfills. *Geotimes* 25: 18–19.

3 LIPPMANN, M.; and SCHLESINGER, R. B. 1979. *Chemical contamination in the human environment.* New York: Oxford University Press.

4 WALDBOTT, G. L. 1978. *Health effects of environmental pollutants.* 2nd ed. St. Louis: C. V. Mosby.

5 PATRICK, R.; BINETTI, V. P.; and HALTERMAN, S. G. 1981. Acid lakes from natural and anthropogenic causes. *Science* 211: 446–48.

6 GUTHRIE, F. E., and PERRY, J. J., eds. 1980. *Introduction to environmental toxicology.* New York: Elsevier.

7 MARTIN, D. H. 1980. Nuclear power. In *Introduction to environmental toxicology,* eds. F. E. Guthrie and J. J. Perry, pp. 77–90. New York: Elsevier.

8 HANSON, W. G. 1967. Cesium-137 in Alaskan lichens, caribou, and Eskimos. *Health Physics* 13: 383–89.

9 MACLEOD, G. K. 1981. Some public health lessons from Three Mile Island: A case study in chaos. *Ambio* 10: 18–23.

10 CAIRNS, J., Jr. 1980. Estimating hazard. *BioScience* 30: 101–7.

11 WHITE, G. F. 1980. Environment. *Science* 209: 183–90.

12 PERRY, J. J. 1980. Oil in the biosphere. In *Introduction to environmental toxicology,* eds. F. E. Guthrie and J. J. Perry, pp. 198–209. New York: Elsevier.

13 OFFICE OF TECHNOLOGICAL ASSESSMENT. 1979. *Pest management strategies in crop protection.* Vol. 1.

14 BROWN, A. W. A. 1978. *Ecology of pesticides.* New York: Wiley.

15 CARSON, R. 1962. *Silent Spring.* Boston: Houghton Mifflin.

16 DURHAM, W. F. 1970. Benefits of pesticides in public health programs. In *Biological impact of pesticides in the environment,* ed. J. W. Gilbert, pp. 153–55. Corvallis: Oregon State University Press.

17 STICKEL, W. H. 1975. Some effects of pollutants in terrestrial ecosystems. In *Ecological research: Effects of heavy metals and organohalogen compounds,* eds. A. D. McIntyre and C. F. Mills, pp. 25–74. New York: Plenum Press.

18 CLESCERI, L. S. 1980. PCBs in the Hudson River. In *Introduction to environmental toxicology,* eds. F. E. Guthrie and J. J. Perry, pp. 227–35. New York: Elsevier.

19 CHANLETT, E. T. 1979. *Environmental protection.* 2nd ed. New York: McGraw-Hill.

20 PIER, S. M. 1975. The role of heavy metals in human health. *Teax Reports on Biology and Medicine* 31: 85–106.

21 WOODWELL, G. M., and REBUCK, A. L. 1967. Effects of chronic gamma radiation on the structure and diversity of an oak-pine forest. *Ecological Monographs* 37: 53–69.

22 SCHNEIDER, W. J. 1970. *Hydraulic implications of solid-waste disposal.* U.S. Geological Survey Circular 601F.

23 COUNCIL ON ENVIRONMENTAL QUALITY. 1973. *Environmental quality.*

24 WALKER, W. H. 1974. Monitoring toxic chemical pollutants from land disposal sites in humid regions. *Ground Water* 12: 213–18.

25 MAGNUSON, E. 1980. The poisoning of America. *Time* 116: 58–69.

26 ALEXANDER, M. 1981. Biodegradation of chemicals of environmental concern. *Science* 211: 132–38.

27 OFFICE OF INDUSTRY RELATIONS. 1974. *The nuclear industry: 1974.*

28 BREDEHOEFT, J. D.; ENGLAND, A. W.; STEWART, D. B.; TRASK, J. J.; and WINGRAD, I. J. 1978. *Geologic disposal of high-level radioactive wastes—Earth science perspectives.* U.S. Geological Survey Circular 779.

29 MICKLIN, P. P. 1974. Environmental hazards of nuclear wastes. *Science and Public Affairs* 30: 36–42.

FURTHER READING

1 ALEXANDER, M. 1981. Biodegradation of chemicals of environmental concern. *Science* 211: 132–38.

2 BOND, R. G., and STRAUB, C. P. 1973–74. *CRC handbook of environmental control.* Cleveland: CRC Press.

3 BUTLER, G. C. 1978. *Principles of ecotoxicology.* (SCOPE Report #12.) New York: Wiley.

4 CHANLETT, E. T. 1979. *Environmental protection.* 2nd ed. New York: McGraw-Hill.

5 HOLDGATE, M. D. 1979. *A perspective of environmental pollution.* Cambridge, England: Cambridge University Press.

6 HORNE, R. A. 1978. *The chemistry of our environment.* New York: Wiley.

7 MUDD, J. B., and KOZLOWSKI, T. T. 1975. *Responses of plants to air pollution.* New York: Academic Press.

8 NATIONAL ACADEMY OF SCIENCES. 1981. *Testing for effects of chemicals on ecosystems.* Washington, D.C.: National Academy of Sciences.

9 OFFICE OF TECHNOLOGY ASSESSMENT. 1979. *Pest management strategies in crop protection.* Vol. 1.

10 SCHLESINGER, W.; REINERS, W.; and KNOPMAN, D. 1974. Heavy metal concentrations and deposition in bulk precipitation in montane ecosystems of New Hampshire, U.S.A. *Environmental Pollution* 6: 39–47.

11 STOKER, H. S., and SEAGER, S. L. 1976. *Environmental chemistry: Air and water pollution.* 2nd ed. Glenview, Ill.: Scott Foresman.

12 WALDBOTT, G. L. 1978. *Health effects of environmental pollutants.* 2nd ed. St. Louis: C. V. Mosby.

STUDY QUESTIONS

1 What kinds of life forms would most likely survive in a highly polluted world? What would their general ecological characteristics be?

2 Some environmentalists argue that there is no such thing as a "threshold" for pollution effects. What is meant by this statement? How would you determine if it were true for a specific chemical and its effect on a specific species?

3 What is a "reversible" effect of a pollutant?

4 What is biomagnification?

5 You are lost in Transylvania while trying to locate Dracula's castle. Your only clue is that the soil around the castle has an unusually high concentration of the heavy metal arsenic. You wander in a dense fog, only able to see the ground a few meters in front of you. What changes in vegetation warn you that you are nearing the castle?

6 Distinguish between acute and chronic effects of pollutants.

7 Design an experiment to test whether tomatoes or cucumbers are more sensitive to lead pollution.

8 Why is it difficult to establish standards for acceptable levels of pollution? In giving your answer, distinguish among the geological, climatological, biological, and social reasons.

9 You are hiking in the Blue Ridge Mountains of Virginia. You notice a blue haze on the hills in the distance. Your companion says, "That haze is produced by trees. They give off chemicals that pollute the air. Chemical pollution is natural—it is as old as trees. If trees pollute the air, so can we—it's perfectly natural." Would you agree or disagree? State your reasons.

10 In what ways is controlling pollution from solid wastes easier than controlling air pollution?

11 A new highway is built through a pine forest. Driving along the highway, you notice that the pines nearest the road have turned brown and are dying. You stop at a rest area and walk into the woods. One hundred meters away from the highway the trees seem undamaged. How could you make a crude dose-response curve from direct observations of the pine forest? What else would be necessary to devise a dose-response curve which would be used in planning the route of another highway?

12 Why is arsenic a poor choice as a pesticide?

13 In what ways are the effects of acid rain effects on *ecosystems?*

14 In your study of acid rain, you add sulphuric acid to a natural pond. After one year you observe no changes. Would you stop the experiment and conclude that acid rain is harmless for that pond?

15 Coal and nuclear power plants both create pollutants. What pollutants do they share in common? Which (coal or nuclear) poses the greater short-term (1 to 10 years) problem? Which poses the greater long-term (100 years or longer) problem?

15

Putting a Value on the Environment: Environmental Ethics and Environmental Economics

THE RELATIONS BETWEEN HUMAN BEINGS AND NATURE: NATURE AS AN IDEA

Every human society has a set of beliefs about nature, the effects of nature on human beings, and the effects of human beings on their natural surroundings. These universal and ancient concerns include an attempt to find order and harmony in nature, a design and purpose for this natural order, and the role of humanity in nature. Environmental studies brings together these ancient concerns with modern scientific and technological knowledge and principles. As the human population has changed, so has the relationship between human beings and their environment. Environmentalism—a concern with the environment and activism to protect and use it wisely—seems a relatively recent interest, but in fact its roots are deep within human history, society, and psychology.

Throughout the history of Western civilization, there have been three central questions asked about people and nature:

1 What is the *condition* of nature undisturbed by human influences?

2 What is the influence of nature on humankind?

3 What is the effect of humanity on nature, and what is humanity's role in nature [1]?

These are social and personal issues of moral, ethical, religious, and metaphysical importance, and they must be interpreted and reexamined in every age by every generation. At many times in human history these three questions have been controversial. In this section we will discuss the history of each of these questions and the various answers that have been given to them (see Fig. 15.1).

WILDERNESS AS A CONCEPT AND A REALITY

The concern with nature undisturbed is a concern with wilderness. Today it is popular to think of wilderness in a positive way, and to believe that it is something to be valued and preserved. But this was not always so. In some nontechnological societies—prior to herding and agriculture—there was no concept of wilderness as separate from one's immediate surroundings. Indeed, everything was habitat [2]. As one American Indian, Chief Luther Standing Bear, said, "We did not think of the great open plain, the beautiful rolling hills and the winding streams with their tangled growth as 'wild.' Only to the white man was nature a wilderness and . . . the land infested with wild animals and savage people'' [2]. It was only after early peoples began to herd animals and plant crops that the idea of uncontrolled or wild plants, animals, and land emerged.

The idea of wilderness as separate from people and their habitat can be found rather early in European civilization. The word itself is derived from the old Anglo-Saxon word *wild(d)ēor* (wild beast) and means literally the place of wild creatures. In the great Anglo-Saxon epic poem *Beowulf*, wilderness was viewed as a place of danger, horror, and discomfort [3]. It was the home of strange, mysterious, and dangerous creatures like the monster Grendel and a place where a brave man could test himself and prove himself a hero, as does Beowulf when he ventures out from the warm hearth of the king's castle to slay the evil Grendel and his mother. In many primitive societies, when civilization had yet to develop much control over the environment, this is the way wilderness was seen: as wild, dangerous, and horrible, and as a place where a person could test himself against the challenges of nature.

This view of wilderness is common in Western civilization, and many authors have written of it in this way. Even in eighteenth-century England we find wilderness similarly described. In a classical work of that century, *Natural History, General and Particular*, Count de Buffon describes nature untouched by human beings as ''melancholy deserts'' which are ''overrun with briars, thorns and trees which are deformed, broken and corrupted''; or wetlands ''occupied with putrid and stagnating water . . . covered with stinking aquatic plants'' that ''serve only to nourish venomous insects, and to harbour impure animals.'' A person in wilderness has to ''watch perpetually lest he should fall victim to [wild animals'] rage, terrified by the occasional roarings and even struck with the awful silence of those profound solitudes'' [4].

To those who see wilderness in this sense—as chaotic, uncontrolled, and dangerous—the role of people is to tame, manage, and order it. It is a human being who ''cuts down the thistle and the bramble, and he multiplies the vine and the rose'' [4]. From this point of view, *nature is disordered, and the role and purpose of humankind is to add the order, harmony, and balance that is lacking.* Sometimes this concept is extended: *Homo sapiens'* purpose on Earth is to add the final touches of order to an almost harmonious world.

This idea of people as the husbanders of nature is in great contrast to one of the predominant views that runs throughout the history of Western civilization and is particularly common in our own time. This is the view that *nature undisturbed is perfectly ordered, balanced, and harmonious* [1]. According to this belief, it is human

(a)

(b)

(c)

(d)

(e)

FIGURE 15.1
The many views of nature and wilderness. (a) *The Unicorn in Captivity*. In this medieval tapestry we see plants illustrated in exquisitely accurate detail surrounding the mythical animal. This mixture of the real and the imagined, of the hard facts of one's habitat and the idealizations that might lie within it, runs through the history of human thoughts about nature. (b) Nature as wild, dangerous, and a place where a person could test himself against the challenges of nature is illustrated fancifully in Rousseau's nineteenth-century painting *The Indian and the Gorilla in the Jungle.* (c) Nature as disordered, and the role of people as adding order, harmony, and balance, are illustrated in the 1872 American painting *American Progress* by John Gast. One can also see in this painting a representation of nature as a commodity to use up or a raw resource to mold into something useful. (d) Nature as harmonious, ordered, and complete without human influence is illustrated by this nineteenth-century painting by Joseph Francis Cropsey, *Autumn on the Hudson River.* This painting can also be seen as illustrating nature as beautiful and a key for an individual to discover himself. (e) Nature as unpredictable and uncaring, and the role of human beings as a struggle to survive against the capriciousness of Mother Nature, are illustrated by the nineteenth-century painting *Sea Tragedy* by Albert Pinkham Ryder. [Photo (a) courtesy of Metropolitan Museum of Art, New York; photos (b) and (d) from the National Gallery of Art, Washington, D.C.; photo (c) from Harry T. Peters, Windholme Farm, Orange, Virginia; photo (e) from Ball State Art Museum, Ball State University, Muncie, Indiana. From Sweeney, 1977.]

391

beings who upset the order and are the great destroyers of nature's balance. This is the viewpoint evident in the writings of many classical philosophers in Greece and Rome. Their writings suggest a belief that physical and biological nature have a perfect order, balance, and harmony. This perfect order suggested that there was a purpose behind the order, that the purpose was divine—from gods or the God—and that the object of this purpose was humankind. Aristotle, who some might call the grandfather of the study of ecology, perceived this order in many aspects of biology as he observed it [5]. Cicero summarized many of the classical beliefs in the century just before the modern era in his book *The Nature of the Gods.* Cicero saw order and purpose in many of the ecological phenomena described in earlier chapters of this text. He saw order in the food habits of animals and in the amazing adaptations of living creatures for their needs. Lacking a theory of biological evolution, Cicero and other classical writers believed these adaptations were part of a divinely clever and purposeful plan. Even the elephant, Cicero observed, "has a trunk, as otherwise the size of his body would make it difficult for him to reach his food" [6]. The order in nature was seen in the interactions among species, as in symbiotic relationships. Cicero describes a shellfish that "by entering into an alliance" with a small fish obtains its food: "When a small fish swims inside the gaping shell, the [fish] with a bite signals it to close it. So two very different creatures combine to seek their food together." Cicero marveled at these adaptations and asked, "What power is it which preserves them all according to their kind," adding, "Who cannot wonder at this harmony of things, at this symphony of nature which seems to will the well-being of the world?"

Assuming there was a purpose behind the order in nature, the next question was, For whom was the world so well ordered? The answer naturally enough had to be human beings, who were intelligent enough alone among all living things to appreciate it. This interpretation of nature, then, is that *nature is ordered, balanced, and harmonious; human beings, like all living things, have a place and purpose in this order; and the divine purpose of nature's order is for human benefit.*

Among Greek and Roman writers, Lucretius was one of the strongest opponents of this view. He argued the opposite: Nature was not made for human benefit; nature gave human beings only a hard life; and one must struggle to survive and obtain the necessities of life against the natural workings of things. He saw nature as capricious. "How many a time the produce of great agonies of toil burgeons and flourishes," wrote Lucretius in *De Rerum Natura*, "and then the sun is much too hot and burns it to a crisp; or sudden cloudbursts, zero frosts, or winds of hurricane force are, all of them, destroyers" [7]. While nature provides every need of nonhuman creatures, human beings alone must struggle for their existence. This view characterizes *nature as unpredictable and uncaring,* and sees *the role of human beings as a struggle to survive against the capriciousness of Mother Nature.* Human influence on nature is small and only affects those living things that bear directly on human life.

It has been argued that people in hunting and gathering societies do not make a distinction between human beings, their habitat, and a separate wilderness. Once technology began—even the most primitive agriculture and herding—then wilderness was seen as separate from home and human habitat [8]. The less control one has over nature, the more likely one would view nature undisturbed by human influence as dangerous, capricious, and an obstacle that one must struggle against to survive. Thus it is not surprising to find such views among societies with less technological control over their environment, and less assurance about the production of the necessities of life, than our own.

The European Renaissance forced a reexamination of the ideas for and against order in nature, and there was a renewal of the controversy over the balance and harmony of nature. The explorations and discovery of new, strange creatures; the development of science and an increased understanding of the processes that were involved in nature; and the increase in technology and the power of civilization over the environment all promoted such a reexamination. The same issues were restated, and the same arguments presented, with new pieces of evidence [9].

With the development of the new physics of Newton, Galileo, and others, people optimistically believed that the workings of the world could be understood from physical laws—that the world and the creatures who lived in it could be understood like a mechanical device. Because they believed nature was a system following inexorable laws of the universe, they could learn to understand and control it. Their view was that *nature is like a machine, and human beings can learn to be nature's engineers—captains of the great Earth ship* [9].

In the nineteenth century, another great change in ideas took place, best known to us through the writings of the Romantic poets. Part of this view is that wildness is to be appreciated, that the power and unpredictable grandeur of nature is beautiful, sublime, and a sign of the power and glory of God. The change from the belief that wilderness was dangerous and therefore bad to a belief that the dangerous was wonderful and magnificent can be traced in the reports of British travelers through the Alps in the eighteenth and nineteenth centuries [9]. At

FIGURE 15.2
The Alps. (Photo courtesy of the Swiss National Tourist Office.)

first, those who traveled through the Alps saw them as places of horrible disorder and danger to be passed through quickly. Soon after, travelers began to talk of the Alps as having a "terrible joy." Finally, Percy Bysshe Shelly saw in the Alps' Mont Blanc the ultimate in powerful sublimity. "Mont Blanc yet gleams on high; /The power is there," he wrote. "Power dwells apart in its tranquillity, / Remote, serene, and inaccessible." In viewing the mountain, Shelley found "The secret strength of things, / Which governs thought, and to the infinite dome of heaven as a law" [10] (Fig. 15.2). The Romantic view was that nature's wild power was grand, and for a person, the experience of wild nature was of great significance and a source of inspiration, peace, and beauty, This is a view of *nature as beautiful in its wildness, and a key for an individual to discover himself and the meaning for existence.*

With the discovery of America, the largest wilderness ever known to Western civilization was opened to exploration. Early settlers saw wilderness as a commodity, a resource to be utilized to improve their own economic well-being. Wilderness was to be conquered and transformed. Progress, as in the famous painting of nineteenth-century America (Fig. 15.1c), was seen as a grand lady subduing the wilderness, driving away the

wild animals and primitive Indians, and bringing with her the wondrous new inventions of technology.

For example, in early New England, the wilderness was forest to be opened up—to be burnt, cut, and transformed into farms. In the region around Mt. Monadnock in New Hampshire, fires lit in the nineteenth century to clear the forest burned so hot over the mountaintop that the very bedrock cracked and trees were eliminated completely from the summit. The once forested summit is now a barren, rocky ridge; closed woodlands have been replaced by a bald summit, which at least provides today a place to stand and see a magnificent view of the New England countryside (Fig. 15.3).

In the Big Woods of Michigan, wilderness was 8 million hectares of white pine, a commodity to be cut to build houses across the eastern and midwestern new nation. Today a single preserve of less than 25 hectares in Hartwick pines exists in central Michigan. The American wilderness was Buffalo Bill killing Indians; it was Daniel Boone; it was the big sky. It was "The People Yes" country of Carl Sandburg's poem [11]—the place for optimism, of big things to conquer. The American western wilderness was the land of the strange and the big, and the land of the tall tale, where the wind blew so strong you had to make a kite of an iron shutter and a chain [12]; where fog

FIGURE 15.3
The summit of New Hampshire's Mt. Monadnock, once forested to the top, was burned in the nineteenth century by fires lit to clear lands in the valleys. Trees and soils were removed by the fires, leaving a barren, rocky summit which has not revegetated. (Photo by D. B. Botkin.)

was so thick you could put shingles on it; where plants grew so fast that "the boy who climbed a cornstalk . . . would have starved to death if they hadn't shot biscuits up to him"; and where a herd of cattle got lost in a hollow redwood tree [11].

Following not far behind the belief in progress and subduing the wilderness for personal benefit was the recognition that industry might kill the soil that was growing the golden corn and ruin the streams that had been full of trout and salmon. At first the country seemed too big a wilderness for anyone to destroy it. The early loggers in Michigan are said to have believed they would never run out of wood; it would take so long to cut it the first time that by the time the virgin forests were cut the trees would have grown back enough to start cutting again.

The view that *wilderness was a commodity to use up or a raw resource to mold into something usable* con-

flicted with a new belief that wilderness and human beings were somehow intimately tied together and that the preservation of the latter required the preservation of the former. The first important American statement of these concerns was made in *Man and Nature* in 1864 by George Perkins Marsh, native of Vermont and U.S. ambassador to Italy and Egypt [13]. Struck by the differences in the soils, forests, and general appearance of the landscape in Europe and North Africa, which had been used by civilized human beings for thousands of years, and the still barely touched wilderness of Vermont and New Hampshire, Marsh proposed that the rise and fall of civilizations were linked to the use and misuse of nature. He suggested that the misuse of farmland contributed to the fall of Rome; as the Romans exhausted the soil near to the city, they were forced to expand their empire to obtain new sources of food. The process repeated itself, leading to an ever-increasing network of transportation and government. The empire finally collapsed when the distances required to transport the food exceeded the capabilities of the technologies of the period.

While modern historians may argue that such a view is oversimplified, Marsh was the first to state the possibility of a life-sustaining dependency of *Homo sapiens* on nature's balance. He became the mid-nineteenth century's classic proponent of the idea that nature undisturbed achieves a permanency of form and substance, a harmony that only human beings destroy. Thus the view that *in wilderness is one of the mechanisms for the survival of Homo sapiens* began in the United States with George Perkins Marsh.

The opening of the American West led to the discoveries of scenic grandeur in Yosemite, Yellowstone, and the Grand Canyon (Fig. 15.4), which were set aside as national parks—monuments of nature's freaks, curiosities, and grandeur. As the historian Alfred Runte has made clear, monumental scenery was the United States' answer to the architectural and sculptural wonders of the older civilizations in Europe [14]. The national parks were not seen as biological or ecological units to preserve as natural living systems, but as places where the peculiar (geysers, rock formations, etc.) could be viewed by the average citizen, and from which a sense of peace, beauty, and even religious experience, such as Shelley had found early in the nineteenth century in the Alps, could be gained. In the late nineteenth century, then, developed the idea of *wilderness as the place of the strange and monumental.* Wilderness exists outside of the needs or uses of ordinary life, as much for entertainment as a circus for some, or as a shrine for spiritual inspiration as a church for others [14].

The first great U.S. conservation movement began in the late nineteenth and early twentieth centuries, stimu-

FIGURE 15.4
Yosemite Valley, a place of scenic grandeur, illustrates the American view of wilderness as a curiosity to be seen and set aside for the future. (Photo courtesy of Douglas Whitesides.)

lated by the nineteenth-century writings of Henry David Thoreau, Ralph Waldo Emerson, and Marsh; by the ideas of the founders of the national parks and national forests,

such as John Wesley Powell, the first to travel down the Grand Canyon's Colorado River in 1869; and by the deep reverence for nature found in the writings of John Muir.

The "land ethic" put forward by Aldo Leopold in 1949 affirms the right of all resources, including plants, animals, and earth materials, to continued existence and, at least in certain locations, their continued existence in a natural state [15]. This ethic effectively changes our role from conquerors of the land to that of citizen and protector of the environment. This new role requires that we revere and love our land and not see the land as solely an economic commodity to be used up and thrown away.

Leopold emphasized the lack of ethics regarding the environment through the story of Odysseus, who, upon returning from Troy, hanged a dozen slave women for suspected misbehavior during his absence. There was no question of ownership; the women were property, and the disposal of property in this case was a matter of expediency, much as it is today. Although since that time ethical values have been applied to many other areas of human behavior, only within this century have moral considerations been extended to the relation between civilization and its physical environment (Fig. 15.5).

Ecological ethics limit social as well as individual freedom of action in the struggle for existence [16]. A land ethic assumes that we are ethically responsible not only to other individuals and society, but also to that larger community consisting of plants, animals, soils, atmosphere, and water—that is, the environment.

There exists a possible dichotomy or source of confusion between an ideal and a realistic land ethic. Giving rights to the plants and animals and landscape might be interpreted as granting to individual plants and animals

FIGURE 15.5
The evolution of ethics. (After Nash, 1977.)

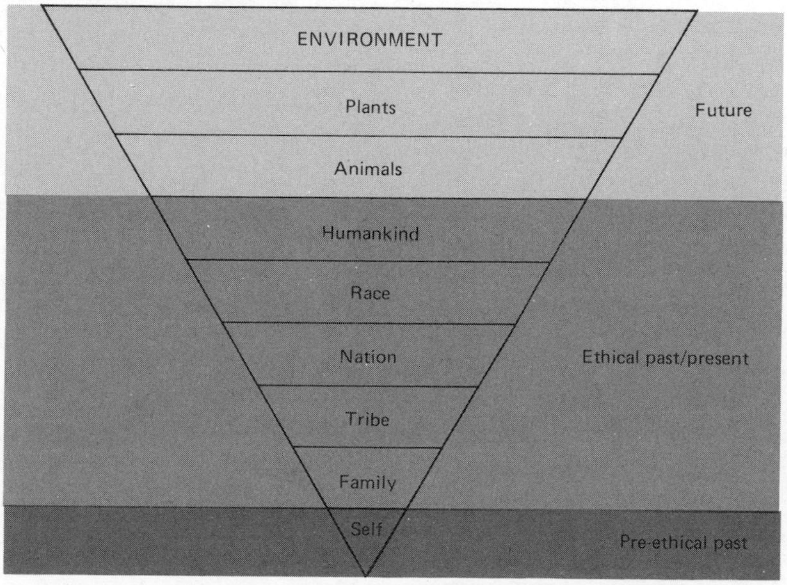

ENVIRONMENT

Plants

Animals

Future

Humankind

Race

Nation

Ethical past/present

Tribe

Family

Self

Pre-ethical past

the fundamental right to live, as in the Eastern Indian religion called Jainism. However, we must eat to live; not being autotrophs, we must consume other organisms. Therefore, although the land ethic assigns rights for animals to survive as species, it does not necessarily assign rights to an individual deer, cow, or chicken for that survival. The same argument may be given to justify the use of stream gravel for construction material or the mining of other resources necessary for our well-being. However, unique landscapes with high aesthetic value or ecosystems that sustain endangered species need to be protected within this ethical framework.

The land ethic places human beings in the role of the husbanders of nature, with a moral responsibility to sustain it for itself and for future generations of human beings. According to this view, *wilderness is of intrinisic value to be maintained for itself and because our own survival depends upon it.*

This chapter began with the statement that ideas about nature must be reinterpreted and reexamined in every age and by every generation. Having seen the development of these ideas throughout history, we can ask, What is the truth about nature for our times? In Chapter 1 we learned that nature undisturbed is not constant in time; it does not have the perfect constancy or balance that has been, and is still, so often attributed to it. In our own times, the controversy about nature undisturbed continues, with some additions. In spite of evidence to the contrary, many still argue that wilderness undisturbed achieves an ecological permanency which only human beings disrupt.

From the viewpoint of the individual, the wilderness, like the white whale Moby Dick, means many different things. In our age wild nature represents unfettered power to be watched in awe or to serve as a place to test oneself. It is also a place of sublime peace and beauty, or a reservoir of mineral and biological resources to be used for our economic betterment. At the end of the twentieth century, wilderness can be found in many different places. As one of this century's historians of wilderness, Roderick Nash, has written, wilderness is in the end a state of mind—wilderness is where you find it [8, 16]. The wilderness that is a source of peace or a place to test oneself might be found by a city dweller in an urban park or a seashore much altered by civilization. Recent incidences of mountain climbers scaling city skyscrapers may be seen as one illustration of this viewpoint. For someone familiar with back country, only the most pristine areas distant from all sounds of modern civilization might serve that same purpose.

From a societal point of view, mounting scientific evidence tells us that we indeed depend on the living processes of nature; ours is truly a living planet whose atmosphere, oceans, and sediments are biological products. Two modern scientists, J. E. Lovelock and L. Margulis, have argued that we should return to the ancient Greek idea of *Gaia,* or the idea of Mother Earth as the sustainer [17]. This idea recognizes that in the end our survival depends on the persistence of ecosystems and biospheric processes which our civilization has already begun to change. The moral and ethical implication of this view is that human beings must husband or manage wisely the biosphere and local ecosystems and their rare and endangered species. Thus, *wilderness and human beings are one; their persistence depends on each other.* Nature is still seen as strongly influencing individuals, but each person in his own way detects this influence to a greater or lesser degree. The nature of nature still remains an area of intensive scientific inquiry—a subject with many questions but few answers—and an important area for us to study.

Once we have adopted a set of beliefs or principles about the environment, how do we put them into practice? How do we plan so that our use of the biosphere is consistent with our desires? In the process of planning, we make use of the knowledge we have of the environment, as presented in the first two sections of this book. Next, we need to consider economic principles, the methods and principles of planning and forecasting, and the possible social mechanisms, including the legal framework, that make planning a reality.

ENVIRONMENTAL ECONOMICS

Once we have decided that something is an environmental "good," how do we put a value on it to compare it with other factors? How do we turn our choices into decisions, that is, how do we insure that individuals and society will act in a way to accomplish environmental goals? What are the options open to a society?

THE ENVIRONMENT AS A "COMMONS": PROFITS, RENTS, AND EXTERNALITIES

A society has a number of social mechanisms to achieve its goals. Laws and legal regulation are one method; use of resources can be limited by setting quotas or by selling a limited number of licenses (as with sport fishing). Or, a society might rely on individual motivation, under the assumption that what an individual finds best for himself will also be best for society. This approach leaves the individual complete freedom of action.

One of the first questions often asked about managing the environment is, Why don't individuals want to act in a way that always protects the environment and maintains biological resources in a renewable state? One way to answer this question is by economic analysis. Economic analysis shows us that the individual's profit motive alone will not always act in the best interest of the environment, at least not when certain *externalities* are brought into account.

The Management of Biological Resources The management of biological resources involves several major economic and social issues. Two of these issues are, How can we increase the production of biological resources without endangering their long-term survival or other aspects of the environment? and, What is the upper limit of world production of each of these resources?

Some biological resources, such as much of the U.S. forests, occur on privately owned lands. Other resources occur on nationally owned lands, such as 38 percent of our nation's forests. Resources that occur in international regions are not controlled by any single nation. Many of the world's fisheries are in this last category.

When biological resources occur in international or national areas with open access, the resources are threatened by what Garrett Hardin has called the "tragedy of the commons" [18]. The **commons** were parts of old English and New England towns where every farmer could graze his cattle. This practice works as long as the number of cattle is low. Hardin points out that each herdsman in such a situation is trying to maximize his gain, and must ask himself continually whether he should add more cattle. The addition has both a positive and negative utility. The positive utility is the benefit the herdsman receives when he sells the cattle; the negative one is the overgrazing caused by the addition of any one animal. Hardin argues that since the benefit to the individual of selling a cow is greater than his individual loss in the deg-

radation of the commons, freedom of action in a commons inevitably brings ruin to all. He says that, without some management or control, all natural resources treated like commons will be inevitably destroyed. The overcrowding of national parks and the pollution of the atmosphere are given as examples of this tragedy [18].

As another example, many have argued for a reduction or elimination of commercial whaling in the interest of conserving the great whales. A question that naturally arises about whaling is, If whalers profit by harvesting whales, why do they not act to protect the resource on which their livelihood depends? Ranchers, after all, do not intentionally kill all their cattle just because they can make money by selling cattle. However, whalers have brought species of the great whales to the brink of extinction. The blue whale was reduced to only an estimated few hundred before harvesting of it was stopped in the 1960s (Fig. 15.6).

Hardin's analysis shows that a fundamental issue in whaling and fishing is the lack of property rights. Ranchers refrain from killing off entire herds because the benefits of maintaining the herds (and the yet unborn) are reaped by them in the future. Fishermen or whalers, however, cannot be assured that they will reap those benefits. The offspring of the fish or whales they do not kill today are not theirs to harvest in the future.

From the example of whaling, one might argue in favor of private ownership of what have been the commons. On the other hand, some argue that private ownership of public goods, like fisheries and forests, is undemocratic and unfair. The issue of resource regulation and ownership has been raised by the "Sagebush Rebellion"—a political movement of western ranchers who want more local control and access to public lands for private cattle grazing. As the human population grows and resources remain constant or decline, greater and greater

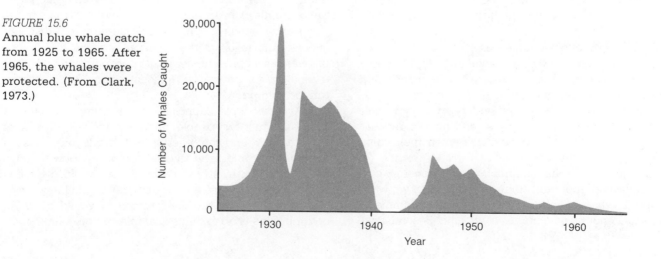

FIGURE 15.6
Annual blue whale catch from 1925 to 1965. After 1965, the whales were protected. (From Clark, 1973.)

competition will occur for their use, and any decisions about political and economic control will be more hotly contested.

Another reason that individuals will tend to overexploit natural resources held in common, according to mathematician Colin Clark, has to do with what economists call **economic rent** [19]. To understand this concept, consider the whales in the ocean and the whale oil, which is a marketable product, as the capital investment of the industry. Whalers have a variety of policies that they might adopt toward harvesting whales. A **policy** is a set of rules to determine which actions will lead to a certain goal. How can whalers get the best return on their "capital"? (Here we need to remind ourselves that the whales, as a biological population, increase only if there are more births than deaths.) Let's examine two extreme policies. If they adopt a *conservative policy,* they will harvest only the net biological productivity each year and maintain the total abundance of whales at its current level; that is, they will stay in the whaling business indefinitely. If they adopt the *maximum immediate profit policy,* they will harvest all the whales now, sell the oil, get out of the whaling business, and invest the profits.

Suppose they adopt the first policy. What is the maximum gain they can expect? Whales, like other large, long-lived creatures, reproduce slowly; a calf born every 3 or 4 years per female is typical. The total net growth of a whale population is unlikely to be more than 5 percent per year. If all the oil in the whales in the ocean today represented a value of $10 million, then the most they could expect to take in each year would be 5 percent of this, or $500,000.

If they adopt the second policy and harvested all the whales, then they could invest the money from the oil. Although investment income varies, even a conservative investment of $10 million would very likely yield more, particularly since this income could be received without the cost of paying a crew, maintaining the ships, buying fuel, marketing the oil, and so on.

Thus, it is quite reasonable—and in fact quite practical, if one only considers direct profit—to adopt the second policy: Harvest all the whales, invest the money, and relax. Whales just are not a highly profitable long-term investment. It is no wonder that there are fewer and fewer whaling companies and countries, and no wonder that countries leave the whaling business when their ships become old and inefficient.

The problems of the commons and of economic rent point out several important things. First of all, if we want to conserve whales, we must think beyond the immediate, direct economic advantages of whaling. Second, policies that seem ethically good to us may not be the most profitable for an individual. In the example of whaling, the economic analysis clarifies how an environmental resource is used. An economist would say that we must be concerned with externalities in whaling. An **externality** is an effect not normally accounted for in the cost-revenue analysis of producers [19].

Other factors to consider in resource use are the relative scarcity of a necessary resource and its price. For example, if a whaler lived on a desert island and whales were the only food he could obtain, then he would have to consider his interest in staying alive for awhile as well as his short-term profit. Of course, even a whale-eating whaler might choose a policy somewhere in between the two extremes. He might decide that his own life expectancy was only 10 years, and he would try to harvest the whales so that they and he would go extinct at the same time. "You can't take it with you" would be his attitude. This would not happen necessarily if ocean property rights existed. A whaler could then sell his rights to future whalers, or mortgage against them, and thus reap the benefits of whales to be caught after his death.

THE DISCOUNT FACTOR

The preceding example reminds us of the old saying, "A bird in hand is worth two in the bush." That is, a profit *now* is worth much more than a profit *in the future.* This economics concept—the future value compared with the present value—is another important one for environmental studies. Economists refer to this concept as the **discount factor.** The discount factor is how much something is worth in the future compared with what it is worth now. Economists observe that the market determines a discount rate that is often, but not always, less than 1. (A discount factor less than 1 means that something promised in the future has less value than something given today.) The market-determined discount factor is the result of the interaction of the consumer's preferences for present and future consumption and of the technology for transferring present consumption into future consumption.

As an example, suppose you were dying of thirst on a desert and met two people; one offered to sell you a glass of water now, and the other offered to sell you a glass of water if you could be at the well tomorrow. How much is each glass worth? If you believed you would die today without water, the glass of water today would be worth all of your money and the glass tomorrow would be worth nothing. This is an extreme example of a discount factor.

In real life, things are rarely so simple and distinct. But we all know we are mortal, so we tend to value personal wealth and goods more if they are available now than in the future. This evaluation is made more complex, however, because we are accustomed to think of the future—to plan for retirement or a nest egg for our children.

Modern concerns with the environment have placed a novel emphasis on the discount factor. Conservationists often argue that we have a debt to future generations and must leave the environment in at least as good a condition as we found it. Such conservationists would argue that the future environment is not to be valued less than the present. In economic terms, this means that the discount factor is 1 (the environment in the future is just as valuable as the environment today) or greater than 1 (the environment in the future is worth more than the environment today).

For example, suppose you are the manager of the whooping cranes and are paid in relation to your success in keeping that species from extinction. The assurance that whooping cranes would be alive 10 years from now would seem to have more value to you than the assurance that they would be alive tomorrow. These different attitudes toward the discount factor pose a dilemma for environmental studies.

First of all, economists would argue that it is difficult, if not impossible, to make a sound economic analysis when the discount factor is greater than 1. Suppose we agree that a whooping crane alive next year is worth twice a live one today, and so on. In 2 years, the whooping crane is 4 times as valuable, and in 3 years, 8 times as valuable. If your salary is proportional to that value, your salary will very quickly become infinite. Clearly, this is not a feasible approach.

Second, many would argue that, rhetoric to the contrary, everybody really does place a higher value on a possession today rather than promised tomorrow. In other words, you would really rather have wilderness today while you are alive to enjoy it than wilderness tomorrow when you might be dead. This situation, of course, would be subject to the technological devices which affect the transfer of present consumption into future consumption.

How to deal with the discount factor in an economic analysis of environmental issues is an unsolved problem. The concept of the discount factor is, however, very important as we seek the environment we desire.

From the preceding discussion we see that effective management of the environment requires a clear understanding of the reasons for overexploiting our resources. That is, we need to know about the problems of the commons and the concepts of the discount factor, economic rent, and externalities.

RISK/BENEFIT ANALYSIS

Our discussion of the economics of whaling and the discount factor illustrates that the riskiness of the future influences the value we place on things now. Another important concept used in environmental economics is that of **risk/benefit analysis.** All of life, including populations, resources, ecosystems, and the biosphere, involves risk. Consider, for example, the effects of pollutants on life. How can we deal with these? How much are we willing to pay to reduce or eliminate a pollutant? The answers depend on the risks involved.

Pollutants have effects on human health, commercially important or essential products such as food crops, wildlife, natural ecosystems, and the biosphere. The ecosystem and biospheric effects may in turn have indirect negative effects on human beings. Pollutants can cause annoyance, injury, or death. Like natural hazards, they can be dangerous to human beings; but unlike natural hazards, they often act in subtle and slow, sometimes almost imperceptible, ways. How do we know if something is indeed dangerous or toxic?

Death is the fate of all individuals, and every activity in life involves risk of injury or death. What then does it mean to save a life by reducing the level of a pollutant?

With some activities, the relative risk is clear. It is much more dangerous to stand in the middle of a busy highway than to stand on the sidewalk. Hang gliding has a much higher mortality rate than hiking. Table 15.1 gives the risk associated with a variety of activities. The effects of pollutants, however, are often more subtle. Populations subject to high levels of certain pollutants have a lower average life expectancy or a higher incidence of certain diseases. But even in such a population any one of us might live a ''normal'' lifespan, or even longer.

The degree of risk is important in our legal processes. For example, the Toxic Substances Control Act states that no one may manufacture a new chemical substance or process a chemical substance for a new use without obtaining a clearance from the U.S. Environmental Protection Agency (EPA). The act establishes procedures to estimate the hazard to the environment and to human health of any new chemical *before* it becomes widespread. The EPA examines the data provided and judges the degree of risk associated with all aspects of the production of the new chemical or the new process, including extraction of raw materials, manufacturing, distribution, processing, use, and disposal. The chemical can be

TABLE 15.1
The risk of death for various activities (given as the chance of death per year).

A. Recreation

Activity		No. of Deaths in 1975	Risk/yr
Football			4×10^{-5}
Automobile racing			1.2×10^{-3}
Horse racing	Averaged over participants		1.3×10^{-3}
Motorcycle racing			1.8×10^{-3}
Powerboating			1.7×10^{-4}
Boxing (amateur)	40 hr/yr engaged in sports		2×10^{-5}
Skiing			3×10^{-5}
Canoeing			4×10^{-4}
Rock climbing (U.S.)			10^{-3}
Sunbathing, mountain climbing (skin cancer risk/curable)		300,000 cases	5×10^{-3}
Fishing (drowning)	Averaged over fishing licenses	343	1.0×10^{-5}
Drowning (all recreational causes) all over U.S.		4110	1.9×10^{-5}
Bicycling (assuming one person per bicycle)		1000	10^{-5}

B. Commonplace Activities[a]

Activity	No. of Deaths in 1974	Risk/yr
Motor vehicle (in 1975)		
Total	46,000	2.2×10^{-4}
Pedestrian (certainly involuntary)	8,600	4×10^{-5}
Home accidents (1975)	25,500	1.2×10^{-5}
Alcohol		
Cirrhosis of the liver (1974)		1.6×10^{-4}
Cirrhosis of the liver (moderate drinker)		4×10^{-5}
Air travel		
One transcontinental trip/yr		3×10^{-6}
Jet-flying professor		10^{-4}
Accidental poisoning		
Solids and liquids	1,274	6×10^{-6}
Gases and vapors	1,518	7×10^{-6}
Inhalation and ingestion of objects	2,991	1.4×10^{-5}
Electrocution	1,157	5×10^{-6}
Falls	16,339	7.7×10^{-5}
Tornadoes (average over several years)	160	5×10^{-7}
Hurricanes (average over several years)	118	4×10^{-7}
Lightning (average over several years)	90	4×10^{-7}
Air pollution		
Total U.S. estimate (sulphates)	30,000	1.5×10^{-4}
Urban U.S. (benzo(α)pyrene)—cancer risk		3×10^{-5}
Vaccination for smallpox (per occasion)		3×10^{-6}
Living for 1 yr downstream of a dam (calculated)		5×10^{-5}

[a]Accuracy approximately 30 percent.

TABLE 15.1 (continued)

C. Radiation (cancer risks)

Activity	Estimated Uncertainty (factor of 3)	Risk/yr
Cosmic ray risks		
One transcontinental flight/yr	3	5×10^{-7}
Airline pilot (50 hr/month at 35,000 ft)	3	5×10^{-5}
Frequent airline passenger	3	1.5×10^{-5}
One summer (4 months) camping at 15,000 ft	3	10^{-5}
Living in Denver compared with New York	3	10^{-5}
Other radiation risks		
Average U.S. diagnostic medical X-rays	3	10^{-5}
Increase in risk from living in a brick building (with radioactive bricks) compared with wood	3	5×10^{-6}
Natural background at sea level	3	1.5×10^{-5}

D. Food, Tobacco, and Drugs (cancer risks)

Activity	Estimated Uncertainty (factor of 3)	Risk/yr
Eating and drinking		
One diet soda (saccharin)	10	10^{-5}
Average U.S. saccharin consumption	10	2×10^{-6}
4 tb peanut butter/day (aflatoxin)	3	4×10^{-5}
One pint milk per day (aflatoxin)	3	10^{-5}
Miami or New Orleans drinking water (chloroform)	5	1.2×10^{-6}
½ lb charcoal broiled steak once a week (benzopyrene) (cancer risk only; heart attack, etc. additional)	10	4×10^{-7}
Alcohol		
Averaged over smokers and nonsmokers	3	5×10^{-5}
Light drinker (one beer/day)	3	2×10^{-5}
Tobacco		
Smoker		
Cancer only	3	1.2×10^{-3}
All effects (including heart disease)	3	3×10^{-3}
Person in room with smoker	10	10^{-5}
Miscellaneous		
Taking contraceptive pills regularly	10	2×10^{-5}

E. Employment-related Activities[a]

Activity	Number of Fatalities (in 1975 unless stated)	Risk/yr
Mining and quarrying (accident only)	500	6×10^{-4}
Coal mining		
Accident (average 1970–1974)	180	1.3×10^{-3}
Black lung disease (1969)	1135	8×10^{-3}

TABLE 15.1 *(continued)*

E. Employment-related Activities[a] *(continued)*

Activity	Number of Fatalities (in 1975 unless stated)	Risk/yr
Agriculture		
Total	2100	6×10^{-4}
Tractor driver (one driver/tractor)		1.3×10^{-4}
Trade	1200	6×10^{-4}
Manufacturing	1500	8×10^{-5}
Service	1800	9×10^{-5}
Government	1100	1.1×10^{-4}
Transportation and utilities	1600	3.3×10^{-4}
Airline pilot		3×10^{-4}
Truck driver (one driver/truck)	400	10^{-4}
Jet-flying consultant and professor		10^{-4}
Steel worker (accident only) (1969–1971)	66	2.8×10^{-4}
Railroad worker (1974) (all accidents excluding grade crossing)	688	1.3×10^{-3}
Firefighters (1971–1972 average)		8×10^{-4}

[a]Accuracy approximately 30 percent.

F. Environmental Toxins

Chemical Name	Production (g)	Potency (mg^{-1} kgd)	Cancer Dose (g)	Total Potential No. of Cancers (yr^{-1})
Trichloroethylene	2.5×10^{11}	0.0001	1.5×10^{7}	1.6×10^{4}
NTA (until 1970)	7.5×10^{10}	0.0002	8×10^{6}	8×10^{3}
Tetrachloroethylene	3.4×10^{11}	0.001	1.5×10^{6}	2×10^{3}
Tetrachloroethane	2×10^{11}	0.004	4×10^{5}	4×10^{4}
Trifluralin	1.2×10^{10}	0.0004	5×10^{6}	2×10^{3}
Tris (or TBP)	1.5×10^{9}	0.01	1.6×10^{3}	8×10^{3}
2,4-Diaminoanisole sulphate	1.5×10^{7}	0.0008	2×10^{6}	6.4
1,2-Dichloroethane	4.5×10^{12}	0.003	6×10^{5}	7×10^{6}
1,4-Dioxane	8×10^{9}	0.001	1.6×10^{6}	5×10^{3}
1,2-Dibromoethane	1.6×10^{11}	0.5	3×10^{3}	5×10^{7}
Dicofol	2×10^{9}	0.003	5×10^{5}	3×10^{3}
Saccharin	3×10^{9}	0.0004	6×10^{6}	4.6×10^{2}
Benzene				
Produced	4.8×10^{12}	0.001	1.8×10^{6}	2.7×10^{6}
Emitted	2.5×10^{11}	0.001	1.8×10^{6}	1.4×10^{5}
Aflatoxin	1.7×10^{4}	1000	1.8	9.4×10^{3}
Aldrin	8×10^{8}	0.5	3.3×10^{3}	2.3×10^{5}
Chlordane	3×10^{9}	0.2	1.0×10^{4}	2.9×10^{5}
Dioxin	4.5×10^{3}	5000	4×10^{-1}	1.3×10^{4}
Dioxin	(Seveso)			
	3.0×10^{2}	5000	4×10^{-1}	8×10^{2}
	(2-4-5T)			
Arsenic	2.0×10^{10}	15	1.0×10^{2}	1.5×10^{8}
DDT	2.0×10^{10}	—	—	—
Parathion	—	0.1	2.0×10^{4}	—
Dieldrin	—	0.4	5.0×10^{3}	—
Heptachlor	—	0.5	4.0×10^{3}	—
Endrin	—	0.9	2.0×10^{3}	—
Benzo(α)pyrene				
Food	3.0×10^{4}	7.0	2.5×10^{2}	1.2×10^{2}
Air	1.0×10^{9}	7.0	2.5×10^{2}	4.0×10^{5}
Nuclear waste	After 10 yr	—	—	8.0×10^{10}
Ingestion	After 500 yr	—	—	1.0×10^{7}

SOURCE: From Wilson, 1980.

banned or restricted in either manufacturing or use if the evidence suggests that it will pose an unreasonable risk of injury to human health or to the environment. But what is "unreasonable" [20]?

The preceding discussion indicates that pollutants increase the risk of injury or death to individuals and the risk of damage to the environment. It is commonly believed that future discoveries will help to decrease the risk, eventually allowing us to attain a zero-risk environment. But the more likely case is that any given society has a socially, psychologically, and ethically acceptable level of risk for any cause of death or injury. While ideally one would like to eliminate all pollutants from the environment, detailed analyses reveal that complete elimination in many cases is either technologically impossible or impossibly expensive (Fig. 15.7). The level of acceptable risk changes over time in society, depending on the risks from other causes, the expense of decreasing the risk, and the social and psychological acceptability of the risk.

Therefore, we must ask several questions. What risk from a particular pollutant is acceptable? How much is a certain reduction in risk from that pollutant worth to us? How much will each of us, as individuals or collectively as a society, be willing to pay for a certain reduction in that risk?

Novel or new risks appear to be less acceptable than long established or "natural" risks. Thus our society tends to pay more to reduce novel risks than to reduce natural or long established ones. For example, in France approximately $1 million is spent to reduce the likelihood of one air traffic death, but only $30,000 is spent for the same reduction in automobile deaths. Some argue that the greater safety of commercial air travel compared with automobile travel is in part a function of the relatively novel fear of flying compared with the more ordinary fear of death from a road accident.

While in an ethical sense it is impossible to put a value on a human life, it is possible to find out how much people are willing to pay for a certain reduction in risk or a certain probability of an increase in longevity. For example, a study by the Rand Corporation considered measures that would save the lives of heart attack victims, such as increasing ambulance services and initiating pretreatment screening programs. According to the study, which identified the likely cost per life saved and the willingness of people to pay, people were willing to pay approximately $32,000 per life saved or $1600 per year of longevity [21].

The willingness to pay would vary with the essentialness and desirability of an activity. For example, many people accept much higher than average risks for sports or recreational activities than they would for risks associated with transportation or employment (Table 15.1).

Although information is very incomplete, it is possible to estimate the cost of extending lives in terms of the dollars per person per year for various actions (Fig. 15.8). For example, based on the *direct* effects on human health, it costs more to increase longevity by a reduction in air pollution than it would to directly reduce deaths by addition of a coronary ambulance system. Such a comparison is useful as a basis for decision making. Clearly, when a society chooses to reduce air pollution, many factors beyond the direct measurable health benefits are considered. We might want to choose a slightly higher risk of death in a pleasanter environment (spend money to clean up the air instead of on increased ambulance services) than increase the chances of living longer in a poor environment. Whether we like it or not, we cannot avoid making choices of this kind. The issue boils down to whether we should improve the quality of life for the living or extend the life expectancy regardless of the quality of life [22].

Although pollution control may involve many dollars, the cost per family in the United States, in terms of reduced purchasing power for other things, has been estimated to be between $30 and $60 per year for a family with a median income. On the other hand, federal air quality standards are estimated to reduce the risk of asthma 3 percent, and the risk to locally exposed adults of chronic bronchitis and emphysema 10 to 15 percent. Air

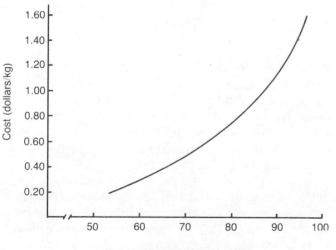

Percent Reduction in Discharge

FIGURE 15.7
The marginal cost of reducing the discharge of a pollutant. Costs rise rapidly as one attempts to remove a higher and higher percentage of a pollutant. This curve for organic effluents from a sugar refinery is characteristic of many pollutants. Such curves suggest that removing 100 percent of a pollutant may be prohibitively expensive. (After Baumol and Oates, 1979.)

FIGURE 15.8
One way to rank the effectiveness of various efforts to reduce pollutants is to estimate the cost of extending a life in dollars per year. This graph shows that reducing sulphur emissions from power plants to the Clean Air Act level (d) would extend a human life 1 year at a cost of about $10,000. Similar restrictions applied to automobile emissions (b,c) would increase lifetimes by 1 day. More stringent automobile controls would be much more expensive (a), while mobile units and screening programs for heart problems would be much cheaper. This graph represents only one step in an environmental analysis. (After Wilson, 1980.)

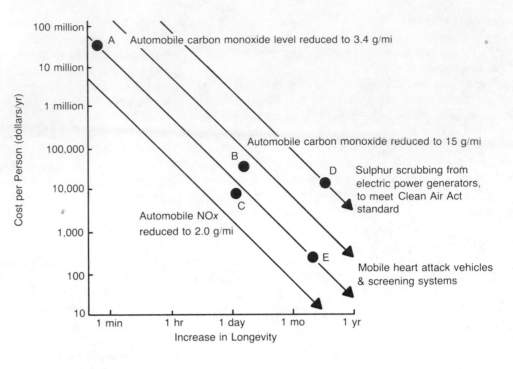

pollution contributes to inflation by reducing the number of productive work days; reducing work efficiency; adding to direct expenditures on health treatments; and incurring costs of restoration of nonhuman environmental damage. Estimates of the total cost of the direct and indirect effects on human health from stationary sources of air pollution are $250 per family per year. On this basis, air pollution control appears not to be inflationary; in fact, it has economic benefits [23].

In summary, the risk associated with a pollutant can be determined by the present levels of exposure and predicted future trends. These trends depend on the production and origin of the pollutant, the pathways it follows through the environment, and the changes it undergoes along these pathways. Dose-response curves establish the risk to a population from a particular level of a pollutant. The relative risks of different pollutants can be determined by comparing the current levels and their dose-response curves.

So far we have described scientific and technological steps to estimate risk, but an acceptable risk is much more than a scientific or technical issue. Once the risk is established, it is then possible to estimate the cost of reducing that risk. The acceptability of a risk involves ethical and psychological attitudes of individuals and of society. Risks that are voluntary appear to be more acceptable than those that are not voluntary. Risks that af-

fect a small portion of a population (such as all employees at nuclear power plants) are usually more acceptable than those that involve all of a society (such as the risk from radioactive fallout). Finally, familiar, long established risks seem to be more acceptable than novel ones.

WHO PAYS AND HOW?

How does a society achieve an environmental goal? For our purposes in this section we will distinguish two such goals: (1) the use of a desirable resource (catching fish, hiking in the wilderness, obtaining wood or fuel, and so on); and (2) the reduction of an undesirable product (minimizing the release of a pollutant).

Any society has several methods to achieve these goals: persuasion, regulation (establishing laws, regulatory agencies, etc.), taxation and subsidies, and licenses (Tables 15.2 and 15.3). In every environmental matter there is a desire on the one hand to maintain individual freedom of choice and on the other to achieve a specific social goal. In ocean fishing, for example, how does a society allow every individual to choose whether or not to fish and yet prevent everyone from fishing at the same time so as to cause extinction of the fish? This interplay between private good and public good is at the heart of environmental issues. Some argue that the market itself will provide the proper control. For example, it can be

TABLE 15.2
Approaches to environmental policy.

POLICY INSTRUMENTS

1. Moral Suasion (publicity, social pressure, etc.)

2. Direct Controls
   a. Regulations limiting the permissible levels of emissions
   b. Specification of mandatory processes or equipment

3. Market Processes[a]
   a. Taxation of environmental damage
      1) Tax rates based on evaluation of social damage
      2) Tax rates designed to achieve preset standards of environmental quality
   b. Subsidies
      1) Specified payments per unit of reduction of waste emissions
      2) Subsidies to defray costs of damage-control equipment
   c. Issue of limited quantities of pollution "licenses"
      1) Sale of licenses to the highest bidders
      2) Equal distribution of licenses with legalized resale
   d. Refundable deposits against environmental damage
   e. Allocation of property rights to give individuals a proprietary interest in improved environmental quality

4. Government Investment
   a. Damage prevention facilities (e.g., municipal treatment plants)
   b. Regenerative activities (e.g., reforestation, slum clearance)
   c. Dissemination of information (e.g., pollution-control techniques, opportunities for profitable recycling)
   d. Research
   e. Education
      1) Of the general public
      2) Of professional specialists (ecologists, urban planners, etc.)

ADMINISTRATIVE MECHANISMS

1. Administering Unit
   a. National agency
   b. Local agency

2. Financing
   a. Payment by those who cause the damage
   b. Payment by those who benefit from improvements
   c. General revenues

3. Enforcement mechanism
   a. Regulatory organization or police
   b. Citizen suits (with or without sharing of fines)

SOURCE: From Baumol and Oates, 1979.
[a]Subsidies and taxes can also be distinguished by using a property-rights framework. Per unit subsidies implicitly confer ownership of the right to pollute on the polluter, and these rights are then purchased by the government via the subsidy. Taxes essentially say that there is public ownership of usage rights which can be purchased from the public through its agent, the government, by private parties upon payment of the tax (price).

argued that people will stop fishing when there is no longer a profit to be made. We have already seen, however, that two factors interfere with this argument: (1) The level of fishing that results in no economic gain may still be a level that causes biological extinction; and (2) Even when one may not make a profit today, there may be an advantage in harvesting the entire resource and getting out of the business.

Use of Resources   Desirable resources can be privately owned or controlled by a single user (like many mineral mines or some forests) or common property. The oceans are common property, as are the fish and mammals who live in them. What is common property may change over time. The establishment of the 200-mile (325-kilometer) limit by many nations has turned some fisheries from completely open common property to national property open only to domestic fishermen.

In fisheries there are four main management options:

1 Establish total catch quotas for the entire fishery.

2 Issue a restricted number of licenses, but allow each licensed fisherman to catch many fish.

3 Tax the catch (the fish brought in) or the effort (the cost of ships, the fuel, etc.).

4 Allocate fishing rights [24].

When total catch quotas are established, the fishery is closed when the catch quota is reached. Whales, Pacific halibut, tropical tuna, and anchovies have been regulated in this way. Although regulating the total catch might be done in a way that helps the fish, it tends to increase the number of fishermen and the capacity of vessels, and the end result is a hardship on fishermen.

Recent economic analysis suggests that taxes which take into account the cost of externalities can work to the

TABLE 15.3
Performance of various policy instruments by specified criteria.

Policy Instrument	Reliability	Permanence	Adaptability to Growth	Resistance to Inflation
Moral suasion	Good[a]	Poor	Good*	Good*
Direct controls				
a. By quota	Fair	Poor	Fair	Excellent
b. By specification of technique	Fair	Poor	Good*	Good*
Fees	Excellent	Excellent	Fair	Fair
Sale of permits	Excellent	Excellent	Excellent	Excellent
Subsidies				
a. Per unit reduction	Fair[d]	Good	Fair	Fair
b. For equipment purchase	Fair	Good	Fair	Fair
Government investment	Good	?	?	?

SOURCE: From Baumol and Oates, 1979.
*Authors' judgment based on little concrete evidence.
[a]For short periods of time when urgency of appeal is made very clear.
[b]Induces contributions from decision makers who are most cooperative, not necessarily from those able to do the job most effectively (most inexpensively).
[c]Tends to allocate reduction "quotas" among firms in cost-minimizing manner, but if the number of emissions permits is too small it will force the community to devote an excessive quantity of resources to environmental protection.
[d]Tends to allocate reduction quotas among firms in cost-minimizing manner, but introduces inefficiency into the environmental protection process by attracting more polluting firms into the subsidized industry, so that aggregate response is questionable.

best advantage of fishermen and fish. Another technique with similar results is a quota allocated to each fishermen which is transferable and salable.

Which of the social methods to choose to achieve the best use of a desirable environmental resource is not a simple question. The answer will vary with the specific attributes of both the resource and the users. However, we can use the tools of economics to determine the methods that will work best within a given social framework.

The Control of Pollutants   In discussing the control of pollutants, the concept of the **marginal cost** is useful. Marginal cost is the cost in dollars per unit of pollutant reduction as a function of the percentage reduction in total discharge. It is generally true that the marginal cost increases rapidly as the percentage reduction increases. For example, the marginal cost of reducing the biological oxygen demand in waste water from petroleum refining increases exponentially. When 20 percent of the pollutants have been removed, the cost of removing an additional kilogram is 5 cents. When 80 percent of the pollutants have been removed, it costs 49 cents to remove an additional kilogram. Extrapolating from this, it would cost an infinite amount to remove all the pollution.

There are two types of direct controls of pollution: (1) setting maximum levels of pollution emission, and (2) requiring specific procedures and processes that reduce pollution. In the first case, a political body could set a maximum level for the amount of sulphur emitted from the smokestack of an industry. In the second, it could restrict the kind of fuel. In fact, many areas have chosen the latter method by prohibiting the burning of high-sulphur coal. The problem with the first approach—controlling emissions—is that careful monitoring is required indefinitely to make certain the allowable levels are not exceeded. Such monitoring is costly and may be difficult to carry out. The disadvantages of the second approach are that the required methodology may impose a financial burden on the producer of the pollutant, restrict the kinds of production methods open to an industry, and become technologically obsolete.

Although the United States has emphasized the use of direct regulation to control pollution, other countries have been successful in controlling pollution by charging effluent fees. For example, charges for effluents into the Ruhr River in West Germany are assessed on the basis of both the quality (concentration of pollutant) and total quantity (total amount of polluted water) emitted into the

TABLE 15.3 (continued)

Incentive for Improved Effort	Economy	Feasibility without Metering	Noninterference in Private Decisions	Political Attraction	
				Actual	Potential
Fair	Poor[b]	Excellent	Excellent	Excellent	—
Poor	Poor	Poor	Poor	Excellent	—
Poor	Poor	Excellent	Poor	Excellent	—
Excellent	Excellent	Poor	Excellent	Poor	Good
Excellent	Excellent[c]	Poor	Excellent	Poor	Good
Excellent	Good	Poor	Excellent	Good	—
Good	Poor	Excellent	Excellent	Good	—
—	?	Excellent	—	Good	—

SOURCE: From Baumol and Oates, 1979.

river. As a result of this practice, plants have introduced water recirculation and internal treatment in order to reduce emissions [25].

In recent years the U.S. government has spent considerable amounts of money in environmentally related programs. Protection and enhancement (rehabilitation of sites, protection of unique natural areas, etc.) cost more than $1 billion per year. Pollution control and abatement cost almost $2 billion per year. Programs for observing and predicting weather, ocean conditions, and environmental disturbances such as earthquakes cost more than $1 billion per year [26].

A "public good," such as clean air, cannot be sold by private sellers. For example, suppose an individual went into the business of cleaning up the air. Since anyone can use the clean air, people will not voluntarily pay a charge for it. Natural ecosystems provide this service to some extent. Forests may absorb particulates, salt marshes may convert toxic compounds to nontoxic forms, and sewage put into streams can be removed by biological activity. We all profit from these public service functions of natural ecosystems, but we have no simple way to put a value on them or even to estimate the amount of pollution removal that takes place.

To summarize, knowledge of the specific resource and an understanding of economics as well as the characteristics of the biosphere are necessary to choose the best methods of using a desirable resource or reducing undesirable pollutants. In all such practices there is a desire to maintain individual freedom of action while insuring a public good. Environmental economics, an important and developing field, provides methods to analyze and understand these issues.

SUMMARY AND CONCLUSIONS

Every human society has a set of beliefs about nature, the effects of nature on human beings, and the effects of human beings on their natural surroundings. In the history of Western civilization there have been a variety of beliefs about nature. Nature has been seen as

1 Wild, dangerous, and horrible, and a place where an individual could test himself against its challenges.

2 Disordered without human beings, requiring people to create a perfect order. Humanity's role in nature is to create the final order, harmony, and balance.

3 Perfectly harmonious, ordered, and in balance without human beings; only disrupted and made unharmonious and imbalanced by people's poor actions. Human beings can learn to act in the right way by understanding the natural order.

4 Made for human beings, and providing for our needs.

5 Resembling a machine whose workings are exact and predictable. People are nature's engineers and can modify nature's operations to their own ends.

6 Unpredictable and capricious, creating a world where people must struggle to survive against natural hazards of all kinds. People create a small amount of order and regularity and predictableness in a chaotic world.

7 Magnificent and beautiful in its wildness, power, and unpredictability, a key for individual self-discovery and a meaning for existence; an expression of the power of God or gods.

8 A commodity to use up. People are the marketers of the commodity.

9 A place of the strange, monumental, and peculiar. People are entertained by nature or inspired by it.

10 A life support system necessary in its wild state for the survival of human beings.

11 Having an instrinsic value in and of itself. People have a moral obligation to preserve it. (This belief is related to #2.)

12 Being one with human beings; the persistence of each depends on the other.

Once we develop a set of values about nature, we attempt to achieve or maintain a desirable condition for nature while maintaining at the same time human activities, freedom of individual action, and a high standard of living. Economics provides a basic framework for the analysis of the policies that will achieve these social goals for the environment. Put another way, "A prerequisite for effective regulation is a clear understanding of the basic reasons for overexploitation" [19].

Two major issues are important for society: (1) the use of a desirable resource (fish in the ocean, oil in the ground, forests on the land); and (2) the minimization of undesirable pollution. Resources may be common property, or they may be privately controlled. The kind of ownership affects the methods available to achieve an environmental goal. From the example of whaling, we learned that there is a tendency to overexploit a common-property resource and to harvest to extinction non-essential resources whose innate growth rate is low. The discount factor can be an important determinant of the level of exploitation.

The relation between risk and benefit affects our willingness to pay for an environmental good.

Societal methods to achieve an environmental goal include persuasion, direct regulation, taxation and subsidies, licensing, and establishment of quotas. All five kinds of controls have been applied to the use of desirable resources. In the United States, regulation and licensing have been commonly used to control the use of desired resources; regulation has been the common method to control pollution. Taxation and the establishment of salable quotas can be shown to be most effective in fisheries management.

REFERENCES

1 GLACKEN, C. J. 1967. *Traces on the Rhodian Shore: Nature and culture in Western thought from ancient times to the end of the eighteenth century.* Berkeley: University of California Press.

2 MCLUHAN, T. C. 1971. *Touch the Earth: A self-portrait of Indian existence.* New York: Promontory.

3 KLAEBER, F., ed. 1950. *Beowulf and the fight at Finnsburg.* Boston: D. C. Heath.

4 LECLERC, G. L. (Count de Buffon). 1812. *Natural history, general and particular.* Translated by W. Smellie. London: C. Wood.

5 EGERTON, F. N. 1975. Aristotle's population biology. *Arethusa* 8: 307–30.

6 CICERO, Marcus Julius. 1972. *The nature of the gods.* Translated by H. C. P. McGregor. Aylesbury, England: Penguin Press.

7 LUCRETIUS (Titus Lucretius Carus). 1968. *De rerum natura.* Translated by R. Humphries. Bloomington: Indiana University Press.

8 NASH, R. 1979. Wilderness is all in your mind. *Backpacker* 31: 39.

9 NICOLSON, M. H. 1959. *Mountain gloom and mountain glory.* Ithaca, N. Y.: Cornell University Press.

10 SHELLEY, P. S. "Mont Blanc." Lines 127, 96–97, 139–41.

11 SANDBURG, C. 1970. "The People Yes." In *The complete poems of Carl Sandburg,* pp. 439–617. New York: Harcourt, Brace, Jovanovich.

12 BOTKIN, B. A. 1944. *A treasury of American folklore.* New York: Crown Publishers.

13 MARSH, G. P. 1967. *Man and nature.* (Edited by D. Lowenthal.) Cambridge, Mass.: Belknap Press. (Originally published in 1864 by Charles Scribner's Sons, New York.)

14 RUNTE, A. 1979. *National parks: The American experience.* Lincoln: University of Nebraska Press.

15 LEOPOLD, A. 1949. *A Sand County almanac.* New York: Oxford University Press.

16 NASH, R. 1977. Do rocks have rights? *Center Magazine* 10: 2–12.

17 LOVELOCK, J. E., and MARGULIS, L. 1973. Atmospheric homeostasis by and for the biosphere: The Gaia hypothesis. *Tellus* 26: 1–9.

18 HARDIN, G. 1968. The tragedy of the commons. *Science* 162: 1243–48.

19 CLARK, C. W. 1973. The economics of overexploitation. *Science* 181: 630–34.

20 CAIRNS, J., Jr. 1980. Estimating hazard. *BioScience* 30: 101–7.

21 SCHWING, R. C. 1979. Longevity and benefits and costs of reducing various risks. *Technological Forecasting and Social Change* 13: 333–45.

22 GORI, G. B. 1980. The regulation of carcinogenic hazards. *Science* 208: 256–61.

23 OSTRO, B. D. 1980. Air pollution, public health, and inflation. *Environmental Health Perspectives* 34: 185–89.

24 CLARK, C. W. 1981. Economics of fishery management. In *Renewable resource management: Lecture notes in biomathematics 40,* eds. T. L. Vincent and J. M. Skowronski, pp. 95–111. New York: Springer-Verlag.

25 BAUMOL, W. J., and OATES, W. E. 1979. *Economics, environmental policy, and the quality of life.* Englewood Cliffs, N. J.: Prentice-Hall.

26 COUNCIL ON ENVIRONMENTAL QUALITY. 1976. *Environmental quality.*

FURTHER READING

1 BAUMOL, W. J., and OATES, W. E. 1979. *Economics, environmental policy, and the quality of life.* Englewood Cliffs, N. J.: Prentice-Hall.

2 CLARK, C. 1976. *Mathematical bioeconomics: The optimal management of renewable resources.* New York: Wiley.

3 EGERTON, F. N. 1973. Changing concepts of the balance of nature. *Quarterly Review of Biology* 48: 322–50.

4 GLACKEN, C. J. 1967. *Traces on the Rhodian Shore: Nature and culture in Western thought from ancient times to the end of the eighteenth century.* Berkeley: University of California Press.

5 LEOPOLD, A. 1949. *A Sand County almanac.* New York: Oxford University Press.

6 MARSH, G. P. 1967. *Man and nature.* (Edited by D. Lowenthal.) Cambridge, Mass.: Belknap Press. (Originally published in 1864 by Charles Scribner's Sons, New York.)

7 NASH, R. 1967. *Wilderness and the American mind.* New Haven, Conn.: Yale University Press.

8 NICOLSON, M. H. 1959. *Mountain gloom and mountain glory.* Ithaca, N. Y.: Cornell University Press.

9 PAGE, T. 1977. *Conservation and economic efficiency.* Baltimore: John Hopkins University Press.

10 PARTRIDGE, E., ed. 1981. *Responsibilities to future generations.* Buffalo, N. Y.: Prometheus Books.

STUDY QUESTIONS

1 What is meant by the phrase "the tragedy of the commons"? Which of the following are the result of this "tragedy": (a) the California condor; (b) the right whale; (c) the high price of walnut wood used in furniture?

2 Explain what is meant by "risk assessment."

3 What is meant by the statement, "Wilderness is a state of mind"?

4 Why might a society that changed from hunting to agriculture also change its view of wilderness?

5 How might the following inventions alter a society's views of wilderness: (a) the magnetic compass; (b) the sailing ship; (c) the steam engine?

6 Is the surface of the moon a wilderness? Explain your answer.

7 It has been said that national parks are the United States' answer to the great sculptures and paintings of Europe. Explain this idea, which is known as "monumentalism."

8 Should cities have national parks? Explain your answer.

9 You are invited by a friend to invest in a walnut plantation. She tells you that walnut is an extremely valuable wood and the price can only go higher as the tree becomes more scarce. You investigate further and discover that walnut is one of the longer-lived trees of eastern North America. Would you join your friend's investment? Why or why not?

10 Cherry and walnut are valuable woods used to make fine furniture. Using the information below, which would you invest in: (a) a cherry plantation; (b) a walnut plantation; (c) a mixed stand of both species; or (d) an unmanaged woodland where you see some cherry and walnut growing?

Species	Longevity	Maximum Size	Maximum Value
walnut	400 years	1 meter	$15,000 per tree
cherry	100 years	1 meter	$10,000 per tree

11 Flying over Los Angeles, you see smog below you. Your neighbor in the next seat says, "That smog looks bad, but eliminating it would save only a few lives. Doing that isn't worth the cost. We should spend the money on other things, like new hospitals." Do you agree or disagree? Give your reasons.

16 Thinking Ahead with Lessons from the Past: Environmental Planning

INTRODUCTION

Oliver Wendell Holmes wrote, "Every year, if not every day, we have to wager our solution upon some prophesy based upon imperfect knowledge." This statement cer-

tainly applies to environmental planning, where we often do not have all the data necessary to make the best possible choices. Even so, remaining buildings from earlier societies suggest that careful attention was paid to the environment. For example, approximately 1000 years ago in Mesa Verde, Arizona, American Indians had some unusual design features in their houses. Set in shallow recesses along the steep sides of mesas, the houses were positioned so that they were shaded from the summer sun by the rocks above but warmed directly by winter sun (Fig. 16.1). The sun in winter was (and is) at a lower angle in the sky and shone directly on the houses. The summer sun in midday was at a much higher angle in the sky and was shaded from the houses by overhanging rocks. Thus, the Mesa Verde Indians' homes took advantage of certain local aspects of the environment (the local rock formations) and certain characteristics of the transmission of the sun's energy [1]. The Mesa Verde Indians built their houses with what we call today a "design with nature" concept [2].

In contrast, many houses built in the United States during the middle of the twentieth century, when central heating and air conditioning were common and fossil fuels cheap, were designed with little concern for their location relative to sunlight. In some tract housing developments, the same design would be repeated on different streets or on different sides of the same street and facing in different compass directions. In like manner, little consideration was given to the physical environment, so many older homes today stand on ancient landslides, active traces of faults, and other unstable areas, such as the sea cliff environment.

(a)

FIGURE 16.1
Map and simplified diagram of Mesa Verde houses.

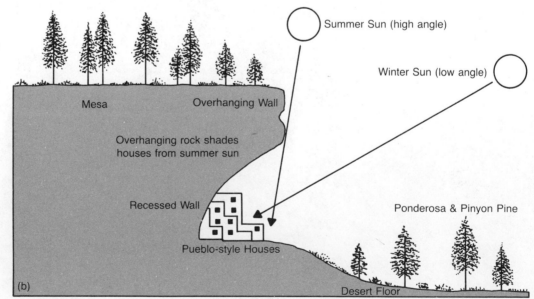

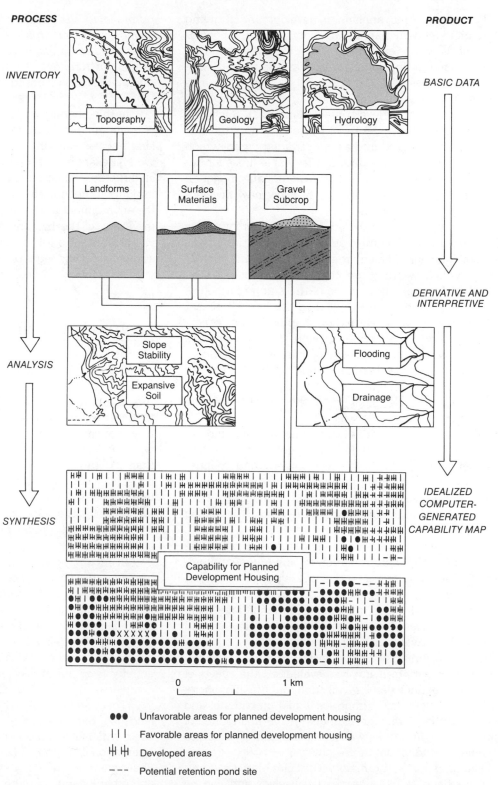

FIGURE 16.2
The modern land planning process is illustrated by this flow chart, which shows the use of many kinds of information about the environment in land-use planning for the Franconia area of Fairfax County, Virginia. The planning process involves inventory, analysis, and synthesis phases. (Modified from Froelich et. al., 1978.)

INVENTORY

Topography

Geology

Hydrology

BASIC DATA

Landforms

Surface Materials

Gravel Subcrop

DERIVATIVE AND INTERPRETIVE

ANALYSIS

Slope Stability

Expansive Soil

Flooding

Drainage

SYNTHESIS

Capability for Planned Development Housing

IDEALIZED COMPUTER-GENERATED CAPABILITY MAP

0   1 km

●●● Unfavorable areas for planned development housing

| | | Favorable areas for planned development housing

╫╫ ╫╫ Developed areas

--- Potential retention pond site

XXX Special engineering studies required prior to development

Fortunately, this unplanned development is giving way to more modern planning concepts involving consideration of the entire environment. For example, the Franconia area in Fairfax County, Virginia, less than 15 kilometers southwest of the District of Columbia, has integrated the concept of designing with nature into planning for housing development (Fig. 16.2). Such features as topography, geology, hydrology, and hazards analysis are integrated into a *land capability map*, which is a computer composite showing favorable areas for development. These maps have had a significant impact on decisions to rezone existing land or allow new development in the Franconia area.

Today we recognize the advantages of careful house siting in relation to nature, as the Mesa Verde Indians practiced long ago. The environment has specific characteristics which we may use to our benefit or our loss. Some decisions, like the choice of the siting of a house, may affect the future no longer than the lifetime of that house. However, as we learned in Chapter 1, *some effects of land use tend to be cumulative, and, therefore, we have an obligation to future generations to minimize the negative effects of land use*. In other words, we need to plan for the future. In this chapter we ask the question, How do we plan? How do we make use of the knowledge of the environment to plan more wisely?

Planning takes place at local, regional (state and national), and global levels. Planning methods are themselves in various stages of development and testing; these methods and techniques are an important component of applied environmental work. Evaluating, planning, and forecasting are all part of the process of dealing with the environment.

Local land-use planning, including site selection, construction, and determination of environmental impact, is now a common practice. At regional, state, and national levels, there is planning for transportation facilities, recreational areas (including national and state parks), wilderness areas, water supply, clean air and rivers, and renewable natural resources.

Global planning today consists primarily of evaluation and forecasting. It involves evaluating the effects of our highly technological civilization on the biosphere, such as the effects of burning fossil fuel on climate, and forecasting the worldwide changes in the abundance of people and resources, as well as the amount of resources available per individual in the future.

Planning has a long history. City planning, for example, is perhaps as old as cities themselves and can be traced back at least to the classical Greek civilization. Here we are reminded of one of environmental studies' fundamental principles: *All technology, from the most primitive to the most recent inventions, causes some change in the environment. However, the Industrial and Scientific revolutions have led to ever more rapid changes in our environment*. As our ability to alter the environment increases, planning becomes more and more important.

In this chapter we will consider the basic principles of environmental planning and describe planning at different scales—urban, regional and national, and global. We will examine the methodology of land-use plans, including the evaluation of impacts and the possibilities for mitigation. Recent attempts to forecast the conditions of the environment at the beginning of the next century and the limits of the biosphere for various human activities will illustrate the role of forecasting in planning. We will also consider the legal processes that have been and are used to accomplish desirable environmental goals.

LAND-USE PLANNING

The use of land in the United States and almost everywhere else in the world has changed during the twentieth century. For example, the amount of land converted from rural to nonrural uses has increased (Fig. 16.3). Although the increase in conversion of rural land to nonagricultural uses appears to be very slow, currently it amounts to about 8100 square kilometers per year—an area more than 2½ times the size of Rhode Island. The intensive conversion of rural land to urban development, transportation networks and facilities, and reservoirs is nearly matched by the extensive conversion of rural land to wildlife refuges, parks, and wilderness and recreation areas (Fig. 16.4).

In recent years the need for land near urban areas has led to the concepts of **multiple** and **sequential land use** rather than permanent, exclusive use. An example of multiple land use is active subsurface mining below urban land. Multiple use is less common than sequential land use, which involves changing use with time. The basic idea of sequential land use is that after a particular activity (e.g., mining or a sanitary landfill operation) is completed, the land is reclaimed for another purpose.

There are several well-known examples of sequential land use. Sanitary landfill sites are often planned to be used, when filled, for other purposes. Part of J. F. Kennedy International Airport in New York City is built on landfill. Other sanitary landfill sites in California, North Carolina, and other locations are being planned so that when each site is completely filled in, the land will be used for a golf course. The city of Denver used abandoned sand and gravel pits for sanitary landfill sites, then converted these

FIGURE 16.3
Changes in land use in the United States from 1900 to 1974 (excluding Alaska and Hawaii). Built-up areas have increased considerably, as has cropland; rangeland has decreased slightly. (From Council on Environmental Quality, 1979.)

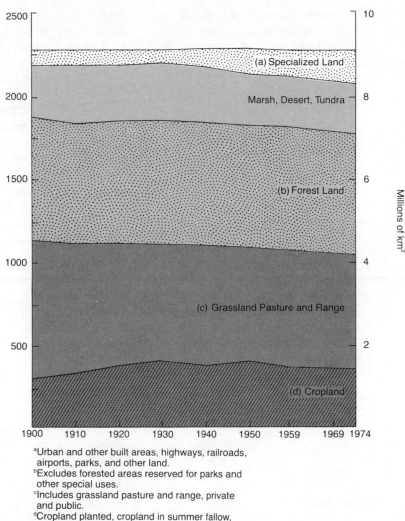

ªUrban and other built areas, highways, railroads, airports, parks, and other land.
ᵇExcludes forested areas reserved for parks and other special uses.
ᶜIncludes grassland pasture and range, private and public.
ᵈCropland planted, cropland in summer fallow, soil-improvement crops, and land being prepared for crops and idle.

FIGURE 16.4
Approximate annual conversion of rural land in the United States to nonagricultural uses. (After Council on Environmental Quality.)

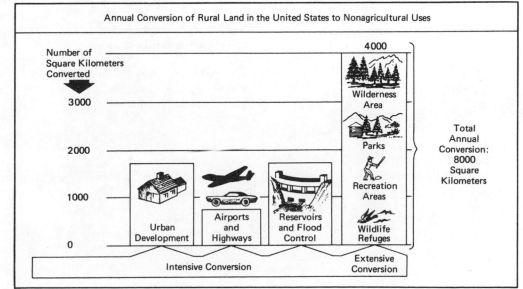

to a parking lot and the Denver Coliseum. Enormous underground limestone mines in Kansas City, Springfield, and Neosho, Missouri, have been profitably converted to warehousing and cold-storage sites, offices, and manufacturing plants. Other possibilities also exist: for example, abandoned surface mines could be used for parking below shopping centers or for chemical or petroleum storage [3].

THE LAND-USE PLAN

There are four basic elements of a land-use plan: (1) a statement of land-use issues, goals, and objectives; (2) a summary of data collection and analysis; (3) a land classi-

fication map; and (4) a report that describes appropriate development for areas of special environmental concern [4]. The preparation of a land-use plan is complex and requires a team approach.

The statement of major land-use issues assists in planning future development for at least a 10-year period and therefore should be prepared in cooperation with citizens and public agencies. The statement includes information on issues such as impact of economic and population trends, housing and services, conservation of natural resources, and protection of important natural environments, and cultural, historical, and scenic resources [4].

Data collection and analysis include the following information: analyses of present population, economy, and

TABLE 16.1
Required and optional data analysis items for the land-use plan.

Element	Required	Optional
1. Present Conditions		
a. Population and economy	Brief analysis, utilizing existing information.	More detailed analyses relating to human resources (population composition, migration rates, educational attainment, etc.) and economic development factors (labor force characteristics, market structure, employment mix, etc.).
b. Existing land use	Mapped at generalized categories.	Mapped with more detailed categories, including more detailed analyses, building inventory, etc.
c. Current plans, policies, and regulations		
1) Plans and policies	1) Listing and summary.	1) Detailed impact analysis of plans and policies upon land-development patterns.
2) Local regulations	2) Listing and description of their enforcement mechanism.	2) Detailed assessment of adequacy and degree of enforcement.
3) Federal and state regulations	3) Listing and summary (to be provided by N. C. Dept. of Natural and Economic Resources).	
2. Constraints		
a. Land potential		
1) Physical limitations	1) Analysis of following factors (maps if information available): — Hazard areas — Areas with soil limitations — Sources of water supply — Steep slopes	1) Detailed analysis and mapping of required items. Analysis and mapping of additional factors: — Water-quality limited areas — Air-quality limited areas — Others as appropriate

TABLE 16.1 (continued)

Element	Required	Optional
2) Fragile areas	2) Analysis of following factors (maps if information available): — Wetlands — Frontal dunes — Beaches — Prime wildlife habitats — Scenic and prominent high points — Unique natural areas — Other surface waters — Fragile areas	2) Detailed analysis and mapping of required items. Analysis and mapping of additional factors.
3) Areas with resource potential	3) Analysis of following factors (maps if information available): — Areas well suited for woodland management — Productive and unique agricultural lands — Mineral sites — Publicly owned forests, parks, fish and game lands, and other outdoor recreational lands — Privately owned wildlife sanctuaries	3) Detailed analysis and mapping of required items Analysis and mapping of additional factors: — Areas with potential for commercial wildlife management — Outdoor recreation sites — Scenic and tourist resources
b. Capacity of community facilities	— Identification of existing water and sewer service areas — Design capacity of water-treatment plant, sewage-treatment plant, schools, and primary roads — Percent utilization of the above	Detailed community facilities studies or plans (housing, transportation, recreation, water and sewer, police, fire, etc.).
3. Estimated Demand a. Population and economy 1) Population	1) 10-yr estimates based upon Dept. of Administration figures as appropriate.	1) Detailed estimate and analysis, adapted to local conditions using Department of Administration model.
2) Economy	2) Identification of major trends and factors in the economy.	2) Detailed economic studies.
b. Future land needs	Gross 10-yr estimate allocated to appropriate land classes.	Detailed estimates by specific land-use category (commercial, residential, industrial, etc.).
c. Community facilities demand	Consideration of basic facilities needed to service estimated growth.	Estimates of demands and costs for some or all community facilities and services.

SOURCE: North Carolina Coastal Resources Commission, 1975.

land use, and current plants, policies, and regulations; constraints, such as land potential or the capacity of community facilities (water and sewer service, schools, etc.); and estimated demand from changes in population and economy, future land needs, and demand for facilities. Table 16.1 lists in detail required and optional data-analysis items for land-use plans submitted under the North Carolina Coastal Management Act of 1974.

The land classification map is the heart of the land-use plan. It serves as a statement of land-use policy and has five aims: (1) to achieve and encourage coordination and consistency between local and state land-use policies; (2) to provide a guide for public investment in land;(3) to provide a useful framework for budgeting and planning construction of facilities, such as schools, roads, and sewer and water systems; (4) to coordinate regulatory policies and decisions; and (5) to help provide guidelines for development of an equitable land tax. An example of a currently used land classification system is shown in Table 16.2.

The report accompanying the land classification map gives special attention to areas of environmental concern—those areas in which uncontrolled or incompat-ible development might produce irreparable damage. Examples of such areas include coastal marshland, estuaries, renewable resources (watersheds or aquifers), and fragile, historic, or natural resource areas (in the North Carolina example, the Outer Banks, barrier islands, etc.). The report should be precise regarding the permissible land uses of these areas [4].

TABLE 16.2
Example of a land classification system.

a. *Developed*—Lands where existing population density is moderate to high and where there are a variety of land uses which have the necessary public services.

b. *Transition*—Lands where local government plans to accommodate moderate- to high-density development and basic public services during the following 10-year period will be provided to accommodate that growth.

c. *Community*—Lands where low-density development is grouped in existing settlements or will occur in such settlements during the following 10-year period, and which will not require extensive public services now or in the future.

d. *Rural*—Lands whose highest use is for agriculture, forestry, mining, water supply, etc., based on their natural resource potential. Also included are lands for future needs not currently recognized.

e. *Conservation*—Fragile, hazardous, and other lands necessary to maintain a healthy natural environment and necessary to provide for the public health, safety, or welfare.

SOURCE: North Carolina Coastal Resources Commission, 1975.

PLANNING FOLLOWING EMERGENCIES

In recent years two types of planning have emerged: projects in which design and environmental impact analysis are integral parts; and emergency projects following catastrophic events, such as hurricanes or volcanic eruptions, which cause widespread damage. Emergency planning is always in response to pressing needs and an influx of millions of dollars in emergency money. Too often the work authorized is overzealous—beyond what is necessary—and is not carefully thought out. As a result, the emergency projects may cause further environmental disruptions. While emergency work is desperately needed, care must be taken when emergency money arrives to insure that it is used for the best purposes possible.

For example, following severe storms and floods that struck Virginia in the aftermath of Hurricane Camille in 1969, emergency federal aid was used to channelize streams in hopes of alleviating future floods. In some instances, local bulldozer operators with little or no instruction or knowledge of streams were contracted to clear and straighten stream channels. Results of this unplanned and unsupervised channel work have been disastrous to many kilometers of streams. Catastrophic storms can seriously damage roads, farms, and homes, in which case emergency channel work is clearly needed, but such emergency work should be confined to stream channels that require immediate attention. Emergency funds should not be considered a license for wholesale modification of any stream in the damaged area at the request of property owners or others.

The eruption and catastrophic landslide/debris avalanche of Mt. St. Helens in 1980 delivered 2.5 cubic kilometers of debris into the Toutle River. Fearing continued downstream sediment pollution, the U.S. Army Corps of Engineers constructed two emergency catch dams to filter the water through rockfill and gravel barriers, thus maintaining the flow of water in the river while trapping the sediment in the catch dams. Unfortunately, the dams were constructed before reliable estimates could be made of the volume of sediment likely to be delivered to the dams. On the north fork of the Toutle River the catch dam had a capacity to hold 0.0065 cubic kilometer of debris. The Corps of Engineers estimated that the sediment load would be 0.01 to 0.02 cubic kilometer per year, so the river would have to be dredged 2 to 3 times per year to remain in service. However, after observing the effects of a small August flood, the U.S. Geological Survey estimated that the sediment yield would be closer to 0.3 to 0.38 cubic kilometer per year, and that the catch dams were about 100 times too small to survive expected winter storms and floods. This prediction was proven true during a storm in late October when the dams failed. Thus, while the idea of catch dams was worthwhile, they were constructed with poor or incomplete planning and were too small.

Even if there is not sufficient time to prepare an environmental impact statement, planning following emergencies should be carefully carried out. Immediate work may be necessary to restore communication and protect property, but those responsible should limit early work to what is absolutely necessary. Other projects should be more carefully evaluated to determine whether they really are necessary and whether they will cause future problems.

SITE SELECTION AND EVALUATION

Site selection is the process of choosing and evaluating a physical environment that will support human activities. It is a task shared by professionals from many aspects of environmental studies, including engineering, landscape architecture, planning, Earth science, biological sciences, social science, and economics, and thus involves a multidimensional approach. The goal of site selection for a particular land use is to insure that the site development is compatible with both the possibilities and the limitations of the natural environment.

PHYSIOGRAPHIC DETERMINISM, OR DESIGNING WITH NATURE

In recent years, a philosophy of site evaluation known as **physiographic determinism** has emerged [5]. The basic thrust of this philosophy is ''design with nature.'' Rather than laying down an arbitrary design or plan for an area, the approach is to find a plan that nature has already provided in the topography, soils, vegetation, and climate [5]. Using this method, planners maximize the amenities of the landscape while minimizing social and economic expenditures whenever possible [6].

Although the philosophy of working in harmony with nature is obviously advantageous, it is often overlooked. People still purchase land for various activities without considering if the land use they have in mind is compatible with the site they have chosen. There are well-known examples of poor siting resulting in increased expense, limited production, or even abandonment of partially completed construction. Construction of a West Coast nuclear power facility was terminated when fractures in a rock (active faults) were discovered and possible serious foundation problems arose (Fig. 16.5). The productivity of a large chicken farm in the southeastern United States was greatly curtailed because the property was purchased before it was determined if there was sufficient groundwater to meet the projected needs. A housing developer in northern Indiana purchased land and built country homes in one of the few isolated areas where bedrock (shale) was at the surface. Thus, septic-tank systems had to be abandoned and a surface sewage treatment facility built. Furthermore, the rock made it much more expensive to excavate for basements and foundations for the homes.

FIGURE 16.5
Nuclear power development was abandoned due to an earthquake hazard at this site at Bodega Bay, California, after expenditure of millions of dollars and years of time in site preparation. This waste of time and money could have been avoided had the data on the fault been available earlier. (From Wallace, 1974.)

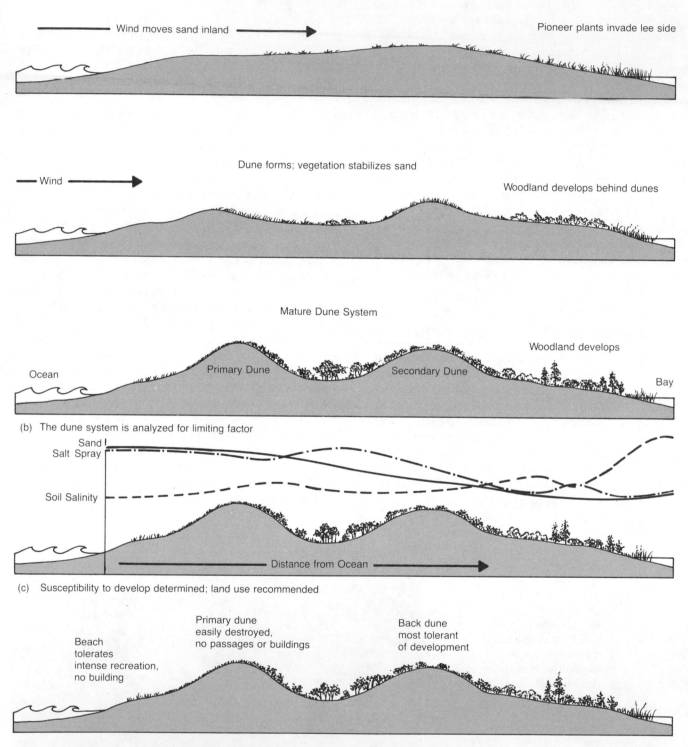

(a) Analysis of geological and ecological processes

Dune forms on a new sand bar by wind moving sand

Wind moves sand inland

Pioneer plants invade lee side

Dune forms; vegetation stabilizes sand

Wind

Woodland develops behind dunes

Mature Dune System

Woodland develops

Ocean

Primary Dune

Secondary Dune

Bay

(b) The dune system is analyzed for limiting factor

Sand
Salt Spray

Soil Salinity

Distance from Ocean

(c) Susceptibility to develop determined; land use recommended

Beach tolerates intense recreation, no building

Primary dune easily destroyed, no passages or buildings

Back dune most tolerant of development

FIGURE 16.6

Designing with nature for a sand dune area. The natural geological and biological processes that maintain the dunes are analyzed first (a); then the important variables are analyzed to determine which are limiting (b). A synthesis of information yields a map of the susceptibility to development and the best land use for each part of the dune (c). (After McHarg, 1971.)

420

The concept of "design with nature," coined by Ian McHarg, is illustrated by his analysis of the development of a sandy beach like that on Cape Cod, Massachusetts, or along the New Jersey coast. First, the geological and ecological processes that form and maintain the dunes are considered. In this case, the dunes are formed by windblown sand. The sand is stabilized by vegetation, and the dune in turn protects the leeward side from salt spray and allows thickets and woodland to develop (Fig. 16.6a). In this stage of analysis two points are recognized: (1) the dunes are *dynamic;* they change over time and depend on an influx of sand; and (2) the *vegetation* plays an important role in stabilizing the dune, and therefore the vegetation must be protected.

In the second stage of analysis, other factors that are important to the vegetation and the dune are measured (Fig. 16.6b). For example, salt spray is lower on the back dune, allowing vegetation to grow that is not resistant to that spray.

Finally, a land-use scheme is developed (Fig. 16.6c). The previous stages of analysis clarify which parts of the dune system are fragile and must be protected and which can tolerate heavy use. For example, the primary dune is easily destroyed because it is stabilized by vegetation that exists in a harsh environment and is easily killed by trampling and other human impacts. On the other hand, as long as the primary dune remains intact, the back dune can tolerate considerable use; its vegetation exists in a less harsh environment and grows back after disruption more rapidly. The back dune is also less subject to disturbance from wind and storms. Conclusion: Protect the primary dune and develop the back dune. Design *with* nature.

LANDSCAPE AESTHETICS: SCENIC RESOURCES

Scenery in the United States has been recognized as a natural and national resource at least since 1864, when the first national park, Yosemite Valley in California, was established. At that time, the purpose of establishing a park was to preserve unique scenic landscapes. Since then, the definition of *scenic value* has broadened. Even "everyday" nonurban landscapes are considered to have scenic value which enhances the total resource base of a region. We now accept that there are varying scenic values relating to other, more tangible resources [7].

Quantitative evaluation of tangible natural resources such as water, forests, or minerals is standard procedure before economic development or management of a particular land area. Water resources for power or other uses may be evaluated by the flow in the rivers and storage in lakes; forest resources may be evaluated by the number, type, and size of the trees and their potential yield of lumber; and mineral resources may be evaluated by estimating the number of tons of economically valuable mineral material (ore) at a particular location. We can make a statement of quality and quantity for each tangible resource compared with some known low quality or quantity. Ideally, we would like to make similar statements about the more intangible resources, such as scenery [8]. Unfortunately, few specific standards are available for comparison.

Landscape evaluation of scenic resources as part of land-use planning or assessing environmental impact generally rests on a rather subjective methodology. Figure 16.6 shows generally how a quantitative evaluation might proceed from a recognition of scenic resources to relative or absolute indices to define their scenic values.

ENVIRONMENTAL IMPACT ASSESSMENT

PHILOSOPHY AND METHODOLOGY

The probable effects of human use of the land are generally referred to as **environmental impact.** This term became widely known after passage of the National Environmental Policy Act (NEPA) in 1969. This act requires that all major federal actions that could possibly affect the quality of the human environment be preceded by an evaluation of the project and its impact on the environment. In order to meet both the letter and the spirit of NEPA, the Council on Environmental Quality has prepared guidelines to assist in preparing environmental impact statements. The steps in the environmental impact assessment process are shown in Table 16.3. The major components of the council's statement, revised in 1979 [9], are

1 A statement of purpose and need for the project.

2 A rigorous comparison of the reasonable alternatives.

3 A succinct description of the environment of the area to be affected by the proposed project.

4 A discussion of the environmental consequences of the proposed project and its alternatives. This discussion must include direct and indirect effects; energy requirements and conservation potential; resource requirements; impacts on urban quality and cultural or historical resources; possible conflicts with state or local land-use plan policies and controls; and mitigation measures. *Mitigations* are actions taken to reduce adverse environmental impacts. They include repair and

restoration of the environment, reduction or elimination of impacts, and compensation to affected parties. The success of mitigation determines the real significance of impacts and the ranking of alternatives.

5 A list of the names and qualifications of the persons primarily responsible for preparation of the environmental impact statement, and a list of agencies to which the statement was sent.

6 An index.

For example, consider a plan to develop a new electric power plant on a major river. The site lies in agricultural land within 50 kilometers of a major city. The *purpose* is the production of electricity, and the description of *need* would include current and projected electrical usage in the area and current and projected production. The comparison of *reasonable alternatives* would include the effects of energy conservation and siting the plant elsewhere. The river and land in the areas that would be affected would be part of the *description of the environment*. The projected future for the site without the proposed project (e.g., the site might be expected to undergo development for residential and commercial use even without the power plant) would be included. The *description of the environmental consequences* would include the project's *effects* (e.g., the power plant would remove water from the river for cooling and return it at a higher temperature) and its *impacts*, both short-term (e.g., fish swim away from the heated water) and long-term (e.g., the reproduction and therefore population size of some fish species are expected to decrease). A *mitigation* could be a plan to stock the river with young fish grown in state fish hatcheries.

Because mitigations determine the real significance of impacts, the professional environmental planner must place considerable emphasis on this aspect. The environ-

mental impact analysis is intended not only to determine the negative effects, but also to outline what positive steps may be taken to promote wise use. In some cases, such as a plan which would affect the wintering grounds of an endangered bird species like the whooping crane, there may be no mitigating actions. In those cases the recommendation would be for preservation and no development.

The environmental impact statement process was criticized during the first 10 years under NEPA because it initiated a tremendous volume of paperwork by requiring detailed reports that tended to obscure important issues. No one knew exactly what had to be included or what might be attacked, and it seemed safer to include more information rather than less. There was a tendency to produce extremely long documents that listed every possible impact and lacked a focus on crucial issues. Therefore, the revised regulations (1979) introduced two other important changes: scoping and record of decision.

Scoping is the process of identifying important environmental issues that require detailed evaluation early in the planning of a proposed project. The **record of decision** is a concise statement by the planning agency of the alternatives considered and which alternatives are environmentally preferable. The agency then has the responsibility to monitor the project to ensure its decision is carried out. This is a very significant point. For example, if in an environmental impact statement for a proposal to construct an interstate highway the agency commits itself to specific designs and locations for the right-of-way in order to minimize environmental degradation, then the contracts authorizing the work must be conditional on incorporation of those designs and locations.

There is no accepted method for assessing the environmental impact of a particular project or action. Furthermore, because of the wide variety of actions—from the hunting of migratory birds to the construction of a large reservoir—there is no one methodology of impact assessment appropriate for all situations. What is important is that those responsible for preparing the statement strive to minimize personal bias and maximize objectivity. The analysis must be scientifically, technically, and legally defensible and prepared by highly objective scientific inquiry [10].

Environmental-impact analysis for major projects often requires the combined efforts of scientists from different disciplines. It is important to remember that the specific function of the task force is to *evaluate;* they are not to decide the issues. The task force provides necessary information to those with authority to make just decisions. At this stage in the development of our environmental awareness, we are still trying to determine the

TABLE 16.3
The steps in environmental impact assessment.

1. Describe the present environment (the *baseline* conditions).
2. Describe the *project*, including purposes and needs.
3. Describe the *effects* of the project.
4. Describe the *impacts*, both *short-term* and *long-term*.
5. Suggest and compare *alternatives*.
6. Provide a *projection* of the future of the site with and without the project.
7. Suggest *mitigating activities*.

kind and amount of information necessary to make good decisions. However, as more and more work is done and decisions are made, the critical elements of environmental information will be better understood, and eventually some will be acknowledged as requirements in the same way as cost and profit data are now required in economic analysis [10].

In keeping with the spirit of NEPA, we must consider environmental consequences of a particular action before it is implemented and identify potential conflicts and problem areas as early as possible. In this way, we can minimize regrettable and expensive environmental deterioration.

CASE HISTORY: THE HIGH DAM AT ASWAN

The High Dam at Aswan and Lake Nasser in Egypt is a dramatic example of what can happen when environmental impacts are ignored or inadequately evaluated. The High Dam project is causing many serious problems, including health problems that could one day be catastrophic, as the following case study by Claire Sterling will demonstrate [11].

The High Dam at Aswan, completed in 1964, is one of the biggest and most expensive dams in the world. It was designed to store a sufficient amount of the Nile's yearly floodwaters to irrigate existing land, reclaim additional land from the desert, produce electricity, and protect against drought and famine.

Unexpected water loss in the reservoir of the Aswan Dam, which may present problems in dry years, occurs because of two factors: a 50-percent error in computing the evaporation loss, and tremendous water losses to underground water systems. The error in computation of evaporation loss, which meant an unanticipated loss of 5 billion cubic meters of water per year, resulted from overlooking the evaporation loss induced by high-velocity winds traveling over the tremendous expanse of water in this very hot, dry region of the Earth. A chain of smaller and less expensive dams would have avoided this waste of water resources. However, even this loss is small compared with the water lost underground [11].

It has long been known that, for hundreds of kilometers upriver from Aswan, the Nile cut across an immense sandstone aquifer that fed the river an incalculable amount of water. When the first and much smaller Aswan Dam was built in 1902, the flow of groundwater was reversed; the water pressure caused by the reservoir forced the water to move elsewhere through numerous fissures and fractures in the sandstone. That smaller dam remained until 1964, when the High Dam was completed. From 1902 until 1964, the Aswan Reservoir stored about 5 billion cubic meters of water per year, but lost 12 billion cubic meters of water per year through reversed groundwater flow. The amount of water escaping from the modern Lake Nasser behind the new dam is unknown, but because the new reservoir is designed to store 30 times more water than the old Aswan Reservoir, and because seepage tends to vary directly with lake depth, the amount of water being lost must be tremendous [11].

It might be argued that in time the clay settling out from the lake will plug fractures in the rock and the lake will fill. However, if the fractures are very large and numerous, the water could essentially escape forever, and Egypt might well end up with less water than it had before the dam was constructed [11].

Unfortunately, the direct water problems with the High Dam and Lake Nasser are only part of the project's total impact. There are five other major environmental consequences of this project. First, the High Dam lacks sluices to transport sediment through the reservoir, so the reservoir traps 134 million tons of the Nile's sediment per year. Historically, the sediment has produced and replenished the fertile soils along the banks of the Nile; no practical, artificial substitute is available to counteract this loss.

Second, the eastern Mediterranean is deprived of the nutrients in the Nile's sediment, resulting in a one-third reduction of the plankton, which is the food base for sardine, mackerel, lobster, and shrimp. An emerging fisheries industry in Lake Nasser may eventually compensate for this setback, but this has not yet developed.

Third, the lake and associated canals are becoming infested with snails that carry the dreaded disease *schistosomiasis* (snail fever). This disease has always been a problem in Egypt, but the swift currents of the Nile floodwaters flushed out the snails each year. The tremendous expanse of waters in Lake Nasser and the irrigation canals are now providing a home for these snails.

Fourth, there is an increased threat of a killer malaria carried by a particular mosquito often found only 80 kilometers from the southern shores of Lake Nasser. Malaria has historically migrated down the Nile to Egypt. The last epidemic in 1942 killed 100,000 Egyptians. Authorities fear that the larger reservoir and the irrigation canal system might be invaded on a more permanent basis by the disease-carrying insects.

Fifth, salinity of soils is increasing at rather alarming rates in middle and upper Egypt. Soil salinity has been a long-standing problem on the Nile delta, but was alleviated upstream by the natural flushing of the salt by the floodwaters of the Nile River. Millions of dollars will have to be spent on fertilizers to counteract the rising soil salinity that threatens the productivity of the land.

On the positive side, the High Dam and lake have converted 2800 square kilometers from natural floodwater irrigation to canal irrigation and allowed double cropping —but at a tremendous cost. In the future, however, tremendous amounts of money will have to be spent on fertilizers to replenish soil nutrients and to control water- and insect-borne disease [11].

CASE HISTORY: THE TRANS-ALASKA PIPELINE

The controversy surrounding the Trans-Alaska Pipeline, which went into the construction phase in 1975 and was completed in 1977, provides a good example of environmental impact analysis. It is hoped that experience gained from this project will set precedents and establish procedures for future impact analyses.

In 1968, vast subsurface reservoirs of oil and gas were discovered near Prudhoe Bay in Alaska. Since the Arctic Ocean is frozen much of the year, it is impractical to ship the oil by tankers, and thus a 1270-kilometer pipeline from Prudhoe Bay to Port Valdez, where tankers can dock and load oil the entire year, was suggested [10,12].

The general route of the Trans-Alaska Pipeline is shown in Figure 16.7. The corridor for the pipeline traverses rough topography, large rivers, areas with extensive permafrost, and areas with a high earthquake hazard. Over 80 percent of the pipeline crosses federal lands. Because of the sensitive nature of the Arctic environment and the certainty of irreversible environmental degradation, a comprehensive impact analysis was required to evaluate both the negative and the positive aspects of the project. A summary of the evaluation published by the U.S. Geological Survey emphasizes the critical aspects of the natural and socioeconomic impacts of the project [10].

The pipeline crosses three major mountain ranges: the Brooks Range (which is the Alaskan continuation of the Rocky Mountains), the Alaska Range, and the tectonically active Chugach Mountains. On the north side of the Alaska Range, the pipeline crosses the Denali fault, an active fault zone that has experienced recent displacement of the ground.

The corridor also traverses a number of large rivers, and it was feared that scour or lateral erosion at meander bends might damage the pipeline. A study to evaluate the river crossings [13] identified areas that might experience excessive bank erosion or scour during the 30-year expected life of the pipeline.

The permafrost (permanently frozen ground) areas crossed by the pipeline presented another potential hazard. Once frozen ground melts, it is extremely unstable. This aspect of the project has been thoroughly engineered, and it is hoped that problems with permafrost will be minimized. Nevertheless, the many kilometers of a large hot-oil pipeline crossing vast areas of permafrost provides sufficient cause for concern.

Analysis of possible physical, biological, and socioeconomic impacts associated with the pipeline established three areas of concern [10]: the construction, operation, and maintenance of the pipeline system, including access roads and highways; development of the oil fields; and operation of the marine tanker system at Port Valdez.

FIGURE 16.7
Approximate corridor for the Trans-Alaska Pipeline.
(After Brew, 1974.)

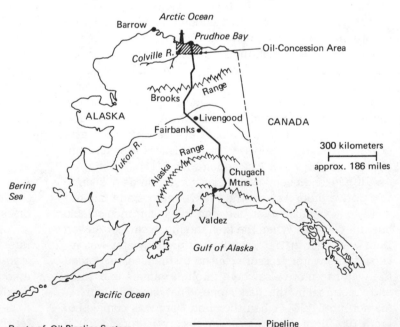

Route of Oil-Pipeline System ———— Pipeline

Furthermore, while some of the effects would cause predictable, unavoidable disruptions, other effects are more speculative, such as the impact of an oil spill caused by a break in the pipeline. These threatened effects are not easily evaluated or predicted [10]. A list of primary and secondary effects of the Trans-Alaska hot-oil pipeline, gas pipeline, and tanker transport system considered by the task force to evaluate environmental impact is shown in Table 16.4.

In addition to these effects, possible indirect linkages between the effects had to be evaluated. For example, an oil spill or an unavoidable release of oil in the tanker operation into the marine environment obviously affects marine resources. The extent of the effects, however, is difficult to evaluate because of the variable or unknown aspects of the hazard, the dynamic hydrologic environ-

ment, and seasonal and other changes in fish and marine resources. So, in general, the task force concluded that the impact linkage data were the largest obstacle to quantitatively determining possible impacts [10].

Based on the environmental analysis, both unavoidable and threatened impacts of the Trans-Alaska Pipeline were determined [10]. There were three main unavoidable effects: (1) disturbances of terrain, fish and wildlife habitats, and human use of the land during construction, operation, and maintenance of the entire project, including the pipeline itself and access roads, highways, and other support facilities; (2) discharge of effluents and oil into Port Valdez from the tanker transport system; and (3) increased human pressures on services, utilities, and many other areas, including cultural changes of the native population.

TABLE 16.4
Primary and secondary impacts associated with the Trans-Alaska Pipeline, arctic gas pipelines, and the proposed tanker system.

A. Primary Effects Associated with Arctic Pipelines:
1. Disturbance of ground
2. Disturbance of water (including treated effluent discharge into water)
3. Disturbance of air (including waste discharged to air and noise)
4. Disturbance of vegetation
5. Solid waste accumulation
6. Commitment of physical space to pipeline system and construction activities
7. Increased employment
8. Increased utilization of invested capital
9. Disturbance of fish and wildlife
10. Barrier effects on fish and wildlife
11. Scenery modification (including erosional effects)
12. Wilderness intrusion
13. Heat transmitted to or from the ground
14. Heat transmitted to or from water
15. Heat transmitted to or from air
16. Heat to or from vegetation
17. Moisture to air
18. Moisture to vegetation
19. Extraction of oil and gas
20. Bypassd sewage to water
21. Human-caused fires
22. Accidents that would amplify unavoidable impact effects
23. Small oil losses to the ground, water, and vegetation
24. Oil spills affecting marine waters
25. Oil spills affecting freshwater lakes and drainages
26. Oil spills affecting ground and vegetation
27. Oil spills affecting any combination of the foregoing

B. Secondary Effects Associated with Arctic Pipelines:
1. Thermokarst development
2. Physical habitat loss for wildlife
3. Restriction of wildlife movements
4. Effects on sports, subsistence, and commercial fisheries
5. Effects on recreational resources
6. Changes in population, economy, and demands on public services in various communities, including native communities, and in native populations and economies
7. Development of ice fog and its effect on transportation
8. Effects on mineral resource exploration

C. Primary Effects Associated with Tanker System:
1. Treated ballast water into Port Valdez
2. Vessel frequency in Port Valdez, Prince William Sound, open ocean, Puget Sound, San Francisco Bay, southern California waters, and other ports
3. Oil spills in any of those places

D. Secondary Effects Associated with Tanker System:
1. Effects on sports and commercial fisheries
2. Effects on recreational resources
3. Effects on population in Valdez and other communities

SOURCE: From Brew, 1974.

The major threatened effect was accidental loss of oil from the oil field, pipeline, or tanker system. Accidental loss of oil from the pipeline could be caused by slope failure (landslides), differential settlement in permafrost areas, stream bed or bank scour at river crossings, or destructive sea waves causing a leak or rupture in the pipeline. Oil loss from tankers could be caused by shipwreck or accidental loss during transfer operations at Port Valdez. Because the potential loss of oil from the pipeline and tanker systems involves a great number of variables, it is difficult to predict how much oil would be lost, but some loss is inevitable. Estimates place maximum oil loss at 1.6 to 6 barrels per day at Valdez for tanker operations, and 384 barrels per day from tanker accidents. The latter, of course, could be either a series of small spills or several large spills at unknown times, locations, and intervals [10].

From before the final impact statement was completed early in 1972 until construction started in 1975, the pipeline generated (and continues to generate) contro-

versy. It is controversial because of the conflicts in balancing the need for resource development and the known or predicted environmental degradation. Although alternative routes and transport systems, including trans-Canada routes and railroads, other Alaska routes, and marine routes, were extensively evaluated (Fig. 16.8), the earlier proposed route to Port Valdez was ultimately approved, perhaps because this route led to the most rapid resource development while maintaining national security. However, no one route is superior in all respects to the others [10].

Comparison of the alternative routes suggests the following. First, all the trans-Alaska routes have less unavoidable adverse impact on the abiotic (nonliving) systems than the trans-Alaska-Canada routes. Second, the trans-Alaska route to the Bering Sea would probably have the least unavoidable impact on terrestrial-biologic and socioeconomic systems. The trans-Alaska-Canada coastal route would be next in minimizing unavoidable impact on these systems. Third, the trans-Alaska-Canada

FIGURE 16.8
Alternative routes for transporting oil from the North Slope of Alaska. (From Brew, 1974.)

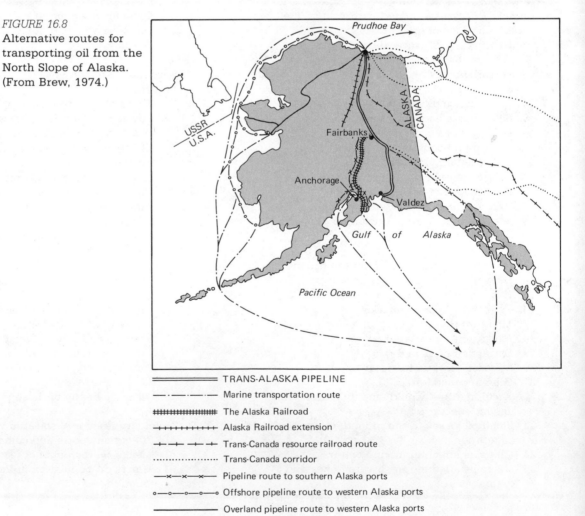

════════	TRANS-ALASKA PIPELINE
— · — · — · —	Marine transportation route
▓▓▓▓▓▓▓▓▓▓	The Alaska Railroad
++++++++++	Alaska Railroad extension
+ — · + — · +	Trans-Canada resource railroad route
··················	Trans-Canada corridor
— x — x — x —	Pipeline route to southern Alaska ports
o — o — o — o	Offshore pipeline route to western Alaska ports
───────────	Overland pipeline route to western Alaska ports

routes would have the least unavoidable impact on marine environments because no direct marine transport of oil is involved [10].

Based on these comparisons and other evaluations and analyses [10], it was concluded that the trans-Alaska-Canada route to Edmonton, Canada, was the route that would cause the least environmental impact. This route would avoid the marine environment and earthquake zones and would enable both an oil and a gas pipeline to be placed in one corridor. Although this example is past history, it emphasizes two things. First, the alternatives considered show the scientists' obligation to state their opinion based on sound scientific information, even though this opinion may be either unpopular or likely to be overridden in the final balancing of alternatives. Second, diverse political maneuvering at all levels can significantly affect which alternative is chosen.

CASE HISTORY: THE M-X MISSILE

One of the largest federal land-use projects ever proposed in the United States was the M-X missile system. The size and complexity of the proposed project and the extensive analyses of environmental impact that have been conducted make it a classic example of the impact analysis process.

As originally conceived, the M-X missile system was to be a land-based intercontinental missile system which would involve several hundred missiles and more than 40,000 shelters. The missiles would be moved among the shelters so that their exact location could not be predicted. Each missile could be concealed in any of 23 shelters. There would be 2 operating bases, 5 support centers, 300 maintenance facilities, and 200 remote surveillance sites.

One of the original proposals was to develop the system in Nevada and Utah. The alternatives included location elsewhere (in Texas and New Mexico); a split location (some in Nevada-Utah, some in Texas–New Mexico) (Fig. 16.9); and the placement of the M-X missiles in existing, fixed silos. Suitable zones for the system had to meet the operational requirements of the system itself, have few effects on the environment, and be able to coexist with other human activities. Thus the system had to avoid population centers, parks and environmentally sensitive areas, mineral resource areas (oil, gas, and ore mines), and so forth. Certain geological requirements of the system itself (including certain limitations on slopes and depth to bedrock and groundwater) also had to be met.

The construction of the M-X system would require 520 square kilometers and about 14,000 kilometers of road margins. In the original plan, shelters would be located in

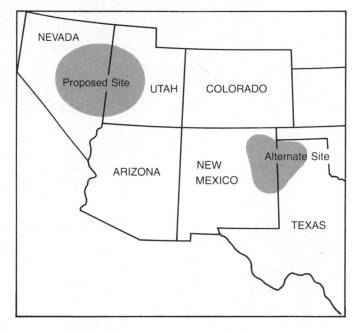

FIGURE 16.9
General location of the proposed M-X missile system as it was originally conceived.

29 valleys and spread over almost 17,000 square kilometers of the Great Basin in Nevada and Utah.

The environmental impact involves both the *construction* and the *operation* of the system. The construction effects would include establishment of camps for workers, construction roads, gravel pits, and recreational activities for the workers. At the peak of construction, the project would involve 19,000 construction workers and 6000 other personnel. Construction was estimated to take 8 years and use 100 million to 160 million cubic meters of water, 400,000 tons of steel, and 1.5 million tons of concrete.

The Great Basin, the original primary proposed location, is part of the desert biome of North America. The area is a desert because of the rain shadow effect of the Sierras to the west in California. It is the "high desert" whose vegetation includes sagebrush, rabbitbrush, and other low and widely scattered shrubs. Its five major vegetation communities are shown in Figures 16.10 and 16.11. Although the area in Nevada and Utah is sparsely populated and appears barren to the casual visitor, it has many rare and fragile habitats and ecosystems.

The Great Basin is a series of north-south mountain ranges alternating with broad, flat valleys (Figs. 16.10 and 16.11). Annual rainfall ranges from 10 to 25 centimeters (4 to 10 inches), and there is snow in the winter. Elevation ranges from 1400 meters (4500 feet) to almost 4000 meters (13,000 feet).

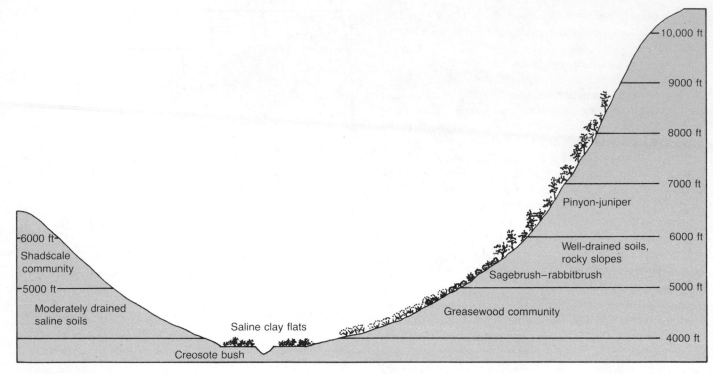

FIGURE 16.10
Vegetation zones in the Great Basin, the proposed location for the M-X missile system. The diagram shows changes in vegetation with elevation and soils.

The proposed M-X system would be the most recent of a long series of human disturbances of the Great Basin. The major current human-caused disturbance of the Great Basin is livestock grazing [14]. The area is used as winter range for about 9 million cattle and sheep. Livestock grazing decreases palatable vegetation, and overgrazing increases annual weeds, rodents, and insect herbivores. Cattle trampling of vegetation leads to decreased aeration of soils and decreases plant growth and storage of water.

Some valley grasslands appear to be maintained by fire. Fire control in this century has reduced the grass abundance. Other disrupting human activities are mining and recreation. Deep-shaft and open-pit mining began in the area in 1859. Mine spoils revegetate slowly, and many are visible today. Much of the recreation, including off-road vehicles, photography, hunting, and rock collecting, is due to the presence of wildlife (bighorn sheep, pronghorn antelope, mule deer, and elk). Predators such as bears, mountain lions, and coyotes, which were once much more abundant, have been controlled.

The following impacts of the M-X missile system have been determined:

1 The system would create ecological islands.

2 Construction would clear vegetation in an area where recovery is slow.

3 Construction could have adverse impacts on communities because of rapid, temporary growth.

4 Construction would compete for a limited water supply.

5 The system would restrict access to exploration for mineral and energy resources in the area.

6 There would be a potential impact on the following natural resources: ground and surface water; air; mammals (bighorn sheep, prairie dog, pronghorn antelope); reptiles (desert tortoise); birds (sage grouse); rare plants; aquatic organisms; and wilderness.

Human resources that could be affected are health, education, housing, public safety, recreation, grazing, native American culture, water and land use, and archeological resources.

Each of these impacts has been intensively studied and mitigations suggested. For example, native vegetation would be degraded even without the project because of other human activities (Fig. 16.11c). The project would remove, at least temporarily, 650 square kilometers (250 square miles) of vegetation—an area four-fifths the size of Grand Canyon National Park. Suggested mitigations include vegetation reclamation program and reestablishment of soils.

FIGURE 16.11
The Great Basin desert has several major vegetation communities. In (b) the juniper-pinyon woodland is in the foreground, and, at lower elevations, the sagebrush-rabbitbrush is in the background. Part (a) shows the shadscale community, which covers much of the area. In an area used to water sheep (c), there are bare areas due to overgrazing and trampling. (Photos courtesy of Walter L. Moore.)

(a)

(b)

(c)

As another example, pronghorn antelopes would lose habitat and food over the short term. Mitigations of this impact include establishment of new habitat and increased policing to reduce poaching.

Large areas of wilderness in the project area, under the jurisdiction of the Bureau of Land Management, add up to the size of the state of Delaware. More than 60 percent of these areas occur within 1.6 kilometers (1 mile) of proposed M-X project features. Roads developed for the project would provide better access to these wilderness areas and thus increase degradation by people. Some have proposed that a buffer zone be established between each wilderness area and the areas affected by the project.

In the case of an impact assessment as large and complex as this one, many groups and organizations become involved. The planning for and evaluation of the proposed project have involved a number of consulting companies, with some making independent assessments of similar issues. In a large project like the M-X, public hearings are part of the impact analysis process. At these hearings, the public is free to speak and respond to preliminary drafts of the impact analysis report.

In summary, as the largest federal land development project ever proposed, the M-X missile system has involved some of the most extensive environmental impact assessments conducted to date. Expertise of many kinds was required—from geologists, engineers, biologists, and those trained in remote sensing and map interpretation to the experts in political science and the social effects on recreation.

AN EXAMPLE OF NATIONAL LAND-USE PLANNING

As pressures on the land from many uses have increased during the last decades, planning at the national level has developed for certain land uses, particularly those that affect very sensitive areas. Wilderness is one such sensitive area. For example, the National Wilderness Preservation Act of 1964 (Fig. 16.12) set aside 3.6 million hectares (9 million acres) of wilderness, which has increased since then to include more than 7 million hectares (17 million acres) in more than 170 areas. These designated wilderness areas occur in lands with many kinds of jurisdiction. For example, the U.S. Forest Service administers more than 100 wilderness areas, totaling more than 6.7 million hectares (16.6 million acres). The National Wild and Scenic Rivers System (Fig. 16.13) is another example of national planning to protect sensitive land areas [15].

The National Wilderness Preservation Act required the U.S. Forest Service to study a number of roadless areas to consider whether they should be made part of the wilderness system. Thirty-four areas were studied under a project called RARE I (Roadless Areas Review and Evaluation). As a result, 5.6 million hectares (12.3 million acres) out of 25.5 million hectares (56 million acres) studied were selected for inclusion in the wilderness system. RARE I was followed by RARE II, which was designed to consider the entire national forest system for planning wilderness areas. The RARE II report is important because it provides maps for the United States showing areas in several different kinds of relatively undisturbed

stages and suggests new additions to the wilderness system [15].

These national plans are part of the attempt to develop multiple and sequential land use. There are other such planning attempts, some of which, like the project called Man and the Biosphere, are international.

GLOBAL FORECASTING

The global effects of our modern technological civilization have been discussed in earlier chapters. The burning of fossil fuels is adding a significant amount of carbon dioxide to the atmosphere; lead used in gasoline fuels has spread to the glaciers of Greenland; and DDT and other pesticides and artificial organic compounds are found in marine organisms inhabiting the Antarctic. We can imagine others, such as a nuclear war or a large-scale release of a toxic compound. Few have yet tried to plan on a global level. However, in the last decade procedures for global forecasting and evaluation have been developed.

In response to these global concerns, an international group of business executives, intellectuals, and government officials founded the Club of Rome, an organization whose objectives were to promote a better understanding of humanity's predicament; to disseminate information about this predicament; and to stimulate development of new attitudes, policies, and institutions to redress the present situation.

Out of the concerns of this organization came the development of computer simulation models—models of global phenomena such as human population growth, the use of the world's resources, and human impacts on the biosphere. The results of the use of these models were reviewed in 1972 in *The Limits to Growth* [16]. This book was controversial in many ways. First of all, its forecasts showed some dire consequences of our current activities. Supporters of *The Limits to Growth* hailed it as a new approach which could help save us from ourselves by forcing us to recognize the true limits to our uses of the biosphere. Opponents condemned it as another example of the "GIGO" rule of computer simulation—garbage in, garbage out. They claimed that it said nothing more than had been said by Malthus hundreds of years before.

Since the publication of *The Limits to Growth*, global forecasting has become an important aspect of modern civilization's attempt to deal with its own global effects. The Council on Environmental Quality was directed by the President to conduct a study of the probable changes in the world's population, natural resources, and environment through the end of the century. This study resulted in *The Global 2000 Report to the President* [17]. This re-

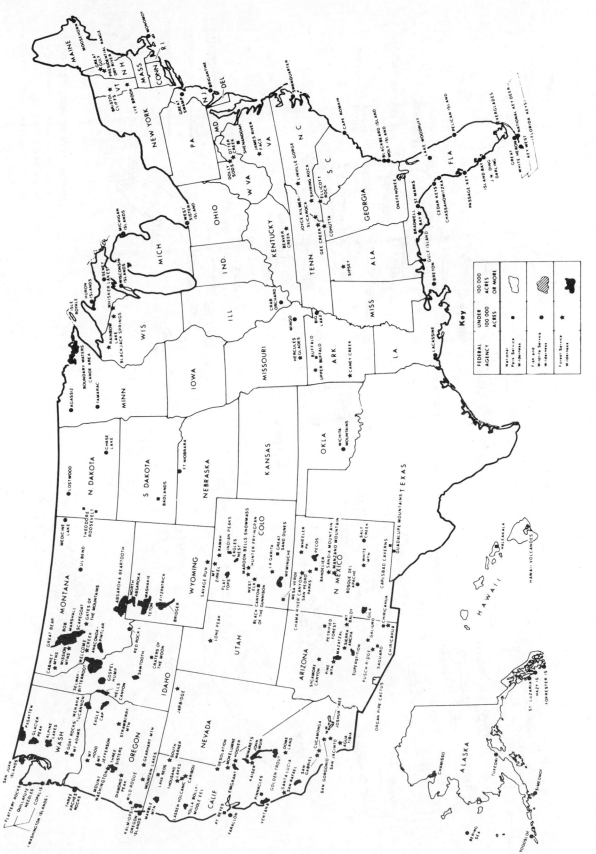

FIGURE 16.12
Map of the National Wilderness Preservation System as of September 1, 1979.
(from U.S. Forest Service, 1980.)

431

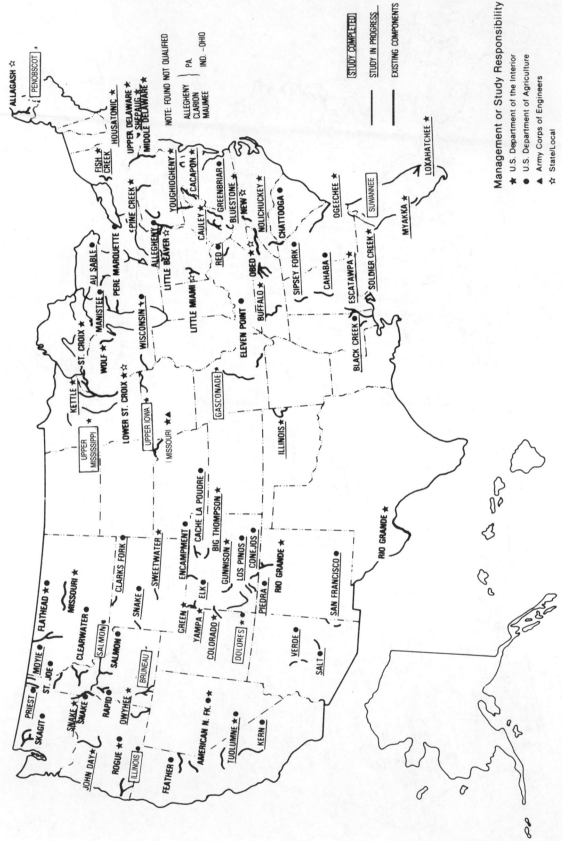

FIGURE 16.13

Map of the National Wild and Scenic Rivers System as of September 1, 1979. (From U.S. Forest Service, 1980.)

432

port used several methods to assess the status of population resources and environment by the end of the century, including a review of several computer models for global forecasting.

The methods of analysis used in global forecasting are illustrated in a general way in Figure 16.14, which shows the two steps in integrating the environment into the process of analysis. The actual models are more complex, involving many sections or compartments. Figure 16.15 shows the basic diagram of the world model of glo-

bal forecasting by M. Mesarovic and E. Pestel. Many factors are considered, from population and food demand to economics, trade, machinery, and education. Such a model can be used to investigate the implications of different assumptions about present and future trends (Fig. 16.16). For example, the projections in Figure 16.16a assume no changes in certain political aspects, while the projections in Figure 16.16b assume isolationist policies. Too many assumptions cloud the meaning of the projections, and we must be careful not to take the projections

FIGURE 16.14
The steps in global forecasting consider a number of environmental factors. The gray box represents the analysis prior to integration with the environmental analysis. (From Council on Environmental Quality and the Department of State, 1980.)

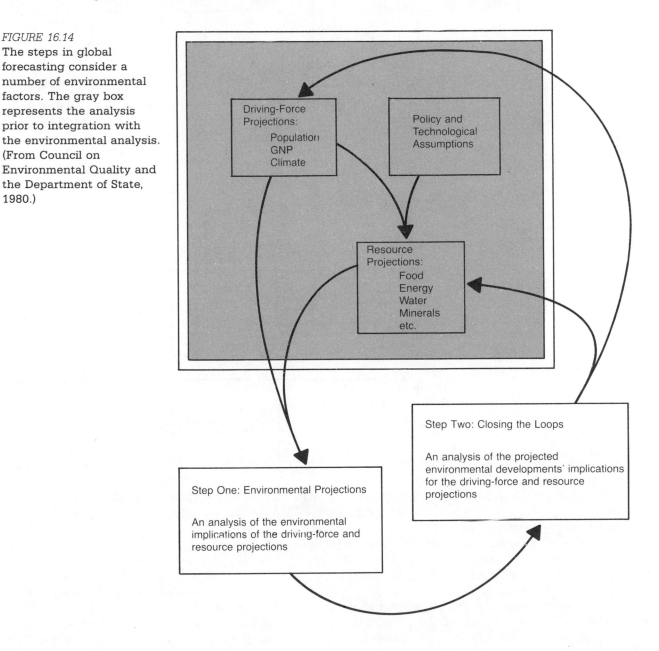

Driving-Force Projections:
Population
GNP
Climate

Policy and Technological Assumptions

Resource Projections:
Food
Energy
Water
Minerals
etc.

Step One: Environmental Projections

An analysis of the environmental implications of the driving-force and resource projections

Step Two: Closing the Loops

An analysis of the projected environmental developments' implications for the driving-force and resource projections

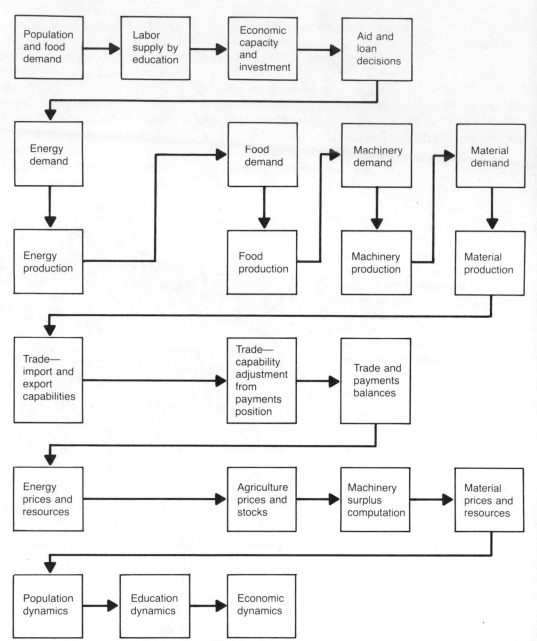

FIGURE 16.15
This flow chart of the Mesarovic-Pestel model for global forecasting shows the sequence in which calculations are made. Other world models involve similar calculations. (From Case Western Reserve University, Systems Research Center, and Council on Environmental Quality and the Department of State, 1980.)

at face value. Important information for such forecasts will come in the future from global remote-sensing data, which will provide large-scale images necessary for a global perspective. In addition, a better understanding of the history of climate and long-term changes in the distribution and abundances of life on the Earth will help us in our attempt to make projections.

Global forecasting remains a controversial activity. It is always difficult; if not impossible, to predict the future. The more we understand about the processes that govern change in the environment, the more likely our forecasts will be helpful. Such forecasts are perhaps most valuable in showing us general effects, or the consequences of what we *assume* governs these processes. In this way the forecasts force us to recognize the meaning of what we think we know and suggest to us where our knowledge is lacking.

The most important impact of the global forecasts is the public attention they have drawn to the issues. *The Global 2000 Report* is important because it is a projection of the future, concluding with an endorsement of a limit to growth, that has the approval of the President of the United States. In addition, *The Global 2000 Report* provides a useful summary of current trends.

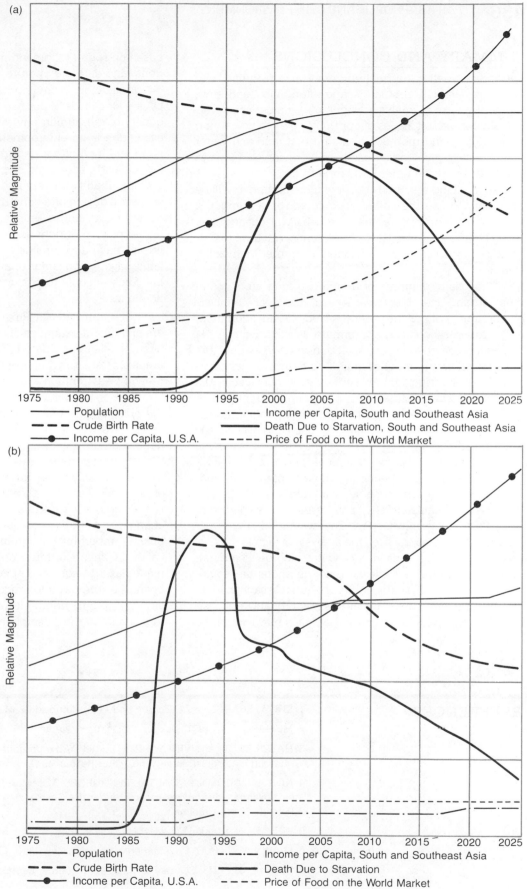

FIGURE 16.16
Projections of future trends from global models are made under different assumptions or "scenarios." Here the model diagram used in Figure 16.15 is used to project trends for a "no change" political policy (a) and an isolationist political policy (b). (From Case Western Reserve University, Systems Research Center, and Council on Environmental Quality and the Department of State, 1980.)

435

SUMMARY AND CONCLUSIONS

Landscape evaluation, including land-use planning, site selection, evaluation of landscape intangibles, and environmental impact analysis, is one of the most controversial environmental issues of our times. We are now trying to develop sound methods to insure that land and water resources are evaluated, used, and conserved in ways consistent with an emerging land ethic.

As our land resources become more limited, it will be necessary to use the same piece of land in a number of ways at the same time, and the uses will change over time. Therefore, future land-use planning will have to emphasize sequential or multiple land use rather than exclusive use.

The basic elements of a land-use plan are a statement of land-use objectives, issues, and goals; analysis and summary of pertinent data; a land classification map; and a report discussing appropriate developments.

In the process of site selection and evaluation, the physical environment is evaluated to determine its capability of supporting human activity and, conversely, the possible effects of human activity on the environment. A philosophy of site evaluation based on physiographic determinism, or ''design with nature,'' serves to balance the traditional economic aspects of site evaluation issues.

Evaluation of scenic resources and other environmental intangibles is becoming more important in landscape evaluation. Because such intangibles are difficult to measure quantitatively, they cannot be easily compared with economic factors. The evaluation of environmental intangibles involves the analysis of factors such as topographic relief, presence of water, degree of naturalness, and diversity. While all these methods are somewhat subjective, the subjectivity is not necessarily bad. In fact, a completely objective analysis is probably impossible to obtain. What is important is that the evaluation be based on sound judgment and that its methods, assumptions, and data are open to review by others.

Evaluation of environmental impact is now required by law for all federal actions that could possibly affect the quality of the human environment; the result of such an evaluation is an environmental impact statement. There is no one method for determining environmental impact for the wide spectrum of possible actions and projects that might affect the environment. The objective of analysis before design and construction is to minimize the possibility of causing extensive environmental degradation. In the past, pollution, loss of resources, or creation of a hazard have accompanied certain projects. These problems have led to unfortunate closings of industries and has forced people to adjust to possible hazards and economic loss.

The Trans-Alaska Pipeline and M-X missile system are significant examples of the way possible impacts and alternatives are evaluated. Furthermore, both of these examples stress the importance of considering the many different environmental aspects of proposed projects.

As our population increases and land is degraded, the pressures for many different uses of the land will increase. We will need to plan at larger scales so that the nation and the biosphere can maintain land with all of the variety of uses required by a modern society. Computer projections, as well as national and global maps, are part of this process. Satellite and aircraft remote-sensing projects on a large scale, long-term studies of climatic change, and studies of the history of climate and life will provide important background information for such forecasts. In an uncertain world with natural and human-made hazards, sound judgment remains, as it has always been, the true basis for good planning. In environmental studies, sound judgment requires knowledge of many topics and an interdisciplinary perspective.

REFERENCES

1 KNOWLES, R. L. 1974. *Energy and form.* Cambridge, Mass.: The MIT Press.

2 MCHARG, I. L. 1971. *Design with nature.* Garden City, N.Y.: Doubleday.

3 HAYES, W. C., and VINEYARD, J. D. 1969. *Environmental geology in town and country.* Missouri Geological Survey and Water Resources, Educational Series No. 2.

4 NORTH CAROLINA COASTAL RESOURCES COMMISSION. 1975. *State guidelines for local planning in the coastal area under the Coastal Area Management Act of 1974.* Raleigh, North Carolina.

5 WHYTE, W. H. 1968. *The last landscape.* Garden City, N.Y.: Doubleday.

6 FLAWN, P. T. 1970. *Environmental geology.* New York: Harper & Row.

7 ZUBE, E. H. 1973. Scenery as a natural resource. *Landscape Architecture* 63: 126–32.

8 LINTON, D. L. 1968. The assessment of scenery as a natural resource. *Scottish Geographical Magazine* 84: 219–38.

9 COUNCIL ON ENVIRONMENTAL QUALITY. 1979. *Environmental quality*.

10 BREW, D. A. 1974. *Environmental impact analysis: The example of the proposed Trans-Alaska Pipeline*. U.S. Geological Survey Circular 695.

11 STERLING, C. 1971. The Aswan disaster. *National Parks and Conservation Magazine* 45: 10–13.

12 NATIONAL RESEARCH COUNCIL, COMMITTEE ON GEOLOGICAL SCIENCES. 1972. *The Earth and human affairs*. San Francisco: Canfield Press.

13 BRICE, J. 1971. *Measurements of lateral erosion at proposed river crossing sites of the Alaskan Pipeline*. U.S. Geological Survey, Water Resources Division, Alaska District.

14 WAGNER, F. H. 1978. Livestock grazing and the livestock industry. In *Wildlife and America*, ed. H. T. Brokaw, pp. 121–45. Washington, D.C.: Council on Environmental Quality.

15 U.S. FOREST SERVICE. 1978. *RARE II. Draft environmental statement roadless area review and evaluation*.

16 MEADOWS, D. H.; MEADOWS, D. L.; RANDERS, J.; and BEHRENS, W. W. III. 1972. *The limits to growth: A report for the Club of Rome's Project on the Predicament of Mankind*. New York: Universe Books (Potomac Associates).

17 COUNCIL ON ENVIRONMENTAL QUALITY AND THE DEPARTMENT OF STATE. 1980. *The global 2000 report to the President: Entering the twenty-first century*.

FURTHER READING

1 COUNCIL ON ENVIRONMENTAL QUALITY AND THE DEPARTMENT OF STATE. 1980. *The global 2000 report to the President: Entering the twenty-first century*.

2 FORRESTER, J. W. 1971. *World dynamics*. Cambridge, Mass.: Wright-Allen Press.

3 MCHARG, I. L. 1971. *Design with nature*. Garden City, N.Y.: Doubleday.

4 MEADOWS, D. L., and MEADOWS, D. H., eds. 1973. *Toward a global equilibrium*. Cambridge, Mass.: Wright-Allen Press.

5 MESAROVIC, M., and PESTEL, E. 1974. *Mankind at the turning point*. New York: Sutton Press.

6 RAU, J. G., and WOOTEN, D. C., eds. 1980. *Environmental impact analysis handbook*. New York: McGraw-Hill.

7 SMITH, N. J. H. 1981. Colonization lessons from a tropical forest. *Science* 214: 755–61.

STUDY QUESTIONS

1 Make a map of your neighborhood. Does your neighborhood seem planned with the environment in mind?

2 Make a plan for a solar energy installation on your home. Where would you locate it, and which direction would you have it face?

3 What is meant by ''mitigation of an environmental impact''?

4 An expert on impact analysis tells you that ''mitigation is the impact. Once you have determined the mitigating factors, you have determined the impact.'' Explain. Do you agree?

5 What are the major steps which must be taken in the assessment of any environmental impact?

6 Discuss the elements of a land-use plan. Which is likely to be the most controversial? Why?

7 Differentiate between sequential and multiple land use. Provide examples of each.

8 Why is planning immediately following disasters loaded with potential problems?

9 Compare and contrast the "design with nature" method of site selection with benefits-cost evaluation (which tries to compare the benefits of a particular project over some time period with its cost).

10 Do you think scenery is a resource? Why or why not?

11 Discuss the major components of an environmental impact statement. Which are the most necessary? Why?

12 What are the major advantages of the scoping process in environmental impact analysis?

17

Artifice as Environment: An Environmental Perspective on Cities

INTRODUCTION

The city of Long Beach, California, which lies just south of Los Angeles, has undergone rapid development in the twentieth century. While more and more buildings were constructed, oil and gas were being pumped out from under the surface and water was being obtained from local wells. The removal of fluids—oil, gas, and water—caused *subsidence,* or settling of the soil. The city settled enough so that it became vulnerable to ocean waters; retaining walls had to be built to keep the ocean back, and treated water was pumped into the ground to reduce the subsidence.

Subsidence was first noticed in 1940. By the mid-1970s, the ground level had subsided 9 meters in the central area and had caused over $100 million in damage. Predictions of an ultimate subsidence of 13.7 meters promoted a community effort to stop it. In a repressuring program started in 1958, tremendous quantities of water were injected into the rocks from which the oil had been pumped out, raising the fluid pressure. This action actually reversed the subsidence, and by 1963 there had been appreciable rebound amounting to as much as 15 percent of the initial subsidence in some parts of the oil field [1].

Another example of subsidence occurred in Mexico City, where a rapid increase in population (from less than 500,000 in 1895 to more than 8 million in 1975) put rapidly increasing demands on local water supplies. Removal of groundwater led to subsidence of as much as 7 meters. The continuous sinking has produced many problems for buildings. The Palace of Fine Arts, constructed in 1934, has subsided approximately 3 meters; and, reportedly, steps that once went up to the first floor now go down to that floor [1].

As the people of Long Beach and Mexico City learned, cities influence and are influenced by their environment. We can ignore these relationships only at our own peril and expense. As we become aware of city-environment relationships, we can develop and maintain our cities wisely so that they function well, are pleasing, and have bright futures.

The relationship between cities and the environment is two-sided. On the one hand, the location, success, and importance of cities are in some ways consequences of the environment. On the other hand, a city affects the local environment; in some ways a city creates its own environment. However, while the environment places certain limits on cities, the environment is only one factor which determines the success or failure of a city.

In a sense, cities are one of the ways that human society is able to pass beyond the direct limitations the environment imposes on nonhuman species. We have built cities north of the Arctic, such as Tromsφ, Norway, which lies at latitude 69° 40′ N. We have also built cities in deserts or near deserts; Los Angeles, for example, is located in a region where local rainfall is too small to support a huge urban population and water is supplied from far away.

In planning for our future, it is useful to consider these relationships between a city and the environment. How will a city change as the environment changes? Is a city's environment healthy or unhealthy? How can we plan a city to make it more livable? These are some of the questions we will discuss in this chapter.

An environmental perspective on cities is particularly important because we are an urban society. In the United States, about 70 percent of the people live on 3 percent of the land area [2]. The world population is also becoming increasingly urban. In 1950, 392 million people lived in cities of 100,000 or larger, and by 1975, 983 million lived in these cities. It is projected that by the year 2000, 2.2 billion will live in cities of these sizes [2]. In 1950, 29 percent of the world's population lived in cities; in 1975, 39 percent. It is projected that 50 percent will live in cities by the year 2000 [3].

An environmental perspective on cities can tell us what is necessary for a city to support life. Although this perspective can tell us where it is easier or more difficult to build and maintain cities, it will not tell us where we *will* build cities or where we will maintain them. Human history can give a city its own momentum which will maintain it beyond a time when simple environmental factors no longer make it the best choice for an urban location.

Like any life-supporting system, a city must maintain a flow of energy, provide necessary material resources, and have ways of removing wastes. These ecosystem functions are maintained in a city by transportation and communication with outlying areas. A city is not a self-contained ecosystem; it depends on other cities and rural areas. A city cannot exist without a countryside to support it. As was said half a century ago, city and country, urban and rural, are one thing, not two things [4].

Thus a city can never be free of environmental constraints, even though its artificial, human constructions give us a false sense of security. As Lewis Mumford, the historian of cities, has written, cities give "the illusion of self-sufficiency and independence and of the possibility of physical continuity without conscious renewal." This is attested to by "the disintegration of one civilization after another" [5].

A city changes the landscape and therefore the relation between the biological and physical aspects of the environment. Areas that had natural soils and ecosystems, readily absorbing rain water, are converted to water-impervious roadways, walkways, and buildings.

Everything is concentrated in a city; this means that all classes of pollutants are concentrated. City dwellers are exposed to more kinds of toxic chemicals in higher concentrations and to more human-produced noise, heat, and particulates than their rural neighbors.

The environment in cities makes life riskier. The age-adjusted mortality rates for adults in the United States indicate a consistent 13 percent excess of cancer mortality in metropolitan counties which include central cities compared with other areas [6]. Deaths from cancer are highest in central cities, intermediate in noncentral city metropolitan counties, and lowest elsewhere. This pattern holds regardless of race or sex [6].

The city of Cleveland, Ohio, shows these patterns. In terms of air pollution, the city can be divided into five zones. Air pollution is heaviest in the central city, particularly along the waterfront where the industries are concentrated. Deaths from cancer are 50 percent greater for people living in the zone of highest air pollution than for those living in the least polluted area.

ENVIRONMENT AND THE LOCATION OF CITIES

Cities are not located at random; their location is strongly influenced by the environment. In ancient Rome, for example, all important cities were located near waterways. As a more modern example, most major cities of the eastern United States occur either at major ocean harbors or at the fall line on major rivers (Fig. 17.1). The **fall line** occurs where there is an abrupt drop in elevation. Because this topographic feature provided good sites for water power, which was an important source of energy in the eighteenth and nineteenth centuries, many eastern cities were established near the fall line.

The fall line was also the farthest inland large river vessels could sail. Also, crossing a river was much easier above the falls than below. In fact, it was only with the development of steel bridges in the late nineteenth century that the spanning of wider regions of the rivers below the falls became practical [7].

River valleys have other advantages for the location of cities. Valley bottoms and flats have water-deposited rich soils which are good for agriculture. The rivers also provided an important early means of waste disposal.

As in the case of the fall line and harbors, cities often are founded at crucial transportation points, growing up around a market, a river crossing, or a fort. Newcastle and Budapest are located at the lowest bridging point on a river; other cities, such as Geneva, are located where a river enters or leaves a major lake. Cities at crucial defensive locations include those with rock outcrops, such as

Edinburgh, Athens, and Salzburg; or a peninsula, such as Istanbul and Monaco. Other cities are located at the confluence of major rivers. For example, St. Louis lies close to where the Missouri and Mississippi rivers meet. Reading, Coblenz, and Khartoum are other cities located at the confluence of two rivers. Cities are often founded close to a mineral resource, such as salt (Salzburg); metals (Kalgoorlie, Australia); and medicated waters and thermal springs (Spa, Bath, Vichy, and Saratoga Springs). A successful city can grow and spread over surrounding terrain so that its original purpose may be obscured to a resident; its original market or fort may have evolved into a minor square or historical curiosity. Cities can change rapidly; areas that are desirable in one decade may be slums in the next and vice versa. These trends sometimes further obscure the original importance of a city.

SITE AND SITUATION

The location of a city is said to be influenced by two factors: **site,** which is the exact spot where the city is located and all of its environmental features; and **situation,** which is the location of the city with respect to other areas, including the areas it serves and the areas it is connected to by transportation routes. A good site includes a good geologic substrate for buildings, including a firm rock base and well-drained, dry land; good nearby supplies of water; local, good agricultural land; abundant timber and other natural resources; a benign local climate; and a location that is easily defended. Venice is an example of a site which developed because it was easily defended, even though it had other disadvantages. Venice was originally a small coastal settlement. Because the original site on the mainland could not be defended, merchants began building warehouses and mansions on islands offshore. Today the negative aspects of the site have come to the fore; the islands are sinking, endangering historic buildings. Good site characteristics occur frequently along foothill slopes, the natural terraces along the bases of mountain ranges; these can provide shelter from bad weather.

An ideal location for a city would have both a good site and a good situation, but this does not always occur. Paris is an example of a city with a good site and a good situation. Paris was founded on an island in the Seine River—a site which was easily defended, had ample water, and where the river was relatively easily crossed. There are several advantages to the situation of Paris: it is located near where several tributaries join the Seine, thus connecting the city to rivers from the northeast, east, and southeast; the Seine flows into the English Channel, which made it possible for vessels to travel from the ocean to the island of Paris. The city is also near the cen-

FIGURE 17.1
The major cities of the eastern and southern United States lie either at the sites of good harbors or along the fall line (shown by the dotted line). The location of cities is thus strongly influenced by the characteristics of the environment. (From *Natural regions of the United States and Canada* by Charles B. Hunt. Copyright © 1974 by W. H. Freeman and Company.)

ter of a large, prosperous farming region. This highly advantageous natural site and situation was improved further by social and technological developments. Canals were dug to improve and extend transportation, thus providing access to the regions of Loire, the Rhône-Saône, the Lorraine industrial area, and the French-Belgian industrial area.

An excellent situation can sometimes compensate for a poor site. For example, New Orleans has a good situ-

ation. Lying at the mouth of the Mississippi, New Orleans is an important transportation center. It is a transfer point for goods moving from the ocean to the interior up the Mississippi, and vice versa. However, the city has a poor site. It is located on the low delta of the Mississippi, which floods frequently and which provides a poor substrate for construction. The slow backwaters and swampy areas provide good breeding areas for mosquitoes but offer little as a local resource. In spite of its poor site, New

Orleans is a major city because it is an important transportation center at the mouth of the Mississippi.

Situation is a product of environmental and social factors. The natural influences that make a situation good are called the **natural nodality.** (A *node* is the meeting point of two or more lines.) For example, the situation of New Orleans was one of a natural transportation center. The social influences that make a situation good are called the **artificial nodality.** Society can make a situation significant by social, political, or technological means. The decision to locate Washington, D.C., at about the center of the north-south distribution of the original thirteen states was primarily a political effect on situation. Thus the site of Washington, D.C. (low, marshy ground along the Potomac River), which was of little utility originally, was excavated and filled. Once developed, the land became valuable, transportation routes developed, and so forth. While one might argue that the modern capital might be more logically located at the geographic center of the United States, the historical significance of Washington—including the massive investment in its government structures—makes such a suggestion almost unthinkable. Moscow and Madrid are other examples of cities that owe their location to political factors.

Mecca, Jerusalem, and Salt Lake City have situations created in part by religious and cultural history, which give these cities an importance beyond that of their natural site or situation.

Site begins as something provided by the environment, but technology can change the characteristics of a site. Modern technology has reduced the importance of site in the location of cities. Technology can also alter situation. For example, Santa Fe, New Mexico, is considered to have an excellent site. When it was bypassed by the transcontinental railroad in the nineteenth century, its importance declined in relation to Albuquerque, where the railroad was located.

The influence of railroads on situation is also reflected in European cities. In Germany, the town of Tangermunde intentionally and successfully prevented the construction within its borders of the railroad from Berlin to Hanover. The railroad junction was located in the nearby village of Stendal, which consequently grew rapidly and became larger than Tangermunde. Once this advantage was lost, it could not be regained. A later branch line to Tangermunde did not help it regain its previous importance [8].

Site and situation are often mixed blessings from an environmental point of view. Those factors that make a location a good situation can make it a poor site, and vice versa. Thus broad river valleys provide a good situation for land and water transportation, but make the site vulnerable to flooding and open to military attack.

Site and situation can change over time. For example, towns and cities that have lost their waterways have lost importance. Bruges, Ghent, and Ravenna are cities whose harbors silted; Ghent responded by improving its artificial situation by building a canal.

THE CITY AS AN ENVIRONMENT

The environment of cities is obviously different from those of lower-density human residential areas and natural areas. Although it is common to think of cities as unpleasant environments, cities have been centers of civilization and have often created an environment believed beautiful by the inhabitants. For example, a writer earlier in the twentieth century remembered earlier days in a city with great nostalgia. "The clicking of the hoofs upon the hard macadam, the rhythmical creaking of the harness, the merry rattle of the lead bars," he wrote, "are delectable sounds" [9]. Some of the ways that urban development proceeds to influence the environment are shown in Figure 17.2.

In this section, we will examine certain aspects of the city environment—its stages of development, its climate and wildlife, and its role as an ecological island. By understanding the special effects the city has on its environment, we can develop an ecological perspective in planning which can help make cities more livable.

URBAN MICROCLIMATE

Cities affect the local climate (Table 17.1), and as the city changes, so does its climate. For example, in the middle of the eighteenth century New York City's Manhattan Island was "generally reckoned very healthy" [10], perhaps because of its nearness to the ocean and its relatively unobstructed ocean breezes. Today, the air pollution and the effects of tall buildings on air flow lead the average visitor to Manhattan to a quite different conclusion.

Although air quality in urban areas is in part a function of the amount of pollutants present or produced, it is also affected by the city's ability to ventilate and thus flush out pollutants. The amount of ventilation depends on several aspects of the urban microclimate.

Cities are warmer than surrounding areas. The observed increase in temperature in urban areas is approximately 1 to 2C° in the winter and 0.5 to 1.0C° in the summer for mid-latitude areas. The temperature increase results from increased production of heat energy; the heat emitted from the burning of fossil fuels and other indus-

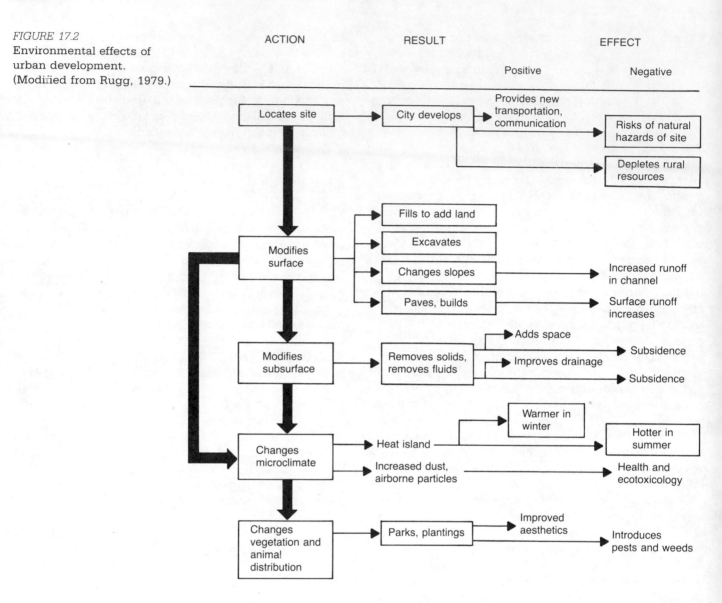

FIGURE 17.2
Environmental effects of
urban development.
(Modified from Rugg, 1979.)

trial, commercial, and residential sources; and from the decreased rate of heat loss, because the dust in the urban air traps and reflects back into the city long-wave (infrared) radiation emitted from the city surface. In the winter, space heating from the city is primarily responsible for heating the local air environment. For example, on Manhattan Island in New York City, the input of heat from industrial, commercial, and residential space heating in the winter has been measured to be about 2.5 times the solar energy that reaches the surface of the city; the annual average, however, is closer to 33 percent of the solar input. In large urban areas characterized by warmer winters, the heat input from artificial sources is much less [11]. Concrete, asphalt, and roofs also tend to act as solar collectors and quickly emit heat, helping to increase the sensible heat in cities [12].

In spite of stories about Chicago as "The Windy City," cities are in general less windy than nonurban areas. Air over cities tends to move more slowly than in surrounding areas because buildings and other structures obstruct the flow of air. Thus, it is not uncommon that wind velocities are reduced by 20 to 30 percent and the number of calm days are 20 percent more abundant in urban areas relative to nearby rural areas [11].

Particulates in the atmosphere over cities are often 10 times or more greater than for surrounding areas. Although these tend to reduce incoming solar radiation by up to 30 percent and thus cool the city, the effect of particulates is small relative to the processes that produce heat in the city [11].

The combination of lingering air and abundance of particulates and other pollutants in the air produces the

TABLE 17.1
Typical climate changes caused by urbanization.

Type of Change	Comparison with Rural Environs
Temperature	
Annual mean	0.5–1.0C° higher
Winter minima	1.0–3.0C° higher
Relative Humidity	
Annual mean	6% lower
Winter	2% lower
Summer	8% lower
Dust Particles	10 times more
Cloudiness	
Cloud cover	5–10% more
Fog, winter	100% more frequent
Fog, summer	30% more frequent
Radiation	
Total on horizontal surface	15–20% less
Ultraviolet, winter	30% less
Ultraviolet, summer	5% less
Wind Speed	
Annual mean	20–30% lower
Extreme gusts	10–20% lower
Calms	5–20% more
Precipitation[a]	
Amounts	5–10% more
Days with 0.2 in.	10% more

SOURCE: After Landsberg in Matthews et al., 1971. From Council on Environmental Quality and the Department of State, 1980.
[a]Precipitation effects are relatively uncertain.

well-known **urban dust dome** and **heat island effect** (Fig. 17.3). Also shown in Figure 17.3 is the general circulation pattern of air moving from the rural or suburban areas toward the inner city, where it flows up and then laterally near the top of the dust dome. This circulation of air often occurs when a strong heat island develops over the city. For example, when the dust dome and heat island have developed during a calm period in New York City, there is an upward flow of air over the heavily developed Manhattan Island accompanied by a downward flow over the nearby Hudson and East rivers, which are green belts and thus are characterized by cool air temperatures [12]. Figure 17.3 also shows the heat profile over the city which delineates the heat island.

The urban dust dome and heat island cause problems for many urban areas. Especially in the summer, long hot spells with poor air quality may lead to social unrest. Although the air quality problems of older cities will be more difficult to minimize, there is some indication that emission of pollutants is on the decline. For new urban areas, certainly we can plan more carefully to minimize both the dust dome and heat island effects.

The water budget is changed by the city's large impervious surfaces and lack of surface water. There is less exchange (evaporation) of water from the surface to the atmosphere, explaining why mid-latitude cities generally record a lower relative humidity (2 percent in the winter to 8 percent in the summer) than the surrounding countryside.

Particulates in the dust dome provide condensation nuclei, and thus urban areas experience 5 to 10 percent more precipitation and considerably more cloud cover and fog than surrounding areas. The formation of fog is particularly troublesome in the winter and may impede air traffic into and out of airports. If the pollution dome moves downwind, then increased precipitation may be reported outside of the urban area. For example, in the mid-1960s, effluent from the south Chicago–northern Indiana industrial complex apparently caused a 30 percent increase in precipitation at La Porte, Indiana, 48 kilometers downwind to the south (Fig. 17.4). La Porte also has almost 2½ times as many hailstones, 38 percent more thunderstorms, and less sunshine than the countryside not directly downwind [13]. La Porte is an extreme example because of certain unique features of its setting. It receives moisture from Lake Michigan and particles, heat, and moisture from the steel mills—all of which contribute to the increased precipitation [14]. The lake produces a dome of cold air that prevents the flow of air away from La Porte. While the La Porte case is extreme, it is not unique,

FIGURE 17.3
Lingering air, an abundance of particulates, and the flow of air over heavily built-up areas create an urban dust dome and a heat island.

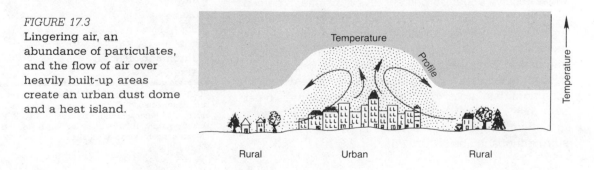

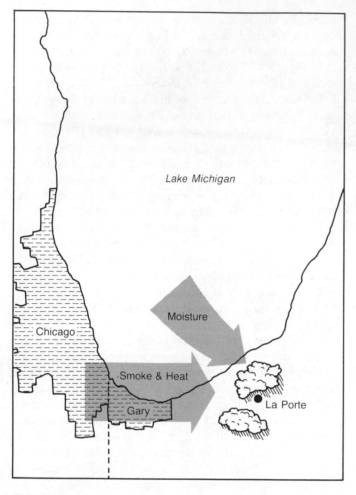

FIGURE 17.4

La Porte, Indiana, is 48 kilometers downwind from the steel mills of Chicago and Gary, Indiana. The mill effluents and the moisture from Lake Michigan increase rain, hail, and thunderstorms in La Porte.

and there is little doubt that particulate matter in the atmosphere from urban sources has altered local weather at numerous locations.

In summary, cities are cloudier, warmer, rainier, and less humid than their surroundings. Cities in middle latitudes receive about 15 percent less sunshine, 5 percent less ultraviolet light during the summer, and 30 percent less ultraviolet light during the winter than nonurban areas. They are 10 percent rainier and 10 percent cloudier and have a 25 percent lower average wind speed, 30 percent more summer fog, and 100 percent more winter fog than nonurban areas. Average relative humidity is 6 percent less in cities than in their environs. The average maximum temperature difference between a city and its surroundings is about 3C° [14].

Cities and Air Pollution　The city dweller is subject to many kinds of pollutants, as demonstrated by the levels of pollutants collected by snow in Ottawa, Ontario (Table 17.2). Snow near windows facing streets had a particularly high concentration of some pollutants, especially particulates.

Air pollution tends to be most intense at city centers. For example, particulate matter in New York City has been consistently higher in Manhattan than in the other boroughs, and has been lowest in Richmond (Staten Island), the least built-up of the boroughs (Fig. 17.5).

TABLE 17.2

Levels of pollutants found in snow deposits in Ottawa, Ontario.

Pollutant and Location	Windows Adjacent to Street (mg/kg)	Snow Disposal Sites (mg/kg)
Suspended solids		
Arterial street	3570	—
Collectors	1920–4020	—
Local	1215–2530	—
Parking lot	1620	—
BOD		108[a,b]
Arterial street	16.6	—
Collectors	13.2	—
Local	5.5	—
Parking lot	5.5	—
Chlorides	0–4500	175–2250
Oils	28.6[a]	28.6[a]
Greases	19.6[a]	19.6[a]
Phosphates		1.5[a]
Arterial street	0.032[a]	—
Collectors	0.087[a]	—
Local	0.065[a]	—
Lead		0.9–9.5
Residential	2[a]	—
Industrial	4.7[a]	—
Commercial	3.7[a]	—
Highway	102.0	—
Cadmium	—	<0.05
Barium	—	<0.50
Zinc	—	0.6
Copper	—	0.19
Iron	—	30.0
Chromium	—	<0.02
Arsenic	—	<0.02

SOURCE: From Richards, J. L., and Associates, Ltd., and Labrecque, Verina and Associates, 1973. Reprinted by permission of the publisher from *Introduction to environmental toxicology*, eds. F. E. Guthrie and J. J. Perry, p. 453. Copyright 1980 by Elsevier North Holland, Inc.
[a]Mean concentration at all sites.
[b]In mg/liter.

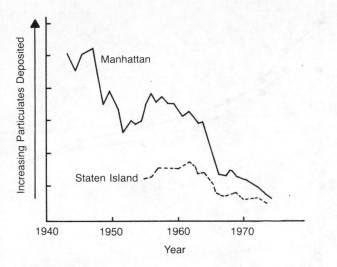

FIGURE 17.5

The deposit of particulates (dust) has been much higher in central, densely urbanized Manhattan than in less urbanized Staten Island. However, the Clean Air Act and technological improvements have brought the two boroughs of New York City to similar levels in recent years. (From Council on Environmental Quality, 1973.)

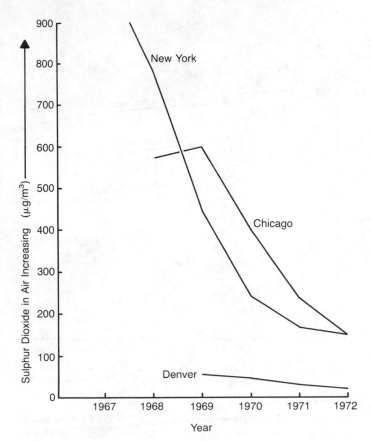

FIGURE 17.6

Air quality in New York and Chicago, two highly industrialized cities with a history of considerably high air pollution, improved greatly between 1967 and 1972. The air's sulphur dioxide concentration was reduced to one ninth of earlier values. Denver, a much less polluted city, also improved. (After Baumol and Oates, 1979.)

Air pollution levels in cities in developed countries have steadily improved, due to increases in regulation of pollution during the last several decades. For example, sulphur oxide concentrations in New York and Chicago decreased markedly from the late 1960s to the early 1970s (Fig. 17.6). While air pollution has improved, the outlook is not bright. As oil and gas become scarce and industrialized societies return to a dependency on coal, air quality in cities is projected to decline [3]. In the United States, if emission standards remain constant but energy production begins to emphasize coal, urban air quality could begin to deteriorate before 1990.

Exposure to air pollutants decreases with increased income; the city's poor live where there is more pollution (Fig. 17.7). For example, in St. Louis, Missouri, those with incomes less than $3000 per year were subject to suspended particulates of 91.3 micrograms per milliliter of air; those with incomes over $30,000 lived where the air had a concentration of 64.9. Those who can afford it move away from pollution.

The potential for air pollution in urban areas is determined by the following factors: the rate of emission of pollutants per unit area; the length downwind that a mass of air may move through an urban area; the average speed of wind; and, finally, the height to which potential pollutants may be thoroughly mixed in the lower atmosphere (Fig. 17.8) [15]. The concentration of pollutants in

the air is directly proportional to the first two factors. That is, as either the emission rate or downwind length of the city increases, so will the concentration of pollutants in the air. On the other hand, city air pollution decreases with increases in two meteorological factors: the height of mixing and the wind velocity. The stronger the wind and the higher the mixing layer, the lower is the pollution. Assuming a constant rate of emission of air pollutants as the column of air moves through the urban area, it will collect more and more pollutants. The inversion layer acts as a lid for the pollutants, but near a geologic barrier such as a mountain, there may be a "chimney effect" in which the pollutants spill over the top of the mountain (Fig. 17.8). This effect has been noticed particularly in the Los Angeles basin, where pollutants may climb several thousand meters, damaging mountain pine trees and other vegetation and spoiling the air of mountain valleys.

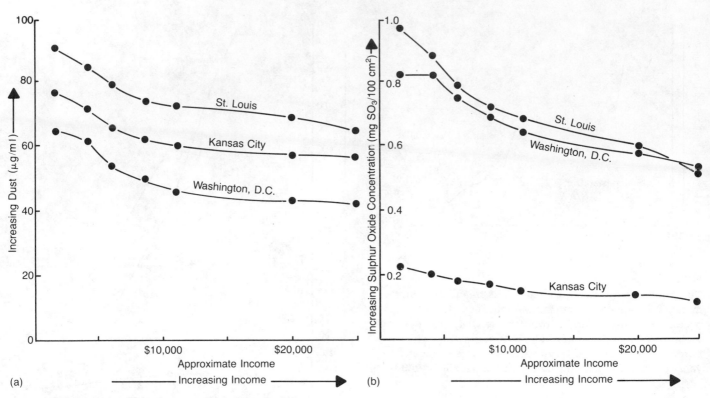

(a)

Increasing Income ⟶

(b)

Increasing Income ⟶

FIGURE 17.7

In urban residential areas, air pollution decreases as income increases. This is true
for particulates (a) and sulphur oxides (b). Although the level differs for each city,
the trend is the same. St. Louis has the highest levels for both forms of pollutants.
Kansas City has the lowest level of sulphur oxides and Washington, D.C., the
lowest level of particulates. (After Baumol and Oates, 1979.)

Planning to reduce air pollution in urban areas can
be helpful. For example, natural ventilation routes in
cities such as valleys and other green belts should not be
blocked by development. Zoning to take advantage of so-

lar energy is also important. Buildings constructed in tiers
for active and passive solar application will reduce the
sensible heat input into the atmosphere from burning
fuels. Interspersed small parks provide natural filters (from

FIGURE 17.8
The higher the wind
velocity and the higher the
mixing layer (shown here
as an inversion layer), the
less is the air pollution. The
higher the emission rate
and the longer the
downwind length of city,
the greater is the air
pollution. The "chimney
effect" allows polluted air
to move over a mountain
down into an adjacent
valley.

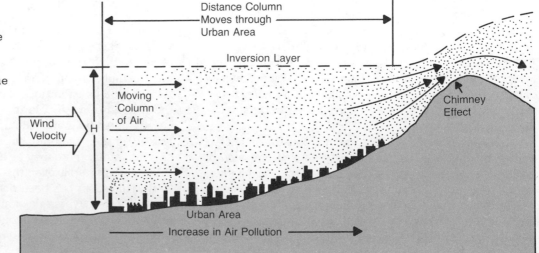

the vegetation) for air pollution and help circulate cooler air toward buildings, where heat tends to build up. Finally, it is hoped that stronger controls for automobile and industrial emissions, such as those set forth in the Clean Air Act amendments of 1977, will further improve urban air quality.

WILDLIFE IN CITIES

Peregrine falcons once hunted pigeons above the streets of Manhattan. Unknown to most New Yorkers, the falcons nested on the ledges of skyscrapers and dived on their prey in an impressive display of predation. The falcons disappeared when DDT and other organic pollutants caused a thinning of their eggshells and a failure in reproduction, but they have been reintroduced recently into the city. In New York City's Central Park approximately 260 species of birds have been observed, and 100 in a

single day. Foxes live in London, feeding on garbage and road kills (animals run over by motor vehicles); shy and nocturnal, they are seen by few Londoners [16].

We do not associate wildlife with cities, but, as these examples show, cities provide homes to many forms of life. Cities are a habitat, albeit artificial. They can provide all the needs for both plants and animals, including physical structure and necessary material resources (food, minerals, water). We can identify ecological food chains in cities (Fig. 17.9).

For some species, the city's structure is sufficiently like their original habitat to be a home. For example, chimney swifts originally lived in hollow trees, but are now common in factory chimneys and other vertical shafts. Their nests are glued to the walls with saliva. A city can easily have more chimneys per square kilometer than a forest has hollow trees.

FIGURE 17.9
(a) An urban food chain based on plants of disturbed places and insect herbivores. (b) An urban food chain based on road kills.

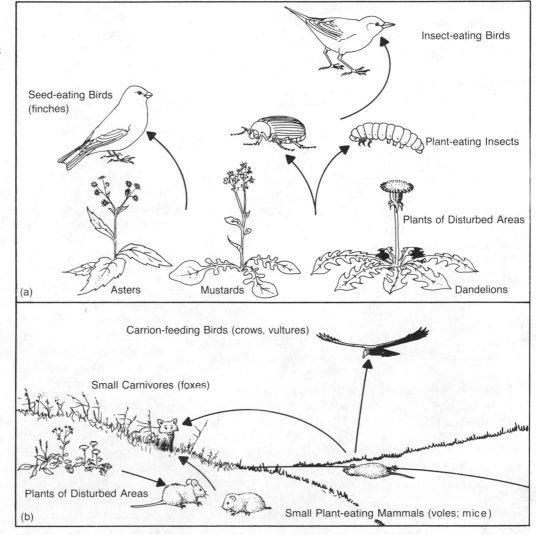

Cities include natural habitats in parks and preserves. Modern parks provide some of the world's best wildlife habitats, and the importance of parks will increase as the truly wild areas are reduced in extent. Jamaica Bay, a park in New York City, was recovered from natural marshes and wetlands. Until the 1960s, the area had been polluted by sewage and had become a wasteland, supporting a few species. After restoration, it is now 15,000 acres (about 7000 hectares) of urban wilderness, with a diverse population of birds. During the spring and fall migration many bird watchers from New York visit the bay to see avocets, dowitchers, sandpipers, and godwits [16]. Jamaica Bay is unusual in that it was planned as a natural park, emphasizing native vegetation and habitats, and has succeeded in attracting native species of animals.

Cities that are major harbors often have many species of marine wildlife at their doorsteps. New York City's waters include sharks, bluefish, mackerel, tuna, shad, striped bass, and nearly 250 other species of fish [16]. Small ponds contain freshwater fish and frogs.

Wild plants that do particularly well in cities are those characteristic of disturbed areas. City roadsides have wild mustards, asters, and other early successional plants. Trees, including exotic flowering trees, are planted for decorative purposes. In this way cities with like climates tend to have, through purposeful introductions, similar kinds of trees and flowers. Cities are becoming islands for such species.

Disturbances in cities promote certain kinds of plants. Curiously, during World War II, many species of wildflowers were found near bombed areas in London that had not been recorded previously. In addition, 342 species of plants were recorded where less than 100 had been recorded before. For example, bracken fern, rare in an English city, became common during that war and persisted afterwards in London [16].

We can divide city wildlife into the following categories: (1) those species that cannot persist in an urban environment and disappear; (2) those that tolerate an urban environment, but do better elsewhere; (3) those that have adapted to urban environments and are abundant there, and are either neutral or beneficial to human beings; and (4) those that are so successful that they are pests.

The last category of urban wildlife, familiar to almost every urban dweller, includes cockroaches, mice, rats, fleas, and pigeons. Pests compete with people for food and spread diseases. Before modern sanitation and medicine, such diseases played a major role in limiting the human population density in cities. Although cities are still less healthy than rural areas in terms of the effects of pollutants on human health, the control of these diseases has been an important improvement.

In summary, cities provide a significant habitat for wildlife of many kinds. As true wilderness decreases, city zoos will become more and more important as habitat for rare and endangered species. Cities today have a greater potential for the enjoyment of wildlife than most city dwellers realize. Perhaps one of the great ironies of modern environmentalism is that, although most of us are city dwellers, few studies of wildlife are done in cities and most activities of conservation take place outside of cities.

A BRIEF ENVIRONMENTAL HISTORY OF CITIES

With this background in mind, we can now appreciate the importance of the relationships between environment, society, technology, and the history of cities. We can divide the history of cities into roughly four stages.

Stage 1: The first towns. The first cities emerged on the landscape thousands of years ago during the New Stone Age with the development of agriculture, which provided the excess of food resources that is necessary to the maintenance of a city. Mumford writes that these *neolithic* villages, "lacking the size and complexity of the city," nevertheless exhibited the essential features of a city: "the encircling mound or palisade, setting it off from the fields; permanent shelters; storage pits and bins, with refuse dumps and burial grounds" [5].

In this first stage, the density of people per square kilometer is much higher than in the surrounding countryside, but the density is still too low to cause rapid, serious disturbance to the land. In fact, the city dwellers and their animals provide waste, which is an important fertilizer for the surrounding farmlands.

In this first stage, the city still relies on local food production. Its size is restricted by the ability of primitive transportation methods to bring food and necessary resources into the city and to remove wastes. Because of such limitations, no medieval town served only by land transportation had a population greater than 15,000 [8]. The size of a city varied with the richness and amount of local agricultural land.

Stage 2: Waterways and roads made possible the first large cities. In the second stage, more efficient transportation made possible the development of much larger urban centers, with a totally urban social core. Boats, barges, canals, and harbor wharfs, and roads, horses, carriages, and carts made possible a city which depended on agriculture far away. Rome, originally dependent on local produce, became a city fed by granaries of Africa and the Near East.

A city in this second stage was limited in size by pedestrian travel. The city could be no larger in area than the distance a worker could walk to work, do a day's work, and walk home. The density of people per square kilometer was limited by architectural techniques and primitive waste disposal. Thus these cities never exceeded a population of one million, and there were few cities of this size, including Rome and some cities in China.

This city contained the seeds of its own destruction. There is a tendency, as Mumford notes, to transform the city center from natural to artificial and to replace grass and soil with pavement, gravel, house, and temple, creating an impression of the dominance of the environment by civilization. Ironically, the very artifice of the city makes it all the more dependent on its rural surroundings for all resources. So although it appears to the inhabitants to be stronger and more independent, the city is actually more fragile. Cities in this stage grow at the expense of the surrounding countryside, destroying the surrounding landscape on which it ultimately depends. As the nearby areas are ruined for agriculture and the transportation network extends, the use, misuse, and destruction of the environment increase. In this stage, many of the classic cities of history were built, prospered, and declined.

Stage 3: The city of the Industrial Revolution. The Industrial Revolution allowed greater modification of the environment and greatly increased the ease and capacity of transportation. Two technological advances that had significant effects on the environment were improved sanitation methods, which led to the control of many diseases, and modern transportation methods, including the improved sailing ship and the steam engine.

The size and density of cities increased rapidly. While the city of Rome in 274 A.D. occupied about 13 square kilometers, modern greater London occupies about 1685 square kilometers and metropolitan New York City more than 6475 square kilometers. Prior to World War II, only 25 cities had populations over 1 million; by 1954 there were 64, 83 in 1960, and more than 100 in 1975.

In the third stage rapid transportation allowed the development of suburbs, and industries intensified pollution of the air and water. These changes increased the urban dwellers' sense of separateness from their "natural" environment. Some attribute the recent interest in hiking, wilderness, and environmental conservation to the ease of transportation, which allows the urban dweller brief experiences with a nonurban landscape, and the increased artificiality of the city itself. In the climax of the third stage, the city became more concentrated and artificial, with skyscrapers reaching nearly a kilometer upward and surface and underground transit systems moving people long distances to and from work. In this stage the urban park achieved a new importance.

Stage 4: Telecommunications and air travel. We are perhaps at the beginning of a fourth stage in the development of cities. With telecommunications (telephones, television, and computers linked by satellite and cable), much communication and commerce (which is much of the life of the city) can be done by people isolated from one another, at home, or long distances apart. The airplane has also freed us even more from the traditional limitation of *situation*. We now have urban areas where previously transportation was poor—in the Arctic (Fairbanks, Alaska) or on islands (Honolulu, Hawaii).

The negative effects of urban sprawl have led many back to the urban centers. The drawbacks of suburban commuting and the destruction of the landscape in suburbs have brought the city center a new appeal. Perhaps, as telecommunications free us from the necessities of certain kinds of commercial travel and related activities, the city can return to a cleaner, more pleasing center of civilization.

This optimistic future for cities requires a continued abundance of energy and material resources, which are certainly not guaranteed, and a wise use of these resources. However, if energy resources are rapidly depleted, modern mass transit may fail; fewer people will be able to live in suburbs and the cities will become crowded; reliance on coal and wood will increase air pollution; and the continued destruction of the land within and near cities could compound transportation problems, making local production of food impossible. The future of our cities depends on our ability to plan and to conserve and use our resources wisely.

CITY PLANNING AND THE ENVIRONMENT

Of all the aspects of planning, city planning has perhaps the longest history. Although many cities in history have grown "like Topsy" without any conscious plan, formal plans for new cities can be traced in modern history as far back as the fifteenth century. Sometimes cities have been designed for specific social purposes, with little consideration of the environment; in other cases, the environment and its effect on city residents have been a major planning consideration.

Among the earliest European planned towns and cities were fortress cities, designed for defense. For example, in 1485 Leon Battista Alberti described the shape, walls, fortification, internal transportation, and water supply of fortified towns in his book *De Re Aedificatoria.*

FIGURE 17.10
The city of Vitry-le-François, planned in 1545, was designed with an emphasis on fortification. (Courtesy of the Library of Congress.)

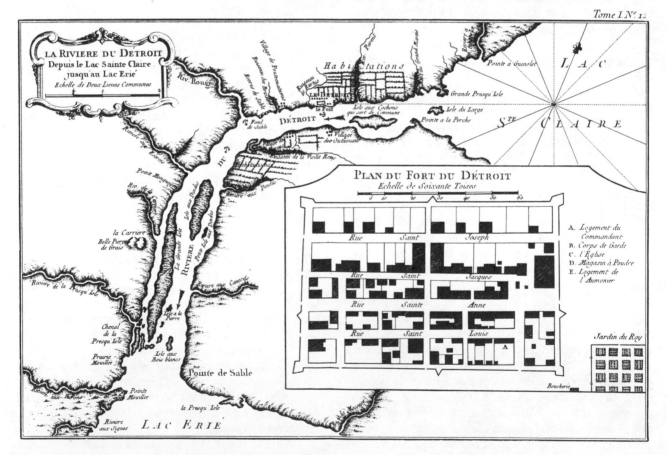

FIGURE 17.11
Plan of Detroit. Like many earlier cities, Detroit was designed for fortification. The fortress design is evident in the 1764 plan. (Courtesy of the Library of Congress.)

However, even in these early plans the aesthetics of the town were considered. Alberti argued that large and important towns should have broad and straight streets; smaller, less fortified towns should have winding streets to increase their beauty. Alberti advocated the inclusion of town squares and recreational areas. Figure 17.10 shows the 1545 plan for Vitry-le-François, which illustrates the emphasis on fortification. The area's topography is used to add to the defense, but the shape of the city is determined primarily by military rather than residential needs [17].

The usefulness of walled cities essentially ended with the invention of gunpowder, and the Renaissance sparked an interest in the "Ideal City." A preference for gardens and parks—emphasizing recreation—developed in the seventeenth and eighteenth centuries, culminating in the plan of Versailles, built for Louis XIV by André Le Nôtre. Versailles includes a number of parks of various sizes as well as tree-lined walks. New York City also reflected this new concept. After a visit to this city in the late eighteenth century, the Swedish naturalist Peter Kalm wrote that trees planted on the main street give them in the summer "a fine appearance" and "afford a cooling shade." During the winter, "I found it extremely pleasant to walk in the town," he wrote, "for it seemed like a garden" [10]. The trees planted at that time were sycamore and locust. Both the fortress town and park town planning influenced the planning of cities in North America.

The first new "town" built in the New World by Europeans was a crude fortress constructed by Christopher Columbus on the island of Espanada from the wreckage of timbers of one of Columbus' ships, the *Santa Maria*. The first city in what is now the United States was St. Augustine, Florida, begun in 1565.

Cities in the United States were planned for many different purposes. Detroit, founded in 1701 by Cadillac, was originally designed as a fortress town on the Detroit River, which connected Lake Erie with Lake St. Clair (Fig. 17.11). The care taken in planning some early American towns is illustrated by Williamsburg, Virginia, established by the Virginia legislators at the beginning of the eighteenth century. The town plan—the most detailed up to that time in the English colonies—specified the size of the town, the site of the capital, and location of principal roads. On the main street, all houses were to be set back 6 feet (2 meters) (Fig. 17.12).

From the earliest towns in North America, urban planners have attempted to provide for defense, transportation, access to necessary resources, needs of residential and commercial buildings, and aesthetics. The importance of aesthetic considerations is illustrated in the plan

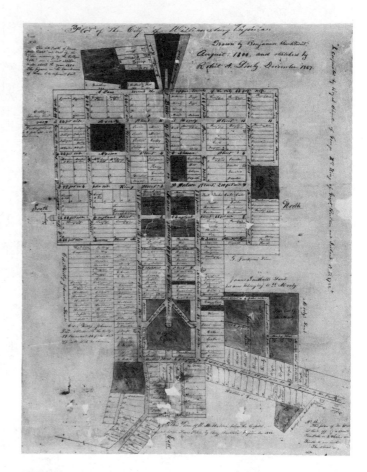

FIGURE 17.12

Plan of Williamsburg, Virginia. Great attention was paid to detail in planning this famous, and now restored, early American city. (Courtesy of Swem Library, College of William and Mary.)

of Washington, D.C., designed by the Frenchman L'Enfant. He mixed a traditional rectangular grid pattern of streets with broad avenues set at angles. The intention was to design a city of beauty (Fig. 17.13).

The most important event in the development of city parks was the planning and construction of Central Park in New York City by Frederick Law Olmsted. Central Park was the first large public park in the United States. In his plan, Olmsted carefully considered the opportunities and limitations of the topography (Fig. 17.14). He pointed out that the topography divided the site into two nearly equal parts—the northern half being more rugged, and the southern portion including flat meadows and a more varied topography. Because of the terrain and the proximity of the southern half of the park to the city center, most of the recreational areas were placed there. Depressed roadways allowed traffic to cross the park without detracting from the vistas as seen by park visitors.

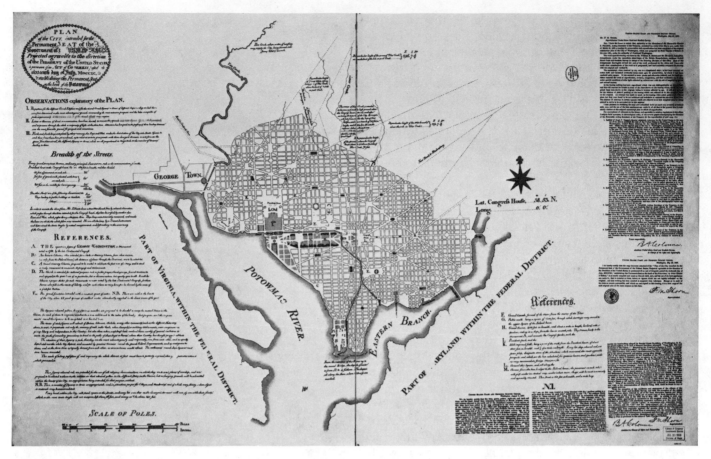

FIGURE 17.13
L'Enfant's plan for Washington, D.C. Aesthetics was an important factor in this 1791 design. (Courtesy of the Library of Congress.)

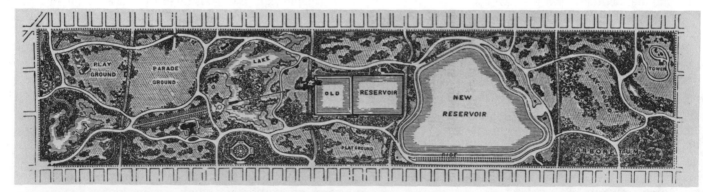

FIGURE 17.14
Olmsted's plan for Central Park, New York City, in 1871. This plan had a great impact on subsequent city planning. Variety, the use of natural topography, and the successful designing for multiple uses make this plan a landmark in the history of city planning. [Reproduced from a volume of the *Central Park Competition Designs* (1898) at Loeb Memorial Library, Harvard Graduate School of Design.]

Central Park's main drive was designed to bring visitors to a vista that would attract their attention further into the park. Large open areas were planned as parade grounds or ball fields; where a marsh existed, a lake was proposed. Olmsted recognized that eventually the time would come when New York would be entirely built up, and that the only area that would remind the city dwellers of the original terrain would be in the great park. Central Park is an example of ''design with nature.'' The potentials and limitations of the site—its geology, hydrology, and vegetation—were carefully considered and influenced the design. The design of Central Park influenced other U.S. city parks, and Olmsted remained a major figure in American planning throughout the nineteenth century[18].

SUMMARY AND CONCLUSIONS

As an urban society, we must recognize the city's relation to the environment. A city influences and is influenced by its environment, and is an environment itself.

Like any life-supporting system, a city must provide for a flow of energy and a cycling of chemical elements necessary for life. Not being self-sufficient, a city must have a source of energy and material resources and must have a sink for waste disposal. These in turn require a transportation network.

Because cities depend on the resources outside, they could only appear when human ingenuity resulted in an excess of production. The history of cities can be divided into four stages: (1) the rise of towns accompanying the beginning of agriculture which depended on local resources and had relatively small negative effects on the environment; (2) the era of great classic cities, in which cities came to dominate their surroundings and improved transportation made possible the use of resources far away; (3) the Industrial Revolution, in which mechanized transportation, modern sanitation, and other technological changes allowed a great increase in the size and density of cities as well as an increase in the pollution of air, water, and soil; and (4) the age of mass telecommunication, computers, and new forms of travel, in which the future of our cities may be decided.

The location of cities is strongly influenced by the environment. It is clear that cities are not located at random, but in places of particular importance or environmental advantage, such as near ocean harbors, at the fall line, near mineral resources, or at crucial transportation points. Their location is a consequence of *site* and *situation*. The site is the exact location where the city is built and the characteristics of that location. The situation is the relationship to other areas. A good site includes firm, solid geological foundations for construction; well-drained soil with abundant water; defensibility; and a moderate climate. A good situation includes good, natural transportation, such as a place where a river is most easily crossed or a transfer point between two modes of transportation. Most of the great cities of history are located at good sites and important situations. However, one can compensate for another. In particular, an important situation can compensate for a poor site.

A city creates an environment that is different from its surrounding areas. Cities change the local climate, and in general are cloudier, warmer, rainier, and less humid than their surroundings. Because cities are ecological islands, they favor certain kinds of animals and plants. Natural habitats in city parks and preserves will become more important as true wilderness decreases.

In general, life in a city is riskier because of higher concentrations of pollutants and pollutant-related diseases. Although pollution levels have improved in cities in recent decades, the trend is expected to worsen in the future. Because there are so many factors that affect a city's potential for air pollution, careful planning is needed to improve air quality.

City planning has a long history. Cities have been planned for many uses, from forts to commercial and political centers. Each major use has shaped the original city design. The fortress city, one of the earliest types of planned towns, and the park town influenced the planning of cities in North America. The importance of aesthetics in planning is demonstrated in L'Enfant's plan of Washington, D.C., and Central Park is an excellent example of the ''design with nature'' concept of planning. Wise city planning helps make a city livable and includes parks, open spaces, and recreational areas.

REFERENCES

1 POLAND, J. F., and DAVIS, G. H. 1969. Land subsidence due to withdrawal of fluids. In *Reviews in engineering geology*, eds. D. J. Varnes and G. Kiersch, pp. 187–269, The Geological Society of America.

2 ROBIE VESTAL, J. 1980. Pollution effects of storm-related runoff. In *Introduction to environmental toxicology*, eds. F. E. Guthrie and J. J. Perry, pp. 450–56. New York: Elsevier.

3 COUNCIL ON ENVIRONMENTAL QUALITY AND THE DEPARTMENT OF STATE. 1980. *The global 2000 report to the President: Entering the twenty-first century.*

4 JEFFERSON, M. 1931. Distribution of the world's city folks: A study in comparative civilization. *Geographical Review* 11: 446–65.

5 MUMFORD, L. 1972. The natural history of urbanization. In *The ecology of man: An ecosystem approach,* ed. R. L. Smith, pp. 140–52. New York: Harper & Row.

6 FORD, A. B., and BIALIK, O. 1980. Air pollution and urban factors in relation to cancer mortality. *Archives of Environmental Health* 35: 350–59.

7 HUNT, C. B. 1974. *Natural regions of the United States and Canada.* San Francisco: W. H. Freeman.

8 LEIBBRAND, K. 1970. *Transportation and town planning.* (Translated by N. Seymer.) Cambridge, Mass.: The MIT Press.

9 BOTKIN, B. A., ed. 1954. *Sidewalks of America.* (Reprinted in 1976.) Westwood, Conn.: Greenwood Press.

10 KALM, P. 1750. *Peter Kalm's travels in North America.* (The English version of 1770.) Translated by A. B. Benson. Reprinted in 1963 by Dover Books, New York.

11 DETWYLER, T. R., and MARCUS, M. G., eds. 1972. *Urbanization and the environment.* North Scituate, Mass.: Duxbury Press.

12 MARSH, W. M., and DOZIER, J. 1981. *Landscape.* Reading, Mass.: Addison-Wesley.

13 GATES, D. M. 1972. *Man and his environment: Climate.* New York: Harper & Row.

14 LYNN, D. A. 1976. *Air pollution — Threat and response.* Reading, Mass.: Addison-Wesley.

15 PITTOCK, A. B.; FRAKES, L. A.; JENSSEN, D.; PETERSON, J. A.; and ZILLMAN, J. W., eds. 1978. *Climatic change and variability: A southern perspective.* (Based on a conference at Monash University, Australia, 7–12 December, 1975.) New York: Cambridge University Press.

16 BURTON, J. A. 1977. *Worlds apart: Nature in the city.* Garden City, N.Y.: Doubleday

17 REPS, J. W. 1965. *The making of urban America.* 2nd ed. Princeton, N.J.: Princeton University Press.

18 MCLAUGHLIN, C. C., ed. 1977. *The papers of Frederick Law Olmsted. Vol. 1: The formative years 1822–1852.* Baltimore: Johns Hopkins University Press.

FURTHER READING

1 BEAUJEU-GARNIER, J., and CHABOT, G. 1967. *Urban geography.* New York: Wiley.

2 BRYSON, R. A., and ROSS, J. E. 1972. The climate of the city. In *Urbanization and the environment,* eds. T. R. Detwyler and M. G. Marcus, pp. 51–68. North Scituate, Mass.: Duxbury Press.

3 BURTON, J. A. 1977. *Worlds apart: Nature in the city.* Garden City, N.Y.: Doubleday.

4 de BLIJ, H. J. 1977. *Human geography: Culture, society and space.* New York: Wiley.

5 FEBVRE, L. 1966. *A geographical introduction to history.* New York: Barnes and Noble.

6 LEGGETT, R. F. 1973. *Cities and geology.* New York: McGraw-Hill.

7 LEIBBRAND, K. 1970. *Transportation and town planning.* (Translated by N. Seymer.) Cambridge, Mass.: The MIT Press.

8 MUMFORD, L. 1961. *The city in history.* New York: Harcourt, Brace and World.

9 ———. 1968. *The urban prospect.* New York: Harcourt, Brace and Jovanovich.

10 ———. 1972. The natural history of urbanization. In *The ecology of man: An ecosystem approach,* ed. R. L. Smith, pp. 140–52. New York: Harper & Row.

11 REPS, J. W. 1965. *The making of urban America.* 2nd ed. Princeton, N.J.: Princeton University Press.

12 ROBINSON, H. 1976. *Human geography.* 2nd ed. Plymouth, England: MacDonald and Evans.

13 WHYTE, W. H. 1968. *The last landscape.* Garden City, N.Y.: Doubleday.

STUDY QUESTIONS

1 How does the environment influence the location of cities?

2 What cities are most likely to become ghost towns in the next 30 years? In answering this question, make use of your knowledge of changes in resources, transportation, and communication.

3 Some futurists picture a world that is one giant biospheric city. Is this possible? Under what conditions would it be possible?

4 Among the ancient Greeks it was said that a city should have no more people than the number which can hear the sound of a single voice. Would you apply this rule today? If not, how would you decide how to plan the size of a city?

5 Standing on top of the Sears Tower in Chicago, Illinois, you overhear someone say, "Planning never works. The most interesting cities just grow. Planned cities are always dull and sterile." He points to large, low-income housing developments far in the distance. How would you respond?

6 You are the manager of Central Park in New York City and receive the following requests. Which would you approve?

(a) A gift of $1 million to plant trees from all of the eastern states.
(b) A gift of $1 million to set aside half the park to be forever untouched, thus producing an urban wilderness.
(c) A gift of the construction of an asphalt jogging track and a gym for physical fitness. The donor says that lack of physical fitness is a major urban health problem.
(d) A request to install an ice skating rink with artificially made ice. Facilities include an elegant restaurant with many views of the park.

7 Your state asks you to *locate* and *plan* a new town. The purpose of the town is to house people who will work at a "wind farm"—a large area of many windmills, all linked to produce electricity. You must locate the site for the wind farm first; then plan the town. Describe how you would proceed and what factors you would take into account.

8 Visit your town center. What changes, if any, would make better use of the environmental location? How could the area be made more livable?

9 In what ways does air travel alter the location of cities? The value of land within a city?

18

Whose Right To Do What?: An Introduction to Environmental Law

INTRODUCTION

In 1841, a suit brought before the U.S. Supreme Court concerned mudflats along the shores of the Raritan River in New Jersey. The mudflats were rich in oysters, and the owner of the land wanted to prevent other people from harvesting them. He claimed sole rights to the oysters on the basis of his land title, which could be traced back to a grant in 1664 from King Charles II of England to the Duke of York. That title conveyed to the duke "all the lands, islands, soils, rivers, harbors, mines, minerals, quarries, woods, harbors, waters, lakes, fishings, hawkings, huntings and fowlings." According to the landowner, the title gave him private rights to the environment (Fig. 18.1) [1].

The Supreme Court rejected his claim, stating that the navigable waters and the land under them were held by the king "in public trust." In English common law, this public trust gave the king responsibilities to protect the lands for the public. Accordingly, the king did not have the power to alter the public uses of fisheries; he could not abridge his obligation to the public. Oysters living in a natural estuary could not be claimed as private property [1].

This landmark case, known as *Martin* v. *Waddell,* not only marked an important turning point in wildlife law in the United States, it epitomized the legal conflicts that have persisted ever since. At the heart of these conflicts lies the question of who owns, controls, and is responsible for the well-being of the environment, as well as the question of who can represent the environment in court. The environment, which seems on the one hand to belong to everyone and on the other to belong to no one, poses unique legal issues.

THE DEVELOPMENT OF ENVIRONMENTAL LAW

Today a land, or environmental, ethic is being incorporated into our legal principles. No longer is the best use of land that which returns the greatest profit. The new land ethic establishes that the human race is part of a land community including trees, rocks, animals, and scenery, and that we are morally bound to assure the community's continued existence. Thus, this ethic affirms our belief that this Earth is our only suitable habitat (Fig. 18.2), and it recognizes the rights of people to breathe clean air, drink unspoiled water, and generally exist in a quality environment.

According to the National Environmental Policy Act of 1969, the national environmental policy is to encourage productive and enjoyable harmony between people and their environment, to promote efforts to eliminate or at least minimize degradation of the environment, and to continue investigating relations between ecological systems and important natural resources. This policy suggests that each generation has the moral responsibility to provide the next generation with healthful, productive, and aesthetically pleasing surroundings. It implies a **doctrine of public trust** which asserts that all public land and, to a lesser extent, some private land are essentially held in trust for the general public.

The Trust Doctrine as a theory is closely related to constitutional issues involving the right of people to a quality environment. Although nowhere in the Constitution does it state this right, some students of law believe that such a right can be inferred from the Ninth Amendment to the Constitution, which recognizes that the list-

FIGURE 18.1
The question of who owns the rights to shellfish in tidal areas along the east coast of North America brought before the U.S. Supreme Court in 1841 the issue of public trust. This case represents one of the first in environmental law in the United States. (Photo by D. B. Botkin.)

FIGURE 18.2
The Earth is indeed our only suitable habitat, and there are few alternatives to maintaining a quality environment. (Photo courtesy of NASA.)

ing of specific rights in the Bill of Rights does not deny the existence of other unlisted rights. That is, there are many rights not specifically listed but always held by the people [2].

Fundamental rights entitled to the protection of the Ninth Amendment are defined as those so basic and important to our society that it is inconceivable that they are not protected from unwarranted interference [3]. It might, therefore, be argued that this amendment encompasses the people's right to have clean air, clean water, and other resources necessary to insure a quality environment.

It is important to recognize that there is a certain amount of friction between the Trust Doctrine, which establishes that particularly significant land resources are held in a public trust, and the Fifth Amendment, which establishes that land may not be taken from an individual without due process of law and just compensation. Interpreting the law and determining how to measure just compensation are two of the problems encountered. The real issue, however, is defining at what point the private use of land infringes on the rights of other people and future generations.

Although the term *environmental law* has only recently been commonly used and there are few new laws in this field, it is fast becoming an important part of our jurisprudence [2]. The many recent works devoted to environmental law, the establishment of environmental law societies, and the teaching of environmental law at leading law schools also attest to its importance [4].

Air, water, and other resources necessary for our survival are often taken for granted. In the United States, we still tend to adhere to the myth of superabundance and to think of resources as inexhaustible. However, with in-

creasing urbanization, our resources are deteriorating, so regulations and laws concerning air and water pollution are necessary. Such regulations are not new; London's air by 1306 was polluted by large amounts of smoke from the burning of coal to the extent that a royal proclamation was issued to curtail the use of coal. Violations of the law were punishable by death [4].

Air pollution was proclaimed a public nuisance as early as 1611, when an English court ruled, in essence, that property owners have the right to have the air they breathe smell clean and be free of pollutants. The case involved a plaintiff who asked for an injunction and damages against his neighbor for raising hogs. The defendant was found to be creating a nuisance, even though he pleaded that the raising of hogs was necessary for his subsistence and that neighbors should not have such delicate noses that they cannot bear the smell of hogs [5]. The decision in this case depended on the English Nuisance Law of 1536, a type of common law still used in the United States. Its main principle is that if other people suffer equally from a particular pollution, an individual cannot bring suit against the polluter. In the case of the hog farmer, it is obvious that the effect of the nuisance varied with proximity to the hog pens [4].

It is beyond the scope of our discussion to consider in detail the processes of law and how they relate to the environment. Suffice it to say that law is a technique for the ordered accomplishment of economic, social, and political purposes, and the most desired legal technique generally is one that most quickly allows ends to be reached. However, law usually serves the dominant interest in a culture, and in our sophisticated culture, the major concerns are wealth and power. The resource base is used to produce goods and services, and our legal system provides the vehicles to insure that productivity [6].

The general trend of law and the courts has been to render judgments based upon the greatest good to the greatest number and to use what is commonly referred to as the **Balancing Doctrine,** which asserts that public benefit from and importance of a particular action should be balanced with potential injury to certain individuals. However, the method of balancing is changing. The courts are now considering possible long-range injury caused by certain activities to large numbers of citizens rather than to just the immediate complainants; this trend is more likely to insure a true balancing of the equities [6].

CASE HISTORIES

The following case histories illustrate, in a legal context, the complex conflicts of interest that arise over the environment and some legal aspects of environmental problems. The particular selection of cases does not imply a judgment of any particular activity, but rather indicates the considerable variability of and possibilities in environmental law.

Storm King Mountain: Hydroelectric Power versus Fish and Scenery   The Storm King Mountain dispute is a classic example of possible conflict between a utility company and conservationists. In 1962, the Consolidated Edison Company of New York announced plans for a hy-

FIGURE 18.3
Storm King Mountain in the Hudson River Highlands (a) has long been known as a scenic area. The highlands of the Hudson were painted often in the nineteenth century, as shown by Samuel Colman's *Storm King on the Hudson* (b). The proposed hydroelectric project in this area raised issues concerning the need for energy and the conservation of a well-known scenic resource. The idealized diagram in (c) shows how the entire Storm King Mountain hydroelectric project might be placed underground. [Part (b) from the National Museum of American Art, Smithsonian Institute, gift of John Gellatly.]

(a)

(b)

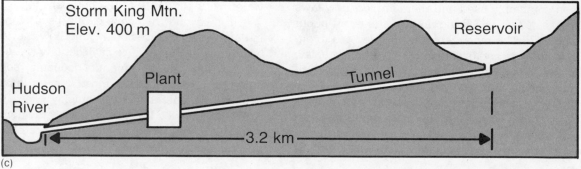

(c)

droelectric project approximately 64 kilometers north of New York City in the Hudson River Highlands, an area considered by many to have unique aesthetic value. Nowhere else in the eastern United States has a major river eroded through the Appalachian Mountains at sea level, giving the effect of a fjord [7]. The early plans for the project called for a powerhouse to be constructed above ground, requiring a deep cut into Storm King Mountain. However, the project was redesigned to site the powerhouse entirely underground, eliminating the cut on the mountain (Fig. 18.3).

The conservationists continued to oppose the project, broadening the issues to include possible damage to fishery industries. They argued that the high rate of water intake from the river (31,200 cubic meters per minute) would draw many fish larvae into the plant, where they would be destroyed by turbulence and abrasion. One study showed that 25 to 75 percent of the annual hatch of striped bass, the most valuable sport fish of the Hudson, might be destroyed if the plant were operating. The fish return from the ocean to tidal water to spawn, and the Hudson River is the only estuary north of Chesapeake Bay where the striped bass spawn. In addition, the proposed plant location is near the lower end of the 13-kilometer reach in which the fish spawn. The project is still being studied, and it is the responsibility of the utility company to show that the project will not do unacceptable damage to the fishery resource [7].

After a decade of legal activity, the Storm King Mountain controversy is still unresolved. The case emphasizes the difficulty in making decisions about multidimensional issues. On the one hand, there is a utility company trying to survive in New York City, where there are extremely high peak-power demands accompanied by high labor and maintenance costs. On the other hand, conservationists are fighting to preserve a beautiful landscape and a fishery resource. Both have legitimate arguments, which makes it difficult to resolve the conflicts.

Fossil Beds versus Land Development   A recent case in Colorado emphasizes the significance of the Trust Doctrine and the Ninth Amendment as they pertain to land management [3]. The conflict surrounded the use of 7.3 square kilometers of land near Colorado Springs. The land is part of the Florissant Fossil Beds, where insect bodies, seeds, leaves, and plants were deposited in an ancient lake bed approximately 30 million years ago. Today, they are remarkably preserved in thin layers of volcanic shale. Unfortunately, the fossils are delicate and, unless protected, tend to disintegrate when exposed. The fossils are considered by many people to be unique and irreplaceable. At the time of the controversy, a bill had been

introduced into Congress to establish a Florissant Beds National Monument. The bill had passed the Senate, but the House of Representatives had not yet acted on it.

While the House was deliberating the bill, a land development company, which had contracted to purchase and develop recreational home sites on 7.3 square kilometers of the ancient lake bed, announced it was going to bulldoze a road through a portion of the proposed national monument site to gain access to the property it wished to develop. A citizens' group formed to fight the development until the House acted on the bill. They tried to obtain a temporary restraining order, which was at first denied because there was no law preventing the owner of the property from using that land in any way desired, providing existing laws were upheld.

The conservationists then went before an appeals court and argued that even though there was no law protecting the fossils, they were subject to protection under the Trust Doctrine and the Ninth Amendment. They argued that protection of an irreplaceable, unique fossil resource was an unwritten right retained by the people under the Ninth Amendment, and that since the property had tremendous public interest, it was also protected by the Trust Doctrine. An analogy used by the plaintiffs was that if a property owner were to find the U.S. Constitution buried on the land and wanted to use it to mop the floor, certainly that person would be restrained. After several more hearings on the case, the court issued a restraining order to halt development, and shortly thereafter the bill to establish a national monument was passed by Congress and signed by the President [3].

The court order prohibiting the destruction of the fossil beds may have deprived a landowner of making the most profitable use of the property, but it did not prohibit any use consistent with protecting the fossils. For instance, the property owner is free to develop the land for tourism or scientific research. While these uses might not result in the largest possible return on the property owner's investment, they probably would return a reasonable profit [3].

SOME MAJOR ENVIRONMENTAL LEGISLATION

To insure wise use of our natural environment and resources, we need ecologically sound, responsible, socially acceptable, economically possible, and politically feasible legislation [3]. To achieve this goal, we need professional people to assist legislators at all levels of government in drafting laws and regulations.

Environmental legislation has already had a tremendous impact on our society. New standards regulating the discharge of pollutants into the environment have placed restrictions on industrial activity. Furthermore, before any new activity is begun by the federal government, the environmental impact of the proposed activity must be evaluated. Many states are following the example of the federal government by passing environmental protection legislation.

A complete description of all federal legislation with environmental implications is beyond the scope of this text. However, a discussion of some of the major acts is valuable in understanding the basics of such legislation. In this section, we will discuss the Refuse Act of 1899, the Lacey Act of 1900, the Migratory Bird Treaty Act of 1918, the National Environment Policy Act of 1969, the Water Quality Improvement Act of 1970, the Endangered Species Act of 1973, the Resource Conservation and Recovery Act of 1976, the Surface Mining Control and Reclamation Act of 1977, and the Clean Water Act of 1977.

THE REFUSE ACT OF 1899

The Refuse Act of 1899 states that it is unlawful to throw, discharge, or deposit any type of refuse from any source, except that flowing from streets and sewers, into any navigable water. Furthermore, the act implies that it is unlawful to discharge refuse into tributaries of navigable water. For all practical purposes, this means that it is against the law to pollute any stream in the United States. However, the Secretary of the Army can allow the discharge of refuse into a stream if a permit is granted.

THE LACEY ACT OF 1900

The Lacey Act of 1900 authorizes the Secretary of Agriculture to take all measures necessary for the ''preservation, distribution, introduction, and restoration of game birds and other wild birds.'' These actions, however, are subject to the laws of the states and territories. The act was a direct response to the rapid destruction of the passenger pigeon, which had been hunted in great numbers and was rapidly becoming extinct. The act also prohibited for the first time interstate transportation of any wild animals or birds that had been killed in violation of a state law. Thus the federal government was given the power to assist states in the enforcement of state game laws.

THE MIGRATORY BIRD TREATY ACT OF 1918

Great Britain and the United States signed a treaty in 1916 establishing a closed season on birds whose migrations brought them across international boundaries. The Migratory Bird Treaty Act of 1918 implemented that treaty within the United States. The act makes it unlawful to hunt, take, capture, or kill any bird protected by the

convention, except as permitted by the Secretary of Agriculture. This act has had the important effect of allowing government restrictions on the taking of wildlife on private property.

THE NATIONAL ENVIRONMENT POLICY ACT OF 1969

The philosophical purposes of the National Environmental Policy Act are to declare a national policy that will encourage harmony with our physical environment; to promote efforts that prevent or eliminate environmental degradation, thereby stimulating human health and welfare; and to improve our understanding of relations between ecological systems and important natural resources. To promote interest and research in and the authority to achieve these purposes, the act established the Council on Environmental Quality. The council is in the Executive Office of the President and is responsible for preparation of a yearly environmental quality report to the nation. It also provides advice and assistance to the President on environmental policies.

Perhaps the most significant aspect of the NEPA Act is that it requires an environmental impact statement before major federal actions are taken that could significantly affect the quality of the human environment. This requirement extends to such activities as the construction of nuclear facilities, airports, federally assisted highways, electric power plants, and bridges; the release of pollutants into navigable waters and their tributaries; and resource development on federal lands, including mining leases, drilling permits, and other uses. During the first 9 years after the act, more than 11,000 environmental impact statements were prepared for major federal projects.

THE WATER QUALITY IMPROVEMENT ACT OF 1970

The Water Quality Improvement Act of 1970 is a comprehensive water pollution control law that essentially gives more power to the federal Water and Pollution Control Act of 1956. The purpose of the latter act was to enhance the quality and value of our water resources and to establish a national policy to prevent, control, and abate pollution of the country's water resources. The 1970 act provides for control of oil pollution by vessels and offshore and onshore oil wells, control of hazardous pollutants other than oil, control of sewage from vessels, research and development methods to control and eliminate pollution of the Great Lakes, research grants to universities and scholarships to students to promote training in water quality control, and projects to demonstrate methods to eliminate and control drainage of acid water from both active and abandoned mines [3].

In addition, the act requires that all federally licensed facilities or activities that involve discharge into navigable water must also obtain a certificate of reasonable assurance that the proposed activity will not violate the state's water quality standards as approved by the Environmental Protection Agency.

THE ENDANGERED SPECIES ACT OF 1973

A major change in the federal government's role in the protection of endangered species, in the legal attitude toward endangered species, and in the legal position of endangered species occurred as the result of the Endangered Species Act of 1973. The act declares that endangered species of wildlife and plants "are of aesthetic, ecological, educational, historical, recreational, and scientific value to the nation and its people" and provides a means whereby the ecosystems upon which endangered species depend may be conserved. Thus the act establishes the need for an ecosystem approach to the management of endangered species. Furthermore, the act states that all federal departments and agencies should seek to conserve endangered species and use their authorities to further the purpose of the act. The act recognizes two groups of protected species: endangered and threatened. *Endangered species* include any in danger of extinction through all or a significant portion of its range. *Threatened species* include any which is likely to become endangered within the foreseeable future through all or a significant portion of its range.

THE RESOURCE CONSERVATION AND RECOVERY ACT OF 1976

The purpose of the Resource Conservation and Recovery Act of 1976 is to control hazardous wastes and protect human health. When the act is fully implemented (in the 1980s), it will provide for "cradle to grave" control of hazardous wastes. At the heart of the act is the goal of identifying hazardous wastes and their life cycles. Once identified, hazardous wastes will be regulated by requiring stringent record keeping and reporting to ensure that wastes do not present a public nuisance or health problem.

THE SURFACE MINING CONTROL AND RECLAMATION ACT OF 1977

The purpose of the Surface Mining Control and Reclamation Act of 1977 is to control the environmental effects of strip mining. The act prohibits mining practices that in the past have led to environmental degradation and requires reclamation of the land after mining. Although the act is greatly improving the way strip mining is con-

ducted, it has been criticized because it does not sufficiently allow for site-specific conditions in the regulations that control mining and reclamation.

THE CLEAN WATER ACT OF 1977

The purpose of the Clean Water Act of 1977 is to clean the nation's water. In particular, the act requires most municipal sewage treatment plants to implement secondary treatment or use the best practicable waste treatment technology. Billions of dollars in federal grants have been awarded to meet these goals. The act clearly encourages the use of innovative and alternative techniques in water treatment and waste disposal, such as land application of sewage sludge, aquifer recharge of treated waste water, and energy recovery. Results of water treatment nationwide have been encouraging. For example, in the 1950s and 1960s the Detroit River was almost devoid of life. As a result of water treatment, people are now catching walleye pike, muskellunge, smallmouth bass, salmon, and trout. Although the Detroit River is still not a really clean river, improvement continues as the discharge of pollutants is reduced.

STATE AND LOCAL ENVIRONMENTAL LEGISLATION

The far-reaching federal legislation has provided an example for state and local governments to follow. In fact, a number of states have enacted legislation analogous to the federal laws. Areas of particular public concern at the state and local level are environmental impact, floodway regulations, sediment control, land use, and water and air quality. The most effective programs and laws, excepting federal activity, will probably be at the local level because acceptance and enforcement, when combined with education and communication, are most easily achieved where environmental degradation is experienced firsthand.

An example of important state and county legislation is the North Carolina Sediment Pollution Control Act of 1973. The act establishes that control of erosion and sedimentation is vital to the public interest and necessary for public health and welfare. The act provides for the creation, administration, and enforcement of a sediment-control program to minimize detrimental effects from sediment pollution in future development.

The Sediment Pollution Control Act of 1973 in North Carolina recognizes the need for local participation, and it encourages local government to draft ordinances consistent with the act. Mecklenburg County, part of the fourth largest urbanizing area in the southeastern United States, recently drafted a sediment-control ordinance that calls for erosion- and sediment-control plans to be submitted

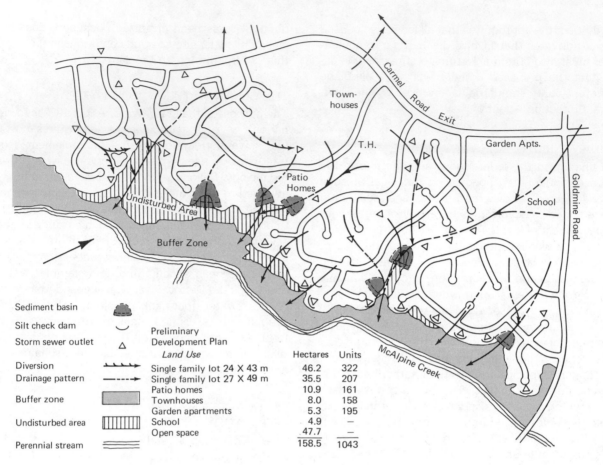

Land Use	Hectares	Units
Single family lot 24 X 43 m	46.2	322
Single family lot 27 X 49 m	35.5	207
Patio homes	10.9	161
Townhouses	8.0	158
Garden apartments	5.3	195
School	4.9	—
Open space	47.7	—
	158.5	1043

FIGURE 18.4

Erosion- and sediment-control plan for a housing development including single-family homes, patio homes, townhouses, garden apartments, a school, and open space. This is a plan responsive to the North Carolina Sediment Pollution Control Act of 1973. (Courtesy of Braxton Williams, Soil Conservation Service.) [10 m = 33 ft; 10 hectares = 25 acres]

before land-disturbing activities are started, if more than one contiguous acre is to be uncovered. These plans are to be approved by the county engineer and reviewed by the Soils and Water Conservation District Office for comments and recommendations. An example of such a plan is shown in Figure 18.4.

WATER LAW

Water is so important that water resources may be the most legislated and discussed commodity in environmental law. Large areas such as New York City and southern California struggle to find sufficient water. In New York, intrastate and legislative contracts were necessary to obtain water for New York City from the Delaware River. California and Arizona have fought for 50 years over Colo-

rado River water, and the issue is only partially resolved [4].

There will always be problems in allocating water resources, but such problems are most severe where water is scarce, such as in the southwestern United States (Figs. 18.5 and 18.6). The situation in California illustrates the types of conflict that arise when a large population with substantial industrial and agricultural activity is concentrated in an area with a natural deficiency of water. Approximately two thirds of the state's water supply comes from the northern third of California, but the greatest need for water is in the southern two thirds of the state, where most of the people live and where most of the industrial and agricultural activity takes place [8].

The deficiency of water in southern California, combined with an almost insatiable demand for water, resulted in the construction of the California aqueduct to move water from the northern and southeastern parts of the state to the Los Angeles area (Fig. 18.6). The legal grounds that allowed California citizens to vote for and pass a state bond of nearly $2 billion for the California aqueduct, which now transports water from the northern part of the state to the southern (Fig. 18.5), was derived from the California Constitution: "The general welfare requires that water resources of the state be put to beneficial use to the fullest extent of which they are capable...that the conservation of such water is to be exercised with a view to the reasonable and beneficial use thereof in the interest of the people and for the public welfare" [4]. From a philosophical viewpoint, a question might be raised, Is the continued development in southern California warranted? Perhaps the people should be located where the water is rather than moving huge quantities of water hundreds of kilometers over rough terrain and active faults at tremendous costs to an area al-

FIGURE 18.5
Map of California showing the major aqueducts supplying water to southern California. (From State of California, Department of Water Resources.)

FIGURE 18.6
California aqueducts in the San Joaquin Valley, California. (Photo courtesy of State of California, Department of Water Resources.)

ready overcrowded and suffering from pollution and other environmental problems. The construction of the aqueduct indicates the price people are willing to pay to support their standard and style of living. Furthermore, it should be pointed out that many of the people in the Feather River and Owens River areas, from where the water transported to Los Angeles is diverted, are not so pleased as those receiving the water.

There is a rather elaborate framework of law surrounding the use of surface water, two major aspects of which are the Riparian Doctrine and the Appropriation Doctrine.

Riparian rights to water are restricted to owners of the land adjoining a stream of standing water. (The word *riparian* comes from the Latin word meaning bank.) Traditional **Riparian Doctrine** is a common law concept which essentially holds that each landowner has the right to make reasonable use of water on his land, provided the water is returned to its natural stream channel before it leaves the property. Furthermore, the property owner has the right to receive the full flow of the stream undiminished in quantity and quality. However, the owner is not entitled to make withdrawals of water that infringe upon the rights of other riparian owners [9].

The Riparian Doctrine was the water law used by most states before 1850, and it is still used in all the states east of the Mississippi and in the first tier of states immediately west of the river (Fig. 18.7). The right to use water is considered as a real property right, but the water itself does not belong to the property owner. Riparian water rights are considered as natural rights and a part of property that enters into the value of land. Water rights may be transferred, sold, or granted to other people [10].

The **Appropriation Doctrine** in water law holds that prior usage is a significant factor. That is, the first to use the water for a beneficial purpose has prior right to it. Furthermore, this right is perfected by use and is lost if beneficial use ceases [10].

Appropriation water law is common in the western part of the United States. Generally, states with the poorest water supply manage their water most closely. Arizona is a good example. With an average precipitation of less than 38 centimeters per year, Arizona must, of necessity, manage its water very closely. The state constitution asserts that riparian water rights are not authorized, and the state's comprehensive water code declares that all water is subject to appropriation. Preferred uses are domestic, municipal, and irrigation. Colorado also has a limited water supply and has declared that all streams are to be considered as public property subject to appropriation [10].

Comparison of the two doctrines suggests that management of water resources is considerably more effective when the Appropriation Doctrine is used, because the Riparian Doctrine requires a judicial decision and therefore is subject to possible variations and interpretations in different courts. As a result, property owners are never sure

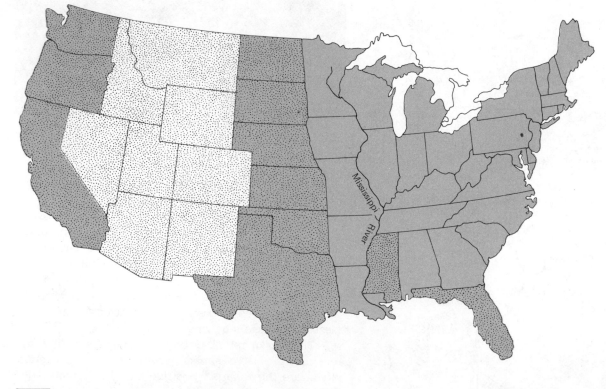

Appropriation

Riparian

Both Appropriation and Riparian

FIGURE 18.7
Surface water laws for the conterminous United States. (Data from Garrity and Nitzschke, 1968.)

of their legal position. Furthermore, the Riparian Doctrine tends to encourage nonuse of water and thus is counterproductive in times of shortage. On the other hand, states with an appropriation system have the power to make and enforce regulations based upon sound hydrologic principles, and this is more likely to lead to effective management of water resources [10].

LAW OF THE SEABED

For more than 20 years, there has been a controversy surrounding mining of the seabed. Industrial nations with international companies interested in mining the seabed are at odds with developing countries who argue that mineral wealth on the sea bottom is a common heritage to all people on Earth, consistent with a U.N. resolution on the issue of seabed resources. Five international consortia of companies have spent hundreds of millions of dollars on research and exploration in preparation for mining nodules containing manganese, cobalt, nickel, and copper, and naturally they want to be sure that their investment is returned. Nevertheless, developing nations, wanting a relatively large share of the mineral wealth, have generally taken a slow-growth-and-development position at each Law of the Sea Conference, which are sponsored by the United Nations.

Because there has been little action toward ratifying a treaty concerning mining of the seabed, industrial nations have threatened to begin mining without an agreement. Some developing nations have responded with threats to cut off exports of oil, metals, and other resources to industrial nations mining the seabed without approval of the nations attending the Law of the Sea Conference. In spite of this, the U.S. Congress has passed legislation authorizing U.S. companies to mine the seabeds. Although the controversy still exists, it is expected that mining plans will continue, and that developing countries will eventually modify their demands in light of the tremendous amount of money being spent by industrial nations on mining technology and exploration.

LAND-USE PLANNING AND LAW

Few environmental topics are as controversial as land-use planning legislation. The controversy is a result of several factors. First, whereas air and water pollution can be measured, evaluated, and possibly corrected, it is more difficult to determine the "highest and best use" of the natural environment rather than its "most profitable use." Second, landowners are fearful that land-use planning

would take away their right to decide what to do with their property. They fear that individual ownership of property would be transferred into a social ownership in which the individual has only a caretaker role. Third, there is considerable concern over federal or state government initiation of or control over local land-use planning.

People in the United States greatly value private ownership of land, and therefore, a law that requires rural landowners to use their lands in a very restricted way is not likely to be accepted. On the other hand, it is increasingly apparent that private ownership of land does not mean that the owner has the right to deliberately or inadvertently degrade the environment. To be effective, land-use planning must give the property owner certain alternative means of land management from which to choose freely.

Should the federal government control land-use planning at all levels? The argument in favor of a federal land-use planning act stresses that such planning is urgently needed to minimize degradation of land resources of statewide or national importance. Those who favor such an act point out that land use is the most important aspect of environmental quality control that remains to be stated as national policy [11].

The federal government's role in land-use planning could take other forms: increased federal money to sponsor continued research and dissemination of information about land-use planning; federal funds to underwrite selected experiments in guiding growth and land use; and finally, if Congress should determine that it is federal responsibility to recommend land-use planning at the state and local levels, then perhaps revenue-sharing measures might be appropriate [12].

A number of states are now considering, or have passed, land-use planning legislation. Table 18.1 summarizes land-management programs in the United States. For example, in 1970 the Maine legislature enacted a site law to maintain the highest and best use of the natural environment. The law was drafted to regulate large developments in such a way that the burden of proof was shifted to the developer; that is, the developer must show that the project will not degrade the land, water, or air. A recent test of the law in court upheld the legislature's policy to regulate land use. This decision was significant because it strongly reaffirmed the right of states to limit property use in such a way as to preserve the environment [13].

In 1970, Vermont also passed a state land-use act. Under this act, to obtain a permit to develop land, a developer must prove to a district environment commission that their project: (1) would not cause undue air or water pollution; (2) could obtain sufficient water; (3) would not

TABLE 18.1
State land-use programs as of 1976.

State	Comprehensive Permit System[a]	Coordinated Incremental[b]	Mandatory Local Planning[c]	Coastal Zone Management[d]	Wetlands Management[e]	Power Plant Siting[f]	Surface Mining[g]	Designation of Critical Areas[h]	Differential Assessment Laws[i]	Floodplain Management[j]	Statewide Shorelands Act[k]
	Type of State Program										
Alabama				X		X	A			X	
Alaska		X		X		X			B		
Arizona		X				X			A	X	
Arkansas						X	A, B		A	X	
California		X		X		X	X		C	X	
Colorado						X	X	X	A	X	
Connecticut		X		X	X	X			B	X	
Delaware		X		X	X				A		X
Florida	X	X	X	X	X	X	A	X	A, C		
Georgia		X		X	X		A, B				
Hawaii	X	X		X		X	X	X	B	X	
Idaho		X					X		A		
Illinois				X		X	A, B		B	X	
Indiana		X		X			A, B		A	X	
Iowa							A, B		A	X	
Kansas							A, B				
Kentucky						X	A, B		B		
Louisiana				X	X						
Maine	X	X	X (LTD)	X	X	X	A	X	B	X	
Maryland		X		X	X	X	A, B	X	B	X	
Massachusetts				X	X	X			B		
Michigan				X			X		C	X	X
Minnesota		X		X	X	X	X	X	B	X	X
Mississippi				X	X					X	
Missouri						X	X		A	X	

SOURCE: Council of State Governments, 1976.
[a]State has authority to require permits for certain types of development.
[b]State-established mechanism to coordinate state land-use-related problems.
[c]State requires local governments to establish a mechanism for land-use planning (e.g., zoning, comprehensive plan, planning commission).
[d]State is participating in the federally funded coastal zone management program authorized by the Coastal Zone Management Act of 1972.
[e]State has authority to plan or review local plans or the ability to control land use in the wetlands.
[f]State has authority to determine the siting of power plants and related facilities.
[g]State has statutory authority to regulate surface mines. (A) State has adopted rules and regulations. (B) State has issued technical guidelines.
[h]State has established rules, or is in the process of establishing rules, regulations, and guidelines for the identification and designation of areas of critical state concern (e.g., environmentally fragile areas, areas of historical significance).
[i]State has adopted tax measure which is designed to give property tax relief to owners of agricultural or open space lands. (A) Preferential Assessment Program—Asssessment of eligible land is based upon a selected formula, which is usually use-value. (B) Deferred Taxation—Assessments of eligible land are based upon a selected formula, which is usually use-value and provides for a sanction, usually the payment of back taxes, if the land is converted to a noneligible use. (C) Restrictive Agreements—Eligible land is assessed at its use-value, a requirement that the owner sign a contract, and a sanction, usually the payment of back taxes if the owner violates the terms of the agreement.
[j]State has legislation authorizing the regulation of floodplains.
[k]State has legislation authorizing the regulation of shorelands of significant bodies of water.

result in excessive soil erosion, highway congestion, or municipal service and educational burden; and (4) would not have undue detrimental effects on the natural scenery, historical sites, or rare and irreplaceable natural areas. In addition, the development would have to conform with existing local or regional land-use plans as well as any subsequently adopted plan [12]. The law also required establishment of an environmental board to hear appeals in the permit procedure and to prepare a state land-use plan consisting of a map showing broad categories of the proper use of all the land in the state. In 1974,

the Vermont legislature stripped the environmental board of responsibility to develop a state land-use plan and created a land-use study commission to pursue the matter.

Several states have passed laws concerning fragile coastal areas. California citizens concerned about coastal areas (Fig. 18.8) voted in 1972 to require a permit for any development taking place in a zone extending from approximately 900 meters inland from the sea's high tide to about 5 kilometers seaward. The law also calls for a master plan for coastal development to be prepared and adopted as state policy.

TABLE 18.1 (continued)

| State | Type of State Program | | | Coastal Zone Management[d] | Wetlands Management[e] | Power Plant Siting[f] | Surface Mining[g] | Designation of Critical Areas[h] | Differential Assessment Laws[i] | Floodplain Management[j] | Statewide Shorelands Act[k] |
	Comprehensive Permit System[a]	Coordinated Incremental[b]	Mandatory Local Planning[c]								
Montana		X	X			X	A, B	X	B	X	X
Nebraska			X			X			B	X	
Nevada		X	X			X		X	B		
New Hampshire				X	X	X			B, C		
New Jersey				X	X	X			B	X	
New Mexico		X				X	A		A		
New York	X	X		X	X	X	X	X	B	X	
North Carolina		X		X	X	X			B	X	
North Dakota						X	A		A		
Ohio				X		X	A		B		
Oklahoma							X		A	X	
Oregon		X	X	X		X	A	X	B		
Pennsylvania				X	X	X	A	X	B		
Rhode Island		X		X	X	X			B		
South Carolina				X		X	A		B		
South Dakota							A	X	A		
Tennessee						X	A, B				
Texas				X	X	X			B		
Utah		X					A		B		
Vermont	X	X				X	X		C	X	
Virginia			X	X	X		A, B		B		
Washington		X		X	X	X	A		B	X	X
West Virginia							A, B			X	
Wisconsin		X		X	X	X	X	X		X	
Wyoming		X	X			X	A		A		

In 1974 North Carolina passed the Coastal Area Management Act, which recognizes the need to preserve for the people an opportunity to enjoy the aesthetic, cultural, and recreational quality of natural coastlines. The act provides a management system to preserve and protect the natural productivity and biological, economic, and aesthetic value of the estuaries, barrier islands, sand dunes, and beaches. The act also requires that development in the coastal area does not exceed the capabilities of the land and water resources.

FIGURE 18.8
California citizens concerned about coastal areas prompted legislation to control development in an area from 900 meters inland to 5 kilometers seaward. This is a photograph of the California coast near San Simeon. (Photo by D. B. Botkin.)

The North Carolina act establishes a cooperative program of coastal management between local and state governments, with local government having the initiative in the planning. The role of state government is to set standards and to review and support local government in its planning program.

WILDLIFE LAW

Ever since the Roman Empire, wildlife has occupied a unique status from a legal point of view. Like the air and the oceans, wildlife belonged to everyone; unlike the air and oceans, wildlife could become the property of anyone who could catch them (Fig. 18.9). In Rome, this legal right to take wildlife was restricted only by private land ownership. A landowner had an exclusive right to the wildlife on his land. However, this right seems to be related more to the Roman concept of land ownership than to their concept of wildlife.

In England, after the Norman conquest in 1066, the king claimed exclusive authority over hunting and fishing. This power was gradually taken by the British Parliament.

In the United States, the first case concerning wildlife to be considered by the Supreme Court was *Martin* v. *Waddell* in 1842, which we discussed in the introduction to this chapter. This case laid the groundwork for public and state ownership of wildlife. This case and Roman and British antecedents reveal the major issues that surround the legal aspects of wildlife: (1) who owns wildlife; (2) who controls the activities that affect wildlife; and (3) who protects wildlife.

In the United States, these issues are addressed in terms of *states' rights versus individual rights* and *states' rights versus federal rights*. In general, during the last century states' rights to control and protect wildlife and its use have prevailed over the rights of individuals, and the right of the federal government has prevailed over the states' rights. International treaties have legal precedence over state and private rights. However, the issues of private, state, and federal rights are by no means resolved.

Other major issues in the United States include a concern with the legal tools that will be used to manage wildlife. The legal tools available include (1) the regulation of the taking of wildlife; (2) the regulation of commerce in wildlife and wildlife products (skins, whale oil, ivory, meat, etc.); (3) the acquisition and management of wildlife habitat; (4) the regulation of land development where it could interfere with the conservation of wildlife; and (5) the appropriation of monies for activities that involve regulation, land acquisition, commerce, and environmental impact. In addition, wildlife law includes laws controlling predators and pesticide usage and international agreements.

The development of federal control of wildlife and the development of laws to protect endangered wildlife species can be traced back to the turn of the century in the United States. As we discussed earlier in this chapter, in 1900 the Lacey Act prohibited the interstate transportation of wild animals and birds killed in violation of state law. The Lacey Act increased the power of the states and gave the federal government power to help the states enforce their laws about wildlife.

FIGURE 18.9
Wildlife has always occupied a unique legal status: like air and oceans, it belongs to everyone; unlike air and oceans, it becomes the property of whoever captures it. Here we see a mountain lion in Zion National Park, protected by the park boundaries. (Photo by Dale Smith, Zion National Park, Springdale, Utah.)

The Migratory Bird Treaty Act passed in 1918 is one of the most important wildlife laws. It prohibited for the first time in the United States the taking of any bird protected by a migratory bird convention signed in 1916 with Great Britain. The most far-reaching effect of this law was that it prevented landowners from killing and capturing wildlife on their own lands.

The protection of wildlife was increased by several laws concerning endangered species, including the Endangered Species Preservation Act of 1966, the Endangered Species Conservation Act of 1969, the Marine Mammal Protection Act of 1972, and the Endangered Species Act of 1973. The 1966 act directed the Secretary of the Interior to "carry out a program in the United States of conserving, protecting, restoring and propagating selected species of native fish and wildlife" and authorized acquisition of land for these purposes. The 1969 act extended the concern to international conservation of wildlife.

The Marine Mammal Protection Act of 1972 added a unique perspective to the protection of wildlife. The primary goal of this law is to maintain the health and stability of marine ecosystems. This is the first act to name the maintenance of ecosystems as its major goal. It is unique because it requires that the U.S. government also promote the health and stability of ecosystems not directly owned by this country. A secondary goal is to promote the optimal sustainable population of marine mammals. The concept of an optimal sustainable population has caused considerable controversy because no one is quite certain what such a population level for a marine mammal (or for any organism, for that matter) might be. The act also recognizes certain population levels, including depleted, threatened, and endangered. If a species is depleted or threatened, it requires conservation even though it is not yet endangered. Thus the act fosters a preventative approach to wildlife conservation.

The Endangered Species Act of 1973 continued this approach and the emphasis on conserving the ecosystems upon which endangered species depend. Under this act, any species listed as endangered is protected from being taken anywhere in the United States, its territorial sea, or on the high seas. *Take* includes wounding and harassing as well as hunting, trapping, or attempting any of these actions. The act also introduced the concept of *critical habitats* (habitats essential to the survival of a species), requiring the government to insure that federal activities do not destroy or modify critical habitats on federal lands. The critical habitat concept has not only been applied to endangered species, but has become important in environmental impact assessment and in land use.

THE USE OF FEDERAL LANDS

THE NATIONAL FORESTS AND NATIONAL PARKS

The lands owned and controlled by the federal government which are important in environmental law are the national forests, Bureau of Land Management lands, and the national parks. The national forests were established by the Forest Reserve Act of 1891. The standards to be used to manage the national forests were set down first in the Forest Reserve Act of 1897, known as the "organic act." The goals stated were the protection of the forest, securing favorable waterflows, and furnishing a supply of timber—in other words, the purpose of national forest management was *multiple use*. Since then, the national forests have been managed by the U.S. Forest Service from a perspective of multiple use.

Wildlife was not specifically mentioned in the Forest Reserve Act as part of the multiple use concept, but in 1928 the case of *Hunt* v. *United States* upheld the power of the Secretary of Agriculture to remove deer that threatened to harm the vegetation in the Kaibab National Forest in Arizona. Eventually, authority was given to the federal government over the states to govern the wildlife in national forests.

The multiple use concept was made explicit in 1969 by the passage of the Multiple Use Sustained Yield Act, which stated that the national forests are established and shall be administered for outdoor recreation, range, timber, watershed, and wildlife, including fish. This act directs the Secretary of Agriculture to manage the national forests for multiple use and sustained yield.

The national parks were first established as unique areas preserving what were seen as freaks of nature or magnificent monuments to nature's grandeur [14]. The more general purposes of the parks were established in 1916 by the National Park Service Act, which lists among its purposes the conservation of scenery, natural and historic objects, and wildlife. The act states that these are to be enjoyed in such a manner as "will leave them unimpaired for the enjoyment of future generations." Recent emphasis in national park management has shifted from the preservation of "monuments" of nature to conservation of natural populations and ecosystems.

The Wilderness Act of 1964 directed that areas administered by federal law "where the Earth and its community of life are untrammeled by man, where man himself is a visitor who does not remain" be preserved for their wilderness characteristics. This act led to the development of a wilderness land system, which we discussed in Chapter 16.

PESTICIDES AND WILDLIFE

The concern with the effects of pesticides on wildlife has had an important effect on environmental law. In particular, as a result of the litigation over DDT and its effects on wildlife, the federal government was granted authority to prohibit people from using a pesticide "in a manner inconsistent with its labeling." An example of this litigation is the case of the *Environmental Defense Fund* (*EDF*) v. *Hardin* in 1969. The 1964 amendments to the Insecticide, Fungicide, and Rodenticide Act of 1947 gave the Secretary of Agriculture the power to prevent the marketing of pesticides that posed a threat to nontarget wildlife. In 1969 the EDF petitioned the Secretary of Agriculture to cancel all pesticides containing DDT, but the Secretary responded by cancelling only certain limited uses of DDT. As a result, the EDF challenged the Secretary in court. This case set certain important precedents, not the least of which was to uphold the right of persons claiming biological harm to human beings and other living things to legally challenge the Secretary.

The case concluded with the court directing that DDT be cancelled. Subsequent proceedings established that DDT would be cancelled except for public health and agricultural pest quarantine use. As a result of this case, the burden of proof of safety was shifted to the manufacturer.

The litigation over DDT also helped to establish the legal process as an important means to influence society's use of the environment. For the first time, many conservationists recognized that the courts could provide a way toward the wise use of our environment [1].

DO TREES HAVE STANDING?

One of the most famous legal environmental controversies in recent years is over who speaks for nature in the courtroom. This has become known as the question, "Do trees have standing?", which was raised in the case of *Sierra Club* v. *Morton*. This case concerned the use of the Mineral King Valley in the Sierras of California (Fig. 18.10). Under the jurisdiction of the U.S. Forest Service, this area had been used for recreation. In 1965 the Forest Service invited bids from private developers for the construction of a ski resort, and Walt Disney Enterprises, Inc., was chosen as the successful bidder. That company planned a $35 million ski development to include ski lifts, restaurants, motels, swimming pools, and parking lots for 14,000 visitors a day. The main development was proposed for 32 hectares (80 acres) of the valley floor.

The Sierra Club brought suit in 1969, claiming that the proposed development violated federal laws governing the preservation of national parks, forests, and game refuges. The case was thrown out because the Supreme Court concluded that the Sierra Club lacked the legal standing to bring the case to court. In a famous dissenting opinion, Justice William O. Douglas argued that trees and other natural objects should have a legal standing. According to the legal doctrine of **standing,** cases must be limited to parties whose allegations of injury are more than theoretical and are not minuscule in light of everyone's common experience. For example, a resident of Chicago cannot bring a legal suit against the city of Los Angeles because the smog affects his relatives who live in that city [15].

The court ruled that the Sierra Club did not have standing; that is, it was not directly aggrieved by the proposal to the point that it had a right to bring the case to court. In his dissenting opinion, Douglas pointed out that inanimate objects sometimes are parties in litigation. For example, a ship has a "legal personability." A corporation is also legally a person, whether "it represents a proprietary, spiritual, aesthetic, or charitable cause." Douglas said, "So it should be as respects valleys, alpine, meadows, rivers, lakes, estuaries, beaches, ridges, groves of trees, swampland, or even air that feels the destructive pressure of modern technology and modern life." He said that "the river as plaintiff speaks for the ecological unit of life that is part of it." In his discussion, Douglas referred to *Should Trees Have Standing? Towards Legal Rights for Natural Objects,* a book by C. D. Stone, a law professor. Stone had argued that just as minors and incompetent persons are represented in court by legal guardians, so should natural objects.

After the Supreme Court's decision to reject the Sierra Club's standing, the club filed another complaint, claiming direct injury to its members. However, that case never came to trial. The development project was abandoned, and therefore the issue of the standing of trees and other natural objects remains unresolved.

Although the case was not brought to trial, the legal procedures were influential in stopping the proposed development. What is perhaps most important is that the case raised the question of legal standing as concerns our environment.

SUMMARY AND CONCLUSIONS

A land ethic is emerging in our national environmental policy. Productive and enjoyable harmony with our physical environment is now encouraged, and legal theories based upon the Trust Doctrine or the Ninth Amendment are being used to argue in court for the right of this generation and future generations to an unspoiled land.

The term *environmental law* has only recently been commonly used, but the subject is becoming increasingly popular among lawyers and promises to be more signifi-

Mineral King ●

(a)

(b)

FIGURE 18.10
A plan to develop the Mineral King Valley in the California Sierras (a) raised the question of the legal standing of trees. In (b) we see a stand of sequoias in the Sierras. (Photo by D. B. Botkin.)

cant in the future. Topics of particular concern to individuals and society are degradation of air, water, scenery, and other natural resources.

Environmental legislation is having a great impact on our society. In addition to regulations controlling emissions of possible pollutants, perhaps the most significant and far-reaching piece of environmental legislation is the National Environmental Policy Act of 1969. The act requires that an environmental impact statement must be completed before the federal government begins any activity that affects the environment. In this chapter we have reviewed other important federal legislation, each of which has added important precedents or concepts to the legal aspects of environmental studies.

Many states are following the federal government's leadership and are developing environmental and land management legislation. Areas of particular concern at the state level are environmental impact, floodway regulation, sediment control, land use, and air and water quality.

Environmental law has influenced our use of wildlife and management of the land and its natural resources. During the history of the United States, the concept of private ownership of wildlife (once it was caught) has changed to the idea that wildlife belongs to everyone and that the state and federal governments have the authority to control our use of living wild resources.

Recent environmental legislation has changed its focus from individual species of wildlife to the conservation of ecosystems, protection of critical habitats, and the establishment of wilderness areas in federal land holdings. Environmental litigation has raised the question of the legal standing of natural objects—a question that is still unresolved.

Water law is of particular importance in regions with water deficiencies. Eastern states generally have what is known as riparian rights, whereby owners of land that adjoins water have the right to reasonable use. In the western states, water law is generally governed by the Appropriation Doctrine, which holds that prior beneficial usage is the key to water rights. In states where water is especially valuable, such as Arizona and Colorado, all water is appropriated on the basis of prior and preferred uses. In comparing the two systems, one can conclude that appropriation of water leads to better management of water resources.

Perhaps the most controversial and potentially significant of all environmental legislation is land-use planning. It is often difficult to determine the highest and best use of land. Also, landowners are fearful that land-use planning will cause them to lose the right to control their property. There is also considerable concern over the role of federal and state governments in initiating local planning.

The argument in favor of land-use planning legislation stresses that planning is urgently needed if degradation of the land is to be controlled, and that land use is the only remaining environmental control that is not stated as national policy.

The significant effect of land-management legislation is that the developer must prove that an activity will not degrade the land. Furthermore, we will have to live with the concept that the most profitable use of land is not always necessarily the best use. This concept comes directly from the Trust Doctrine and an emerging land ethic.

REFERENCES

1 COUNCIL ON ENVIRONMENTAL QUALITY, ENVIRONMENTAL LAW INSTITUTE. 1977. *The evolution of national wildlife law.*

2 LANDAU, N. J., and RHEINGOLD, P. D. 1971. *The environmental law handbook.* New York: Ballantine Books.

3 YANNACONE, V. J., Jr.; COHEN, B. S.; and DAVIDSON, S. G. 1972. *Environmental rights and remedies 1.* San Francisco: Bancroft-Whitney.

4 COATES, D. R. 1971. Legal and environmental case studies in applied geomorphology. In *Environmental geomorphology,* ed. D. R. Coates, pp. 223–42. Binghampton: State University of New York.

5 JUERGENSMEYER, J. C. 1970. Control of air pollution through the assertion of private rights.In *Environmental law,* ed. Research and Documentation Corporation, pp. 17–46. Greenvale, N. Y.: Research and Documentation Corporation.

6 MURPHY, E. F. 1971. *Man and his environment: Law.* New York: Harper & Row.

7 CARTER, L. J. 1974. Con Edison: Endless Storm King dispute adds to its troubles. *Science* 184: 1353–58.

8 CARGO, D. N., and MALLORY, B. F. 1974. *Man and his geologic environment.* Menlo Park, Calif.: Addison-Wesley.

9 RESEARCH AND DOCUMENTATION CORPORATION. 1970. Private remedies for water pollution. In *Environmental law,* ed. Research and Documentation Corporation, pp. 47–69. Greenvale, N. Y.: Research and Documentation Corporation.

10 PETTYJOHN, W. A. 1972. Legal approach to water rights. In *Water quality in a stressed environment,* ed. W. A. Pettyjohn, pp. 255–76. Minneapolis: Burgess.

11 HEALY, M. R. 1974. National land use proposal: Land use legislation of landmark environmental significance. *Environmental Affairs* 3: 355–95.

12 MCCLAUGHRY, J. 1974. The land use planning act—An idea we can live without. *Environmental Affairs* 3: 595–626.

13 WAINWRIGHT, J. K., Jr. 1974. Spring Valley: Public purpose and land use regulation in a "taking" context. *Environmental Affairs* 3: 327–54.

14 RUNTE, A. 1979. *National parks: The American experience.* Lincoln: University of Nebraska Press.

15 MCGINNES, M. 1980. *Principles of environmental law.* Santa Barbara, Calif.: Rainbow Bridge Press.

FURTHER
READING

1 BALDWIN, M. F., and PAGE, J. K., Jr. 1970. *Law and the environment*. New York: Walker.

2 COUNCIL ON ENVIRONMENTAL QUALITY, ENVIRONMENTAL LAW INSTITUTE. 1977. *The evolution of natural wildlife law*.

3 LANDAU, N. J., and RHEINGOLD, P. D. 1971. *The environmental law handbook*. New York: Ballantine Books.

4 MCGINNES, J. M. 1980. *Principles of environmental law*. Santa Barbara, Calif.: Rainbow Bridge Press.

5 MURPHY, E. F. 1971. *Man and his environment: Law*. New York: Harper & Row.

6 RODGERS, W. H., Jr. 1979. *Cases and materials on energy and natural resources law*. St. Paul, Minn.: West.

7 SAX, J, L. 1971. *Defending the environment: A strategy for citizen action*. New York: Knopf.

8 STONE, C. D. 1974. *Should trees have standing? Towards legal rights for natural objects*. Los Altos, Calif.: William Kaufmann.

9 YANNACONE, V. J., Jr.; COHEN, B. S.; and DAVIDSON, S. G. 1972. *Environmental rights and remedies 1*. San Francisco: Bancroft-Whitney.

STUDY
QUESTIONS

1 Do rocks have rights?

2 You are asked to negotiate between tuna fishing fleets and a conservation group trying to save porpoises. Porpoises are caught in the tuna nets and die accidentally. What information do you need? What would you suggest?

3 Introduced goats which are *feral* (meaning they have gone wild) on an island off the coast of California are destroying rare native plants and reducing them to less than one tenth of their original abundances. These plants have no economic value. It costs $2000 to transport goats off the island. As the manager of the island with a budget of $15,000 per year, what goal are you required to seek legally? How would you try to accomplish this?

4 Should our laws lead us to a risk-free society?

5 In general, why are water pollution laws easier to enforce than air pollution laws?

6 Air pollution laws have led industries to build taller stacks and to reduce the emission of dark smoke. What are the benefits of these changes? What are the drawbacks of these changes?

7 Wild horses roam parts of the American West. What protection are these animals afforded by the U.S. Endangered Species Act?

8 A fisheries manager claims that certain fish in Canadian lakes are endangered by acid rain created by industry in the United States. She wants to bring a legal suit to stop the acid rain. Are there any legal precedents for this suit? Are courts the most effective means to solve this problem?

9 What is the Trust Doctrine, and why is it important in environmental law?

10 Why has the Storm King Mountain case been so controversial? What does it have in common with the development on the fossil beds in Colorado?

11 Discuss the relative merits of the Riparian Doctrine and Appropriation Doctrine in the use and conservation of surface water.

12 Why is a law of the seabed taking so long to develop? What do you think could be done to speed up the process?

13 Are we more likely to achieve cleaner air by legally requiring pollution abatement equipment, or by charging a fee according to the amount of pollution emitted?

APPENDICES

APPENDIX A:
THE KINDS OF LIVING THINGS

The millions of species on the Earth are classified into five major groups called **kingdoms**. These kingdoms form a family tree of species with similar characteristics. The five kingdoms are (1) **prokaryotes**, which include bacteria and blue-green algae; (2) the **protists**, which are primarily single-celled organisms; (3) **fungi**; (4) **plants**; and (5) **animals**. Table A.1 lists the major characteristics of each kingdom. The prokaryotes are generally believed to be the most ancient because their cells lack a **nucleus** and other features of the other kingdoms. The appearance of prokaryoticlike forms in the oldest fossils also suggests that the prokaryotes are more primitive than the other kingdoms. Prokaryotes, however, carry out certain chemical transformations, such as the fixation of nitrogen, which are not carried out by organisms in the other kingdoms. The protists include some single-celled organisms which are autotrophs (make their own food) and others which are heterotrophs, such as amoebae.

Biologists believe that the species within a kingdom are more closely related to each other than to species in other kingdoms. According to the theory of evolution, species have evolved from others through the process of **natural selection**. Individuals in a population differ from one another in inherited, or **genetic**, traits. If individuals with one kind of trait have more offspring than others, then natural selection is said to occur with regard to that trait. As a consequence, this trait becomes more and more common in the genetic material, or the **gene pool**. Changes in the environment can change which genetic traits are favored and thus lead to natural selection. Genetic changes can also occur by **mutation**, a change that occurs in the chemical characteristics of the genetic material. Mutations may occur at random or may be caused by the effects of ionizing radiation. Certain chemicals seem to increase the rate of mutation. When a chemical substance or form of radiation causes mutations, it is said to be **mutagenic** or is referred to as a **mutagen**. If two populations become isolated from breeding with one another for a long enough time, they can undergo genetic changes which make interbreeding no longer possible. The two populations therefore can no longer exchange genetic material and are by definition said to be different species. By this process, it is believed that the five kingdoms evolved over the Earth's history. The family tree of connections is shown in Figure A.1.

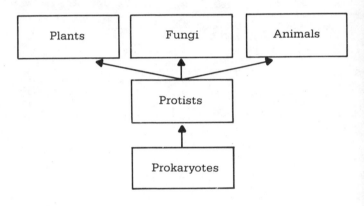

FIGURE A.1
The family tree of the five kingdoms.

TABLE A.1
The five kingdoms.

Kingdom	Cell Nucleus	Chlorophyll	Cell Specialization	Nervous System	Nutrition	Oxygen Requirement	Examples
Prokaryotes	None	Some kinds, but not in chloroplasts	Rare	None	Autotrophs and heterotrophs	Some kinds require oxygen	Bacteria; blue-green algae
Protists	All	Some kinds, in chloroplasts	In a few kinds	Primitive	Autotrophs and heterotrophs	All	Amoebae; dinoflagellates; euglena
Fungi	All	None	In most, but limited	None	Heterotrophs	Most	Mushrooms; molds; yeasts
Plants	All	Almost all; in chloroplasts	All	None	Autotrophs	All	Water lilies; redwoods; moss; wheat
Animals	All	None	All	All	Heterotrophs	All	Crabs; termites; goldfish; elephants; human beings

SOURCE: Modified from Margulis and Schwartz, 1982.

APPENDIX B:
COMMON CONVERSION FACTORS

AREA
1 square mi = 640 acres, 1 acre = 43,560 ft^2 = 4046.86 m^2 = 0.4047 ha
1 ha = 10,000 m^2 = 2.471 acres

	in.^{2}	ft^2	mi^2	cm^2	m^2	km^2
1 in.^{2} =	1	—	—	6.4516	—	—
1 ft^2 =	144	1	—	929	0.0929	—
1 mi^2 =	—	27,878,400	1	—	—	2.590
1 cm^2 =	0.155	—	—	1	—	—
1 m^2 =	1,550	10.764	—	10,000	1	—
1 km^2 =	—	—	0.3861	—	1,000,000	1

LENGTH
1 yard = 3 ft, 1 fathom = 6 ft

	in.	ft	mi	cm	m	km
1 inch (in.) =	1	0.083	1.58×10^{-5}	2.54	0.0254	2.54×10^{-5}
1 foot (ft) =	12	1	1.89×10^{-4}	30.48	0.3048	
1 mile (mi) =	63,360	5,280	1	160,934	1,609	1.609
1 centimeter (cm) =	0.394	0.0328	6.2×10^{-6}	1	0.01	1.0×10^{-5}
1 meter (m) =	39.37	3.281	6.2×10^{-4}	100	1	0.001
1 kilometer (km) =	39,370	3,281	0.6214	100,000	1,000	1

VOLUME

	in.^{3}	ft^3	yd^3	m^3	qt	liter	barrel	gal (U.S.)
1 in.^{3} =	1	—	—	—	—	0.02	—	—
1 ft^3 =	1,728	1	—	0.0283	—	28.3	—	7.480
1 yd^3 =	—	27	1	0.76	—	—	—	—
1 m^3 =	61,020	35.315	1.307	1	—	1,000	—	—
1 quart (qt) =	—	—	—	—	1	0.95	—	0.25
1 liter (l) =	61.02	—	—	—	1.06	1	—	0.2642
1 barrel (oil) =	—	—	—	—	168	159.6	1	42
1 gallon (U.S.) =	231	0.13	—	—	4	3.785	0.02	1

MASS AND WEIGHT

1 pound = 453.6 grams = 0.4536 kilogram = 16 ounces
1 gram = 0.0353 ounce = 0.0022 pound
1 short ton = 2000 pounds = 907.2 kilograms
1 long ton = 2240 pounds = 1008 kilograms
1 metric ton = 2205 pounds = 1000 kilograms
1 kilogram = 2.205 pounds

ENERGY AND POWER

1 kilowatt-hour = 3413 Btus = 860,421 calories

1 Btu = 0.000293 kilowatt-hour = 252 calories = 1055 joules

1 watt = 3.413 Btu/hr = 14.34 calorie/min

1 calorie = the amount of heat necessary to raise the temperature of 1 gram (1 cm^3) of water 1 degree Celsius

1 quadrillion Btu = (approximately) 1 exajoule

1 horsepower = 7.457×10^2 watts

1 joule = 9.481×10^{-4} Btu = 0.2389 cal = 2.778×10^{-7} kilowatt-hour

TEMPERATURE

$$F = \frac{9}{5}C + 32$$

F is degrees Farenheit.
C is degrees Celsius (centigrade).

Farenheit		Celsius
	Freezing of H$_2$0	
32	(Atmospheric Pressure)	0
50	——————	10
68	——————	20
86	——————	30
104	——————	40
122	——————	50
140	——————	60
158	——————	70
176	——————	80
194	——————	90
212	Boiling of H$_2$O (Atmospheric Pressure)	100

OTHER CONVERSION FACTORS

1 ft^3/sec = 0.0283 m^3/sec = 7.48 gal/sec = 28.32 liter/sec

1 acre-foot = 43,560 ft^3 = 1233 m^3 = 325,829 gal

1 m^3/sec = 35.32 ft^3/sec

1 ft^3/sec for one day = 1.98 acre-feet

1 m/sec = 3.6 km/hr = 2.24 mi/hr

1 ft/sec = 0.682 mi/hr = 1.097 km/hr

1 atmosphere = 14.7 lb(in.$^{-2}$) = 2116 lb(ft^{-2}) = 1.013×10^5 N(m^{-2})

APPENDIX C: GEOLOGIC TIME SCALE

Era	Approx. Age in Millions of Years (Radioactivity)	Period (System)[a]			
		Recent (Holocene)		*Man?*	
	2	Pleistocene		Neogene	
	6		Pliocene		
	25		Miocene		
Cenozoic	36	Tertiary	Oligocene	Paleogene	
			Eocene		*Mammals*
	58		Paleocene		
	65				
	135	Cretaceous			
Mesozoic	181	Jurassic		*Flying reptiles*	*First bird*
		Triassic	*Dinosaurs*		
	220–230				
		Permian		*First reptiles*	
	280				
		Pennsylvanian	Carboniferous		
		Mississippian			*First insect fossils*
	345				
		Devonian	*First amphibian*		
	405				
		Silurian		*First land plant fossils*	
	425				
		Ordovician	*First vertebrate fossil fish*		
	500				
		Cambrian	*Oldest tribolites*		
	570–600				
Proterozoic[b]		Vendian	*Oldest skeletonized invertebrates*		
	680				
		Riphean			
			Upper Riphean	*Algal megafossils; increasing abundance of algae*	
	900				
			Middle Riphean	*Eukaryotes; many cyanobacteria*	
	1300				
			Lower Riphean		
Early Proterozoic	2000	*OXYGENIC CRISES*			
		Gunflint iron formation (Canada); bacteria; increasing abundance of stromatolites			
	2800	*Stromatolites*			
Archean	3500	*Earliest autotrophs; oldest stromatolites and microfossils*			
	3800	*Oldest terrestrial rocks; no fossil record*			
Hadean	4000				

[a]*Period* refers to a time measure; *system* refers to the rocks deposited during a period.
[b]Proterozoic to Hadean modified from Awramik, 1981.

GLOSSARY

A soil horizon Uppermost soil horizon sometimes referred to as the *zone of leaching*.

Absorption Taking up, incorporating, or assimilating of a material.

Acid rain Rain made artificially acid by pollutants, particularly oxides of sulphur and nitrogen. (Natural rainwater is slightly acid due to the effect of carbon dioxide dissolved in the water.)

Active solar energy systems Direct use of solar energy which requires mechanical power; usually consists of pumps and other machinery to circulate air, water, or other fluids from solar collectors to a heat sink where the heat may be stored.

Adaption The evolutionary process by which populations become better able to exist under prevailing environmental conditions. Also used to mean specific genetic traits that increase the likely persistence of a population or individual.

Adsorption Attachment of gas molecules or molecules in solution to the surface of solid materials with which they come in contact.

Aerobic Characterized by the presence of free oxygen.

Age structure (of a population) A population divided into groups by age. Sometimes the groups represent the actual number of each age in the population; sometimes the group represents the percentage or proportion of the population of each age.

Aggregate Any hard material (crushed rock, sand, gravel, or other material) that is added to cement to make concrete.

Alkaline soils Soils in arid regions that contain a large amount of soluble mineral salts (primarily of sodium) which in the dry season may appear on the surface as a crust or powder.

Alluvium Unconsolidated sediments, including sand, gravel, and silt, deposited by streams.

Anaerobic Characterized by the absence of free oxygen.

Anthracite A type of coal characterized by a high percentage of carbon and low percentages of volatiles which provides a high heat value. Anthracite often forms as a result of metamorphism of bituminous coal.

Appropriation Doctrine, water law Holds that prior usage of water is a significant factor. The first to use the water for beneficial purposes is prior in right.

Aquiclude Earth material that retards the flow of groundwater.

Aquifer Earth material containing sufficient groundwater that the water can be pumped out. Highly fractured rocks and unconsolidated sands and gravels make good aquifers.

Area (strip) mining Type of strip mining practiced on relatively flat areas.

Artesian Referring to a groundwater system in which the groundwater is isolated from the surface by a confining layer and the water is under pressure. Groundwater that is under sufficient pressure that it will flow freely at the surface from a spring or well.

Asbestos Fibrous mineral material used as insulation. It is suspected of being either a true carcinogen or a carrier of carcinogenic trace elements.

Ash, volcanic Unconsolidated volcanic debris, less than 4 mm in diameter, physically blown out of a volcano during an eruption.

Autotroph An organism that produces its own food from inorganic compounds and a source of energy. There are photoautotrophs (photosynthetic plants) and chemical autotrophs.

Avalanche A type of landslide involving a large mass of snow, ice, and rock debris that slides, flows, or falls rapidly down a mountainside.

B soil horizon Intermediate soil horizon sometimes known as the *zone of accumulation*.

Balancing Doctrine Asserts that public benefits and the importance of a particular action should be balanced with the potential injury to certain individuals. The method of balancing is changing now, and courts are considering possible long-range injury caused by certain activities to large numbers of citizens other than the immediate complainants.

Barrier island Island separated from the mainland by a salt marsh. It generally consists of a multiple system of beach ridges and is separated from other barrier islands by inlets that allow the exchange of seawater with lagoon water.

Basalt A fine-grained extrusive igneous rock. It is one of the most common igneous rock types.

Bauxite A rock composed almost entirely of hydrous aluminum oxides. It is a common ore of aluminum.

Bed material Sediment transported and deposited along the bed of a stream channel.

Bentonite A type of clay that is extremely unstable. Upon wetting, it expands to many times its original volume.

Biogeochemical cycle The cycling of a chemical element through the biosphere; its pathways, storage locations, and chemical forms in the atmosphere, oceans, sediments, and lithosphere.

Biogeography The geography of living things; the study of the distribution of living things and their history, origin, migrations, and the causes of these.

Biomagnification Also called *biological concentration*. The tendency for some substances to concentrate with each trophic level. Organisms preferentially store certain chemicals and excrete others. When this occurs consistently among organisms, the stored chemicals increase as a percentage of the body weight as the material is transferred along a food chain or trophic level. For example, the concentration of DDT is greater in the herbivores than in plants, and greater in plants than in the nonliving environment.

Biomass The amount of living material; or the amount of organic material contained in living organisms, both as live and dead material, as in the leaves (live) and stemwood (dead) of trees.

Biome A kind of ecosystem. The rain forest is an example of a biome; rain forests occur in many parts of the world, but are not all connected with each other.

Biosphere That part of a planet where life exists. On the Earth it extends from the depths of the oceans to the summit of mountains, but most life exists within a few meters of the surface.

Biota A general term for all the organisms of all species living in an area or region up to and including the biosphere, as in "the biota of the Mojave Desert" or "the biota in that aquarium."

Biotite A common ferromagnesian mineral, a member of the mica family.

Birth rate The rate at which births occur in a population, measured either as the number of individuals born per unit of time, or the percentage increase in the population per unit of time.

Bituminous coal Very common type of coal characterized by relatively high carbon content and small amounts of volatiles. Sometimes called *soft coal.*

Blowout Failure of an oil, gas, or disposal well resulting from adverse pressures that may physically blow part of the well casing upward. May be associated with leaks of oil, gas, or, in the case of disposal wells, harmful chemicals.

BOD (biological oxygen demand) A measure of the amount of oxygen necessary to decompose organic materials in a unit volume of water. As the amount of organic waste in water increases, more oxygen is used, resulting in a higher BOD.

Body burden The amount or concentration of a toxic chemical, especially radionuclides, in an individual.

Braided river A river channel characterized by an abundance of islands that continually divide and subdivide the flow of the river.

Breeder reactor A type of nuclear reactor that actually produces more fissionable (fuel) material than it uses.

Brine Water that has a high concentration of salt.

C soil horizon Lowest soil horizon sometimes known as the *zone of partially altered parent material.*

Calcite Calcium carbonate ($CaCO_3OH$); common carbonate mineral that is the major constituent of the rock limestone. It weathers readily by solutional processes, and large cavities and open weathered fractures are common in rocks containing the mineral calcite.

Calorie The quantity of heat required to raise the temperature of one gram of water from 14.5° to 15.5°C.

Capillary action The rise of water along narrow passages, facilitated and caused by surface tension.

Carbonate A compound or mineral containing the radical (CO_3^{--}). The common carbonate is calcite.

Carcinogen Any material that is known to produce cancer in humans or other animals.

Carnivore Organisms that kill and feed on other animals. Contrast with *herbivore* and *detritivore.*

Carrying capacity The maximum population size that can exist in a habitat or ecosystem without detrimental effects to either that population or to the habitat or ecosystem.

Channelization An engineering technique to straighten, widen, deepen, or otherwise modify a natural stream channel.

Chaparral A dense scrubland found in areas with Mediterranean climate (a long warm dry season and a cooler rainy season).

Circum-Pacific belt One of the three major zones where earthquakes occur. This belt is essentially the Pacific plate. It is also known as the *ring of fire,* as many active volcanoes are found on the edge of the Pacific plate.

Clay May refer to a mineral family or to a very fine-grained sediment. It is associated with many environmental problems, such as shrinking and swelling of soils and sediment pollution.

Climate The representative or characteristic conditions of the atmosphere at particular places on the Earth. *Climate* is the average or expected conditions over long periods; *weather* is the particular conditions at one time in one place.

Climax stage (of ecological succession) The final stage of ecological succession, and therefore an ecological community which continues to reproduce itself over time; or a stage in ecological succession during which an ecological community achieves the greatest biomass or diversity. (The first definition is the classical definition.)

Coal A sedimentary rock formed from plant material that has been buried, compressed, and changed.

Cohort All the individuals in a population born during the same time period. Thus, all the people born during the year 1980 represent the world human cohort for that year.

Commensalism A relationship between two kinds of organisms in which one benefits from the relationship and the other is neither helped nor hurt.

Community, ecological A group of populations of different species living in the same local area and interacting with one another. A *community* is the living portion of an ecosystem.

Competition When different individuals, populations, or species both compete for the same resource(s), and the presence of one has a detrimental effect on the other. Sheep and cows eating grass in the same field are competitors.

Competitive exclusion principle The idea that two populations of different species with exactly the same requirements cannot persist indefinitely in the same habitat—one will always win out and the other will go extinct. Which one wins depends on the exact environmental conditions. Referred to as a "principle," the idea has some basis in observation and experimentation.

Cone of depression A cone-shaped depression in the water table caused by withdrawal of water at rates greater than the rates at which the water can be replenished by natural groundwater flow.

Conglomerate A detrital sedimentary rock composed of rounded fragments, 10 percent of which are larger than 2 mm in diameter.

Connate water Water that is no longer in circulation or in contact with the present water cycle; generally saline water trapped during the deposition of sediments.

Continental drift The movement of continents in response to sea-floor spreading. The most recent episode of continental drift supposedly started about 200 million years ago with the breakup of the supercontinent Pangaea.

Continental shelf Relatively shallow ocean area between the shoreline and the continental slope that extends to approximately a 600-foot water depth surrounding a continent.

Contour (strip) mining Type of strip mining used in hilly terrain.

Convection The transfer of heat involving the movement of particles; for example, the boiling of water in which hot water rises to the surface and displaces cooler water which moves toward the bottom.

Convergent evolution See *Evolution, convergent.*

Convergent plate boundary Boundary between two lithospheric plates in which one plate descends below the other (subduction).

Corrosion A slow chemical weathering or chemical decomposition that proceeds from the surface in. Objects such as pipes experience corrosion when buried in soil.

Cost-benefits analysis A type of analysis that compares benefits and costs of a particular project, the most desirable being those for which the benefits-to-cost ratio is greater than one.

Creep A type of downslope movement characterized by slow flowing, sliding, or slipping of soil and other earth materials.

Crystallization Processes of crystal formation.

Death rate The rate at which deaths occur in a population, measured either as the number of individuals dying per unit time, or the percentage of a population dying per unit time.

Debris flow Rapid downslope movement of earth material often involving saturated, unconsolidated material that has become unstable due to torrential rainfall.

Decomposer An organism that obtains its energy and nutritional requirements by feeding on dead organisms; or, a feeder on dead organisms.

Deep-well disposal Method of waste disposal that involves pumping waste into subsurface disposal sites such as fractured or otherwise porous rock.

Denitrification The conversion of nitrate to molecular nitrogen by the action of bacteria—an important step in the nitrogen cycle.

Density-dependent population effects Factors whose effects on a population change with population density. The term is usually restricted to apply to population growth, reproduction, and mortality. For example, during a famine the mortality rate increases. In this case the food supply can be said to have a density-dependent population effect.

Density-independent population effects Changes in the size of a population due to factors which are independent of the population size. For example, certain climatic factors, which are not affected by the size of a specific population, can affect the entire population. A storm which will knock down all trees in a forest, no matter how few or how many there are, is a density-independent population effect.

Desertification The process of creating a desert where there was not one before. Farming in marginal grasslands, which destroys the soil and prevents the future recovery of natural vegetation, would be an example of desertification.

Diamond A very hard mineral composed of the element carbon.

Discharge The quantity of water flowing past a particular point on a stream. Generally is measured in cubic feet per second (cfs).

Divergent plate boundary Boundary between lithospheric plates characterized by the production of new lithosphere. Found along oceanic ridges.

Dose dependency The effects of a certain substance on a particular organism depend upon the dose or concentration of the substance.

Drainage basin The area that contributes surface water to a particular stream network.

Drainage net The system of stream channels that coalesce to form a stream system.

Dredge spoils Solid material, such as sand, silt, clay, rock, or other material deposited from industrial and municipal discharges, that is removed from the bottom of water bodies to improve navigation.

Earthquake Natural shaking or vibrating of the Earth in response to the breaking of rocks along faults. The earthquake zones of the Earth generally correlate very well with lithospheric plate boundaries.

Ecology The science of the study of the relationships between living things and their environment.

Ecosystem An ecological community and its local nonliving environment.

Efficiency The ratios of output to input. With machines, usually the ratio of work or power produced to the energy or power used to operate or fuel them. With living things, efficiency may be defined as either the useful work done or the energy returned in a useful form compared with the energy taken in.

Efficiency, first-law The simple ratio of the output of work to input of energy.

Efficiency, second-law The ratio of the minimum quantity of available work needed to perform a particular task to the actual quantity of available work used to perform the task.

Effluent Any material that flows outward from something. Examples include waste water from hydroelectric plants and water dischargeed into streams from waste-disposal sites.

Effluent stream Type of stream where flow is maintained during the dry season by groundwater seepage into the channel.

Electromagnetic spectrum The collection of all the possible wavelengths of electromagnetic energy, considered as a continuous range, including long wavelength (used in radio transmission), infrared, visible, ultraviolet, X-rays, and gamma rays.

Endemic Confined to a given region, such as an island or a country. The whooping crane is endemic to North America.

Energy An abstract concept referring to the ability or capacity to do work.

Entropy A measure of the amount of energy in a system which is unavailable for useful work. As the disorder of a system increases, the entropy in a system also increases.

Environment All of the factors (living and nonliving) that actually affect an individual organism or population at any point in the life cycle. *Environment* is also sometimes used to denote a certain set of circumstances surrounding a particular occurrence (environments of deposition, for example).

Environmental geology The application of geologic information to environmental problems.

Environmental impact The effects of some action on the environment, particularly by human beings.

Environmental impact statement A written statement that assesses and explores possible impacts associated with a particular project which may affect the human environment.

The statement is required by the National Environmental Policy Act of 1969.

Environmental law A field of law that is growing rapidly and becoming a significant part of our jurisprudence.

Ephemeral Temporary or very short-lived. Characteristic of beaches, lakes, and some stream channels that change rapidly (geologically).

Epicenter The point on the surface of the Earth directly above the focus (area of first motion) of an earthquake.

Equilibrium A point of rest. A system which does not tend to undergo any change of its own accord, but remains in a single, fixed condition, is said to be in *equilibrium*. Compare with *steady state*.

Equilibrium, stable A condition in which a system will remain if undisturbed and to which it will return when displaced.

Erosion The process by which natural forces such as wind and water wear away a surface.

Eukaryote Organism whose cells have cell nuclei and certain other characteristics which separate it from the *prokaryotes*. The eukaryotes include higher plants and animals and many single-celled organisms.

Eutrophic Referring to bodies of water having an abundance of the chemical elements required for life.

Evaporite Sediments deposited from water as a result of extensive evaporation of seawater or lake water. Dissolved materials left behind following evaporation.

Evolution, convergent Unrelated, isolated kinds of organisms can evolve under similar environments to have similar adaptations, as in their morphology and physiology. For example, cactus in the New World and euphorbia in Africa both have green stems and reduced leaves, which are adaptations to desert climate.

Exotic species Species introduced into a new area by human action.

Exponential growth Growth in which the rate of increase is a constant percentage of the current size. Also called *geometric growth*. The graph of exponential growth is sometimes referred to as the J-curve.

Extinction Disappearance of a life form from existence. Usually applied to a species.

Extrusive igneous rocks Igneous rock that forms when magma reaches the surface of the Earth; a volcanic rock.

Fault A fracture or fracture system that has experienced movement along opposite sides of the fracture.

Fault gouge Crushed and altered rock found along fault planes. Is produced as differential motion grinds and crushes rocks in fault zones.

Feldspar The most abundant family of minerals in the Earth's crust. They are silicates of calcium, sodium, and potassium.

Ferromagnesian minerals Minerals containing iron and magnesium. These minerals are characteristically dark in color.

Fission The splitting of an atom into smaller fragments with the release of energy.

Floodplain Flat topography adjacent to a stream in a river valley that has been produced by the combination of overbank flow and lateral migration of meander bends.

Fluvial Concerning or pertaining to rivers.

Flux rate In energy flow and chemical cycling, the amount transferred between two compartments in a unit of time.

Fly ash Very fine particles (ash) resulting from the burning of fuels such as coal.

Foliation Property of metamorphic rock characterized by parallel alignment of the platy or elongated mineral grains. Foliation is environmentally important because it can affect the strength and hydrologic properties of rock.

Food chain The chain of who eats whom, beginning with autotrophs.

Forest An area with trees forming a dense canopy, with branches of adjacent trees touching or nearly touching.

Fossil fuels Fuels such as coal, oil, and gas that have formed by the alteration and decomposition of plants, animals, etc., from a previous geologic time.

Fracture zone A fracture system that may or may not be active and may or may not have an alteration zone along the fracture planes. Fracture zones are environmentally important because they greatly affect the strength of rocks.

Fusion, nuclear Combining of light elements to form heavy elements with the release of energy.

Gasification Method of producing gas from coal.

Geochemical cycle The pathways of chemical elements in geological processes, including the chemistry of the lithosphere, atmosphere, hydrosphere, and biosphere.

Geochemistry The study of the distribution and migration of chemical elements in Earth processes.

Geologic cycle The formation and destruction of earth materials and the processes responsible for these events. The geologic cycle includes the following subcycles: hydrologic, tectonic, rock, and geochemical.

Geometric growth See *Exponential growth*.

Geomorphology The study of landforms and surface processes.

Geothermal energy The useful conversion of natural heat from the interior of the Earth.

Glacier A landbound mass of moving ice.

Gneiss A coarse-grained, foliated metamorphic rock in which there is banding of the light and dark minerals.

Gravel Unconsolidated, generally rounded fragments of rocks and minerals greater than 2 mm in diameter.

Gross production (biology) Production before respiration losses are subtracted.

Groundwater Water found beneath the Earth's surface within the zone of saturation.

Growth rate The net increase in some factor per unit time. In ecology, the growth rate of a population is sometimes measured as the increase in numbers of individuals or biomass per unit time, and sometimes as a percentage increase in numbers of biomass per unit time.

Gypsum An evaporite mineral, $CaSO_4 \cdot 2H_2O$.

Half-life The time required for half of a substance to disappear; the average time required for one half of a radioisotope to be transformed to some other isotope; the time required for a toxic chemical to be converted to some other forms.

Halite A common mineral, NaCl (salt).

Hematite An important ore of iron, a mineral (Fe_2O_3).

Heterotroph An organism that feeds on other organisms.

High-value resource Materials such as diamonds, copper, gold, and aluminum. These materials are extracted wherever they are found and transported around the world to numerous markets.

Humus Organic material in soil.

Hydrocarbon Organic compound consisting of carbon and hydrogen.

Hydrofracturing Pumping of water under high pressure into subsurface rocks to fracture the rocks, thereby increasing their permeability.

Hydrograph A graph of the discharge of a stream with time.

Hydrologic cycle Circulation of water from the oceans to the atmosphere and back to the oceans by way of evaporation, runoff from streams and rivers, and groundwater flow.

Hydrology The study of surface and subsurface water.

Hydrothermal ore deposit A mineral deposit derived from hot water solutions of magmatic origin.

Igneous rocks Rocks formed from the solidification of magma. They are *extrusive* if they crystallize on the surface of the Earth and *intrusive* if they crystallize beneath the surface.

Impermeable Earth materials that retard greatly or prevent the movement of fluids through them.

Infiltration The movement of surface water into rocks or soil.

Influent stream Type of stream that is everywhere above the groundwater table and flows in direct response to precipitation. Water from the channel moves down to the water table, forming a recharge mound.

In-stream water use Refers to water uses that occur in the stream itself, such as the generation of electric power, the support of fish, and recreational activities for people.

Integrated Pest Management (IPM) The coordinated use of several techniques of pest control done in an ecologically and economically sound manner to achieve a stable crop production. A goal of IPM is to maintain pest damage below an economically significant level while minimizing hazards to people, other living things, and the abiotic environment.

Inversion, atmospheric Occurs in the lower atmosphere when a warm body of air overrides a cooler one. Inversions are often associated in urban areas with air pollution events.

Island arc A curved group of volcanic islands associated with a deep-oceanic trench and subduction zone (convergent plate boundary).

Juvenile water Water derived from the interior of the Earth that has not previously existed as atmospheric or surface water.

Karst topography A type of topography characterized by the presence of sinkholes, caverns, and diversion of surface water to subterranean routes.

Kinetic energy The energy of motion. For example, the energy in a moving car due to the mass of the car traveling at a particular velocity.

Land ethic A set of ethical principles which affirm the right of all resources, including plants, animals, and earth materials, to continued existence and, at least in some locations, continued existence in a natural state.

Landslide Specifically, rapid downslope movement of rock and/or soil. Also used as a general term for all types of downslope movement.

Land-use planning Complex process involving development of a land-use plan to include a statement of land-use issues, goals, and objectives; a summary of data collection and analysis; a land-classification map; and a report that describes and indicates appropriate development in areas of special environmental concern.

Laterite A soil formed from intense chemical weathering in tropical or savannah regions.

Lava Molten material that is produced from a volcanic eruption, or a rock that forms from solidification of molten material.

Law of the Minimum (Liebeg's Law of the Minimum) The concept that the growth or survival of a population is directly related to the life requirement which is in least supply and not to a combination of factors.

LD-50 The lethal dose for 50 percent of a population. The dose of a toxin which causes death in 50 percent of a population, or which can be expected on the average to cause death in 50 percent of the population.

Leaching Process of dissolving, washing, or draining earth materials by percolation of groundwater or other liquids.

Liebeg's Law of the Minimum See *Law of the Minimum*.

Light-water reactor A nuclear fission reactor that uses water as the primary coolant. Light-water reactors are sometimes called *burner reactors*, because they consume fuel, primarily uranium-235.

Lignite A type of low-grade coal.

Limestone A sedimentary rock composed almost entirely of the mineral calcite.

Limonite Rust; hydrated iron oxide.

Lithosphere Outer layer of the Earth approximately 100 kilometers thick of which the plates that contain the ocean basins and the continents are composed.

Littoral Pertaining to the nearshore and beach environments.

Macronutrients Chemical elements required in large amounts by all forms of life.

Magma A naturally occurring silica melt, a good deal of which is in a liquid state.

Magnetite A mineral and important ore of iron, Fe_3O_4.

Magnitude, earthquake A number on a logarithmic scale which refers to the amount of energy released by an earthquake.

Manganese oxide nodule Nodules of manganese, iron with secondary copper, nickel, and cobalt, which cover vast areas of the deep-ocean floor.

Marble Metamorphosed limestone.

Maximum sustainable yield (MSY) The maximum usable production of a biological resource which can be obtained in a specified time period. The MSY level is the population size which results in maximum sustainable yield.

Meanders Bends in a stream channel that migrate back and forth across the floodplain, depositing sediment on the inside of the bends, forming point bars, and eroding the outsides of bends.

Meteoric water Water derived from the atmosphere.

Methane A gas, CH_4, the major constituent of natural gas.

Mica A common rock-forming silicate mineral.

Microclimate The climate of a very small, local area. For example, the climate under a tree, near the ground within a forest, or near the surface of streets in a city.

Micronutrients Chemical elements required in very small amounts by at least some forms of life. Boron, copper, and molybdenum are examples of micronutrients.

Mineral Naturally occurring inorganic material with a definite internal structure and physical and chemical properties that vary within prescribed limits.

Mudflow A mixture of unconsolidated materials and water that flows rapidly downslope or down a channel.

Mutualism See *Symbiosis.*

National Environmental Policy Act of 1969 (NEPA) Act declaring that it is national policy that harmony between people and their physical environment be encouraged. Established the Council on Environmental Quality. Established requirements that an environmental impact statement be completed prior to major federal actions significantly affecting the quality of the human environment.

Net production (biology) The production that remains after utilization. In a population, net production is sometimes measured as the net change in the numbers of individuals. It is also measured as the net change in biomass or in stored energy. In terms of energy, it is equal to the gross production minus the energy used in respiration.

Neutron A subatomic particle having no electric charge and found in the nuclei of atoms. Neutrons are crucial in sustaining nuclear fission in a reactor.

Niche The ''profession'' or role of an organism or species. Sometimes expressed as all of the environmental conditions under which the individual or species can persist. The *fundamental niche* is all the conditions under which a species can persist in the absence of competition; the *realized niche* is the set of conditions as they occur in the real world with competitors.

Nonrenewable resource A resource that is cycled so slowly by natural Earth processes that once used, it is essentially not going to be made available in any useful time framework.

Nuclear reactor A device in which controlled nuclear fission is maintained. The major component of a nuclear power plant.

Oil shale An organic-rich shale containing substantial quantities of oil that may be extracted by conventional methods of destructive distillation.

Optimal sustainable population (OSP) The population level which results in an optimal sustainable yield; or the population level that is in some way ''best'' for that population, its ecological community, its ecosystem, or the biosphere.

Optimal sustainable yield (OSY) The largest yield of a renewable resource which can be achieved over a long time period without decreasing the ability of the population or its environment to support the continuation of this level of yield.

Ore An earth material from which a useful commodity can be extracted profitably.

Organic compound A compound of carbon; originally used to refer to the compounds found in and formed by living things.

Overburden Earth materials (spoil) that overlie an ore deposit. Most frequently refers to material overlying or extracted from a surface (strip) mine.

Parasitism In most general terms, a relationship between individuals or populations of two different species in which one benefits and the other suffers some loss of vigor, growth, reproduction, longevity, or survival. Usually restricted to disease-host or pest-host relationships.

Passive solar energy system Direct use of solar energy through architectural design to enhance or take advantage of natural changes in solar energy that occur throughout the year without requiring mechanical power.

Pathogen Any material that may cause disease; for example, microorganisms, including bacteria and fungi.

Pebble A rock fragment between 4 and 64 mm in diameter.

Pedology The study of soils.

Permafrost Permanently frozen ground.

Permeability A measure of the ability of an earth material to transmit fluids such as water or oil.

Photosynthesis Synthesis of sugars from carbon dioxide and water by living organisms using light as energy. Oxygen is given off as a by-product.

Physiographic province A region characterized by a particular assemblage of landforms, climate, and geomorphic history.

Placer deposit A type of ore deposit found in material transported and deposited by agents such as running water, ice, or wind; for example, gold and diamonds found in stream deposits.

Plate tectonics A model of global tectonics which suggests that the outer layer of the Earth known as the *lithosphere* is composed of several large plates which move relative to one another. Continents and ocean basins are passive riders on these plates.

Point bar Accumulation of sand and other sediments on the inside of meander bends in stream channels.

Pollutant In most general terms, any factor that has a harmful effect on living things or their environment.

Pool Common bed form produced by scour in meandering and straight channels.

Population A group of individuals of the same species living in the same area or interbreeding and sharing genetic information.

Population dynamics The study of changes in population sizes and the causes of these changes.

Population regulation See *Density-dependent population effects* and *Density-independent population effects*.

Porosity The percentage of void (empty space) in an earth material such as soil or rock.

Potable water Water that may be drunk safely.

Potential energy Energy that is stored. Examples include the gravitational energy of water behind a dam; chemical energy in coal, fuel oil, and gasoline; and nuclear energy (in the forces that hold atoms together).

Power The time rate of doing work.

Primary treatment (of waste water) Removal of large particles and organic materials from waste water through screening.

Production, ecological The amount of increase in organic matter, usually measured per unit area of land surface or unit volume of water, as in grams per square meter (g/m^2). Production is divided into *primary* (that of autotrophs) and *secondary* (that of heterotrophs). It is also divided into *net* (that which remains stored after use) and *gross* (that added before any use).

Production, primary The production by autotrophs.

Production, secondary The production by heterotrophs.

Productivity, ecological The *rate* of production. That is, the amount of increase in organic matter per unit time (for example, grams per meter squared *per year*). See *Production*.

Prokaryote A kind of organism that lacks a true cell nucleus and has other cellular characteristics which distinguish it from the *eukaryotes*. Bacteria and blue-green algae are prokaryotes.

Protocooperation A symbiotic relationship which is beneficial, but not obligatory, to both species.

Pyrite Iron sulphide, a mineral commonly known as fool's gold. Environmentally important because in contact with oxygen-rich water, it produces a weak acid that may pollute water or dissolve other minerals.

Quartz Silicon oxide, a very common rock-forming mineral.

Quartzite Metamorphosed sandstone.

Radioactive waste Type of waste produced in the nuclear fuel cycle. Generally classified as high-level or low-level.

Radioisotope Any isotope that is radioactive. An example is uranium-235 or carbon-14.

Reclamation, mining Restoring of land used for mining to other useful purposes, such as agriculture or recreation, after mining operations are concluded.

Renewable resource A resource such as timber, water, or air that is naturally recycled or recycled by human-induced processes within a time framework useful for people.

Reserves Known and identified deposits of earth materials from which useful materials can be extracted profitably with existing technology and under present economic and legal conditions.

Resources Includes reserves plus other deposits of useful earth materials that may eventually become available.

Respiration The complex series of chemical reactions in organisms which makes energy available for use. Water, carbon dioxide, and energy are the products of respiration.

Retorting A method of recovering shale oil which involves crushing and heating of the rock to release the oil. May be attempted in place (that is, in the ground) or at a prescribed location.

Riffle A section of stream channel characterized at low flow by fast, shallow flow. Generally contains relatively coarse bedload particles.

Riparian rights, water law Each landowner has the right to make reasonable use of water on his land, provided the water is returned to the natural stream channel before it leaves his property. Furthermore, the property owner has the right to receive the full flow of the stream undiminished in quantity and quality.

Riverine environment Land area adjacent to and influenced by a river.

Rock (engineering) Any earth material that has to be blasted in order to be removed.

Rock (geologic) An aggregate of a mineral or minerals.

Rock cycle A group of processes that produce igneous, metamorphic, and sedimentary rocks.

Saline Salty; characterized by a high salinity.

Salinity A measure of the total amount of dissolved solids in water.

Salt dome A structure produced by the upward movement of a mass of salt. Frequently associated with oil and gas deposits on the flanks of a dome.

Sand Grains of sediment having a size between 1/16 and 2 mm in diameter. Often sediment composed of quartz particles of the above size.

Sand dune A ridge or hill of sand formed by wind action.

Sandstone A detrital sedimentary rock composed of sand grains that have been cemented together.

Sanitary landfill A method of solid-waste disposal not producing a public health problem or nuisance. Confines and compresses the waste and covers it at the end of each day with a layer of compacted, relatively impermeable material such as clay.

Savannah An area with trees scattered widely among dense grasses.

Scenic resources The visual portion of an aesthetic experience. Scenery is now recognized as a natural resource with varying values.

Schist A coarse-grained metamorphic rock characterized by a foliated texture of the platy or elongated mineral grains.

Schistosomiasis Snail fever, a debilitating and sometimes fatal tropical disease.

Sea wall An engineering structure constructed at the water's edge to minimize coastal erosion by wave activity.

Secondary enrichment A weathering process of sulphide ore deposits which may concentrate the desired minerals.

Secondary treatment (of waste water) Use of biologic processes to degrade waste water in a treatment facility.

Seismic Referring to vibrations in the Earth produced by earthquakes.

Septic tank A tank that receives and temporarily holds solid and liquid waste. Anaerobic bacterial activity breaks down the waste. The solid wastes are separated out, and liquid waste from the tank overflows into a drainage system.

Sewage sludge Solid material that remains after municipal waste-water treatment.

Shale A sedimentary rock composed of silt- and clay-sized particles; the most common sedimentary rock.

Silicate minerals The most important group of rock-forming minerals.

Silt Sediment between 1/16 and 1/256 mm in diameter.

Silviculture The farming of trees.

Sinkhole A surface depression formed by the solution of limestone or the collapse over a subterranean void such as a cave.

Sinuous channel Type of stream channel (not braided).

Soil The top layer of a land surface where the rocks have been weathered to small particles. Soils are made up of inorganic particles of many sizes, from small clay particles to large sand grains. Many soils also include dead organic material.

Soil (in engineering) Earth material that can be removed without blasting.

Soil (in soil science) Earth material modified by biological, chemical, and physical processes such that the material will support rooted plants.

Soil horizons Layers in soil (A, B, C) that differ from one another in chemical, physical, and biological properties.

Soil survey A survey consisting of a detailed soil map and descriptions of soils and land-use limitations. Generally prepared by the Soil Conservation Service in cooperation with local government.

Solar cell (photovoltaic) Device that directly converts light into electricity.

Solar collector Device for collecting and storing solar energy. For example, home water heating is done by flat panels consisting of a glass cover plate over a black background upon which water is circulated through tubes. Short-wave solar radiation enters the glass and is absorbed by the black background. As long-wave radiation is emitted from the black material, it cannot escape through the glass, so the water in the circulating tubes is heated up, typically to temperatures of 38° to 93°C.

Solar energy Collecting and using energy from the sun directly.

Solar pond Shallow pond filled with water and used to generate relatively low-temperature water.

Solar power tower A system of collecting solar energy that delivers the energy to a central location where the energy is used to produce electric power.

Solid waste Material such as refuse, garbage, and trash.

Species A group of individuals capable of interbreeding.

Spoils, mining Banks or piles that are accumulations of overburden removed during mining processes and discarded on the surface.

Steady state When input equals output in a system, there is no net change and the system is said to be in a steady state. A bathtub with water flowing in and out at the same rate maintains the same water level and is in steady state. Compare with *equilibrium*.

Storm surge Wind-driven oceanic waves.

Stress Force per unit area. May be compression, tension, or shear.

Strip mining A method of surface mining.

Subduction A process in which one lithospheric plate descends beneath another.

Subsidence A sinking, settling, or otherwise lowering of parts of the crust of the Earth.

Subsurface water All of the waters within the lithosphere.

Succession, ecological The process of the development of an ecological community or ecosystem, usually viewed as a series of stages: early, middle, late, mature (or climax), and sometimes post-climax. Primary succession is an original establishment; secondary succession is a reestablishment.

Surface water Waters above the surface of the lithosphere.

Surface wave One of the types of waves produced by earthquakes. These waves generally cause most of the damage to structures on the surface of the Earth.

Suspended load Sediment in a stream or river carried off the bottom by the fluid.

Symbiosis A living together of individuals of two different species which is beneficial to both. Obligatory symbiosis is called *mutualism;* nonobligatory symbiosis is called *protocooperation.*

Tar sands Naturally occurring sand, sandstone, or limestone that contains a very viscous petroleum.

Tectonic Referring to rock deformation.

Tectonic cycle The processes which change the Earth's crust, producing external forms such as ocean basins, continents, and mountains.

Tertiary treatment (of waste water) Advanced form of waste-water treatment involving chemical treatment or advanced filtration. An example is chlorination of water.

Texture, rock The size, shape, and arrangement of mineral grains in rocks.

Thermal (heat) energy The energy of the random motion of atoms and molecules.

Thermodynamics, First Law of A law that states that energy is always conserved. That is, energy is never created or destroyed but only changes form.

Thermodynamics, Second Law of A law that states that in any real process, energy always tends to go from a more usable form to a less usable form. That is, as energy transformations take place, the system goes from a state that is relatively ordered to one that is relatively disordered, or more random, and in which less of the energy is available for useful work.

Tidal energy Electricity generated by tidal power.

Toxic Harmful, deadly, or poisonous.

Trophic level In an ecological community, all the organisms which are the same number of food chain steps from the primary source of energy. For example, in a grassland, the green grasses are on the first trophic level; grasshoppers on the second; birds that feed on grasshoppers on the third; and so forth.

Tropical cyclone Severe storm generated from a tropical disturbance. Called **typhoons** in most of the Pacific Ocean and **hurricanes** in the Western Hemisphere.

Tsunami A giant sea wave produced when ocean water is vertically displaced by submarine volcanic or earthquake activity. Some are also generated by large landslides in the marine environment. Are sometimes incorrectly referred to as *tidal waves*.

Tundra The treeless land area in alpine and arctic areas, characterized by plants of low stature and including bare areas without any plants and covered areas with lichens, mosses, grasses, sedges, and small flowering plants, including low shrubs.

Turnover rate The fraction of the amount of energy stored which is added and lost (turned over) during some unit of time.

Unconfined aquifer Type of aquifer in which there is no impermeable layer restricting the upper surface of the zone of saturation.

Unified soil classification system A classification of soils, widely used in engineering practice, based upon the amount of coarse particles, fine particles, or organic material.

Uniformitarianism The principle that processes which operate today operated in the past. Therefore, observations of processes today can explain events which occurred in the past and leave evidence, for example, in the fossil record or in geologic formations.

Water budget Inputs and outputs of water for a particular system (a drainage basin, region, continent, or the entire Earth).

Water table The surface that divides the zone of aeration from the zone of saturation. The surface below which all the pore space in rocks is saturated with water.

Weathering Changes that take place in rocks and minerals at or near the surface of the Earth in response to physical, chemical, and biological changes. The physical, chemical, and biological breakdown of rocks and minerals.

Work (physics) Force times the distance through which it acts.

Zone of aeration The zone or layer above the water table in which some water may be suspended or moving in a downward migration toward the water table or laterally toward a discharge point.

Zone of saturation Zone or layer below the water table in which all the pore space of rock or soil is saturated.

Acknowledgments

ALGERMISSEN, S. T., and PERKINS, D. M. 1976. U.S. Geological Survey Open File Report 76–416.

AMERICAN CHEMICAL SOCIETY, SUBCOMMITTEE ON ENVIRONMENTAL IMPROVEMENT, in *Cleaning our environment—The chemical basis for action*, American Chemical Society, Washington, D.C., 1969, p. 107.

AMERICAN NUCLEAR SOCIETY. 1976. *Nuclear power and the environment: Questions and answers*. Hinsdale, Ill.: American Nuclear Society.

ANTHES, R. A.; CAHIR, J. J.; FRASER, A. B.; and PANOFSKY, H. A. 1981. *The atmosphere*. 3rd ed. Columbus, Ohio: Charles E. Merrill.

AUSTRALIA, INQUIRY INTO WHALES AND WHALING. 1979. *The whaling question: The inquiry by Sir Sidney Frost of Australia*. San Francisco: Friends of the Earth.

AWRAMIK, S. M. 1981. The pre-Phanerozoic biosphere—three billion years of crises and opportunities. In *Biotic crises in ecological and evolutionary time*, ed. M. H. Nitecki, pp. 83–102. New York: Academic Press.

BAKER, V. R. 1976. Hydrogeomorphic methods for the regional evaluation of flood hazards. *Environmental Geology* 1: 261–81.

BAUMOL, W. J., and OATES, W. E. 1979. *Economics, environmental policy, and the quality of life*. Englewood Cliffs, N. J.: Prentice-Hall.

BELL, F. W. 1978. *Food from the sea: The economics and politics of ocean fisheries*. Boulder, Colo.: Westview Press.

BILLINGS, W. D. 1970. *Plants, man, and the ecosystem*. 2nd ed. Belmont, Calif.: Wadsworth.

BOCKSTOCE, J. R., and BOTKIN, D. B. 1980. *The historical status and reduction of the western Arctic bowhead whale* (Balaena mysticetus) *population by the pelagic whaling industry, 1848–1914*. New Bedford, Conn.: Dartmouth Historical Society.

BOLIN, B.; DEGENS, E. T.; KEMPE, S.; and KETNER, P., eds. 1979. *The global carbon cycle*. New York: Wiley.

BORNE, A. G. 1964. Birth of an island. *Discovery* 25: 16.

BORRIE, W. D. 1970. *The growth and control of world population*. London: Weidenfeld and Nicolson.

BOWEN, H. J. M. 1966. *Trace elements in biochemistry*. New York: Academic Press.

BREW, D. A. 1974. *Environmental impact analysis: The example of the proposed Trans-Alaska Pipeline*. U.S. Geological Survey Circular 695.

BROBST, D. A., and PRATT, W. P., eds. 1973. *United States mineral resources*. U.S. Geological Survey Professional Paper 820.

BUDYKO, M. I. 1974. *Climate and life*. (English edition edited by D. H. Miller.) New York: Academic Press.

CALIFORNIA ENERGY COMMISSION. 1980. *Comparative evaluation of nontraditional energy resources*. Sacramento: California Energy Commission.

CHAIKEN, E. I.; POLONCSIK, S.; and WILSON, C. D. 1973. Muskegon sprays sewage effluents on land. *Civil Engineering* 43: 49–53.

CHANLETT, E. T. 1979. *Environmental protection*. 2nd ed. New York: McGraw-Hill.

CLARK, C. W. 1967. *Population growth and land use*. New York: Macmillan, St. Martin's Press.

——. 1973. The economics of overexploitation. *Science* 181: 630–34.

COMMITTEE FOR THE GLOBAL ATMOSPHERIC RESEARCH PROGRAM. 1975. *Understanding climatic change: A program for action*. Washington, D.C.: National Academy of Sciences.

CORNING, R. V. 1975. *Virginia Wildlife*, February, pp. 6–8.

COUNCIL OF STATE GOVERNMENTS, TASK FORCE ON NATURAL RESOURCES AND LAND USE INFORMATION AND TECHNOLOGY. 1976. *Land use policy and program analysis*. Lexington, Ky.: Council of State Governments

COUNCIL ON ENVIRONMENTAL QUALITY. 1970. *Ocean dumping: A national policy*.

——. 1973. *Environmental quality*.

——. 1978. *Progress in environmental quality*.

——. 1979. *Environmental quality*.

COUNCIL ON ENVIRONMENTAL QUALITY and THE DEPARTMENT OF STATE. 1980. *The global 2000 report to the President: Entering the twenty-first century*. Vols. 1 and 2.

COX, C. B.; HEALEY, I. N.; and MOORE, P. D. 1973. *Biogeography*. New York: Halsted Press.

CRANDELL, D. R., and MULLINEAUX, D. R. 1975. Techniques and rationale of volcanic hazards. *Environmental Geology* 1: 23–32.

CRANDELL, D. R., and WALDRON, H. H. 1969. Volcanic hazards in the Cascade Range. In *Geologic hazards and public problems, conference proceedings,* eds. R. Olsen and M. Wallace, pp. 5–18. Office of Emergency Preparedness, Region 7.

CUSHING, D. 1975. *Fisheries resources of the sea and their management.* New York: Oxford University Press.

DAVIS, F. F. 1972. Urban ore. *California Geology,* May, pp. 99–112.

DAVIS, M. B. 1981. Quaternary history and the stability of forest communities. In *Forest succession: Concepts and applications,* eds. D. C. West; H. H. Shugart; and D. B. Botkin, pp. 132–53. New York: Springer-Verlag.

DELWICHE, C. C., and LIKENS, G. E. 1977. Biological response to fossil fuel combustion products. In *Global chemical cycles and their alterations by man,* ed. W. Stumm, pp. 73–88. Berlin: Abakon Verlagsgesellschaft.

DESMOND, A. 1962. How many people have ever lived on Earth? *Population Bulletin* 18: 1–19.

DESSENS, J. 1962. Man-made tornadoes. *Nature* 193: 13.

———. 1964. Man-made thunderstorms. *Discovery* 25: 40.

DEWEY, J. F. 1972. Plate tectonics. *Scientific American* 22: 56–58. Copyright © 1972 by Scientific American, Inc. All rights reserved.

DRAKE, C. L. 1972. Future considerations concerning geodynamics. *American Association of Petroleum Geologists Bulletin* 56: 260–68.

DUNCAN, D. C., and SWANSON, V. E. 1965. *Organic-rich shale of the United States and world land areas.* U.S. Geological Survey Circular 523.

EDMONSON, W. T. 1975. Fresh water pollution. In *Environment: Resources, pollution, and society,* ed. W. M. Murdoch, pp. 250–71. Sunderland, Mass.: Sinauer.

EHRLICH, P. R.; EHRLICH, A. H.; and HOLDREN, J. P. 1977. *Ecoscience: Population, resources, environment.* 3rd ed. San Francisco: W. H. Freeman.

EISENBERG, R. M. 1966. The regulation of density in a natural population of the pond snail *Lymnaea elodes. Ecology* 47: 889–906.

EKDAHL, C. A., and KEELING, C. D. 1973. Atmospheric carbon dioxide and radiocarbon in the natural carbon cycle: 1. Quantitative deductions from records at Mauna Loa Observatory and at the South Pole. In *Carbon and the biosphere,* eds. G. M. Woodwell and E. V. Pecan, pp. 51–85. Brookhaven National Laboratory Symposium No. 24. Oak Ridge, Tenn.: Technical Information Service.

EVANS, D. M. 1966. Man-made earthquakes in Denver. *Geotimes* 10: 11–18.

FERRIANS, O. J., Jr.; KACHADOORIAN, R.; and GREENE, G. W. 1969. *Permafrost and related engineering problems in Alaska.* U.S. Geological Survey Paper 678.

FISKE, R. F., and KOYANAGI, R. Y. 1968. U.S. Geological Survey Professional Paper 607.

FOOD AND AGRICULTURE ORGANIZATION OF THE UNITED NATIONS. 1978. *Mammals in the seas.* Report of the FAO Advisory Committee on Marine Resources Research, Working Party on Marine Mammals. FAO Fisheries Series 5, vol. 1. Rome: FAO.

FOSTER, R. J. 1979. *Physical geology.* 3rd ed. Columbus, Ohio: Charles E. Merrill.

FOWLER, C. W.; BUNDERSON, W. T.; CHERRY, M. R.; RYEL, R. J.; and STEELE, B. B. 1980. *Comparative population dynamics of large mammals: A search for management criteria.* National Technical Information Service Publication PB80–178627. Washington, D.C.: U.S. Department of Commerce.

FOWLER, J. M. 1978. Energy and the environment. In *Perspectives on energy,* 2nd ed., eds. L. C. Ruedisili and M. W. Firebaugh, pp. 11–28. New York: Oxford University Press.

FOXWORTHY, G. L. 1978. Nassau County, Long Island, New York—Water problems in humid country. In *Nature to be commanded,* eds. G. D. Robinson and A. M. Spieker, pp. 55–68. U.S. Geological Survey Professional Paper 950.

FREED, V. H.; CHIOU, C. T.; and HAGUE, R. 1977. Chemodynamics: Transport and behavior of chemicals in the environment—A problem in environmental health. *Environmental Health Perspectives* 20: 55–77.

FROELICH, A. J.; GARNAAS, A. D.; and VAN DRIEL, J. N. 1978. Franconia area, Fairfax County, Virginia: Planning a new community in an urban setting—Lehigh. In *Nature to be commanded,* eds. G. D. Robinson and A. M. Spieker, pp. 69–89. U.S. Geological Survey Professional Paper 950.

GARRELS, R. M.; MACKENZIE, F. T.; and HUNT, C. 1975. *Chemical cycles and the global environment.* Los Altos, Calif.: William Kaufmann.

GARRITY, T. A., and NITZSCHKE, E. T. 1968. *Water law atlas: A water law primer.* Circular 95. Socorro, N.M.: State Bureau of Mines and Mineral Resources.

GATES, D. M. 1965. Heat, radiant and sensible. *Meteorological Monographs* 6: 1–26.

———. 1980. *Biophysical ecology.* New York: Springer-Verlag.

GIDDINGS, J. C. 1973. *Chemistry, man, and environmental change: An integrated approach.* San Francisco: Canfield Press.

GODFREY, P. J., and GODFREY, M. M. 1973. Comparison of ecological and geomorphic interactions between altered and unaltered barrier island systems in North Carolina. In *Costal geomorphology,* ed. D. R. Coates, pp. 239–58. Binghampton: State University of New York.

GRAHAM, H. E. 1955. Fire whirlwinds. *Bulletin of the American Meteorological Society* 36: 99.

GRAUNT, J. 1662. *Natural and political observations made upon the bills of mortality.* London: Roycraft. (Reprinted in 1973 with introduction by P. Laslett by Gregg International Publishers, Ltd., Germany.)

HANSON, W. G. 1967. Cesium-137 in Alaskan lichens, caribou, and Eskimos. *Health Physics* 13: 383–89.

HAUB, C. 1981. *1981 world population data sheet.* Washington, D.C.: Population Reference Bureau.

HERSEY, J. R. 1946. *Hiroshima.* New York: Knopf.

HIDY, G. M., and BROCK, J. R. 1971. *Topics in current aerosol research.* New York: Pergamon Press.

HUBBERT, M. K. 1962. *Energy resources.* Report to the Committee on Natural Resources of the National Academy of Sciences–National Research Council. Washington, D.C.: National Academy of Sciences–National Research Council.

HUNT, C. B. 1974. *Natural regions of the United States and Canada.* San Francisco: W. H. Freeman.

HUTCHINSON, G. E. 1950. Survey of contemporary knowledge of biogeochemistry. 3. The biogeochemistry of vertebrate excretion. *Bulletin of the American Museum of Natural History* 96: 481–82.

———, ed. 1970. Ianula: An account of the history and development of the Lago di Monterosi, Latium, Italy. *Transactions of the American Philosophical Society* 60: 1–178.

———. 1973. Eutrophication. *American Scientist* 61: 269–79.

———. 1978. *An introduction to population ecology.* New Haven, Conn.: Yale University Press.

INDIANA STATE BOARD OF HEALTH.

JANERICK, D. T.; BURNETT, W. S.; FECK, G.; HOFF, M.; NASCA, P.; POLEDNAK, A. P.; GREENWALD, P.; and VIANNA, N. 1981. Cancer incidence in the Love Canal area. *Science* 212: 1404–7.

JENSEN, W. A., and SALISBURY, F. B. 1972. *Botany: An ecological approach.* Belmont, Calif.: Wadsworth.

JORDAN, P. A.; BOTKIN, D. B.; and WOLF, M. I. 1971. Biomass dynamics in a moose population. *Ecology* 52: 147–52.

KASPERSON, R. E. 1977. Water re-use: Need prospect. In *Water re-use and the cities,* eds. R. E. Kasperson and J. X. Kasperson, pp. 3–25. Hanover, N. H.: University Press of New England.

KATES, R. W., and PIJAWKA, D. 1977. Reconstruction following disaster. In *From rubble to monument: The pace of reconstruction,* eds. J. E. Haas; R. W. Kates; and M. J. Bowden. Cambridge, Mass.: The MIT Press.

LANDSBERG, H. 1947. Fire storms resulting from bombing conflagrations. *Bulletin of the American Meteorological Society* 28: 72.

LANGER, W. L. 1964. The Black Death. *Scientific American* 210: 114–21.

LAVIGNE, D. M.; BARCHARD, W.; INNES, S.; and ORITSLAND, N. A. 1976. Pinniped bioenergetics. ACMRR/MM/SC/112. Rome: Food and Agriculture Organization of the United Nations (FAO).

LAWS, R. M.; PARKER, I. S. C.; and JOHNSTONE, R. C. B. 1975. *Elephants and their habitats.* Oxford, England: Clarendon Press.

LECLERC, G. L. (Count de Buffon). 1812. *Natural history, general and particular.* (Translated by W. Smellie.) London: C. Wood.

LEOPOLD, L. B. 1968. *Hydrology for urban land planning.* U.S. Geological Survey Circular 559.

LIKENS, G. E.; BORMANN, F. H.; PIERCE, R. S.; EATON, J. S.; and JOHNSON, N. M. 1977. *The biogeochemistry of a forested ecosystem.* New York: Springer-Verlag.

LINSLEY, R. K., Jr.; KOHLER, M. A.; and PAULHUS, J. L. H. 1975. *Hydrology for engineers.* 2nd ed. New York: McGraw-Hill.

LIPPMANN, M., and SCHLESINGER, R. B. 1979. *Chemical contamination in the human environment.* New York: Oxford University Press.

LOS ALAMOS SCIENTIFIC LABORATORY. 1978. LASL 78-24.

LOTKA, A. J. 1956. *Elements of mathematical biology.* New York: Dover.

LOVINS, A. B. 1979. *Soft energy paths: Towards a durable peace.* New York: Harper & Row.

MACARTHUR, R. H. 1958. Population ecology of some warblers of northeastern coniferous forests. *Ecology* 39: 599–619.

MACARTHUR, R. H., and CONNELL, J. H. 1966. *The biology of populations.* New York: Wiley.

MACARTHUR, R. H., and WILSON, E. O. 1963. An equilibrium theory of insular zoogeography. *Evolution* 17: 373–87.

MACKINTOSH, N. A. 1965. *The stocks of whales.* London: Fishing News Books, Ltd.

MADER, G. G.; SPANGLE, W. E.; BLAIR, M. L.; MEEHAN, R. L.; and BILODEAU, S. W. 1980. *Land-use planning after earthquakes.* Portola Valley, Calif.: William Spangle and Associates.

MARGULIS, L., and SCHWARTZ, K. V. 1982. *The five kingdoms.* San Francisco: W. H. Freeman.

MARGULIS, L., and LOVELOCK, J. E. 1974. Biological modulation of the Earth's atmosphere. *Icarus* 21: 471–89.

MARSH, W. M., and DOZIER, J. 1981. *Landscape.* Reading, Mass.: Addison-Wesley.

MARTIN, P. S. 1973. The discovery of America. *Science* 179: 969–74.

MARUYAMA, M. 1963. The second cybernetics: Deviation-amplifying mutual causal processes. *American Scientist* 51: 164–79.

MATTHEWS, W. H., et al., eds. 1791. *Man's impact on the climate.* Cambridge, Mass.: The MIT Press.

MATHEWSON, C. C., and CASTLEBERRY, J. P., II. (Undated). *Expansive soils: Their engineering geology.* College Station: Texas A & M University.

MATRAS, J. 1973. *Populations and societies.* Englewood Cliffs, N. J.: Prentice-Hall.

MCHARG, I. L. 1971. *Design with nature.* Garden City, N.Y.: Doubleday.

MILLER, P. C.; BRADBURY, D. E.; HAJEK, E.; LA MARCHE, V.; and THROWER, N. J. W. 1977. Past and present environment. In *Convergent evolution in Chile and California,* ed. H. A. Mooney, pp. 27–72. Stroudsburg, Pa.: Dowden, Hutchinson & Ross.

MUNIZ, I. P.; LEIVESTAD, H.; GJESSING, E.; JORANGER, E.; and SVALASTOG, D. 1976. *Acid precipitation: Effects on forest and fish.* SNSF Project. IR 13/75. As, Norway: Government of Norway.

MUROZIMI, M.; CHOW, T. J.; and PATTERSON, C. 1969. Chemical concentration of pollutant lead aerosol; terrestrial dusts; and sea salts in Greenland and Antarctic snow strata. *Geochim. Cosmochim. Acta* 33: 1247–94.

MURPHY, J. E. 1966. *Sulfur content of United States coals.* U.S. Bureau of Mines Information Circular 8312.

NASH, R. 1977. Do rocks have rights? *The Center Magazine* 10: 2–12.

NATIONAL OCEANIC AND ATMOSPHERIC ADMINISTRATION (NOAA). 1977. Earth's crustal plate boundaries: Energy and mineral resources. *California Geology* 30: 108–9.

NATIONAL RESEARCH COUNCIL, ADVISORY COMMITTEE ON THE BIOLOGICAL EFFECTS OF IONIZING RADIATIONS. 1972a. *The effects on populations of exposure to low levels of ionizing radiation.* Washington, D.C.: National Academy of Sciences–National Research Council.

NATIONAL RESEARCH COUNCIL, COMMITTEE ON GEOLOGICAL SCIENCES. 1972b. *The Earth and human affairs.* San Francisco: Canfield Press.

NEIBURGER, M; EDINGER, J. G.; and BONNER, W. D. 1973. *Understanding our atmospheric environment.* San Francisco: W. H. Freeman.

NEWMAN, J. R. 1980. *Effects of air emissions on wildlife resources.* U.S. Fish and Wildlife Service, Biological Services Program, National Power Plant Team. FWS/OBS-80/40. Washington, D.C.: U.S. Fish and Wildlife Service.

NICHOLS, D. R., and BUCHANAN-BANKS, J. M. 1974. *Seismic hazards and land-use planning.* U.S. Geological Survey Circular 690.

NOBLE, E. R., and NOBLE, G. A. 1976. *Parasitology: The biology of animal parasites.* 4th ed. Philadelphia: Lea & Febiger.

NORTH CAROLINA COASTAL RESOURCES COMMISSION. 1975. *State guidelines for local planning in the coastal area under the Coastal Area Management Act of 1974.* Raleigh, North Carolina.

ODÉN, S. 1976. The acidity problem—An outline of concepts. *Water, Air, and Soil Pollution.* 6: 137–66.

OFFICE OF INDUSTRY RELATIONS. 1974. *The nuclear industry: 1974.*

OFFICE OF TECHNOLOGY ASSESSMENT. 1980. *Energy from biological processes.* OTA-E-124.

PARIZEK, R. R., and MYERS, E. A. 1968. Recharge of groundwater from renovated sewage effluent by spray irrigation. *Proceedings of the Fourth American Water Resources Conference,* pp. 425–43.

PARK, C. F., Jr., 1968. *Affluence in jeopardy: Minerals and the political economy.* San Francisco: Freeman, Cooper, and Co.

PARK, T. 1948. Experimental studies of interspecies competition. 1. Competition between populations of the flour beetles, *Tribolium confusum* Duval and *Tribolium castaneum* Herbst. *Ecological Monographs* 18: 265–308.

PATRICK, C. H. 1977. Trace substances and electric power generation: The need for epidemiological studies to determine health problems. In *Conference on trace substances in environmental health,* pp. 63–69. Columbia, Mo.: Conference on Trace Substances in Environmental Health.

PAYNE, R. 1968. Among wild whales. *The New York Zoological Society Newsletter.*

PEARL, R. 1932. The influence of density of population upon egg production in *Drosophila melanogaster. Journal of Experimental Zoology* 63: 57–84.

PERSSON, R. 1974. *World forest resources.* Stockholm: Royal College of Forestry of Sweden.

PIERROU, U. 1976. The global phosphorus cycle. In *Nitrogen, phosphorus, and sulfur—Global cycles,* eds. B. H. Svensson and R. Soderlund, pp. 75–88. Stockholm: Ecological Bulletin.

PIMENTEL, D.; TERHUNE, E. C.; DYSON-HUDSON, R; ROCHEREAU, S.; SAMIS, R.; SMITH, E. A.; DENMAN, D.; REIFSCHNEIDER, D.; and SHEPARD, M. 1976. Land degradation: Effects on food and energy resources. *Science.* 194: 149–55

PRESS, F. 1975. Earthquake prediction. *Scientific American* 232: 14–23.

RICHARDS, J. L., AND ASSOCIATES LTD., and LABRECQUE, VERINA AND ASSOCIATES. 1973. Snow Disposal Study for the National Capital Area Technical Discussion. Report for the Committee on Snow Disposal, Ottawa, Ontario, Canada.

ROTTY, R. M. 1976. *Energy and the climate.* Institute for Energy Analysis, Oak Ridge Associated Universities, Oak Ridge, Tennessee.

RUGG, D. S. 1979. *Spatial foundations of urbanization.* 2nd ed. Dubuque, Iowa: William C. Brown.

SCHLÜTER, O. 1952. *Die Siedlungsräume Mitteleuropas in frühgeschichtlicher Zeit,* Part I. Hamburg, Germany.

SHEFFER, V. B. 1951. The rise and fall of the reindeer herd. *Scientific Monthly* 73: 356–62.

———. 1976. Exploring the lives of whales. *National Geographic* 150: 752–66.

SKINNER, S. P.; GENTRY, J. B.; and GIESY, J. P., Jr. 1976. Cadmium dynamics in terrestrial food webs of a coal ash basin. In *Environmental chemistry and cycling processes symposium, sponsored by U.S. Energy Research and Development Administration and the University of Georgia, 28 April–May 1976, Augusta, Georgia.*

SLOBODKIN, L. B. 1980. *Growth and regulation of animal populations.* New York: Dover Press. (First published in 1961 by Holt, Rinehart and Winston, New York.)

SLOSSON, J. E., and KROHN, J. P. 1977. Effective building codes. *California Geology* 30: 136–319.

SNEKVIK, E. 1970. *Norwegian directorate for game and freshwater fish.* Unpublished report.

SOIL CONSERVATION SERVICE, SOIL SURVEY STAFF. 1975. *Soil taxonomy.* Agricultural handbook No. 436.

STEINHART, J. S.; HANSON, M. E.; GATES, R. W.; DEWINKEL, C. C.; BRIODY, K.; THORNSJO, M.; KAMBALA, S. 1978. A low-energy scenario for the United States: 1975–2000. In *Perspectives on energy,* 2nd ed., eds. L. C. Ruedisili and M. W. Firebaugh, pp. 553–80. New York: Oxford University Press.

STODDARD, C. H. 1978. *Essentials of forestry practice.* 3rd ed. New York: Wiley.

STRAHLER, A. N. 1973. *Introduction to physical geography.* 3rd ed. New York: Wiley.

SULLIVAN, P. M.; STANCZYK, M. H.; and SPENDBUE, M. J. 1973. *Resource recovery from raw urban refuse.* U.S. Bureau of Mines R. I. 7760.

SWEENEY, J. G. 1977. *Themes in American painting.* Grand Rapids, Mich.: Grand Rapids Art Museum.

SYMONS, G. E. 1968. Water re-use: What do we mean? *Water and Waste Engineering* 5: 40–41.

TAYLOR, R. J., et al. 1973. Convective activity above a large-scale brush fire. *Journal of Applied Meteorology* 12: 1144.

THOMAS, W. L., Jr., ed. 1974. *Man's role in changing the face of the Earth.* 9th ed. Chicago: The University of Chicago Press.

THORARINSSON, S., and VONNEGUT, B. 1964. Whirlwinds produced by the eruption of Surtsey volcano. *Bulletin of the American Meteorological Society* 45: 440.

TRIMBLE, S. W. 1969. Culturally accelerated sedimentation on the middle Georgia Piedmont. Master's thesis, Univesity of Georgia, Athens, Georgia.

U.S. BUREAU OF THE CENSUS. 1979. *Statistical abstract of the United States.* 100th ed. Washington, D.C.: U.S. Department of Commerce.

U.S. BUREAU OF MINES. 1971. *Strippable reserves of bituminous coal and lignite in the United States.* U.S. Bureau of Mines Information Circular 8531.

U.S DEPARTMENT OF COMMERCE. 1970. *Some devastating North Atlantic hurricanes of the 20th century.*

U.S. DEPARTMENT OF ENERGY. 1978. *The United States magnetic fusion energy program.* DOE/ET-0072.

———. 1979a. *Annual report to Congress.*

———. 1979b. *Environmental development plan for magnetic fusion.* DOE/EDP-0052.

———. 1980. *Magnetic fusion energy.* DOE/ER-0059.

U.S. ENERGY AND RESEARCH DEVELOPMENT ADMINISTRATION. 1976. *Advance nuclear reactors: An introduction.* ERDA-76-107.

———. 1980a. *Guidelines for public reporting of daily air quality—Pollutant Standard Index.*

———. 1980b. *Acid rain.*

U.S. FOREST SERVICE. 1980. *An assessment of the forest and range land situation in the United States.*

U.S. GEOLOGICAL SURVEY. 1975. *Mineral resource perspectives 1975.* Professional Paper 940.

U.S. SECRETARY OF THE INTERIOR. 1975. *Mining and mineral policy, 1975.*

———. 1979. *Mining and mineral policy, 1979.*

———. 1980a. *Principles of a resource/reserve classification for minerals.* U.S. Geological Survey Circular 831.

———. 1980b. *Mining and mineral policy, 1980.*

VAN DER TAK, J.; HAUB, C.; and MURPHY, E. 1979. Our population predicament: A new look. *Population Bulletin* 34: 1–8.

WAGNER, A. A. 1957. *The use of the Unified Soil Classification System by the Bureau of Reclamation.* London: International Conference on Soil Mechanics and Foundation Engineering.

WALDBOTT, G. L. 1978. *Health effects of environmental pollutants.* 2nd ed. St. Louis: C. V. Mosby.

WALLACE, A. R. 1876. *The geographical distribution of animals.* New York: Hafner. (Reissued in 1962 in two volumes.)

WALLACE, R. E. 1974. *Goals, strategies and tasks of the earthquake hazard reduction program.* U.S. Geological Survey Circular 701.

WALMAN, M. G. 1967. A circle of sedimentation and erosion. *Geografiska Annaler* 49A, p. 386.

WALTER, H. 1973. *The vegetation of the Earth.* New York: Springer-Verlag.

WATER RESOURCES COUNCIL. 1971. *Regulation of flood hazard areas to reduce flood losses.*

———. 1978. *The nation's water resources, 1975–2000.*

WAY, D. S. 1978. *Terrain analysis.* 2nd ed. Stroudsburg, Pa.: Dowden, Hutchinson & Ross.

WHITE, D. F., and WILLIAMS, D. L. 1975. Introduction. In *Assessment of geothermal resources of the United States—1975,* eds. D. F. White and D. L. Williams, pp. 1–4. U.S. Geological Survey Circular 726.

WHITE, G. F. 1980. *Environment. Science* 209: 183–90.

WHITE, G. F., and HAAS, J. E. 1975. *Assessment of research on natural hazards.* Cambridge, Mass.: The MIT Press.

WHITTAKER, R. H., and LIKENS, G. E. 1973. Carbon in the biota. In *Carbon and the biosphere,* eds. G. M. Woodwell and E. V. Pecan, pp. 281–302. Brookhaven National Laboratory Symposium No. 24. Oak Ridge, Tenn.: Technical Information Service.

WHITTLE, K. J.; HARDY, R.; HOLDEN, A. V.; JOHNSTON, R.; and PRENTREATH, R. J. 1977. Occurrence and fate of organic and inorganic contaminants in marine animals. *Annals of the New York Academy of Science* 298: 47–79.

WILCOX, B. A. 1980. Insular ecology and conservation. In *Conservation biology,* eds. M. E. Soulé and B. A. Wilcox, pp. 95–117. Sunderland, Mass.: Sinauer.

WILLIAMSON, S. J. 1973. *Fundamentals of air pollution.* Reading, Mass.: Addison-Wesley.

WILSON, R. 1980. Risk/benefit analysis for toxic chemicals. *Ecotoxicology and Environmental Safety* 4: 370–83.

WOODWELL, G. M., and REBUCK, A. L. 1967. Effects of chronic gamma radiation on the structure and diversity of an oak-pine forest. *Ecological Monographs* 37: 53–69.

WRIGHT, R. F.; DALE, T.; GJESSING, E. G.; HENDREY, G. R.; HENRIKSEN, A.; JOHANNESSEN, M.; and MUNIZ, I. P. 1976. Impact of acid precipitation on freshwater ecosystems in Norway. *Water, Air, and Soil Pollution* 6: 438–99.

ZISWILER, V. 1967. *Extinct and vanishing species.* New York: Springer-Verlag.

Index